THE KIND OF MOTION
WE CALL HEAT

STUDIES IN STATISTICAL MECHANICS

VOLUME VI

EDITORS:

E. W. MONTROLL
University of Rochester, New York
and
J. L. LEBOWITZ
Yeshiva University, New York

NORTH-HOLLAND PUBLISHING COMPANY
AMSTERDAM · NEW YORK · OXFORD

THE KIND OF MOTION WE CALL HEAT

A HISTORY OF THE KINETIC THEORY OF GASES
IN THE 19th CENTURY

BOOK 2

Statistical Physics and Irreversible Processes

STEPHEN G. BRUSH
University of Maryland, College Park, USA

1976

NORTH-HOLLAND PUBLISHING COMPANY
AMSTERDAM · NEW YORK · OXFORD

© NORTH-HOLLAND PUBLISHING COMPANY, 1976

All Rights Reserved. No part of this publication may be reproduced, stored in a retrieval system or transmitted, in any form or by any means, electronic, mechanical, photocopying, recording or otherwise, without the prior permission of the Copyright owner.

ISBN North-Holland: 0 7204 0482 7
ISBN American Elsevier: 0 444 11131 x

Published by
NORTH-HOLLAND PUBLISHING COMPANY
AMSTERDAM · NEW YORK · OXFORD

Sole distributors for the U.S.A. and Canada:

AMERICAN ELSEVIER PUBLISHING COMPANY, INC.
52 VANDERBILT AVENUE
NEW YORK, N.Y. 10017

PRINTED IN GREAT BRITAIN

To my parents

Preface

These two books bring together the contents of 15 articles (and parts of three others) on aspects of the history of kinetic theory and statistical mechanics, published in various places during the last two decades. The introductory chapter and the bibliographical survey at the end, as well as several other sections, have been added to provide a reasonably comprehensive and coherent picture of the entire subject, though without attempting to go beyond the boundaries of the 19th century or to cover in depth the numerous applications of the theory even within that century. All chapters have been revised in the light of my own current research and the secondary literature up to June 1975. The organization of the books is explained in §1.1.

The project originated as a paper for Thomas Kuhn's seminar at Harvard in 1954, and I have profited from his advice and criticism on numerous occasions since. My colleague C. W. F. Everitt has generously made available his extensive fund of information on James Clerk Maxwell and the Victorian scientific community, and is the co-author of §5.5. Several chapters have been improved and some serious errors avoided thanks to the technical and editorial scrutiny of C. Truesdell, whose encouragement has been especially appreciated. Elliott Montroll first urged me to put together the books in the present form. Other colleagues have contributed valuable suggestions and information: S. R. de Groot, W. Flamm, E. W. Garber, D. ter Haar, F. C. Haber, R. Hahn, Sir Harold Hartley, E. Hiebert, Gerald Holton, H. Kangro,

Martin J. Klein, J. L. Lebowitz, P. E. Liley, E. A. Mason, E. Mendoza, A. Michels, R. G. Olson, N. H. Robinson, J. S. Rowlinson, J. V. Sengers, H. I. Sharlin, and J. D. van der Waals Jr.

During the extended period of research and writing of which these books are the outcome, financial support has been furnished by a Rhodes Scholarship, a postdoctoral fellowship and three research grants awarded by the National Science Foundation, and a contract with the National Aeronautics and Space Administration. The necessary time, facilities, and intellectual environment were provided by my studies at Harvard College and Oxford University, and my employment at the Lawrence Radiation Laboratory, Livermore; Harvard University (Project Physics, the Department of Physics and the Department of History of Science); and the University of Maryland, College Park (Department of History and Institute for Fluid Dynamics and Applied Mathematics). The collections and staff of the Royal Society of London, the Library of Congress, and the libraries of Harvard University, the University of California at Berkeley and the University of Maryland at College Park have been essential to my work.

Permission to reprint or extract from copyrighted material has been granted by the American Association of Physics Teachers, American Institute of Physics, British Society for History of Science, Cambridge University Library, Fratelli Fabbri Editore, D. Reidel Publishing Company, the Royal Society of London, Charles Scribner's Sons, The Society of the Sigma Xi, Springer-Verlag, Taylor & Francis Ltd., University of Pennsylvania Press, and Brigadier Wedderburn-Maxwell.

Contents

Preface   vii
Contents   ix

BOOK 1

Part A: *Introductory Survey*

1 THE KINETIC THEORY IN THE HISTORY OF PHYSICS   3
   1.1 Organization of this book   4
   1.2 Air pressure   8
   1.3 The kinetic theory in the 18th century   19
   1.4 Absolute temperature and the thermal expansion of gases   23
   1.5 Opinions on the nature of heat before 1840 27
   1.6 The Second Scientific Revolution 1800–1950 35
   1.7 Nature philosophy and the reaction against materialism   51
   1.8 Triumphs and failures of the kinetic theory 69
   1.9 *Fin de Siècle* and the resurrection of atomism   90

Part B: *Personalities*

2 HERAPATH   107
   2.1 Introduction   107

	2.2	Herapath's 1820 paper	110
	2.3	The Royal Society controversy	115
	2.4	Railways	121
	2.5	The *Mathematical Physics*	127
	2.6	Herapath's reputation after the revival of the kinetic theory	130
3	WATERSTON		134
	3.1	Biography	134
	3.2	His 1845 paper	146
	3.3	Theory of sound	149
	3.4	Rayleigh's discovery of Waterston's paper	156
	3.5	Later references	158
4	CLAUSIUS		160
	4.1	Joule's revival of Herapath's theory	160
	4.2	Krönig's 1856 paper	165
	4.3	Clausius and the elementary kinetic theory	168
	4.4	The mean free path	177
	4.5	Later writings on kinetic theory	181
5	MAXWELL		183
	5.1	Possible source of the statistical approach	183
	5.2	Gas viscosity	189
	5.3	Specific heats and the hypothetico-deductive method	194
	5.4	Reception of the kinetic theory in the 1860's	198
	5.5	Maxwell, Osborne Reynolds, and the radiometer (with C. W. F. Everitt, co-author)	210
6	BOLTZMANN		231
	6.1	The distribution law	231
	6.2	Transport equation and H-theorem	235
	6.3	Reversibility and recurrence paradoxes	238
	6.4	Statistical mechanics and ergodic hypothesis	240
	6.5	Other contributions	243
	6.6	Defense of the atomic viewpoint	244
	6.7	Bibliography	247
7	VAN DER WAALS		249
	7.1	The states of matter	249
	7.2	Biographical note	251
	7.3	Earlier research on the gas–liquid transition	256

7.4	Kinetic theory of non-ideal gases	264
7.5	The van der Waals force	269
7.6	Recent speculations about states of matter	270

8 MACH — 274
- 8.1 The background for Mach's opinions on atomism — 277
- 8.2 Mach's early pro-atomism (1862) — 282
- 8.3 Mach turns against atomism (1872) — 285
- 8.4 Miscellaneous remarks (1882–95) — 287
- 8.5 Reconciliation of mechanistic and phenomenological physics (1896–1900) — 290
- 8.6 Mach "sees" an atom (1903) — 294
- 8.7 The unrepentant sinner (1910) — 295
- 8.8 Summary and concluding remarks — 296

INDEX — xv

BOOK 2

Part C: Problems

9 THE WAVE THEORY OF HEAT — 303
- 9.1 Heat and radiation in the 19th century — 303
- 9.2 Radiant heat and the decline of the caloric theory — 307
- 9.3 Ampère's theory — 314
- 9.4 Reception of the wave theory, 1831–45 — 318
- 9.5 Transition from wave theory of heat to thermodynamics — 325
- 9.6 Disappearance of the wave theory of heat after 1850 — 329

10 FOUNDATIONS OF STATISTICAL MECHANICS 1845–1915 — 335
- 10.1 Waterston's equipartition theorem — 336
- 10.2 Clausius's postulate about internal motions — 339
- 10.3 Maxwell's velocity distribution — 342
- 10.4 Equalization of kinetic energy by collisions — 343
- 10.5 The effect of forces on the distribution law: The "Boltzmann factor" — 347
- 10.6 Equilibrium of a column of gas under gravitational forces — 349
- 10.7 Approach to equilibrium and the problem of irreversibility — 351
- 10.8 The paradox of specific heats — 353
- 10.9 Validity of the equipartition theorem — 356
- 10.10 The ergodic hypothesis of Boltzmann and Maxwell — 363

	10.11	Digression on the history of mathematics	378
	10.12	Proof of the impossibility of ergodic systems	383
11	**INTERATOMIC FORCES AND THE EQUATION OF STATE**	386
	11.1	The Newtonian program	386
	11.2	The equation of state in the caloric theory	397
	11.3	Interatomic forces in early kinetic theories	401
	11.4	The virial theorem	404
	11.5	The van der Waals equation	406
	11.6	Later discussions of the equation of state for hard spheres	411
	11.7	Second virial coefficient for continuous force law	416
	11.8	Gibbs' statistical mechanics	419
12	**VISCOSITY AND THE MAXWELL–BOLTZMANN TRANSPORT THEORY**	422
	12.1	Introduction	422
	12.2	Mean-free-path formulae for viscosity	426
	12.3	Maxwell's transport theory	432
	12.4	Boltzmann's equation and the H-theorem	443
	12.5	Hilbert's work on the Boltzmann equation	447
	12.6	Chapman's transport theory	449
	12.7	Enskog's solution of the Boltzmann equation	456
	12.8	Viscosity of dense gases	460
	Appendix 12A: Biographies of Chapman and Enskog	463
13	**HEAT CONDUCTION AND THE STEFAN–BOLTZMANN LAW**	469
	13.1	Introduction	469
	13.2	The Dulong–Petit law of cooling	478
	13.3	Heat conduction in gases, before Maxwell	483
	13.4	Maxwell's kinetic theory of heat conduction	489
	13.5	Experimental tests of Maxwell's theory	499
	13.6	The temperature of the sun	507
	13.7	The Stefan–Boltzmann law	513
	13.8	The three modes of heat transfer	524
	Appendix 13A: Leslie's analysis of heat transfer	534
	Appendix 13B: Derivation of the T^4 law	539
14	**RANDOMNESS AND IRREVERSIBILITY**	543
	14.1	Introduction: the world-machine and cosmic history	543
	14.2	The cooling of the earth	551

CONTENTS   xiii

14.3	The Second Law of Thermodynamics and the concept of entropy	566
14.4	The introduction of statistical ideas in kinetic theory	583
14.5	Boltzmann's statistical theory of entropy	598
14.6	Molecular disorder	616
14.7	The recurrence paradox	627
14.8	Toward quantum theory: Planck's irreversible radiation processes	640

15 BROWNIAN MOVEMENT   655
 15.1 Robert Brown's observations and interpretations thereof   657
 15.2 Miscellaneous observations and qualitative explanations, 1840–78 663
 15.3 Criticisms of the molecular-impact theory   667
 15.4 Einstein's theory of Brownian movement   672
 15.5 Smoluchowski's theory of Brownian movement   686
 15.6 Perrin's experiments and the reality of atoms   693

Part D: Bibliography

16 THE LITERATURE OF KINETIC THEORY   705
 16.1 Quantitative aspects of the history of kinetic theory   705
 16.2 Journals and abbreviations   712
 16.3 Bibliography of research publications on the kinetic theory of gases, 1801–1900 721

INDEX   xv

PART C

Problems

CHAPTER 9

The Wave Theory of Heat

9.1 Heat and radiation in the 19th century*

Research on thermal "black-body" radiation played an essential role in the origin of the quantum theory at the beginning of the 20th century. This is a well-known fact, but historians of science up to now have not generally recognized that studies of radiant heat were also important in an earlier episode in the development of modern physics: the transition from caloric theory to thermodynamics. During the period 1830–50, many physicists were led by these studies to accept a "wave theory of heat," although this theory subsequently faded into obscurity.

According to the wave theory of heat,[1] heat is the vibrations of an ethereal fluid that fills all space, and which transmits vibrational motion from one atom to another. While this theory is in some respects similar to post-1850 conceptions – a similarity which, as we will see, was helpful in facilitating the transition between them – it differs in two significant respects. First, it denies that atomic vibrations alone could account for the phenomena of heat; the role of the ether is essential. Second, it is not assumed that atoms in a gas can move freely through space, as in the modern kinetic theory; they are still constrained to vibrate around fixed equilibrium positions. These two features helped to preserve the

---

* Reprinted with a few additional notes from *Brit. J. Hist. Sci.* 5, 145 (1970), by permission of the British Society for the History of Science.

continuity with older ideas about heat and the structure of gases, but were gradually de-emphasized after 1845.

For a contemporary description of the wave theory of heat, it seems appropriate to turn to the article on Heat in the 8th edition of the *Encyclopedia Britannica*, published in 1856, since this article is one of the few pieces of evidence that have been offered to support the common but erroneous view that scientists accepted the caloric theory until the middle of the 19th century. Now it must be noted first of all that the main part of this article was actually published in the 7th edition, so that it really indicates views that were prevalent before 1842.[2] Second, anyone who reads past the first two paragraphs will see that the author, T. S. Traill, does *not* still believe in the caloric theory of heat. It is true that he reports that the caloric theory has in the past been generally accepted, and criticizes the mechanical theory (based on *atomic* vibrations) as "vague and unsatisfactory." But, after noting that the mechanical theory cannot account for phenomena such as the radiation of heat through empty space, he says:

> It is possible, however, to modify this theory, by supposing that heat is produced not merely by the motions of the particles of the heated substance, but by the vibrations or undulations of a very subtle matter existing in all bodies. This will approximate the vibratory theory to that which has been generally considered as its antagonist, will accord well with some recently discovered facts, and will assimilate the vibratory hypothesis of heat to the undulations now so generally received as explanatory of the phenomena of light, to which heat has so intimate a relation.... These views lead us to the conclusion that the phenomena of caloric are owing to the movements of a subtle fluid, the particles of which are strongly repellent of each other, and have an affinity for those of all other bodies, different in force according to each kind of matter.

The above quotation provides most of the clues needed to unravel the history of the wave theory of heat. Before proceeding with our detailed account, however, we shall first outline the main steps in the development of 19th-century ideas about heat, in order to show the wave theory in a wider context.

(1) Some of the pre-19th century ideas are similar to the wave theory,[3] but these lie outside the scope of the present account; we begin with the situation at the beginning of the 19th century, when both heat and light were supposed to be fluid substances, probably particulate.

§9.1] HEAT AND RADIATION IN THE 19TH CENTURY 305

(2) Widely publicized, and frequently referred to in the first half of the 19th century, were the experiments of Rumford, Davy, and others, showing that heat lacks weight and can be generated in unlimited quantities by mechanical processes such as friction. But these experiments did not by themselves persuade most scientists to abandon the material ("caloric") theory of heat developed by Black, Lavoisier, Laplace, and others.

(3) In the period 1800–35, experiments on radiant heat by William Herschel, John Leslie, Macedonio Melloni, James Forbes, and others showed that radiant heat has most if not all of the properties of light. This led to a widespread belief that heat and light are essentially the same phenomenon, *i.e.* superficially different manifestations of the same physical agent. The adjective "radiant" was easily dropped, so that the problem of the nature of heat was reduced to the problem of the nature of light.

(4) Between 1815 and 1830 the Young–Fresnel wave theory of light replaced the Newtonian particle ("emission") theory, as attention was focused on properties such as interference and polarization.

(5) Hence, as the logical conjunction of (3) and (4) reinforced by (2), the wave theory of heat was adopted after 1830. This did not require a sharp break with the caloric theory. One could first identify caloric with ether, then assume that heat consists in the *vibrations* rather than the *amount* of this fluid, thereby preserving many of the explanations of the older theory, with only verbal modifications.

(6) Between 1842 and 1850, interest in steam engines and in the thermal effects of electromagnetic phenomena led to the enunciation of the principle of conservation of energy and the establishment of modern thermodynamics based on the idea of a "mechanical equivalent of heat." By this time the caloric theory was almost dead, and the wave theory of heat had already made it seem natural to treat heat as a form of mechanical energy. Some of the early statements of the mechanical theory of heat were clearly inspired by the wave theory.

(7) With the revival of the kinetic theory of gases (1848–70), the essential role of the ether in transmitting vibrations from one atom to another was eliminated in dealing with ordinary thermal properties of matter ("sensible heat"). The irrelevance of the wave theory to thermodynamics is already foreshadowed by 1850.

(8) In spite of (7), the wave theory of heat did not die out, since there was continuing interest in radiant heat throughout the rest of the 19th century. Since physicists still believed in the existence of an ether with mechanical properties, the question "why doesn't the ether take its share of vibrational energy corresponding to thermal equilibrium

with matter" remained unsolved up to 1900. This question was seen retrospectively as a "crisis in classical physics" (the "ultraviolet catastrophe" posthumously baptized by Paul Ehrenfest in 1910) though as Martin Klein has shown,[4] it was not viewed as such by Planck at the time he developed the quantum theory.

(9) Maxwell's electromagnetic theory (1866–73) indicated that heat radiation could be viewed as a special type of electromagnetic waves ("infrared radiation") which produces thermal effects when absorbed by matter. This led to a reinterpretation of step (3), making the qualifying adjective "radiant" essential: the nature of *heat* is not necessarily the same as the nature of *radiant* heat.[5]

(10) Boltzmann and Wien, in the 1880's showed that heat radiation could be treated with some success by combining thermodynamic concepts with Maxwell's energy-momentum relation for electromagnetic radiation and the Döppler principle. It was by following this phenomenological approach, rather than by worrying about the equipartition problems involved in the wave theory of heat, that Planck was led to his distribution law in 1900.

This chapter is primarily concerned with steps (5) and (6). Up to now, the significance of these steps has been hidden by the modern distinction between radiant heat and ordinary heat (cf. step (9)) and by two myths about the history of 19th-century physics: the first (now largely discredited), that the mechanical theory of heat was already established by Rumford and Davy at the beginning of the 19th century; the second (still prevalent), that most scientists accepted the caloric theory until it was replaced by thermodynamics. We shall see that both myths originated at least in part in the writings of the same person, William Thomson.

The establishment of the wave theory of light, on the one hand, and of the principle of conservation of energy and thermodynamics, on the other, are generally regarded as two separate events in the history of 19th-century physics. However, I think they should be seen as successive and closely related stages of the same transformation of physical theory, in which explanations of phenomena were increasingly based on *motion* rather than on *matter* [see above, pp. 40–54].

Notes for §9.1

1. More often called the "undulatory" theory in the 19th century; but the modern terminology, already well established even in historical writings on the "wave" theory of light, seems preferable. It should be noted that by some modern criteria the wave

theory of heat is not a distinct "theory" since it did not lead to quantitative predictions different from those of the caloric or mechanical theories, but instead could be viewed as a combination of the two (see discussion in text, below). However, physicists at the time did not seem to be worried by this circumstance, and usually presented it as a distinct alternative to the caloric theory. Another version of the wave theory of heat, which was perhaps more popular before the 19th century, asserted that heat itself is the vibrations of material particles, but that these can be transmitted from one particle to another by ether vibrations. This version would appear to be excluded by the postulate that heat and light are qualitatively identical and are vibrations of the same medium.

2. T.S. T[raill], "Heat", *Encyclopedia Britannica*, 7th ed. (Edinburgh, 1842), XI, 180–97. I am indebted to Mr. V. A. Stenberg, Director of Research at the *Britannica*, for providing me with a copy of this article. The 7th edition does not seem to be available in any major American library, which may partly account for the mistaken idea that this article first appeared in 1856. In the 8th edition the article is reprinted with an additional section at the end referring to the work of Joule, Rankine, and Thomson on the mechanical equivalent of heat; this work is said to support the "mechanical or dynamical theory of heat" of Rumford and Davy, thereby contradicting the remarks at the beginning of the article.

3. For our purposes the most important source is Isaac Newton, *Opticks*, 4th ed. (London, 1730), Qu. 18, which suggests that heat is transmitted through a vacuum by ether vibrations; but this seems to be contradicted by remarks in Qu. 28.

4. M. J. Klein, *Arch. Hist. Exact Sci.* **1**, 459 (1962).

5. "I do not think that radiant heat is heat at all as long as it is radiant" – Maxwell to P. G. Tait, 3 August 1868.

"The motion which we call light, though still more minute and rapidly alternating than that of sound, is, like that of sound, perfectly regular, and therefore is not heat. What was formerly called Radiant Heat is a phenomenon physically identical with light.... Now, the motion which we call heat can never of itself pass from one body to another unless the first body is, during the whole process, hotter than the second. The motion of radiation, therefore, which passes entirely out of one portion of the medium and enters another, cannot be properly called heat." Maxwell, *Nature* **11**, 357 (1875); *Papers* II, 418 (quotation from p. 437).

The statement of S. P. Langley [*Am. J. Sci.*, Jan. 1889] that until the work of Draper in 1872 "it was almost universally believed that there are three different entities in the spectrum, represented by actinic, luminous, and thermal rays" was disputed by Rayleigh, who said that in the period 1850–60 nearly all the leading workers in the field, with the exception of Brewster, held the modern view of radiation. Rayleigh, *Phil. Mag.* [5] **27**, 265 (1889).

9.2 Radiant heat and the decline of the caloric theory

The caloric theory of heat was at the height of its popularity around 1825, and its adherents included many of the leading scientists, especially in Paris. It was the "keystone of anti-phlogistic chemistry," as Lilley has pointed out,[1] and was used in explaining phenomena such as thermal expansion, specific heats, changes of state, latent heat, and the heat evolved in chemical reactions, even though one could not say that the majority of scientists considered the theory firmly established.[2] The chief opponents of the caloric theory at this time were Rumford and Davy, but the caloric theorists had so far been able to combat their arguments fairly successfully.[3] Cajori has cited a number of American and German authors who favored the caloric theory during the period 1800–30;[4] his list could be extended without difficulty. Far from being a merely qualitative explanatory principle or crutch for the imagination, the theory could be presented, as Brown has shown, in a logically coherent, semiquantitative manner to critical students of physics.[5] Among the quantitative accomplishments of permanent value arising from the caloric theory may be mentioned the Laplace–Poisson calculation of the speed of sound (based on the ratio of specific heats of a gas)[6] and Fourier's theory of heat conduction.[7] Looking at the situation in 1825, an observer might well jump to the conclusion that the caloric theory would be firmly entrenched for another generation at least. Yet in less than a decade its credibility had been seriously undermined, and the weight of scientific opinion had shifted in another direction.

One of the arguments for the materiality of heat at the beginning of the 19th century was the fact that heat can apparently travel through empty space without any accompanying movement of matter; hence it cannot be simply molecular motion.[8] This conclusion was perfectly reasonable; but those who used the argument to bolster the caloric theory did not suspect that it would eventually prove treacherous. The increasing use of the phrase "radiant caloric" (*calorique rayonnant*) by writers on caloric theory reflects the growing interest among physicists of the period in the phenomena of radiant heat. This was becoming a very active area of research, beginning with Scheele, Pictet, and Prevost in the late 18th century,[9] and continuing with major discoveries by William Herschel and others around 1800.[10] But from our present viewpoint, the decisive contribution was that by the Italian physicist Macedonio Melloni (1798–1854).[11] With the initial help of his compatriot Leopoldo Nobili (1784–1835) in designing a very sensitive thermopile or "*thermomultiplicateur*" for detecting heat from distant sources, Melloni was able to establish around 1830–32 a number of properties of radiant heat. We shall not attempt to sort out the discoveries of Melloni from those of his predecessors and followers; what is important is the fact that with the help of a favorable report on his work published by the French Academy of Sciences, and the award of the Rumford Medal of the Royal Society of London in 1835, these discoveries gained wide publicity in the scientific world.[12]

For those who did not follow Melloni's work in detail, the most significant result was simply that radiant heat shares all the qualitative properties of light: reflection, refraction, diffraction, polarization, interference, etc.[13] This meant that heat and light must be fundamentally the same, even though quantitative differences in such properties as wavelength might lead to different effects on the human sense organs. Melloni himself was slower to accept this conclusion than other scientists, but he eventually adopted it in 1842,[14] and by 1847 he had become a strong advocate of the identity of heat and light.[15]

Before 1820, evidence for the identity of heat and light was at the same time evidence for the materiality of heat, since the particle theory of light was still generally accepted. This was a handicap for scientists such as Davy who advocated both the particle theory of light and the mechanical theory of heat, and required some awkward contortions to make the two theories seem compatible.[16] On the other hand, as soon as one rejected the particle theory of light, it was quite natural to reject the caloric theory of heat as well. Thus Thomas Young wrote in 1802:

> It was long an established opinion, that heat consists in vibrations of the particles of bodies, and is capable of being transmitted by undulations through an apparent vacuum (Newt. Opt. Qu. 18). This opinion has been of late very much abandoned, Count Rumford, Professor Pictet, and Mr. Davy, are almost the only authors who have appeared to favour it; but it seems to have been rejected without any good grounds, and will probably very soon recover its popularity.[17]

Young made the connection between heat waves and light waves a little more explicit in his *Lectures on Natural Philosophy* (1807):

> It was Newton's opinion, that heat consists in a minute vibratory motion of the particles of bodies, and that this motion is communicated through an apparent vacuum, by the undulations of an elastic medium, which is also concerned in the phenomena of light . . . . It is easy to imagine that such vibrations may be excited in the component parts of bodies, by percussion, by friction, or by the destruction of the equilibrium of cohesion and repulsion . . . .[18]

As in his attempts to establish the wave theory of light, Young operated on the principle that if the name of Newton can be firmly associated with a theory, physicists will eventually accept it.

In the context of early 19th-century ideas about heat and light, Fresnel's success in establishing the wave theory of light can be seen as a major turning-point, not only in the history of optics but also in the prehistory of thermodynamics. Thus, as I have argued in §1.6, Fresnel's work and especially the dramatic confirmation of Poisson's disbelieving prediction of the "bright spot" behind a circular disk, helped to launch the Second Scientific Revolution.

In 1883 John Tyndall, a leader in the later generation of investigators of radiant heat, recalled or at least supposed that

> Long before the death of Melloni, what the Germans call "Die Identitäts Frage," that is to say, the question of the identity of light and heat, agitated men's minds and spurred their inquiries ... the undulatory theory of light being once established, soon made room for the undulatory theory of radiant heat.[19]

But the interaction between theories of light and heat was not entirely in one direction; according to a recent article on Fresnel,

> His earliest statement in favor of light as a form of motion (in a letter of 5 July 1814) envisioned the possibility of referring heat, light, and electricity to the modifications of a single, universal fluid. ... His determination to overhaul optical theory was sparked by a dissatisfaction with the caloric view of heat and an appreciation of the analogies between heat and light.[20]

It is conceivable that if heat had been convincingly shown to be a form of motion before 1820, the equation could have been turned around and the wave theory of light might have been adopted *before* Fresnel presented his theory and its experimental proofs; but that seems unlikely in view of the strong position of the corpuscular theory of light at the time.

Robert Fox, in his comprehensive history of the caloric theory of gases and in an earlier article on Dulong and Petit, has cited several examples of scientists favorable to vibrational or wave conceptions of heat in the 1820's.[21] But he sees this as a period of confusion in which the "Laplacian orthodoxy" was replaced by a general mood of positivism rather than by a specific alternative theory. My impression is that the wave theory, though not yet developed in quantitative detail (in fact it was never to be worked out in much detail), was already being seen as the most likely alternative if the caloric theory were to be abandoned.

Although most writers on heat in the 1820's still accepted the caloric theory, some of them recognized the importance of the radiant heat studies and began to give equal prominence to the wave theory, concluding that the issue could not be decided yet.[22] Moreover, as Gay-Lussac pointed out, the development of a theory of heat in space ("*calorique du vide*") would require a significant conceptual change in the description of heat: whereas in describing heat in a material substance, "whether one considers the caloric as a body, or as a motion, one can measure its quantity; but in an empty space ... one can only conceive of caloric in motion."[23]

But a few years later Sadi Carnot, one of the founders of thermodynamics, wrote the following in his manuscript notes:

> We may be allowed to express here a hypothesis concerning the nature of heat.
>
> At present, light is generally regarded as the result of a vibratory movement of the etherial fluid. Light produces heat, or at least accompanies the radiant heat and moves with the same velocity as heat. Radiant heat is therefore a vibratory movement. It would be ridiculous to suppose that it is an emission of matter while the light which accompanies it could only be a movement.
>
> Could a motion (that of radiant heat) produce matter (caloric)?
>
> Undoubtedly no; it can only produce a motion. Heat is then the result of a motion.
>
> Then it is plain that it could be produced by the consumption of motive power and that it could produce this power.[24]

Though he had accepted the caloric theory (with some reservations) in his *Reflexions* of 1824, it seems clear from the above passage that Carnot has accepted the wave theory of light shortly afterwards, and by thinking about radiant heat has been led to the mechanical theory of heat. Here was a possible route to thermodynamics, which other scientists *might* follow; a few of them did.

Notes for §9.2

1. S. Lilley, *Actes 5ᵉ Cong. Int. Hist. Sci.* (Lausanne, 1947), p. 130; *Arch. Int. Hist. Sci.*, p. 630 (1940). This is confirmed by the contemporary evidence I have seen; for example, E. S. Fischer, Professor of Mathematics and Natural Philosophy at Bonn, wrote that chemists were unanimous in adopting the caloric theory. *Elements of Natural Philosophy* (Boston, 1827, trans. from Biot's French trans.), p. 57.

2. In reading the literature of the early 19th century, I have been impressed by the great caution and open-mindedness with which many scientists presented their views on heat, particularly some of the writers who are usually labeled as supporters of the caloric theory. It was very common to say that most of the phenomena can be explained equally well by considering heat as a substance or as a quality (or "mode of motion"); even if the former view was to be adopted for the sake of convenience, it was not to be regarded as firmly established beyond any doubt; furthermore, even if heat is really a fluid substance, it may have some association with molecular motion. This point has been emphasized by E. Mendoza in his introduction to *Reflections on the Motive Power of Fire by Sadi Carnot and other papers on the Second Law of Thermodynamics by É. Clapeyron and R. Clausius* (New York: Dover, 1960), p. xvi. As examples he cites Lavoisier and Laplace, *Mémoire sur la Chaleur* (Paris, 1780) and Lamé's *Cours de Physique* (Paris, 1836). I will discuss Lamé's opinions below–I do not think they can be put in the same category with those of Lavoisier and Laplace–but as additional evidence one can consult D. Olmsted, *Am. J. Sci.* **11**, 349 (1826); *ibid.*, **12**, 1 (1827), esp. p. 355; *ibid.*, **12**, 359 (1827), esp. p. 363; Joseph Black, *Lectures on the Elements of Chemistry* (Philadelphia, 1807), **1**, 29, 33; E. Bellone, *Physics* **13**, 376 (1971).

Thenard, in what Maurice Crosland has called "the standard textbook of chemistry in France for nearly a quarter of a century [*The Society of Arcueil* (London: Heinemann, 1967), p. 330] accepted the caloric theory yet admitted that since caloric seems to be weightless its real existence is dubious: see L. J. Thenard, *Traité de chimie élementaire, théorique, et pratique* (Paris, 5th ed., 1827), **1**, 35. Many other examples can be found in the recent book by Robert Fox, *The Caloric Theory of Gases from Lavoisier to Regnault* (Oxford: Clarendon Press, 1971).

3. S. C. Brown, *Proc. Am. Phil. Soc.* **93**, 316 (1949).
4. F. Cajori, *Isis* **4**, 483 (1922). It was probably Cajori who first used the *Britannica* article of Traill as evidence for the survival of the caloric theory into the 1850's: see *A History of Physics* (New York: Dover Pubs., 1962, rept. of 1929 ed.), p. 122.
5. S. C. Brown, *Am. J. Phys.* **18**, 367 (1950).
6. See T. S. Kuhn, *Isis* **49**, 132 (1958); B. S. Finn, *Isis* **55**, 7 (1964); M. Crosland, *The Society of Arcueil*, pp. 302–7.
7. A convenient edition which includes most of Fourier's work on heat conduction is *The Analytical Theory of Heat*, English trans. by A. Freeman (New York: Dover Pubs., 1955). Among recent historical studies may be mentioned F. Bureau, *Bull. Acad. Sci. Bruxelles* **39**, 1116 (1953); E. Bellone, *Physics* **9**, 301 (1967); I. Grattan-Guinness, *J. Inst. Math. Applics.* **5**, 230 (1969); J. R. Ravetz and I. Grattan-Guinness, *DSB* **5**, 93 (1972); I. Grattan-Guinness, *Joseph Fourier 1768–1830* (Cambridge, Mass.: MIT Press, 1972). J. Herivel, *Joseph Fourier* (Oxford: Clarendon Press, 1975).

Fourier himself, as is well known, denied that he adopted any particular hypothesis about the nature of heat. However, he did accept the similarity of heat and light (see p. 32 in Freeman's translation). According to Grattan-Guinness, while in his 1809 paper Fourier did not discuss the nature of heat "he presumably had in mind here an interpretation of heat flow in terms of component oscillatory waves, corresponding to the vibrations of mechanical systems around a position of equilibrium to which he had referred in his paper on statics." *Joseph Fourier 1768–1830*, p. 145; see also *ibid.*, p. 406 for another reference to the "wave" behavior of heat.

8. W. Henry, *Manchester Mem.* **5**, 603 (1802). J. Murray, *Elements of Chemistry* (Edinburgh, 1801), p. 163. The argument can be found in the literature as late as 1830,

e.g. J. Bostock's article "Heat" in Brewster's *Edinburgh Encyclopedia* (Edinburgh, 1830), **10**, 690.
9. E. S. Cornell, *Ann. Sci.* **1**, 217 (1936); A. Wolf, *A History of Science, Technology and Philosophy in the 18th Century* (London: George Allen & Unwin, 2nd ed., 1952), **1**, 206–12.
10. See E. S. Barr, *Infrared Physics* **1**, 1 (1961); D. J. Lovell, *Isis* **59**, 46 (1968); Baden Powell, *B.A. Rep.* **2**, 259 (1832); S. P. Langley, *Pop. Sci. Mon.* **34**, 212, 385 (1889).
11. See E. S. Barr, *Infrared Physics* **2**, 67 (1962); G. Tedesco, *N. Cim.* [10] **2** (Suppl.), 501 (1955); G. Dascola, *N. Cim.* [10] **2** (Suppl.), 518 (1955).
12. The importance of the discoveries of James David Forbes in this field has been urged by P. G. Tait, in *Life and Letters of James David Forbes*, by J. C. Shairp, P. G. Tait, and A. Adams-Reilly (London, 1873), p. 401; and also by H. S. Williams, *A History of Science* (New York: Harper, 1904), **III**, 275, 278.
13. As soon as Faraday reported his discovery of the rotation of polarization of light by a magnet, French physicists looked for and found the same effect with radiant heat. F. de la Provostaye and P. Desains, *Ann. Chim. Phys.* [3] **27**, 232 (1849).
14. In 1835, Melloni was reluctant to accept the complete identity of heat and light (as suggested by Ampère, §9.3), arguing that light and radiant heat "proceed from two distinct causes." But in a footnote he added that "These two causes themselves are, perhaps, but different effects of a single cause." He insisted that light and heat are "two essentially distinct modifications which the aethereal fluid suffers in its mode of existence." M. Melloni, *C.R. Paris* **1**, 503 (1835); *Ann. Chim. Phys.* [2] **60**, 418 (1836); English trans. in *Taylor's Sci. Mem.* **1**, 388 (1837); German trans. in *Ann. Phys.* [2] **36**, 486 (1836).
15. M. Melloni, *Arch. Sci. Phys.* **5**, 238 (1847); English trans. in *Phil. Mag.* [3] **32**, 262 (1848) (quotation from pp. 274–75). For further discussion see his book *La Thermochrôse* (Naples, 1850), long footnote on pp. 272–76.
16. H. Davy, *Elements of Chemical Philosophy* (Philadelphia, 1812), pp. 46–53, 120. [This and other books cited below were originally published in Britain but I have cited the American edition when it was the only one easily available.] In 1820 Davy told the Royal Society that "the question of the materiality of heat will probably be solved at the same time as that of the undulatory hypothesis of light, if, indeed, the human mind should ever be capable of understanding the causes of these mysterious phenomena." *Collected Works* (London, 1840), 7, 11.
17. T. Young, *Phil. Trans.* **92**, 12 (1802) (quotation from p. 32).
18. T. Young, *A Course of Lectures on Natural Philosophy and the Mechanical Arts* (London, 1807; New York: Johnson Reprint Corp., 1971), **1**, 654.
19. J. Tyndall, lecture 16 March 1883, in *The Royal Institution Library of Science, Physical Science* (New York: American Elsevier, 1970), **3**, 268 (quotation from p. 271).
20. R. H. Silliman, *DSB* **5**, 165 (1972); *Hist. Stud. Phys. Sci.* **4**, 146 (1974). Cf. his remarks in his first optical paper, in *Oeuvres* (Paris, 1866), **1**, 10, 12.
21. *The Caloric Theory of Gases*; *Brit. J. Hist. Sci.* **4**, 1 (1968–69), especially the letter from Dulong to Berzelius (1820).
22. J. B. Biot, *Précis élémentaire de physique expérimentale*, 2nd ed. (Paris, 1821), **II**, 627, reprinted with little or no change in the 3rd ed. (Paris, 1824), **II**, 625. Biot discusses the similarity of light and heat, and concludes that whether caloric has a material existence "like light," or consists in vibrations propagated through a medium, the results would be the same. But still he leans toward the former view.

Johann Carl Fischer concluded a long article on heat, with much emphasis on radiant heat experiments, with the statement that whether heat is a substance or motion still remains to be decided: see *Physikalisches Wörterbuch* (Göttingen, 1827), X, 633–99. Thomas Thomson, in *An Outline of the Sciences of Heat and Electricity* (London, 1830), claimed that the wave theory of heat was held by "the greater number of the French and German chemists of the last century" but concluded (p. 335) that the problem of heat is still insoluble, though he leaned slightly toward the vibrational theories.

23. J. L. Gay-Lussac, *Ann. Chim. Phys.* [2] **13**, 304 (1820). (I am indebted to Stuart Pierson for bringing this reference to my attention.) A detailed discussion of the historical background of this paper of Gay-Lussac may be found in the recent article by P. Costabel, *Arch. Int. Hist. Sci.* **21**, 3 (1968). Further remarks by Gay-Lussac on this subject, in his *Leçons de Physique* (Paris, 1828), pp. 242–43, are quoted by W. L. Scott, *The Conflict between Atomism and Conservation Theory 1644–1860* (New York: Elsevier, 1970), p. 218.

24. Quotation from the translation in Mendoza's edition, cited in note 2. According to Mendoza, most of these notes were written around the time of the original composition of the *Reflexions*: see E. Mendoza, *Arch. Int. Hist. Sci.* **12**, 377 (1959).

9.3 Ampère's theory

By 1830 the wave theory of heat was being seriously considered as an alternative to, or modification of, the caloric theory. But the first extended discussion of it seems to have been Ampère's paper published in 1832.[1] A subsequent article, going over much of the same ground, appeared in 1835 and received wide publicity,[2] so much so that later writers sometimes simply mentioned "Ampère's theory" without giving a specific reference. The combination of Ampère's prestige, and the developments in radiant heat and the wave theory of light, was sufficient to elevate the wave theory of heat to a prominent place in the scientific world for the next decade.

Some remarks of a general character, found in Ampère's 1825 paper on electrodynamics, will indicate the drift of his earlier thinking about the nature of heat:

> The principal advantage of formulae which are derived in this way from general facts gained from sufficient observations for their certitude to be incontestable, is that they remain independent, not only of the hypotheses which may have aided in the quest for these formulae, but also independent of those which may later be adopted instead. [This refers to his formulation of the laws of electrodynamics.] The expression for universal attraction from the laws of Kepler is completely independent of the hypotheses which some writers have advanced to justify the mechanical cause to

which they would ascribe it. The theory of heat is founded on general facts which have been obtained by direct observation; the equation deduced from these facts, being confirmed by the agreement between the results of calculation and of experiment, must be equally accepted as representative of the true laws of heat propagation by those who attribute it to the radiation of calorific molecules as by those who take the view that the phenomenon is caused by the vibration of a diffuse fluid in space; it is only necessary for the former to show how the equation results from their way of looking at heat and for the others to derive it from general formulae for vibratory motion; doing so does not add anything to the certitude of the equation, but only substantiates the respective hypotheses. The physicist who refrains from committing himself in this respect, acknowledges the heat equation to be an exact representation of the facts without concerning himself with the manner in which it can result from one or other of the explanations of which we are speaking; and if new phenomena and new calculations should demonstrate that the effects of heat can in fact only be explained in a system of vibrations, the great physicist who first produced the equation and who created the methods of integration to apply it in his research, is still just as much the author of the mathematical theory of heat, as Newton is still the author of the theory of planetary motion, even though the theory was not as completely demonstrated by his works as his successors have been able to do in theirs.[3]

Thus Ampère conceded to Fourier the title "Newton of heat" while, in effect, claiming for himself the title "Newton of electrodynamics." Fortunately his positivist disdain for hypotheses about the nature of heat did not last very long. Nevertheless, when he did decide to publish his ideas about the wave theory of heat, his admiration for Fourier's theory forced him to recognize at the outset a major difficulty in using the same hypothesis to explain both the transmission of radiant heat through space and the conduction of heat through material bodies:

> instead of a vibratory motion propagated in undulations or waves in such a manner that every wave leaves at rest the fluid which it sets in motion at the instant of its passage, we have a motion propagated gradually in such a manner that the part which originally was the hottest, and consequently the most agitated (explaining the phenomena of heat by the theory of vibratory

motions), although losing heat by degrees, preserves, however, more than the parts to which it is communicating heat.

In modern terms, the problem was to reconcile the propagation of heat by waves (second-order differential equation in time) in free space, with its propagation as described by Fourier's heat conduction equation (first-order time derivative) in matter. But Ampère thought he could answer this and other possible objections to the theory.

Ampère postulated that the total *vis viva* of the system is conserved, *vis viva* being defined as[4] $\Sigma mv^2 + 2\int\Sigma F \cdot dx$. In equilibrium arrangements the integral term is zero, while it is positive for all positions near equilibrium, this being a condition for the stability of the equilibrium positions. If the atoms vibrate while immersed in a fluid, they will gradually lose *vis viva* to it; if initially one atom is vibrating and the others are at rest, then the fluid will transfer some *vis viva* to these others. However, the total *vis viva* of all the atoms will decrease as waves are propagated through the fluid out of the system, unless we suppose it to be enclosed in a container of vibrators which are maintained in a state of vibration at a constant *vis viva*. Then eventually all the vibrators will approach the same *vis viva* (though they never reach exact equality). If we assume that the rate of flow of *vis viva* between groups of atomic vibrators is proportional to the difference of the *vires viva* of the groups, then we obtain Fourier's heat conduction equation. This of course will be true only if this difference of *vires viva* is proportional to the difference of temperatures. (The modern reader can hardly restrain himself from putting into Ampère's mouth such phrases as "the *vis viva* itself is assumed to be proportional to the absolute temperature" when he reads this paper!)

Does Ampère reject the caloric theory? Well, not quite, for he says:

> We find manifestly the same result by considering the subject as we have just enunciated it, according to the system of emission [*i.e.* the material theory–S.G.B.] or according to that of vibrations, substituting for the quantity of caloric in the first system, the *vis viva* of the vibratory motions of the molecules in the second. It was in order to render the analogy between the propagation of heat in bodies and that of sonorous vibrations from solid to solid, through the medium of air, more easy of comprehension that I supposed in this explanation that the molecules of bodies do not transmit their vibratory motions one to another [immediately– word added by translator in *Philosophical Magazine*]; that in the change of form of a molecule, whatever may remain, at the

> distance at which it is situated from the neighboring molecule, of the attractive and repulsive forces of the atoms of which the two molecules are composed, is susceptible of experiencing any changes which tend to make the atoms of the second molecules vibrate. But this manner of considering the subject requiring calculations which I have not made, I have not thought proper to insist on the development of the consequences of this idea. My object in these considerations is only to demonstrate how the vibrations by which heat is propagated in bodies may follow a law entirely different from that of the vibrations of sound, of light and of radiant heat . . . .

In this remarkable paragraph we find juxtaposed three ideas about the nature of heat: first, that it really makes no difference whether heat is matter or motion, since in principle the same phenomena can be explained either way–this idea is clearly on the way out, though Ampère (writing in Paris, the stronghold of the caloric theory) makes a polite bow to it; second, heat involves vibrations of atoms transmitted always by vibrations of the ether–this idea seems to be most convenient for mathematical or analogical reasoning at the moment; and third, one might be able to dispense with the intervening ether entirely in treating ordinary heat within a material body, using instead only atomic vibrations–this idea is to be held in reserve pending further calculations. If I may suggest a historical analogy, Ampère's paper is strongly reminiscent of Einstein's "Heuristic point of view about the creation and conversion of light" (1905) in which, ostensibly without attempting to overthrow the wave theory of light, he proposed that certain phenomena could be more satisfactorily explained by a particle theory.[5]

Almost as an afterthought, Ampère rejected quite firmly a doctrine that had dominated atomic speculation during the preceding half century:

> Now, it is clear that if we admit the phenomena of heat to be produced by vibrations, it is a contradiction to attribute to heat the repulsive force of the atoms requisite to enable them to vibrate.

Notes for §9.3

1. *Bibliotheque Universelle, Arch. Sci. Phys.* **49**, 225 (1832).
2. A. M. Ampère, *Ann. Chim. Phys.* [2] **58**, 432 (1835); *Bibliotheque Universelle, Arch. Sci. Phys.* **58**, 26 (1835); English trans. in *Phil. Mag.* [3] **7**, 342 (1835).

3. R. A. R. Tricker, *Early Electrodynamics* (Oxford: Pergamon Press, 1965), pp. 157–58.
4. "... the summation of the products of the masses of all its molecules by the squares of their velocities at a given moment, adding double the integral of the sum of the products of the forces multiplied by the differentials of the spaces described, in the direction of those forces, by each molecule." This is clearly just twice the sum of the kinetic and potential energies of the system; Ampère refers to the two terms as explicit and implicit *vis viva*, respectively.
5. Ludvig Colding concluded from reading Ampère's 1835 paper that "the author regards heat as a motion of atoms." *Vid. Selsk. Skr.* [5] **2**, 167 (1850); English trans. in *Ludvig Colding and the Conservation of Energy Principle*, by P. F. Dahl (New York: Johnson Reprint Corp., 1972), see p. 55. D. S. L. Cardwell says Ampère's paper "foreshadowed some of the important points in the dynamical theory of heat, and it earns for Ampère a place as one of the founders of that theory." *From Watt to Clausius* (Ithaca: Cornell University Press, 1971), p. 218.

9.4 Reception of the wave theory, 1831–45

Among the early supporters of the wave theory of heat we find C. Matteucci, an Italian physicist (later a prominent politician);[1] August de la Rive, Swiss physicist who had provided facilities for some of Melloni's experiments;[2] Dionysius Lardner, British encyclopedist;[3] Mrs. Mary Somerville, British science writer;[4] possibly David Brewster, British physicist;[5] James Forbes, British physicist who made important experimental contributions to the study of radiant heat;[6] Gabriel Lamé, Professor of Physics at the École Polytechnique in Paris, best known for his work on elasticity theory;[7] William Whewell, British scientist-historian-philosopher;[8] and several others.[9]

Equally significant, perhaps, in indicating the state of scientific opinion are statements by the few remaining caloric theorists such as Poisson who seem to realize that the wave theory is now the most popular even though they still refuse to accept it themselves.[10] Thus, as early as 1834, William West in England wrote:

> I am aware that the once prevalent doctrine of the materiality of caloric and electricity has given way before the conclusions deduced from certain optical phenomena; but ....[11]

Someone like Robert Hunt who lived through the 1840's without being converted to the wave theory had to be quite defensive about retaining his preference for the caloric theory:

> ... notwithstanding the array of talent which stands forward in [the undulatory theory's] support, we must not allow ourselves to

be deceived by the deductions of its advocates, or dazzled by the brilliancy of their displays of learning.

Radiant heat appears to move in waves; but that calorific action is established by any system of undulation, is a deduction without a proof; and the thermic phenomena of matter are more easily explained by the hypothesis of a diffusive subtile fluid.[12]

Hunt's position was logically unassailable but historically obsolete.

Even the terms in which the issue is stated are favorable to the wave theory; no longer is it a question (as it had been 20 to 30 years earlier) of whether heat is substance or quality; it is now a question of whether one is to accept the "emission" theory or the "undulation" theory. Both sides agree that heat and light must be considered together, and those who still maintain the emission theory of heat often seem to be compelled to maintain also the emission theory of light.[13]

It might appear that the considerable interest shown in molecular theories such as that of Mossotti[14] contradicts our assertion that most scientists accepted the wave theory of heat after 1830. Mossotti's molecular model – a nucleus surrounded by an atmosphere of fluid particles which attract the nucleus but repel each other – is quite similar to that used by caloric theorists, who identified the fluid particles with caloric. But careful examination of later papers on this subject shows that this identification was gradually forgotten.[15] An especially interesting case is the theory of Philip Kelland (Professor of Mathematics at Edinburgh University) who actually did state that the atmosphere of an atom is composed of "caloric" particles; yet then went on to suggest that heat is transmitted by the *vibrations* of the particles of caloric.[16] This example shows how a scientist may adhere to the terminology of the caloric theory yet abandon its substance; perhaps one should translate "caloric fluid" as "ether" in all works written after 1830.

Another apparent exception to the general adoption of the wave theory was the persistent idea that heat (caloric) is a "repulsive principle" or "repulsive force." (Despite Ampère's remark quoted at the end of the last section, even some of the wave theorists retained this idea). It has been suggested that the identification of heat as a force rather than as matter could have played a role in the decline of the caloric theory, and there is indeed some evidence of this as an influence of *Naturphilosophie*.[17] But the idea was common in the 18th century *before* the rise of the caloric theory, and remained popular long after that theory was dead.[18] It also inspired occasional attempts to detect a force between macroscopic heated bodies.[19] Though the "repulsive power of heat" affected the language of various theories, it did not help very much to discriminate between them.

Laplace died in 1827, Fourier in 1830, and Poisson in 1840. Who was left to defend the caloric theory? Only a few minor scientists such as Pouillet, Pinaud, and Soubeiran in France,[20] Comstock and Grund in America,[21] and a handful of others.[22] The Academy of Sciences in Paris did continue to resist the newer ideas, despite the influence of Ampère and Lamé; the *Comptes Rendus* records several papers on the nature of heat, of which only the title or a brief summary was allowed to be published; and some of them probably contained views favorable to the mechanical theory.[23] At least we know that one of Mayer's papers was refused publication until after a translation of Joule's work had appeared in 1848.[24]

By 1842 William Robert Grove, lecturing at the London Institution, could assume that his audience already knew that the wave theory of heat was considered the most satisfactory, although he personally thought it was superfluous to assume a peculiar ethereal fluid: ordinary matter diffused through space would be sufficient to transmit the vibrations.[25] Similarly Mohr in Germany[26] and Joseph Henry in America[27] were acknowledging that they had been led to a wave or vibratory conception of heat by the facts of *radiant* heat, even though they preferred to minimize the importance of the ether. Melloni, as we have already noted, declared his allegiance to the wave theory in 1842 and more definitely in 1847.[28] Berzelius, skeptical in 1839, accepted it by 1845.[29] In 1845, the physiologist Ernst Brücke published a critical review of the evidence against the identity of heat and light, in connection with his studies on the physical properties of the eye; he apparently wanted to believe in this identity and to accept the wave theory of heat, though there were still some obstacles.[30] A few writers, while acknowledging that the material theory of heat was probably wrong, continued to use it for the sake of "simplicity" in explaining the phenomena.[31] This attitude was criticized by a French textbook writer, Bailly, who insisted that since the wave theory had been established by the latest results of scientific research it must also be used in teaching about heat.[32] Bailly also recognized the possibility of a third theory, in which heat would be attributed to the vibrations of bodies rather than to the ether, but he asserted that such a theory is refuted by direct observations and has been generally abandoned.

Although there is thus abundant evidence for the popularity of the wave theory during the 1840's,[33] no one seems to have gone beyond Ampère's work and tried to deduce quantitative consequences different from those of the caloric theory (Ampère himself died in 1836). The most famous prediction of the wave theory of heat is the so-called Rayleigh–Jeans law, which was not published until 1900, and was not believed to be valid even by Rayleigh.[34] Aside from this, the primary significance of the theory for the subsequent development of physics was the part it played in the discovery of energy conservation and thermodynamics, which we must now examine.

Notes for §9.4

1. C. Matteuci, *Bibl. Univ.* **50**, 1 (1832).
2. [A. de la Rive], *Bibl. Univ.* **51**, 243 (1832), **59**, 144 (1835) [see esp. p. 154], **60**, 279 (1835).
3. D. Lardner, *Treatise on Heat* [The Cabinet Cyclopedia] (London, 1833), pp. 394–98.
4. Mrs. Somerville, *On the Connection of the Physical Sciences* (Philadelphia, 1834), p. 195, *Mechanism of the Heavens* (London, 1831), lvii.
5. See editorial note in *Phil. Mag.* [3] **7**, 157 (1835); the editors at that time were Brewster, Richard Taylor, and Richard Phillips.
6. J. D. Forbes, *Trans. R. S. Edinburgh* **13**, 131 (1835) [see esp. p. 147]; *Phil. Mag.* [3] **7**, 246 (1836). Letter from Forbes to R. L. Ellis, 20 December 1840, in *Life and Letters of James David Forbes* by J. C. Shairp et al. (London, 1873), p. 151.
7. G. Lamé, *Ann. Chim. Phys.* **55**, 322 (1833), **57**, 211 (1834). In the textbook cited by Mendoza, *op. cit.* (§9.2, note 2), Lamé clearly prefers the wave theory to the emission theory, though he states (perhaps to avoid offending Poisson) that it is not necessary to decide between them [*Cours de Physique de l'École Polytechnique* (Paris, 1836), 297–98]. We may infer something about how much weight Lamé's opinion might carry from the assessment in a recent article by J. W. Herivel, *Brit. J. Hist. Sci.* **3**, 109 (1966). In addition to providing much useful information about the situation in Paris which is relevant to the background of the wave theory of heat, Herivel points out that in the period 1850–70, "and for that matter in the decade immediately preceding 1850, [there was] no *creative* French theoretical physicist remotely of the calibre of Thomson or Clausius, let alone Maxwell." In a footnote he specifies that "creative" is to be taken "as opposed to a competent, and even original, theoretical physicist such as G. Lamé (1795–1870)" (*ibid.*, p. 115). Stretching this just a bit, we could say that the best physicist in France was a supporter of the wave theory of heat, even if some of the others opposed it.
8. W. Whewell, *History of the Inductive Sciences* (London, 1837), **2**, 180–84. Aside from its value as contemporary evidence for the acceptance of the wave theory of heat, Whewell's work is almost the only publication on the *history* of science which discusses this theory. Rosenberger mentions Ampère's theory but states incorrectly that Ampère was the only scientist who attributed both heat and light to vibrations of the same ether [F. Rosenberger, *Die Geschichte der Physik* (Braunschweig, 1887–90), **3**, 230–33]. Historians of science who have mentioned the wave theory more recently are cited in §9.6, note 6.

Four years earlier Whewell had seemed somewhat favorable to the caloric theory in his Bridgewater Treatise: "If heat be a fluid; if to cool be to part with this

fluid, as many philosophers suppose...." *Astronomy and General Physics considered with Reference to Natural Theology* (London, 1833), p. 162.
9. [J. C. Poggendorff] editorial note added to the German trans. of a paper by Melloni, in *Ann. Phys.* [2] **38**, 2 (1836). [Anon.] *Magazine of Popular Science* **1**, 145 (1836). Pinault, *Traité Élémentaire de Physique* (Paris, 2nd ed., 1836), p. 292. A. Baumgartner, *Anfangsgründe der Naturlehre* (Wien, 1837), pp. 131–32 [this book contains one of the first uses of the word *Thermodynamik*, defined here as the study of the motion of heat]. T. Webster, *The Elements of Physics* (London, 1837), p. 280. C. C. Person, *Éléments de Physique* (Paris, 1837), **2**, 224–25. M. A. Gaudin, *Bibl. Univ.* **52**, 139 (1833), cited by T. M. Cole, Jr., *Isis* **66**, 344 (1975).
10. S. D. Poisson, *Théorie Mathématique de la Chaleur* (Paris, 1835); *Memoire et notes formant un Supplement*... (Paris, 1837), p. 27f.
11. W. West, *Phil. Mag.* [3] **5**, 110 (1834).
12. R. Hunt, *The Poetry of Science or Studies of the Physical Phenomena of Nature* (Boston, 1850, from the 2nd London ed.), p. 65. This is presumably the same Robert Hunt who was Keeper of Mining Records at the Museum of Practical Geology and wrote an article on "The Science of the Exhibition" published in *The Crystal Palace Exhibition, Illustrated Catalogue* (London, 1851; New York: Dover Pubs., 1970) where one finds on p. x the statement: "It must always be borne in mind that no physical power can be produced without a change of material somewhere...."
13. Poisson, *op. cit.* (note 10). J. Barton, *Phil. Mag.* [3] **10**, 342 (1837). R. Hare, *A Compendium of the Course of Chemical Instruction in the Medical Department of the University of Pennsylvania*, new section added to the 4th ed. (Philadelphia, 1840), pp. 75–76. J. Johnston, *A Manual of Chemistry* (Philadelphia, 1842), p. 59. H. W. Brandes, *Vorlesungen über die Naturlehre* (Leipzig, 1844), pp. 471, 558. L. Gmelin, *Handbook of Chemistry* (London, 1848, trans. from German ed. of 1843), pp. 163, 167, 212.
14. O. F. Mossotti, *Sur les forces qui régissent la constitution intérieure des corps* (Turin, 1836); English trans. in *Taylor's Sci. Mem.* **1**, 448 (1837). Other works on this subject are reprinted in his *Scritti* (Pisa: Domus Galilaeana, 1942–55). The editors of *Phil. Mag.* said that his "mutual identification of the attractive forces of electricity, aggregation, and gravitation" constituted "one of the most remarkable discoveries of the present area in science" (see **10**, 320 (1837)).
15. J. Challis, *Phil. Mag.* [3] **7**, 89 (1836). P. Kelland, *Trans. Camb. Phil. Soc.* **6**, 235 (1837), **7**, 25 (1839); *Phil. Mag.* [3] **21**, 124, 202, 263, **22**, 8 (1842). S. Earnshaw, *Trans. Camb. Phil. Soc.* **7**, 97 (1839). This last paper by Earnshaw contains the famous "Earnshaw theorem" in electrostatics which was used as an argument against all static atomic models based on the equilibrium of some arrangement of charged particles, around 1900; see *Am. J. Phys.* **27**, 418 (1959).

See also C. Babbage, *The Ninth Bridgewater Treatise, A Fragment* (London, 1837, 2nd ed., 1838, rept. by Cass, 1967), pp. 180–85; T. Exley, *Phil. Mag.* [3] **11**, 496 (1837); P. Cooper, *Phil. Mag.* [3] **10**, 355 (1837); R. L. E., *Phil. Mag.* [3] **19**, 384 (1841). Various theories of this kind are summarized by I. Todhunter, *A History of the theory of elasticity and of the strength of materials from Galilei to the present time* (Cambridge, 1886; New York: Dover Pubs., 1960), **1**.
16. P. Kelland, *Theory of Heat* (Cambridge, 1837), III, 104, 145, 181–82. Kelland published a further explanation of his views in a note added to a new edition of Thomas Young's *Course of Lectures on Natural Philosophy* (London, 1845), p. 506. He states that although recent experiments on polarization and conduction do show

the wave nature of heat, most other phenomena such as latent heat cannot be explained by a purely wave theory as yet. The facts "appear to demand a corpuscular theory, wholly or partly accompanied by transverse vibrations. The hypothesis which I have advanced [in *Theory of Heat*] is, that heat is due to the existence of repulsive atoms which penetrate all material substances; so that expansion arises from the accumulation of such atoms; but that the transmission of heat is partly effected by transverse pulses.... Solar heat is transmitted altogether by such transverse pulses, so that its intensity is measured by the intensity of the pulses, whilst the heat of a fire is perhaps due in part to normal ones, or, which is the same thing, to a flow of atoms impelling by their repulsion those which are in advance of them." (I am indebted to Dr. Charles Weiner for this reference.)

The problem of explaining latent heat on the wave theory was also raised by R. Kane, *Elements of Chemistry* (New York, 1842), **46**, 103.

17. L. P. Williams, *Hist. Sci.* **1**, 1 (1962). R. Kargon, *Centaurus* **10**, 253 (1964). H. C. Oersted, *Recherches sur l'identite des forces chimiques et électriques* (Paris, 1813), pp. 159–200.
18. See Grove's lecture of 1843, cited below in note 25, p. 53; W. and R. Chambers, *Elements of Natural Philosophy* (New York, 1849), p. 31; W. Petrie, *Edinburgh New Phil. J.* **51**, 120 (1851); S. E. Coues, *Outlines of a System of Mechanical Philosophy* (Boston, 1851), p. 26; Z. Colburn, *An Inquiry into the Nature of Heat* (London, 1863); A. Cazin, *Smithsonian Institution Report*, p. 231 (1868).
19. A. Fresnel, *Ann. Chim. Phys.* [2] **29**, 57, 107 (1825); Baden Powell, *Phil. Trans.* **124**, 485 (1834); *Phil. Mag.* [3] **12**, 317 (1838). R. Addams, *Phil. Mag.* [3] **6**, 415 (1835). W. Crookes, *Phil. Trans.* **164**, 501 (1874). The Crookes paper shows the connection between this earlier research on the repulsion of heat, possibly associated with an emission theory, and the radiometer fad of the 1870's (§5.5).
20. [C.] Pouillet, *Elements de physique expérimentale et de météorologie* (Paris, 2nd ed., 1832), pp. 237–38; *C.R. Paris* **24**, 915 (1847). Pouillet was Professor of Physics at the École Polytechnique; his estimate of the temperature of the sun is mentioned in §13.6. He did not change his statements about heat in the 1847 and 1856 editions of his textbook, but later German editions (1847, 1852, etc.) edited by J. Müller dropped the support for caloric theory.

E. Soubeiran, *Précis élémentaire de physique* (Paris, 2nd ed., 1846). A. Pinaud, *Programme d'un cours élémentaire de physique* (Paris, 5th ed., 1848), p. 113.
21. J. L. Comstock, *Elements of Chemistry* (New York, 10th ed., 1834), p. 11; *ibid.* (New York, 29th ed., 1839), p. 11; also in the 1852 ed. F. J. Grund, *Elements of Natural Philosophy* (Boston, 2nd ed., 1835), p. 186.
22. N. Arnott, Elements of Physics (Philadelphia, 1841), p. 281 [Arnott was at the Royal College of Physicians in London]. F. A. Clemens, *Grundriss der Naturlehre* (Königsberg, 1839), **2**, 75–76. J. P. Cooke, *Elements of Chemical Physics* (Boston, 1860), p. 430.
23. Babinet, *C.R. Paris* **7**, 781 (1838) [cf. *ibid.* **63**, 581, 662 (1866)]. Couche, *C.R. Paris* **18**, 312 (1844). Paget, *C.R. Paris* **19**, 1406 (1844). Briot, *C.R. Paris* **24**, 877 (1847).
24. *C.R. Paris* **23**, 220, 544 (1846), **27**, 385 (1848).
25. W. R. Grove, *A Lecture on the Progress of Physical Science since the opening of the London Institution* (London, 1842), p. 27; see also "On the correlation of physical forces" (1843) in *The Correlation and Conservation of Forces*, ed. E. L. Youmans (New York, 1865), p. 55.
26. F. Mohr, *Ann. Chem. Pharm.* **24**, 141 (1837); English trans. in *Phil. Mag.* [5], **2**, 110

(1876). In his note at the end of this translation, P. G. Tait asserted that this paper "contains, in a considerably superior form, almost all that is correct in Mayer's paper." See also [C.] F. Mohr, *Allgemeine Theorie der Bewegung und Kraft, als Grundlage der Physik und Chemie* (Braunschweig, 1869), which includes a reprint of the 1837 paper; R. E. Oesper, *J. Chem. Ed.* **4**, 1357 (1927).

27. J. Henry, *Proc. Am. Phil. Soc.* **4**, 287 (1846), **6**, 84 (1851); also his report on the interference of heat rays, *Proc. Am. Phil. Soc.* **4**, 285 (1846) and another paper on heat in *Am. J. Sci.* **5**, 113 (1848). In the last paper cited he said, "The facts with regard to heat as well as light therefore show that the theory of undulation is not an imagination, but the expression of a *law*." Henry met Melloni in Paris in 1837 and this encounter may have stimulated his interest in radiant heat: see *Edinburgh New Phil. J.* **26**, 300 (1839) and T. Coulson, *Joseph Henry* (Princeton: Princeton University Press, 1950), p. 122.

28. See §9.2, note 15.

29. J. J. Berzelius, *Traité de Chimie*, "nouvelle edition entierement refondue d'après la 4me edition allemande, publiée in 1838" (Bruxelles, 1839), **1**, 35; *Traité de Chimie Minérale, Végetale et Animale* (Paris, 2nd French ed., 1845), **1**, 35.

30. E. Brücke, *Ann. Phys.* [2] **65**, 593 (1845). The identity of light and heat was rejected by L. Moser, *Ann. Phys.* [2] **58**, 105 (1843).

31. E. Peclet, *Traité élémentaire de physique* (Bruxelles, 4th ed., 1838), pp. 234–35. J.-M.-M. Peyré, *Cours de Physique* (Paris, 2nd ed., 1840), p. 256. J. Persoz, *Introduction à l'étude de la chimie moléculaire* (Paris and Strasbourg, 1839), p. 218. C. Despretz, *Traité élémentaire de physique* (Paris, 3rd ed., 1832), p. 77. J. Johnston, *A Manual of Chemistry* (Middletown, 1840), **13**, 57. J. Daniell, *An Introduction to the study of chemical phenomena* (London, 1843), p. 208–9.

32. C. Bailly, *Nouveau manuel complet de physique* (Paris, 1841), pp. 204–7.

33. [A.] Becquerel, *Traité de Physique* (Paris, 1842), I, 163. Mrs. Somerville, *On the connection of the physical sciences* (London, 7th ed., 1846), p. 258. John W. Draper, *Textbook on Natural Philosophy* (New York, 1847), p. 253. E. Peclet, *Traité élémentaire de physique* (Bruxelles, 1847), p. 361. J. Müller, *Principles of Physics and Meteorology*, trans. from German (London, 1847), p. 497 [Müller argues that because radiant heat consists in ether vibrations, *therefore* sensible heat must consist in vibrations of the material parts of bodies; many others followed this line of argument at least implicitly]. G. Bird, *Elements of Natural Philosophy* (London, 1848), p. 487. C. H. D. Buys-Ballot, *Scheets eener physiologie van het onbewerktuigte ryk de natuur* (Utrecht, 1849), as summarized by Rosenberger, *op. cit.* (note 8), pp. 538–40. Thomas Graham, *Elements of Chemistry*, 2nd American ed. based on the 2nd English ed. of 1850 (Philadelphia, 1852), p. 96.

The authors of many of the textbooks cited in note 4, §9.6 below, probably held similar views in the 1840's. I have not attempted to track down all the first editions, since the evidence already obtained seems to be sufficient to establish the point.

An elaborate critical review of opinions about the nature of heat, with references to a number of works published in the early 19th century, may be found in Muncke's article "Wärme" in *Gehler's Physikalisches Wörterbuch* (Leipzig, 1841) **10**, 1st Abt.

34. Rayleigh, *Phil. Mag* [5] **49**, 539 (1900), reprinted with a note, dated 1902, on Planck's formula, in his *Scientific Papers* (Cambridge University Press, 1903; New York: Dover Pubs., 1964), **4**, 483. Rayleigh's intention in this paper was not to deduce a distribution function for black-body radiation as a rigorous consequence of classical

physics, but to improve Wien's distribution by using the assumption that equipartition applies *only to low frequencies.* In this way he obtained the formula $\theta k^2 \, dk$ (θ = absolute temperature, k = wave number), which if integrated over all k would of course diverge; but Rayleigh explicitly stated that for large k one must introduce an exponential factor $\exp(-ck/\theta)$. Thus he recognized that equipartition could not apply to high frequencies, but did not by any means imply that this was to be regarded as a failure of classical physics. On the physical basis for Rayleigh's assumption here, see my remarks at the end of §1.8.

9.5 Transition from wave theory of heat to thermodynamics

In his classic article on the history of energy conservation, Thomas S. Kuhn lists 12 co-discoverers of the principle.[1] Four of them—Mayer, Joule, Colding, and Helmholtz—are considered the primary discoverers because they not only announced the general principle but also provided concrete quantitative applications. Four others—Carnot, Marc Seguin, Karl Holtzmann, and G. A. Hirn—computed a mechanical equivalent of heat but did not bother to make a general statement about the convertibility of *all* forms of energy. A third quartet—C. F. Mohr, W. R. Grove, Faraday, and Liebig—did make such a general statement but failed to develop the numerical aspects of energy conversions. Kuhn argues convincingly that the "simultaneous" nature of this discovery—all but two published their work between 1837 and 1847, probably independently—implies the existence of some common factors in the environment of early 19th-century science, factors not present earlier. He identifies three such factors: the development of a quantitative bookkeeping approach in steam-engine technology; discoveries of many conversion processes linking electricity, magnetism, and heat; and speculations of *Naturphilosophie* suggesting the basic unity of all forces in nature. Having examined some of the writings of these 12 men, I propose to add a fourth factor: investigations of radiant heat, and in particular the wave theory of heat.

The views of Carnot on radiant heat have already been quoted. The case of C. F. Mohr is also fairly clear-cut since he states:

> The phenomena of heat have been till now almost exclusively explained in textbooks by the assumption of a heat-substance. The discoveries of Melloni have made this view inapplicable to the phenomena of Radiant Heat; they require the assumption of vibrations similar to those of the Undulatory Theory of Light. The Propagation, Transmission, and Polarization of Radiant Heat have been completely explained by these assumptions; and, with such

facts to guide us, it is certainly no mere idle speculation to attempt to extend this view to the phenomena of common or stationary heat.... Heat is thus no longer a particular kind of matter, but an oscillatory motion of the smallest parts of bodies.[2]

Grove, in his 1842 lecture cited above, and in his general statement of energy conservation a year later, indicated his qualified acceptance of the wave theory of heat.[3]

Helmholtz, in his 1847 memoir on the conservation of force, concluded that heat must be explained in terms of motion, preferably by a wave theory such as Ampère's.[4] The importance of the wave theory of heat in his thinking can be better judged from an article on physiological heat which he wrote for a medical encyclopedia two years earlier; here, Helmholtz wrote:

> Recently, especially through the complete equality of the laws of heat radiation with those of light, not only the similarity but indeed the identity of both agents has been made probable, and we are thereby led to a wave theory of heat, as to a wave theory of light. Moreover, it is found that heat can actually be generated by various other natural forces, without the occurrence of such changes in the molecular properties of the body to which one might attribute the liberation of latent heat. In particular, first, heat is liberated by the annihilation of mechanical force in the friction of solids against solids, or solids against fluids; second, by the equalization of electric tension, which can again be produced by rubbing or by the motion of magnets.... Thus the possibility of a material theory of heat disappears, since the conservation of quantity would be the most necessary consequence of such a theory, and we are forced to consider heat as well as light to be motion. The relation between free and latent heat discussed above in the language of the material theory would still remain unchanged, if in place of quantity of substance we put quantity of motion, according to the basic laws of mechanics; there is only a difference when we are concerned with the creation of heat motion by other forces of motion and we have to determine the equivalent amount of heat produced by a definite quantity of mechanical or electrical force.[5]

Colding, in 1850, pointed out that Ampère's theory of heat is in harmony with his own version of the conservation principle:

> [Ampère] has proposed the hypothesis that while all rays of light

and heat advance through the aether in the form of waves, the conduction of heat in bodies depends on the vibrations of atoms and their propagation from particle to particle. While the author thus regards heat as a motion of atoms, he compares the quantity of heat contained in bodies with the living force of atoms, and proceeds to show that the general equations for the propagation of heat in a body must also be valid for the propagation of the living force. As I believe I have demonstrated in the above that the internal activity of a body must necessarily be equal to the living force possessed by its particles, it also necessarily follows from this that the proposed principle, applied to the propagation of heat in bodies, far from being in conflict with nature, does in fact lead to truth demonstrated by experience.[6]

* * *

For the other discoverers of energy conservation, the influence of the wave theory was much weaker. Joule, in 1845, made a brief reference to it, suggesting that Davy's idea of rotating molecules might be revived; in order to apply that theory to radiation,

we have only to admit that the revolving atmospheres of electricity possess, in greater or less degree, according to circumstances, the power of exciting isochronal undulations in the ether which is supposed to pervade space.[7]

But Joule was soon to discard Davy's theory of molecular motion in favor of Herapath's (*i.e.* the kinetic theory of gases), and in his later writings on the nature of heat he implies and sometimes even explicitly states that radiant heat is irrelevant to thermodynamics.[8]

Faraday, in lectures on heat at the Royal Institution in 1845, reviewed Melloni's experiments on radiant heat and endorsed the "analogy that Melloni has drawn between the various rays of light and those of heat" but did not commit himself to any specific theory of the nature of heat.[9] In his speculations on ray-vibrations, disclosed (somewhat unwillingly) the following year, he preferred to discard the conventional ideas about the ether in favor of his lines of force. Thus "radiation" (both luminous and calorific) might consist in vibrations of lines of force; but this did not seem to entail any particular consequences for the nature of heat.[10]

Mayer, as has often been noted, was somewhat contemptuous of all attempts to reduce heat to motion, preferring to think of it as a "force" of equivalent status to other forces; his attitude is best illustrated for our purposes by the following remark which he published in 1851:

> We are taught by history that ... the most sagacious hypotheses concerning the state and nature of a peculiar "matter" of heat, concerning a "thermal aether," whether at rest or in a state of vibration, concerning "thermal atoms," supposed to exercise their functions in the interstices between the material atoms, or other hypotheses of like nature, have not availed to solve the problem.[11]

These examples (taken with my failure to find enthusiasm for the wave theory of heat among those co-discoverers concerned primarily with the engineering aspects of heat and work) suggest that the speculations about radiant heat did contribute something to the climate of scientific opinion that favored the emergence of energy conservation in the 1840's; but, like the other factors mentioned by Kuhn, they were neither necessary nor sufficient in leading to that discovery. That the wave theory of heat was a partial but not a *sufficient* basis for thermodynamics is shown by the case of W. J. M. Rankine, the Scottish engineer-physicist who was one of the three founders of thermodynamics (with Clausius and Thomson). Rankine tells us that the object of his researches on the hypothesis of molecular vortices was:

> to deduce the laws of elasticity and of heat as connected with elasticity, by means of the principles of mechanics, from a physical supposition consistent with and connected with the theory which deduces the laws of radiant light and heat from the hypothesis of undulations. Those researches were commenced in 1842 ...

but put aside for several years for lack of experimental data, then resumed when Regnault's experiments were published.[12] Rankine continued to develop his own version of the wave theory of heat, though it was not recognized as such by his contemporaries, and was generally ignored after the revival of the kinetic theory of gases.

Notes for §9.5

1. T. S. Kuhn, in *Critical Problems in the History of Science*, ed. M. Clagett (Madison: University of Wisconsin Press, 1959), p. 321.
2. See §9.4, note 26. In his textbook on *Heat* (London, 1884, rept. with corrections,

1904), p. 247, P. G. Tait discussed the experiments showing the identity of light and radiant heat, and remarked: "It is curious to notice that the original speculations of Mohr, of date 1837, as to the true nature of heat were mainly based on these discoveries."
3. See §9.4, note 25.
4. H. von Helmholtz, *Ueber die Erhaltung der Kraft* (Berlin, 1847); see Brush, *Kinetic Theory* 1, 108–9.
5. H. von Helmholtz, art. "Wärme, physiologisch," *Encyklopädisch Handwörterbuch der medicinischen Wissenschaften* (1845), as reprinted in his *Wissenschaftliche Abhandlungen* (Leipzig, 1882–95), 2, 680; quotation trans. by S. G. B. from pp. 699–700.
6. See §9.3, note 4.
7. J. P. Joule, *Phil. Mag.* [3] **26**, 369 (1845).
8. J. P. Joule, *B.A. Rep.* **18**, 21 (1848) [transition to Herapath theory, no mention of radiant heat]; *Phil. Trans.* **140**, 61 (1850) [radiant heat and similar subjects "do not exactly come within the scope of the present memoir"]. Further indication of Joule's ambivalent attitude toward the wave theory of heat is found in an undated draft manuscript at Manchester University: "Fresh arguments were however constantly adduced in favor of the vibratory hypothesis and the labours of Forbes and others added new proofs of the real nature [the word 'character' is deleted] of heat [phrase 'when in the year 1843' deleted]. To these I need not advert at any length [phrase 'but will proceed to the researches made by' deleted] as the subject of radiation of heat is [phrase 'not necessarily connected with our subject' deleted] an exceedingly complicated one and would occupy too much time nor is the proof derived from the phenomena of radiation a decisive one . . ." (from papers held at the Department of History of Science and Technology, The University of Manchester Institute of Science and Technology; a microfilm copy was kindly provided by Dr. A. J. Pacey).
9. M. Faraday, *Magazine of Science* **6**, 126, 131, 139, 151, 215 (1845).
10. M. Faraday, *Phil. Mag.* [3] **28**, 447 (1845).
11. J. R. Mayer, *Bemerkungen über das mechanische Aequivalent der Wärme* (Heilbronn and Leipzig, 1851); *Phil. Mag.* [4] **25**, 493 (1863) (quotation from p. 498).
12. W. J. M. Rankine, *Trans. R. S. Edinburgh* **20**, 147 (1850).

9.6 Disappearance of the wave theory of heat after 1850

The wave theory of heat might have been the starting point for the new kinetic theory of gases, but the circumstances were unfavorable. In 1845, a Scottish scientist, J. J. Waterston, submitted a paper to the Royal Society of London, containing a comprehensive development of the kinetic theory (§3.2). Waterston's paper began with the remark:

> Of the physical theories of heat that have claimed attention since the time of Bacon, that which ascribes its cause to the intense vibrations of the elementary parts of bodies has received a considerable accession of probability from the recent experiments

of Forbes and Melloni. It is admitted that these have been the means of demonstrating that the mode of its radiation is identical with that of light in the quantities of refraction and polarization. The evidence that has been accumulated in favour of the undulatory theory of light has thus been made to support with a great portion of its weight a like theory of the phenomena of heat . . . ..[1]

But the Royal Society referees (Baden Powell and John William Lubbock) did not think Waterston's paper deserved publication, and it remained unknown in the Royal Society Archives until 1891 when Lord Rayleigh disinterred it (§3.1). When Joule, Clausius, and Maxwell revived the kinetic theory they based their assumptions on the mechanical theory of heat but tended to treat molecular thermal motion completely apart from radiant heat. Moreover, as Waterston himself had pointed out, in order to accept the kinetic theory of gases, it was necessary to assume that molecules can move freely through empty space (except when they collide with each other or with solid objects) so that any kind of energy exchange with an ether has to be ignored.[2] Thus the role of ether vibrations had to be eliminated from the theory of gas properties, even though it was still important in spectroscopy.

As Louis Soret in Geneva pointed out in 1854, the wave theory of heat is completely consistent with thermodynamics, and at that time there seemed to be no reason why the two theories could not peacefully coexist.[3] Indeed, many books and papers by minor scientists continued to use or refer to the wave theory as if it were still acceptable for several decades after 1850.[4] However, the leading physicists of this period–Joule, Thomson, Helmholtz, Maxwell, Boltzmann, etc.–seemed to ignore it. (Clausius alluded to it in a popular lecture in 1857, as mentioned in §4.3, while in the same year Joseph Henry mentioned in the Agricultural Report of the Commissioner of Patents that heat and light are both transmitted by vibrations of the atoms of the etherial medium.[5])

While there may have been good reasons for dropping the wave theory of heat at this particular stage of physics, it is still rather puzzling that it has been so completely forgotten in works on the history of physics. With the exception of a few 19th-century historians,[6] almost all accounts state or imply that the caloric theory was accepted until it was replaced by thermodynamics around 1845–50.[7] Sometimes this myth is combined with the other one, and it is stated that although Rumford and Davy "really" established the mechanical theory, the caloric theorists obstinately persisted in their error for another 40 or 50 years, until Mayer and Joule finally persuaded other scientists to accept a truth that should have been obvious in 1800.[8] Without becoming involved in an extensive digression on the historiography of 19th-century physics, I would like to suggest one possible origin for both myths.

In 1849, William Thomson wrote a remarkable paper on heat, in which he referred to "the ordinarily-received, and almost universally-acknowledged, principles with reference to 'quantities of caloric' and 'latent heat'."[9] Thomson claims that the principle of conservation of *heat* has been accepted "by almost everyone who has been engaged on the subject" except Joule; and so generally is this principle admitted "that its application in this case has never, so far as I am aware, been questioned by practical engineers." The paper is of course remarkable mainly because it reveals Thomson on the brink of abandoning the caloric theory himself, but I think it also displays an amazing ignorance of the current state of opinion among physicists on the nature of heat.

How could William Thomson have been unaware of the fact that most physicists had accepted the wave theory of heat by 1849, if they had not already adopted the mechanical theory? We know, thanks to the work of Elinor Barber and Robert Merton that Thomson made at least 32 discoveries "which he eventually found . . . had also been made by others"[10] – a record that could hardly be compiled by a scientist who bothered to read the literature before plunging ahead with his own research. If we assume that Thomson was not familiar with any works other than those he explicitly mentions, then we would conclude that his knowledge of theories of heat was gained primarily from Fourier, Philip Kelland (whose ambiguous views have been mentioned above), Carnot, Clapeyron, and some anonymous engineers. This is probably overstating the case, but I do think it is quite fair to say that Thomson's statement about the status of caloric theory in physics in 1849 was simply wrong. (It may have been accurate for engineering.)

Having published a paper that later convinced one group of readers that the caloric theory was generally accepted up to 1849, Thomson wrote two years later a paper which seems to have convinced another group of readers that it had been demolished 40 years earlier. During this interval someone has told him about the wave theory of heat, and of course the famous incident with Joule at the British Association has finally had its effect. After quoting Davy, Thomson says:

The Dynamical Theory of Heat, thus established by Sir Humphry

Davy, is extended to radiant heat by the discovery of phenomena, especially those of the polarization of radiant heat, which render it excessively probable that heat propagated through vacant space, or through diathermane substances, consists of waves of transverse vibrations in an all-pervading medium.[11]

He then refers to Mayer's and Joule's discoveries which "would so afford, if required [!], a perfect confirmation of Sir Humphrey Davy's views."[11] Presumably contemporary physicists only bothered to read the later "correct" paper, and thus learned that the caloric theory had been demolished in 1800; whereas historians went back to the earlier paper for evidence as to views about the nature of heat just before the adoption of thermodynamics. Both were misled.

Notes for §9.6

1. J. J. Waterston, *Phil. Trans.* **183**, 5 (1893 [read 1846]); reprinted in *The Collected Scientific Papers of John James Waterston*, ed. J. S. Haldane (Edinburgh: Oliver & Boyd, 1928).
2. Waterston's *Papers*, pp. 278-79.
3. L. Soret, *Arch. Sci. Phys.* **26**, 33 (1854). Soret quotes Joule's remark [*op. cit.*, note 7, §9.5] about heat waves excited by rotating molecules, not realizing that Joule has since dropped his interest in radiant heat. See also G. von Quintus Icilius, *Experimental-Physik* (Hannover, 1855) who accepts the wave theory of heat and implies that it is compatible with the mechanical theory; J. Jamin, *Cours de Physique de l'École Polytechnique* (Paris, 1859), **2**, 248, 436.
4. Z. Allen, *Philosophy of the Mechanics of Nature* (New York, 1852), **41**, 344; *Solar Light and Heat* (New York, 1879), **28**, 68. C. G. Greisz, *Lehrbuch der Physik* (Wiesbaden, 1853), p. 390. Johann Müller, *Grundriss der Physik und Meteorologie* (Braunschweig, 4th ed., 1853), p. 460. L. Soret, *op. cit.* (note 3). A. Brown, *The Philosophy of Physics* (Redfield, N.Y., 1859), pp. 215-25, 273-77. Quintus Icilius, *op. cit.* (note 3). P. A. Daguin, *Traité élémentaire de Physique* (Toulouse and Paris, 1855, 1861), **1**, 626, **2**, 9-10, and similar remarks in the 4th ed. (Paris, 1878). A. Ganot, *Traité élémentaire de Physique* (Paris, 6th ed., 1856), p. 210; English trans. of Ganot's book, *Elementary Treatise on Physiks* (New York, 12th ed., 1886), p. 260. F. Redtenbacher, *Das Dynamiden-System* (Mannheim, 1857). Laurens P. Hickok, *Rational Cosmology* (New York, 1858), p. 175f. Julius Wenck, *Die Physik* (Leipzig, 1858), p. 342. B. Silliman, *First Principles of Physics* (Philadelphia, 1859), p. 303. J. Jamin, *op. cit.* (note 3). A Mousson, *Die Physik* (Zürich, 1860), **2**, 4. J. Hogg, *Elements of experimental and natural philosophy* (London, 1861), p. 236. Balard, *Revue [des Cours] Scientifique* **1**, 78 (1864). Marié-Davy, art. "Chaleur" in the Privat-Deschanel and Focillon *Dictionnaire Général des Sciences* (Paris, 1864), pp. 430-36. W. A. Norton, *Am. J. Sci.* **38**, 61, 207 (1864), and several subsequent papers. John Tyndall, *Fortnightly Review* **3**, 129 (1865). A. Cazin, *Revue Scientifique* **2**, 431 (1865). C. Puschl, *Das Strahlungsvermögen der Atome* (Wien, 1869). J. Challis, *Notes on the*

Principles of Pure and Applied Calculation (Cambridge, 1869). D. Olmsted, *An Introduction to Natural Philosophy* (New York, 4th ed., 1870), p. 310. E. Saigey, *The Unity of Natural Phenomena*, trans. from French (Boston, 1873), p. 106. H. Hudson, *Phil. Mag.* [4] **42**, 341 (1871) [assumes longitudinal heat waves]; *English Mechanic* **19**, 121 (1874). Favé, *Les Mondes* **41**, 336 (1876). A. Guillemin, *Le Monde Physique* (Paris, 1884), **4**, 6. Joannis, *La Grande Encyclopédie* (Paris, 1886–1902), **10**, 239–43. A. R. V. Miller-Hauenfels, *Richtigstellung der in bisheriger Fassung unrichtigen mechanischen Wärmetheorie und Grundzüge einer allgemeinen Theorie der Aetherbewegungen* (Wien, 1889). Rudolf Mewes, *Dinglers Polytechnisches Journal* (Stuttgart), **317**, 758, 800 (1902), **318**, 42, 75 (1903).

The above is not to be regarded as merely a list of cranks or third-rate scientists; many of these men may have had considerable influence through their teaching positions and the use of their textbooks.

5. See Henry's *Scientific Writings* [published in *Smithsonian Miscellaneous Coll.* **30** (1886)], **2**, 96. [S. Newcomb] *North Amer. Rev.* **93**, 373 (1861).
6. Whewell and Rosenberger, cited in note 8, §9.4. See also H. T. Buckle, *History of Civilization in England* (New York: Appleton, 1907), **II**, 384.
7. When I first wrote this chapter in 1969, I found only one modern writer who gave a reasonably accurate (though greatly abbreviated) statement on this subject: T. W. Chalmers, *Historic Researches* (New York: Charles Scribner's Sons, 1952), pp. 28–29. I overlooked the paper by R. J. Morris, Jr., *Proc. Oklahoma Acad. Sci.* **42**, 195 (1962) and the brief remarks by H. S. Williams, *A History of Science* (New York: Harper, 1904), **3**, 215, 275, 278; and I learned that Charles Weiner had reached some of the same conclusions in his unpublished Ph.D. Dissertation at Case Institute of Technology, *Joseph Henry's Lectures on Natural Philosophy: Teaching and Research in Physics, 1832–1847* (1965). There is also a brief discussion by T. N. Gornshtein, *Voprosy Ist. Est. Tekhn.* **20**, 55 (1966). Subsequently the wave theory of heat was mentioned by four other historians: W. L. Scott, *The Conflict between Atomism and Conservation Theory 1644 to 1860* (New York: Elsevier, 1970), p. 219; D. S. L. Cardwell, *From Watt to Clausius* (Ithaca: Cornell University Press, 1971), pp. 217–18; R. Fox, *The caloric theory of gases* (Oxford: Clarendon Press, 1971), pp. 96, 104, 105, 116, 213, 244, 277–78; P. F. Dahl, *Ludvig Colding and the Conservation of Energy Principle* (New York: Johnson Reprint Corp., 1972), p. xxii. (Dahl says only that the "final decay of the caloric theory [was] largely a result" of the experiments of Melloni and Forbes.)
8. Much of the blame for propagating the second myth is Tait's according to Cardwell, *From Watt to Clausius*, p. 284. Tait of course wanted to take the credit for energy conservation away from Mayer and give it to Joule and Mohr.
9. W. Thomson, *Trans. R.S. Edinburgh* **16**, 541 (1849). J. P. Joule, in a paper read to the Royal Society on 21 June 1849, stated that "the scientific world [was] preoccupied with the hypothesis that heat is a substance" but it was not clear that he thought this was still true in 1849; see Joule's *Scientific Papers* (London, 1884), **1**, 302.
10. R. K. Merton, *Proc. Am. Phil. Soc.* **105**, 470 (1961).
11. W. Thomson, *Trans. R.S. Edinburgh* **20**, 261 (1851). *Proc. R.S. London* **8**, 152 (1856).
12. T. S. Kuhn, note 98 of the paper cited in §9.5 (note 1), has called attention to this curious statement, and asks: "But if Davy had established the dynamical theory in 1799 and if the rest of conservation follows from it, as Kelvin implies, what had Kelvin himself been doing before 1852?" In Thomson's article on "Heat" for the 9th edition of the *Encyclopedia Britannica* (Edinburgh and New York, 1880), **11**,

495–526 (replacing Traill's article quoted at the beginning of this chapter), he gave a classic statement of the "combined myth" mentioned in the text above: "It is remarkable that, while Davy's experiment alone sufficed to overthrow the hypothesis that heat is matter, and Rumford's, with the addition of just a little consideration of its relations to possibilities or probabilities of inevitable alternative, did the same, fifty years passed before the scientific world became converted to their conclusion,– a remarkable instance of the tremendous efficiency of bad logic in confounding public opinion and obstructing true philosophic thought." The article does not mention the wave theory of heat.

CHAPTER 10

Foundations of Statistical Mechanics 1845–1915 *

One of the recurring problems in 19th-century theoretical physics was: Can we formulate a consistent molecular model, within the framework of classical Newtonian mechanics, from which we will be able to compute the observable properties of matter? It was in the kinetic theory of gases that one found the most direct relation between certain simple observable properties and a simple molecular model. There, too, was encountered a basic discrepancy between theory and experiment that could eventually be resolved only by resorting to quantum theory: the anomalous specific heats of polyatomic molecules. Of course there were many other experimental results that could not be understood on the basis of classical mechanics, so why did this one cause such anguish? It must have been because the theory that failed in this particular case was so successful in other cases. In trying to resolve the difficulty, physicists were forced to examine their basic postulates and methods of calculation, with an unusual concern for clarity and rigor that eventually brought them into contact with contemporary developments in mathematics. Thus was created the subject: ergodic theory, which eventually passed out of physics almost entirely and became a new branch of mathematics. We shall trace here the early history of these attempts to establish the foundations of

---

* Reprinted with minor revisions from *Archive for History of Exact Sciences* **4**, 145 (1967), by permission of Springer-Verlag.

statistical mechanics, breaking off at the point when modern mathematics and modern physics again parted company and went off in different directions. Although the final separation did not take place until more than a decade later, it was already clear by 1900 that physicists did not look at such problems in the same way as mathematicians. In that year the physicist S. H. Burbury, referring to the idea of a system which passes through all phases on the energy surface, remarked: "I think we have no cage for such a bird."[1]

Note for §10.0

1. S. H. Burbury, *Phil. Mag.* [5] **50**, 584 (1900), quote from p. 588.

10.1 Waterston's equipartition theorem

Waterston's work has been discussed in ch. 3, so here I will only briefly recall his contribution to our present topic. At the age of 19, he published a paper in which he discussed the properties of a system of small colliding cylindrical particles, arguing that these could generate a gravity-like force between larger bodies immersed in the system. Some of the ideas developed in this paper were later utilized in his kinetic theory, in particular the idea that collisions could result in a transfer of energy from the rectilinear to the rotatory modes of motion.[1] This idea is of course at the basis of the equipartition theorem in statistical mechanics.

In a book on the physiology of the central nervous system, published anonymously in 1843, Waterston stated some of the basic principles of the kinetic theory, such as that the pressure of a system of moving particles is proportional to their density and to the square of their velocity. He states that "the proportion of the whole rectilineal to the whole rotatory momentum of the medium is probably constant, and might be found perhaps by calculation." He suggests that increase of temperature in gases might correspond to increase of molecular kinetic energy, and throws out several other suggestions about the consequences of the kinetic theory of gases.[2] It is not clear whether he thinks mv or mv^2 is equilibrated, or whether he realized the inconsistency of his statements implying that *both* are.

The confusion is cleared up in his paper "On the physics of media that are composed of free and elastic molecules in a state of motion," submitted to the Royal Society in December 1845 but not published until 1892. Here he states quite clearly that "in mixed media the mean square velocity is inversely proportional to the specific weight of the molecules." Since this conclusion was printed in an abstract of his paper given at a British Association meeting,[3] Waterston clearly deserves priority of discovery of the equipartition theorem. He also attempted to derive the ratio of specific heats at constant pressure and at constant volume (now usually denoted by γ) but because of a numerical error he obtained the result $\gamma = \frac{4}{3}$ instead of $\frac{5}{3}$. The former value was in fairly good agreement with the experimental data then available, so that he failed to encounter the discrepancy which plagued later kinetic theorists. Waterston also recognized the inconsistency between the idea that the atoms move freely through the medium, and the idea that an all-pervading ether transmits the waves caused by vibrations of the parts of the atoms; one would have expected the atoms to experience some resistance in moving through such an ether. Waterston simply stated the difficulty but drew no conclusions from it.

Waterston stated very firmly at the beginning of his paper that he intended to discuss the properties of "a hypothetical condition of matter," that is, "a hypothetical medium, which we have carefully to refrain from assimilating to any known form of matter until, by synthetical reasoning, circumstantial evidence has been accumulated sufficient to prove or render probable its identity." His medium is composed of

> ... a vast multitude of small particles of matter, perfectly alike in every respect, perfectly elastic as glass or ivory—but of size, form and texture that requires not to be specified further than that they are not liable to change by mutual action—to be enclosed by elastic walls or surfaces in a space so much greater than their aggregate bulk as to allow them freely to move amongst each other in every direction. As all consideration of attractive forces is left out at present, it is obvious that each particle must proceed on a straight line until it strikes against another, or against the sides of the enclosure; that it must then be reflected and driven into another line of motion, traversing backwards and forwards in every direction, so that the intestine condition of the multitude of these that form the medium may be likened to the familiar appearance of a swarm of gnats in a sunbeam.

He continued with an additional pair of hypotheses, which suggested not only conservation of total kinetic energy (in modern terminology) but also something like an ergodic hypothesis:

> The quality of perfect elasticity being common to all the particles, the original amount of *vis viva*, or living, acting force, of the whole multitude must for ever remain the same. If undisturbed by external action it cannot, of itself, diminish or increase, but must for ever remain as unchanged as the matter that is associated with it and that it endows with activity. Such is the case if we view the whole mass of moving particles as one object, but each individual of the multitude must at every encounter give or receive, according to the ever changing angle and plane of impact, some portion of its force, so that, considered separately, they are for ever continually changing the velocity and direction of their individual motions; striking against and rebounding from each other, they run rapidly in their zig-zag conflict through every possible mode of concurrence, and *at each point of the medium we may thus conceive that particles are moving in every possible direction and encountering each other in every possible manner during so small an elapsed interval of time that it may be viewed as infinitesimal in respect to any sensible period.* The medium must in this way become endowed with a permanent state of elastic energy or disposition to expand, uniformly sustained in every part and communicating to it the physical character of an elastic fluid.
>
> The simplicity of this hypothesis facilitates the application of mathematics in ascertaining the nature and properties of such media, and the study acquires much interest from the analogies that it unfolds. For if the reasoning is correct, the physical laws common to all gases and vapours–those laws, namely, that concern heat and pressure–do actually belong to such media, and may be synthetically deduced from the constitution which has now been assigned to them.

In the passage just quoted, Waterston has indicated as clearly as possible the motivation for the ergodic hypothesis, as well as an intuitive "physical" justification for it. His hypothesis is not actually the same as the one later proposed by Maxwell and Boltzmann, since it refers only to the velocities of individual particles and pairs of particles during a short time period, rather than to the possible combinations of all velocities and positions of the entire gas throughout a very long time period; but for a low-density gas in which only binary collisions are important, Waterston has assumed all he needs to assume.

Notes for §10.1

1. J. J. Waterston, *Phil. Mag.* **10**, 170 (1831); reprinted in *The Collected Scientific Papers of John James Waterston*, ed. J. S. Haldane (Edinburgh: Oliver & Boyde, 1928), p. 531.
2. J. J. Waterston, *Thoughts on the Mental Functions* (Edinburgh, 1843); reprinted in his *Papers* **3**, 167, 183.
3. J. J. Waterston, *B. A. Rep.* **21**, 6 (1851); *Papers*, p. 318.
4. *Papers*, pp. 215–16.

10.2. Clausius's postulate about internal motions

Rankine had criticized Waterston's theory on the basis that the total heat of a gas could not be accounted for by translational motion of particles, and he proposed to remedy this defect with his own theory of rotating molecular vortices.[1] Clausius (though unaware of Waterston's work) noted the same difficulty, and suggested instead that the molecules of gases are composed of two or more atoms.[2] In general, the total kinetic energy will be distributed in some way between the translational motion of the whole molecules and the internal vibratory or rotatory motion of the constituents of the molecules. Clausius asserts that the translatory motion of the whole molecules *will always have a constant relation* to the internal motions of the constituents. His justification for the assertion is the following:

> Conceive a number of molecules whose constituents are in a state of motion, but which have no translatory motion. It is evident the latter will commence as soon as two molecules in contact strike against each other in consequence of the motion of their constituents. On the other hand, if the constituents of a number of molecules in a state of translatory motion were motionless, they could not long remain so, in consequence of the collisions between the molecules themselves, and between them and fixed sides or walls. It is only when all possible motions have reached a certain relation towards one another, which relation will depend upon the constitution of the molecules, that they will cease mutually to increase or diminish with each other.

From the wording of this paragraph it might appear that Clausius only wants to assert that the relation between translational and internal motion will reach a constant value at some particular temperature and volume, and thereafter it will not change with *time*. But he immediately goes on to apply his postulate to changes with *temperature*. He infers that the specific heat of an ideal gas will be constant when the temperature changes, because according to kinetic theory the absolute temperature is proportional to the translational kinetic energy, and according to his postulate the total kinetic energy will be proportional to the translational kinetic energy.

We shall call the statement that internal kinetic energy has a constant relation to translational kinetic energy at all temperatures the "Clausius postulate." While it can hardly be said that Clausius has proved its validity by the argument quoted above, the postulate does acquire a certain amount of plausibility from the fact that the specific heats of gases are constant over fairly large temperature ranges, at least when there is no significant amount of dissociation or ionization of the molecules.

In the same paper Clausius suggested some further reasons why one might expect the idealized kinetic-theory model to apply to real gases. First of all, he disposed of Krönig's assumption that the molecules move only in directions perpendicular to the walls of the container, and showed that the pressure is the same with oblique impacts in random directions. Furthermore,

> it is not actually necessary that a molecule should obey the ordinary laws of elasticity with respect to elastic spheres and a perfectly plane side, in other words, that when striking the side, the angle and velocity of incidence should equal those of reflexion, yet, according to the laws of probability, we may assume that there are as many molecules whose angles of reflexion fall within a certain interval, e.g. between 60° and 61°, as there are molecules whose angles of incidence have the same limits, and that, on the whole, the velocities of the molecules are not changed by the side. No difference will be produced in the final result, therefore, if we assume that for each molecule the angle and velocity of reflexion are equal to those of incidence.

We observe that Clausius is calling on "the laws of probability" to justify an assumption that might better be founded on arguments of symmetry and reversibility. In fact, if we have to deal with an unsymmetrical or non-equilibrium situation, in which there is a preferred direction of motion in the gas, then the possibility that the angle of incidence is not equal to the angle of reflection can become quite important. This is the case in rarefied gases, and in particular in the Crookes radiometer, where non-specular reflection of gas molecules at a surface plays a role in producing an observable effect, the rotation of

§10.2]   CLAUSIUS'S POSTULATE ABOUT INTERNAL MOTIONS   341

the vanes (§5.5). But such considerations still lay in the future when Clausius wrote his paper, and they are still ignored in many modern elementary texts.

Despite what seem to be lapses of logical rigor from the modern viewpoint, it must be recognized that Clausius was taking a big step forward in recognizing that some restrictive conditions must be assumed in order that the kinetic-theory model may correspond to a real gas. In addition to the assumptions already mentioned, Clausius proposed the following three conditions which must be satisfied in order that the ideal gas laws may be strictly fulfilled:

(1) The space actually filled by the molecules of the gas must be infinitesimal in comparison to the whole space occupied by the gas itself.
(2) The duration of an impact, that is to say, the time required to produce the actually occurring change in the motion of a molecule when it strikes another molecule or a fixed surface, must be infinitesimal in comparison to the interval of time between two successive collisions.
(3) The influence of the molecular forces must be infinitesimal.

The third condition is interpreted to mean: (a) when all the molecules are at their mean distances, the attractive forces vanish compared to the expansive force due to the motion; (b) the part of the path of a molecule during which these forces can appreciably alter its motion is small compared to the entire path. (The latter condition is of course similar to (2) but refers to the attractive molecular forces as distinguished from the repulsive or "hard-core" forces, in modern language. The distinction was much more significant for those 19th-century physicists who conceived of atoms as matter occupying a definite volume than it is for us today.)

Clausius concedes that such a model will, strictly speaking, describe only an ideal gas—that is, one whose pressure-volume-temperature relations are exactly in agreement with what he calls Mariotte's and Gay-Lussac's laws. Yet most of the significant research in kinetic theory later in the 19th century was to involve properties for which the influence of impacts and molecular forces is definitely not small. Deviations from the ideal gas laws, transport coefficients, and the approach to equilibrium all depend directly on the effects of collisions and intermolecular forces. Even the Clausius postulate itself requires that all the molecules must suffer at least one, and probably several collisions, before the stable ratio of translational to internal motion is reached. There was clearly a need for a set of assumptions that would permit collisions to play a role.

Notes for §10.2

1. W. J. M. Rankine, *Trans. R. S. Edinburgh* **20**, 565 (1853); *Proc. Glasgow Phil. Soc.* **5**, 126 (1864); *Trans. R. S. Edinburgh* **25**, 557 (1869).
2. R. Clausius, *Ann. Phys.* [2] **100**, 353 (1857); English trans. reprinted in Brush, *Kinetic Theory* **1**, 111.
3. R. Clausius, in Brush, *Kinetic Theory* **1**, 114.

10.3. Maxwell's velocity distribution

Charles Gillispie has recently described the intellectual background of probability theory in the early 19th century in connection with the introduction of statistical methods into kinetic theory.[2] In particular, he calls attention to a review by John Herschel of a book on statistics by Quetelet. In his review, which appeared in the *Edinburgh Review* for July 1850, Herschel attempts to give a proof of the law of errors that is both mathematically rigorous and intelligible to the general educated public. The remarkable thing about Herschel's proof is that it is very similar to the derivation of the distribution law for molecular velocities which Maxwell gave in his first paper on kinetic theory in 1860. Herschel postulates that when, for example, a ball is dropped from a height with the intention that it shall fall on a given mark, the probability of any deviation from the mark must be a function of the sum of the squares of its deviations resolved in any two perpendicular directions. Since the probability of a compound event (a deviation of a certain magnitude in some direction) is the product of the probabilities of the two constituent independent events (deviations in the two perpendicular directions), we can require that the probability function have the property that the product of such functions of two independent events is equal to the same function of their sum. "But it is shown in every work on algebra that this property is the peculiar characteristic of, and belongs only to, the exponential or antilogarithmic function. This, then, is the function of the square of the error, which expresses the probability of committing that error."[3]

Maxwell asserts that the velocity components of a molecule in different directions can be treated as independent random variables. His only justification for this assumption was the fact that, if two elastic spheres collide, all directions of rebound are equally likely. He apparently believed that this fact would ensure, not only that all directions of motion are equally probable in the gas, but also that the probability distribution for each component of the velocity is independent of the values of the other components. His first proof of the distribution law depended on these two assumptions. Thus, he writes the number of particles whose velocity components lie between x and $x + dx$, y and $y + dy$, z and $z + dz$ as $Nf(x)f(y)f(z)\,dx\,dy\,dz$, where f always stands for the same function, and N is the total number of particles. But since there is no preferred direction of motion, this product must be equal to a function of the magnitude of the velocity,

$$f(x)f(y)f(z) = \phi(x^2 + y^2 + z^2)$$

"Solving this functional equation, we find $f(x) = C\,e^{Ax^2}$, $\phi(r) = C^3\,e^{Ar^2}$," where A must be negative and C is determined by normalization.

This first proof of Maxwell's distribution law, which mystified his contemporaries though the result seemed to be correct, may have been simply copied from a book on statistics or even from the article by Herschel which we mentioned above. In any case, it served its purpose of suggesting the probable form of the velocity distribution of molecules in a gas, though it was soon recognized as being unsatisfactory by physicists (including Maxwell himself) who were not prepared to take randomness as a basic postulate in developing their molecular theories (cf. §§14.4–8). We shall now discuss some attempts that were made to justify the distribution law on "physical" (*i.e.* deterministic) grounds.

Notes for §10.3

1. See §5.1 for more detailed discussion and references for this topic.
2. C. C. Gillispie, in *Scientific Change*, ed. A. C. Crombie (New York: Basic Books, 1963), p. 431; *Proc. Am. Phil. Soc.* **116** (1), 1 (1972).
3. J. Herschel, *Edinburgh Review* **92**, 1 (1850); see §14.4 for further discussion and references to related works.

10.4. Equalization of kinetic energy by collisions

Maxwell now turns, in the same paper of 1860, to the problem: "Two systems of particles move in the same vessel; to prove that the mean *vis viva* of each particle will become the same in the two systems." For this proof he calls on the result which he has just previously obtained: that if the velocities in each system are distributed according to the law of errors, then "the distribution of relative velocities [of a pair of molecules, one from each system] is regulated by the same law as that of the velocities themselves, and that the mean relative velocity is the square root of the sum of the squares of the mean velocities of the two systems." Maxwell considers a collision between a particle of the first system, having mass P and the mean velocity p of the particles in that system, and a particle of the second system, having mass Q and the mean velocity q of the particles in the second system. He assumes that their relative velocity is just the mean relative velocity according to the theorem cited, namely $\sqrt{p^2 + q^2}$. This is equivalent to assuming that the directions of motion of the two particles are perpendicular before the collision. He then shows that if the direction of relative velocity after the collision is perpendicular to the direction of motion of the center of mass of the two particles, then the velocities after collision will satisfy the equation

$$Pp'^2 - Qq'^2 = \left(\frac{P-Q}{P+Q}\right)^2 (Pp^2 - Qq^2)$$

Hence the difference in kinetic energies of the two particles "is diminished at every impact in the same ratio, so that after many impacts it will vanish, and then

$$Pp^2 = Qq^2 \ldots"$$

Hence after many impacts, Maxwell concludes, the kinetic energies of all the particles will tend to become equal.

It seems amazing to me that Maxwell should have thought he was proving a tendency toward equalization of kinetic energies by this argument, or that any of his contemporaries who bothered to examine the argument in detail would have accepted it. All Maxwell has done is to pick out one very special kind of collision for which the kinetic energies become more nearly equal and then claim that the same result will follow for *all* collisions. This is an instance of a very common fallacy in statistical reasoning: to assume that if a certain member of a population is just average with respect to property A, then any other property B, computed for that one member, will be equal to the average value of B for the entire population. In other words, one can interchange the operation of averaging with any other analytical operation. This type of shortcut is expecially tempting when, as in gas

§10.4] EQUALIZATION OF KINETIC ENERGY BY COLLISIONS

theory, one has to deal with an effectively infinite population, and one knows that the distributions of some properties, at least, are very sharply peaked around the average value.

Maxwell also considers the collision of "two perfectly elastic bodies of any form" and by similar arguments concludes that "the final state, therefore, of any number of systems of moving particles of any form is that in which the average *vis viva* of translation along each of the three axes is the same in all the systems, and equal to the average *vis viva* of rotation about each of the three principal axes of each particle." But since the ratio of specific heats, γ, is known by experiment to be approximately 1.408, the ratio of the total kinetic energy to the kinetic energy of translation should be $\beta = 2/[3(\gamma - 1)] = 1.634$, instead of $\beta = 2$ according to the equipartition theorem for rotating particles. So, despite the fact that systems of hard elastic particles seem to have many properties similar to those of real gases, the existence of an equipartition theorem proves "that a system of such particles could not possibly satisfy the known relation between the two specific heats of all gases." And here Maxwell ended his first paper on the kinetic theory.

In his masterpiece of 1866 on "The Dynamical Theory of Gases" – a work 50 years ahead of its time – Maxwell returned briefly to the problem of deriving the velocity distribution. He admits that his former assumption "that the probability of a molecule having a velocity resolved parallel to x lying between certain limits is not in any way affected by the knowledge that the molecule has a given velocity resolved parallel to y" is an assumption that "may appear precarious." So he proposed to assume that it is the velocities of two colliding molecules, rather than the velocity components of a single molecule, that are statistically independent.[1]

Suppose, says Maxwell, that two molecules have velocities a and b before they collide; and that as a result of the collision these velocities are changed to a' and b', respectively. Then the number of such encounters in unit time in a volume element dV in velocity space will be

$$f_1(a)f_2(b)(dV)^2 F\,de$$

where "F is a function of the relative velocity and of the angle θ [θ = half the angle through which the relative velocity vector is turned by the collision – S.G.B.], and de depends on the limits of variation within which we class encounters as of the same kind."

Maxwell simply states this expression for the number of encounters as if it were self-evident, and goes on to write down a similar expression for the number of encounters between molecules whose original velocities lie in a differential element dV around a' and b', and whose final velocities lie in an element dV around a and b

$$f_1(a')f_2(b')(dV)^2 F'\,de$$

where F' is asserted to be the same function of its arguments that F is of its arguments.

When the number of pairs of molecules which change their velocities from (a, b) to (a', b') is equal to the number that change from (a', b') to (a, b), then, Maxwell says, "the final distribution of velocity will be obtained, which will not be altered by subsequent exchanges." So one simply equates the two expressions, cancels out $F\,de$ on one side with $F'\,de$ on the other, and "solves" the functional equation

$$f_1(a)f_2(b) = f_1(a')f_2(b')$$

subject to the condition that the total kinetic energy is unchanged by the collision

$$M_1 a^2 + M_2 b^2 = M_1 a'^2 + M_2 b'^2$$

The solution is once again the familiar exponential,

$$f_1(a) = C_1\,e^{-a^2/\alpha^2}, \qquad f_2(b) = C_2\,e^{-b^2/\beta^2}$$

where

$$M_1 \alpha^2 = M_2 \beta^2$$

and with the constants C_1 and C_2 determined by normalization.

Maxwell seems to be quite certain that this is at least a possible form for the final velocity distribution, since it is "not altered by the exchange of velocities among the molecules by their mutual action." But he feels it necessary to offer a further justification for his assertion that this is the *only* stable distribution, which runs as follows:

> if there were any other, the exchange between velocities represented by OA and OA' [the vectors representing a and a', resp.–S.G.B.] would not be equal. Suppose that the number of molecules having velocity OA' increases at the expense of OA. Then since the total number of molecules corresponding to OA' remains constant, OA' must communicate as many to OA'', and so on till they return to OA.

Hence if OA, OA', OA'', etc. be a series of velocities, there will be a tendency of each molecule to assume the velocities OA, OA', OA'', etc. in order, returning to OA. Now it is impossible to assign a reason why the successive velocities of a molecule should be arranged in this cycle, rather than in the reverse order. If, therefore, the direct exchange between OA and OA' is not equal, the equality cannot be preserved by exchange in a cycle. Hence the direct exchange between OA and OA' is equal, and the distribution we have determined is the only one possible.[2]

Notes for §10.4

1. J. C. Maxwell, *Phil. Trans.* **157**, 49 (1867), reprinted in Brush, *Kinetic Theory* **2**, 23.
2. J. C. Maxwell, in Brush, *Kinetic Theory* **2**, 48.

10.5 The effect of forces on the distribution law: The "Boltzmann factor"

Shortly after Maxwell's 1866 memoir was published, Ludwig Boltzmann turned his attention to the problem of the distribution of kinetic energy in a system of mass-points. Boltzmann had just started his career in theoretical physics with a study of the relation between the Second Law of Thermodynamics and the principles of mechanics,[1] but he had not yet come to see the importance of statistical concepts in establishing such a relation. Now, in 1868, began a fruitful dialogue between Boltzmann and Maxwell that was to clarify (though not solve) many of the basic problems of kinetic theory and statistical mechanics.[2]

Boltzmann complained that Maxwell's derivation, which we reviewed in the preceding section, was hard to understand because of its brevity. He therefore devoted the first part of his 44-page memoir to filling in and illustrating with concrete examples the steps that Maxwell had glossed over, such as the nature of the function F. Boltzmann did not, however, challenge the validity of Maxwell's statistical assumption about the independence of velocities of colliding molecules, though later he made this assumption more explicit.[3]

In the second part of this paper, Boltzmann considers the following problem:

"Along a line OX there moves an elastic sphere with mass M, which is attracted to O by a force which is a function only of the distance of its center from O. It is continually being bombarded by elastic spheres of mass m with various velocities at irregular time intervals, such that if one looks at the bombarding spheres far away from O on the line OX, then the number of spheres with velocity between c and $c + dc$ found on the average in unit length is a definite function of c, $N\varphi(c)\,dc$.

The potential of the force with which M is attracted to O is $\chi(x)$; then, as long as the motion is not disturbed by collisions, we have

$$(9) \qquad \frac{MC^2}{2} = \chi(x) + A$$

where C is the velocity of the sphere M and x is the distance of its centre from O. The nature of the collision is completely determined by the three quantities x, A, and c. The time during which, in the course of unit time, the constant A of equation (9) lies between the values A and $A + dA$ will be $\Phi(A)\,dA\ldots$"

The problem is to determine $\Phi(A)$.

Here, in the context of a rather specialized problem, Boltzmann arrives at the generalization of Maxwell's velocity-distribution law for the case of particles affected by forces: the so-called *Boltzmann factor* which is now used in almost all practical applications of statistical mechanics to physico-chemical problems. He found that the total fraction of time during which A is between A and $A + dA$, and also x is between x and $x + dx$, is proportional to $e^{-hA}\,dA\,dx$

$$\Phi(A) \propto (\text{const.}) \exp[h(\chi(x) - MC^2/2)]$$

The Boltzmann factor is an exponential function of the total energy of a particle at a given point in space with a given velocity, that is, the sum of its potential energy (which usually depends only on position) and its kinetic energy (which depends only on velocity).

Notes for §10.5

1. L. Boltzmann, *Wien. Ber.* **53**, 195 (1866), reprinted in his *Wissenschaftliche Abhandlungen* (Leipzig: Barth, 1909), 1, 9.
2. L. Boltzmann, *Wien. Ber.* **58**, 517 (1868); *Abhandlungen* 1, 49.
3. L. Boltzmann, *Vorlesungen über Gastheorie*, Teil I (Leipzig, 1896); English trans. by S. G. Brush, *Lectures on Gas Theory* (Berkeley: University of California Press, 1964), see pp. 40–42 and elsewhere. [Further discussion in §14.6.]

10.6 Equilibrium of a column of gas under gravitational forces

If a column of air is in equilibrium under the influence of gravitational forces, will its temperature change with height? This problem gave Maxwell considerable trouble while he was working on his 1866 memoir. He tells us that the original manuscript sent to the Royal Society contained an equation, no. (147), which implied that a column of air would assume a temperature varying with height. At various times he had decided that the temperature might either increase or decrease with height, but finally discovered that these conclusions were the result of calculational errors. Maxwell asserts that any variation of temperature with height is inconsistent with the Second Law of Thermodynamics, though his own argument on this point seems to show only that if there is such a variation, it must be the same for all substances.[1]

Maxwell repeated his conclusion that temperature is independent of height in his textbook on the *Theory of Heat*,[2] but it was disputed by Frederick Guthrie in a letter to *Nature* in 1873.[3] The ensuing correspondence stimulated Maxwell to work out a new concise derivation of the generalized Maxwell–Boltzmann distribution law, from which the thermal equilibrium of a column of gas followed as a special case.[4] Guthrie's original objection was shown to be based on his erroneous assumption that all molecules at a given height have the same velocity, and Guthrie conceded this point, but later raised another objection. He cited Maxwell's equation

$$n_1 n_2 = n'_1 n'_2$$

where n_1 and n_2 are the numbers of particles colliding with velocities v_1 and v_2, respectively, and the primed quantities apply after the collision. (The symbol f was used in Maxwell's earlier paper, discussed in our §10.4.) Guthrie said:

> This reasoning does not seem convincing... if the number of particles in a given element is a function of its velocity in direction and magnitude, then although the average of the numbers in each direction is maintained, it does not follow that the average numbers of particles having the velocities v_1 and v_2 are directly restored from the particles v'_1 and v'_2. All that can be assumed is, that the average number of particles in a given element of space is maintained from the particles in that and the remaining elements. Just as in the case of an equilibrium of trade, the average course of exchange with respect to a given country is at par; but we cannot

therefore safely assume that the same is the case relatively to any other individual country.

There are several other points in Mr. Maxwell's communication which seem to me to require fortification, but the subject has already assumed so technical a form that it would perhaps be uninteresting to your readers to point them out. My impression is that the whole subject is still somewhat beyond the grasp of strict mathematical reasoning, and is still open to experimental investigation.[5]

Here indeed was a promising opportunity for thrashing out some of the fundamental problems of statistical mechanics before a general audience, on terms which at least some of them (the economists if not the physicists) could understand. But at the time (1874), the kinetic theory enjoyed the general confidence (if not the deep comprehension) of most British and European scientists,[6] and there were not yet any serious reasons for doubting the validity of the concept of statistical equilibrium, aside from the problem of specific of molecules. Maxwell was therefore allowed to get away with this rejoinder to Guthrie:

> This question is treated at length in my paper On the dynamical theory of gases (Phil. Trans., 1866). It is there shown that if the average course of exchange is in a cycle from A to B, B to C, C to A, an equal reason may be given why it should be in the opposite cycle A to C, C to B, B to A, and thus it is shown that the exchange is at par between each pair of states separately. For a far more elaborate theoretical treatment of the subject Prof. Guthrie is referred to the papers of Prof. Ludwig Boltzmann in the Vienna Transactions since 1868. I fear we must delay the experimental investigation for some time, till we are able to count the molecules in a given space, to observe their velocities, and to repeat these operations millions of times in a second.[7]

Notes for §10.6

1. J. C. Maxwell, in Brush, *Kinetic Theory* **2**, 85.
2. J. C. Maxwell, *Theory of Heat* (London, 2nd ed., 1872), p. 300. (The passage appears on p. 320 in later editions.)
3. F. Guthrie, *Nature* **8**, 67 (1873). I am indebted to Dr. C. W. F. Everitt for calling to my attention the Guthrie–Maxwell correspondence.
4. J. C. Maxwell, *Nature* **8**, 85 (1873); F. Guthrie, *Nature* **8**, 486 (1873); J. C. Maxwell,

Nature **8**, 527, 537 (1873). Only the last of these papers is reprinted in Maxwell's *Scientific Papers*.
5. F. Guthrie, Nature **10**, 123 (1874).
6. See §5.4.
7. J. C. Maxwell, note added at the end of Guthrie's letter cited in note 5.

10.7 Approach to equilibrium and the problem of irreversibility[1]

Boltzmann, in 1872, tried to approach the problem of thermal equilibrium from another point of view.[2] Instead of assuming, as Maxwell had done, that thermal equilibrium already exists, and simply looking for analytical conditions on the distribution function which will maintain stable equilibrium, Boltzmann started out by assuming that the gas is not in equilibrium, and attempted to show that the effect of collisions will be to bring about equilibrium. Somehow Boltzmann hit on the idea of defining a functional of the velocity-distribution function,

$$E = f \log f,$$

and showing that, subject to certain assumptions, E must decrease as a result of collisions between particles unless f is the Maxwell distribution function. This has become known as Boltzmann's H-theorem, and has been discussed exhaustively in many articles and books. Since in this chapter we are not primarily concerned with the time-evolution of non-equilibrium systems, we shall not have much to say about the H-theorem except to summarize a few relevant points:

(a) When f is the Maxwellian distribution and the system is assumed to be in thermal equilibrium, $-E$ is simply proportional to the entropy. If one is willing to define entropy as being proportional to $-E$ for non-equilibrium states, then the H-theorem can be interpreted as a generalized Second Law of Thermodynamics: entropy always increases unless a system is in equilibrium.

(b) In deriving the H-theorem, Boltzmann used the same assumption introduced earlier by Maxwell: the velocities of two molecules before they collide are statistically independent.

(c) the fact that $-E$ is proportional to the entropy in the equilibrium state suggests that there is a connection between entropy and probability; in particular, entropy is a measure of disorder or randomness, and the equilibrium state is the most random distribution. Here I am simply reporting the interpretation that one finds in textbooks on statistical mechanics; it would make more sense to say that entropy is a measure of our own *information* about the positions and velocities of the particles. If we know only the total number of particles and the total energy, then the Maxwell–Boltzmann distribution corresponds to specifying the minimum amount of information.

(d) It was pointed out by Lord Kelvin, and later by Loschmidt, that if molecular collisions are governed by Newtonian mechanics, then any given sequence of collisions can run backwards just as well as forwards. Kelvin and Boltzmann argued that this reversibility on the molecular level was compatible with macroscopic irreversibility, simply because the overwhelming majority of possible states of the gas as a whole would be disordered states described by the Maxwell distribution.

(e) Poincaré, in 1890, published his recurrence theorem, which asserts that any bounded system of mass points, moving according to Newtonian mechanics, must eventually return to any initial configuration (specified by positions and momenta) within any specified degree of accuracy, an infinite number of times. Of course the recurrence time may be so long that it is never observable–in the simplest case, it would be simply the time required to go through all possible configurations. The important point, however, is that a deterministic mechanical system cannot get stuck in a final state, as one might have expected from the H-theorem. Poincaré and Zermelo used the recurrence theorem as an argument against the validity of the kinetic theory of gases, since they thought that the generalized Second Law of Thermodynamics must be an absolute principle of physics; any theory that permits exceptions to it would have to be rejected. There followed a debate between Boltzmann and Zermelo, in which Boltzmann described a possible form for the time-variation of entropy (his so-called H-curve) which would be compatible both with short-term irreversibility and long-term recurrences. Boltzmann's arguments seem to have satisfied most modern physicists, even though (as far as I know) no one has ever proved rigorously that any realistic model of a physical system does exhibit such an H-curve.

The outcome of the discussions on the H-theorem was that physicists realized that some stochastic element must be built into the basic postulates of kinetic theory, in order to make practical calculations; yet as soon as this was done, the theoretical system would have qualitatively different properties from the corresponding deterministic mechanical system, with respect to irreversibility. At least this is true for classical mechanical systems, with which we are primarily concerned in this paper.

As soon as one is willing to admit such a stochastic element, he can take advantage of a very direct rigorous method for deriving the equilibrium Maxwell distribution. This is the method of equal *a priori* probabilities in phase space, introduced by Boltzmann.[3] One simply assumes that all possible ways of distributing a finite amount of kinetic energy among a finite number of mass points are equally probable – or rather, to be more precise, one uses the appropriate weighting factor corresponding to equal distribution in momentum rather than kinetic energy. An exact formula can be derived by combinatorial analysis or n-dimensional geometry in the finite case, and the Maxwell distribution comes out very simply as a limit when the number of molecules goes to infinity. No assumptions need be made about collisions among molecules, or about the way in which the system approaches equilibrium.

Notes for §10.7

1. For detailed discussion and references see §§14.5–7.
2. L. Boltzmann, *Wien. Ber.* **66**, 275 (1872), reprinted in his *Wissenschaftliche Abhandlungen* **1**, 316; English trans. in Brush, *Kinetic Theory* **2**, 88.
3. L. Boltzmann, *Wien. Ber.* **58**, 517 (1868), **78**, 7 (1878); J. C. Maxwell, *Trans. Camb. Phil. Soc.* **12**, 547 (1879).

10.8 The paradox of specific heats

We have already mentioned that Maxwell regarded the discrepancy between experimental specific heats of gases and the theoretical value for diatomic molecules as a serious objection to the validity of the kinetic-theory model. According to kinetic theory, the ratio of the specific heat at constant pressure to that at constant volume is given by the theoretical formula

$$\gamma = \frac{n+2}{n}$$

where n is the number of degrees of freedom of a molecule. By the 1860's it had been decided by chemists and physicists that most molecules are diatomic in the gaseous state. If one regarded a molecule as a pair of mass-points bound together by some kind of force, then there should be six degrees of freedom, and γ should be equal to $1\frac{1}{3}$. However, for most of the common gases, experiments indicated that γ is about 1.4 or 1.41.[1]

It was also known by this time that molecules must have some kind of internal structure in order to account for the absorption and emission of spectral lines. In that case, there would have to be a much larger number of degrees of freedom and the theoretical value of γ would be closer to one, that is, even farther from the experimental value. Maxwell and Clausius took the position that this discrepancy showed that kinetic theory was not applicable to properties of gases that involve the internal structure of molecule; but that should not prevent one from applying it to other properties of gases, such as viscosity and diffusion, which do not seem to involve internal structure.

A new light was thrown on the subject by the experiments of Kundt and Warburg in 1875. They found that the specific heat ratio of mercury vapor is almost precisely $1\frac{2}{3}$, the theoretical value that would be appropriate for a monatomic molecule with three degrees of freedom.[2] Later, in the 1890's, Rayleigh and Ramsay found that the noble gases, argon, helium, and krypton, also have specific heat ratios equal to $1\frac{2}{3}$. In the absence of chemical evidence, this was regarded as the best argument at the time in favor of their being monatomic.[3]

The fact that monatomic gases seem to have only three degrees of freedom, as far as specific heat measurements are concerned, despite the fact that they also have an internal structure as indicated by spectral lines, encouraged kinetic theorists to devise mechanical models for diatomic molecules. Boltzmann pointed out that a system of two mass-points rigidly connected together, or an ellipsoid of revolution, would act as if it had only five degrees of freedom instead of six, since collisions of such molecules would not change the amount of rotation around the axis. The theoretical specific heat for such a system would be $\frac{7}{5} = 1.4$, in good agreement with experimental values.[4]

Boltzmann's model for diatomic molecules was accepted by many physicists later in the 19th century as being the best available.[5] Maxwell, however, did not accept it, and published a trenchant criticism in his 1877 review of H. W. Watson's *Treatise on the Kinetic Theory of Gases*.[6] Since this review was omitted from the supposedly complete collection of Maxwell's scientific papers published in 1890 (and from subsequent reprints of this collection), Maxwell's disapproval may not have been generally known. Maxwell objected to the notion that a body could be both rigid and elastic at the same time. This was an old objection to atomic theory: How could an atom be elastic (so that kinetic energy and momentum would be conserved in collisions) and also unchangeable in form (so that it can be an ultimate atom

§10.8] PARADOX OF SPECIFIC HEATS 355

in the philosophical sense)? By the late 19th century, physicists no longer took seriously the requirement that an atom be rigid and unstructured just because it was an atom. Boltzmann's reason for assuming rigidity was that collisions of molecules must not be able to set up internal vibrations, except at the very high temperatures found in electrical discharges.

As Maxwell remarked,

> It will not do to take a body formed of continuous matter endowed with elastic properties, and to increase the coefficients of elasticity without limit till the body becomes practically rigid. For such a body, though apparently rigid, is in reality capable of internal vibrations, and these of an infinite variety of types, so that the body has an infinite number of degrees of freedom....
>
> But Boltzmann's molecules are not absolutely rigid. He admits that they vibrate after collisions, and that their vibrations are of several different types, as the spectroscope tells us. But still he tries to make us believe that these vibrations are of small importance as regards the principal part of the motion of the molecules. He compares them to billiard balls, which, when they strike each other, vibrate for a short time, but soon give up the energy of their vibration to the air, which carries far and wide the sound of the click of the balls.
>
> In like manner, the light emitted by the molecules shows that their internal vibrations after each collision are quickly given up to the luminiferous ether.
>
> If we were to suppose that at ordinary temperatures the collisions are not severe enough to produce any internal vibrations, and that these occur only at temperatures like that of the electric spark, at which we cannot make measurements of specific heat, we might, perhaps, reconcile the spectroscopic results with what we know about specific heat.
>
> But the fixed position of the bright lines of a gas shows that the vibrations are isochronous, and therefore that the forces which they call into play vary directly as the relative displacements, and if this be the character of the forces, all impacts, however slight, will produce vibrations.
>
> Besides this, even at ordinary temperatures, in certain gases, such as iodine gas and nitrous acid, absorption bands exist, which indicate that the molecules are set into internal vibration by the incident light.

The molecules, therefore, are capable, as Boltzmann points out, of exchanging energy with the ether.

But we cannot force the ether into the service of our theory so as to take from the molecules their energy of internal vibration and give it back to them as energy of translation. It cannot in any way interfere with the ratio between these two kinds of energy which Boltzmann himself has established. All it can do is to take up its own due proportion of energy according to the number of its degrees of freedom.

We leave it to the authors of the "Unseen Universe" to follow out the consequences of this statement.

Notes for §10.8

1. See J. R. Partington, *An Advanced Treatise on Physical Chemistry* (London: Longmans, Green and Co., 1949), **1**, 839; J. R. Partington and W. G. Shilling, *The specific heats of gases* (London: Benn, 1924).
2. A. Kundt and E. Warburg, *Ber. D. Chem. Ges.* **8**, 945 (1875), reprinted in *Ann. Phys.* [2] **157**, 353 (1876).
3. W. Ramsay, *C.R. Paris* **120**, 1049 (1895); J. W. S. Rayleigh and W. Ramsay, *Phil. Trans.* **186**, 187 (1896); W. Ramsay and M. W. Travers, *Proc. R.S. London* **63**, 405, 437 (1898). Ramsay, *Contemporary Review* **74**, 681 (1898); *Science* [2] **8**, 768 (1898); *Living Age* **220**, 23 (1899); *Smiths. Rep.* 277 (1898).
4. L. Boltzmann, *Wien. Ber.* **74**, 553 (1877), English trans. in *Phil. Mag.* [5] **3**, 320 (1877). A similar suggestion was made by R. H. M. Bosanquet, *Phil. Mag.* [5] **3**, 271 (1877).
5. See A. Crum Brown, *Nature* **32**, 352 (1885); S. H. Burbury, *Nature* **51**, 127 (1894); O. Lodge, *Phil. Mag.* [6] **2**, 241 (1901); books cited in note 1.
6. J. C. Maxwell, *Nature* **16**, 242 (1877), not reprinted in Maxwell's *Papers*. See also *Nature* **11**, 357, 374 (1875). Maxwell's objection to Boltzmann's hypothesis is discussed in the light of modern ideas about "falsifiability" of hypotheses by A. F. Chalmers, *Brit. J. Phil. Sci.* **24**, 164 (1973).

10.9 Validity of the equipartition theorem

Peter Guthrie Tait, who was one of the authors of the "Unseen Universe,"[1] waited until several years after Maxwell's death before taking up the challenge to investigate the partition of energy among gas molecules. In 1885, Tait produced the first of his series of memoirs, "On the foundations of the kinetic theory of gases," in which he attempted to give a comprehensive exposition of the subject as well as to continue the study of many problems left unsolved by Maxwell. In doing this, Tait was bound to enter territory already explored by Boltzmann. It is somewhat unfortunate that Tait could not find time to examine all of Boltzmann's writings; in many cases he published results already stated by Boltzmann, though buried in the middle of the latter's lengthy and difficult papers.

Tait proposed a set of conditions that must be satisfied in order that the "equilibrium" distribution may actually be attained. He first treated two systems of spheres of different kinds, for which he states the conditions:

(A) That the particles of the two systems are thoroughly mixed.
(B) That in any region containing a very large number of particles, the particles of each kind separately acquire and maintain the error-law distribution of speeds....
(C) That there is perfectly free access to collision between each pair of particles, whether of the same or of different systems; and that, in the mixture, the number of particles of one kind is not overwhelmingly greater than that of the other kind.[2]

Burbury[3] immediately objected that "Tait postulated rather more than is necessary. There may, as it appears to me, be a set of spheres which never collide with each other, and one other sphere which collides with the first set, and it will be found that this single sphere will knock the others into 'the special state' [Maxwellian distribution]."

Boltzmann[4] also criticized Tait's paper, and protested against Tait's statement that his (Boltzmann's) deductions were "rather of the nature of playing with symbols than of reasoning by consecutive steps." He cited some of his previous work on the theory of transport properties, involving collisions between molecules of different kinds, in support of his assertion that Tait's condition (B) is unnecessary: the equipartion theorem holds even if the two systems do not initially have Maxwellian velocity distributions. Moreover, he claims that

> it is not even necessary to assume that the molecules of the first kind are in collision amongst themselves. The only assumptions are: that both the molecules of the first and also those of the second kind are uniformly distributed over the whole space; that throughout they behave in the same way in all directions; and that the duration of impact is short in comparison with the time between two impacts.

Boltzmann then gives a more detailed proof of this statement based on his paper of 1875.[5] He also supports Burbury's assertion that a single molecule of another kind can produce equilibrium in a system of molecules that do not collide with each other.

Tait then admitted that his conditions (A) and (B), "though enunciated separately, are regarded as *consequences* of (C), which is thus my sole assumption for the proof of Clerk-Maxwell's theorem." However, he was not willing to accept Burbury's assertion as having any physical relevance, because it would take much too long for a single molecule to collide with all the molecules in a gas:

> Assuming the usual data as to the number of particles in a cubic inch of air, and the number of collisions per second, it is easy to show ... that somewhere about 40 000 *years* must elapse before it would be so much as *even betting* that Mr. Burbury's single particle (taken to have twice the diameter of a particle of air) had encountered, at least once, each of the $3 \cdot 10^{20}$ very minute particles in a single cubic inch.

The justification for this statement was provided by an investigation performed at Tait's request by his mathematical colleague Arthur Cayley.[6] Tait was also unsatisfied with Boltzmann's arguments, and still maintained that there must be a large number of particles of each kind, and that particles of each kind must collide among themselves, in order that the equipartition theorem hold.

In July 1887, William Burnside read a paper to the Royal Society of Edinburgh on the "Partition of Energy between the Translatory and Rotational Motions of a set of Non-homogeneous elastic spheres." He proposed a system of smooth elastic spheres "whose centers of figure and centers of inertia do not coincide." Following Tait's methods, he claimed to prove that

> The average energies of rotation of a sphere about each of the principal axes are equal, and the whole average energy of rotation of a sphere is twice the average energy of translation.[7]

This was the first of the "test-cases" against the equipartition theorem (not counting Loschmidt's paradoxes) but it deserves little discussion since Burnside's assertion was immediately shown to be erroneous by Boltzmann.[8] According to Boltzmann, Burnside's result

> rests merely on the fact that he left out some infinitesimal terms of the same order of magnitude as those he included, by assuming that the frequency of collisions is not changed by the eccentric position of the center of gravity of the molecule. On taking

account of this circumstance more precisely, one finds on the contrary that Burnside's example is in complete agreement with the general theorem....

This test-case was also criticized by Watson and Burbury, and Burnside withdrew it in 1892.[9]

Lord Kelvin had proposed some further test cases in 1891.[10] Asserting that the so-called elastic spheres discussed in kinetic theory should really be considered as Boscovich point-atoms, he described a system consisting of hollow shells, each with a "globule" inside; the shell and the globule interact only when they touch. The shell and its enclosed globule he calls a double molecule or doublet, and the globule an atom, and he considers the equilibrium of a large number of these doublets and atoms. Since the globules enclosed in shells cannot interact directly with other globules, Tait's condition (C) is violated. In variations of this model, there may be several globules inside each shell; or an internal globule may be connected with the shell by massless springs so that it vibrates harmonically about the center. The latter system had already been proposed as a model for vibratory molecules imbedded in ether in Kelvin's *Baltimore Lectures*.[11] Although he discussed these models at length, Kelvin succeeded in doing no more than raising some doubts as to the validity of equipartition in these cases. But in another paper in 1892,[12] he claimed to have found "A decisive test-case disproving the Maxwell–Boltzmann doctrine regarding distribution of kinetic energy." His new model is a system of

> three bodies, A, B, C, all movable only in one straight line, KHL: B being a simple vibrator controlled by a spring so stiff that when, at any time, it has very nearly the whole energy of the system, its extreme excursions on each side of its position of equilibrium are small: C and A, equal masses; C, unacted on by force except when it strikes L, a fixed barrier, and when it strikes or is struck by B: A, unacted on by force except when it strikes or is struck by B, and when it is at less than a certain distance, HK, from a fixed repellent barrier, K, repelling with a force, F, varying, according to any law, or constant, when A is between K and H, but becoming infinitesimally great when (if at any time) A reaches K, and goes infinitesimally beyond it.

Kelvin then claims to prove that "The average kinetic energy of A is less than the average kinetic energy of C!"

Kelvin's test cases touched off a large amount of controversy, in which Watson, Burbury, Boltzmann, Poincaré, Bryan, and Rayleigh participated.[13] The outcome seemed to be a general agreement that most of Kelvin's test cases did not prove any violation of the equipartition theorem, but, on the other hand, that one could not be sure the theorem was always valid in systems of a finite number of particles.

In the course of this discussion, Burbury developed an interesting variation on the Maxwell–Boltzmann theory. He challenged the assumption, which Boltzmann had made somewhat more explicit,[14] that the velocity distributions of two molecules before a collision are independent of their relative positions. Burbury doubted that such a condition can actually be continually true in a gas of interacting molecules. Instead, he assumed that "molecules very near each other in space have on average a certain velocity in common" and developed a kinetic theory on this basis.[15]

At the meeting of the British Association at Oxford in 1894, G. H. Bryan presented a report on the foundations of thermodynamics, with special reference to the equipartition theorem. Boltzmann attended this meeting, and there was a discussion following the presentation of Bryan's report in which Boltzmann replied to various criticisms of the kinetic theory. According to a note which Bryan sent to *Nature* shortly after this meeting, Boltzmann's statements "are now widely circulated and quoted as being an authoritative admission that the Kinetic Theory of Gases is nothing more than a purely mathematical investigation, the results of which are not in accord with physical phenomena; in short, a mere useless mathematical plaything."[16] Boltzmann was asked specifically about the specific heats of gases, and according to Bryan,

> What I understood Prof. Boltzmann to imply was that his investigations treated the matter purely from a mathematical standpoint, but that the values he obtained by regarding the molecules of a gas as rigid bodies, *viz.* 1.6 for smooth spheres, 1.4 for smooth solids of revolution and 1.3 for solids of any other form, accorded on the whole *very fairly* with the results of experiment.

As for the objection that the kinetic theory does not explain the spectra of gases, Bryan remarked that if spectra were due to the vibrations of atoms within molecules than monatomic gases would not have any optical properties at all. But, said Bryan,

> the electromagnetic theory of light entirely relieves the kinetic theory from the burden which has been imposed on it by its opponents, since if (for example) we regard the molecules of a gas

as perfectly conducting hard spheres, spheroids, or other bodies moving about in a dielectric "vacuum" (*i.e.* space devoid of ordinary matter), we shall be able to account for the spectra by means of electromagnetic oscillations determined by surface-harmonics of different orders without interfering with the assumptions required for explaining the specific heats of gases.

E. P. Culverwell, replying to Bryan's letter in *Nature* quoted above, disagreed with the latter's account of Boltzmann's remarks. According to Culverwell, Boltzmann's reply was not to a special question on the specific heats of gases,

> but was in answer to a very vigorous, if somewhat general, onslaught of Prof. Fitzgerald, who simply stated that it appeared evident from the spectra of gases and other considerations, that the energy could not be equally divided among all the degrees of freedom of the coordinates, and said what he wanted to know from Prof. Boltzmann was *when the theory became inapplicable, what assumptions became invalid*?

Fitzgerald asked why the ether, the solar system, and indeed the whole universe were not subject to equipartition. Boltzmann's reply to this, according to Culverwell, was that "the theory as it left his hands was a mathematical theorem, a piece of pure mathematics, and that it was for *physicists* to say how far it applied to gases."[17] But both Culverwell and Bryan seemed to agree that Boltzmann was not a pure mathematician to the extent of being totally uninterested in the relation between his theorem and the real world; and that Boltzmann's work was motivated by the hope that a quantitative agreement between theory and experiment would eventually be established.

Bryan himself gave another reply to Fitzgerald's criticism, saying that if there were collisions between solar systems, then these systems would eventually tend to assume the Maxwell distribution; as for the ether, it was up to the physicists to "give us a clear and definite statement as to *what are the coordinates and momenta of the ether, and how transference of energy takes place between these and the molecules.*" Since the test case in which the molecules are solids of revolution shows that equipartition does not take place between all degrees of freedom, Bryan suggested that something similar may be involved in the interaction between molecules and the ether.[18] Again, "If a gaseous ether will satisfy the requirements of physicists, then the Boltzmann–Maxwell law is undoubtedly applicable to the ether. If not, the ether falls entirely beyond the scope of our investigation."[19]

Boltzmann finally decided to speak out for himself, and wrote a letter to *Nature* in 1895 in which he proposed to answer two questions:

(1) Is the Theory of Gases a true physical theory as valuable as any other physical theory?
(2) What can we demand from any physical theory?

In this letter Boltzmann takes a swipe at Boscovich's atomic theory, and notes that now in Germany

> every special theory is oldfashioned, while in England interest in the Theory of Gases is still active; vide, among others, the excellent papers of Mr. Tait, of whose ingenious results I cannot speak too highly, though I have been forced to oppose them in certain points.

Boltzmann suggests that the ether does not have time to come into thermal equilibrium with gas molecules in the time available for our experiments.

> The possibility of the transference of energy being so gradual cannot be denied, if we also attribute to the ether so little friction that the Earth is not sensibly retarded by moving through it for many hundreds of years.

Similarly, there is not sufficient time for internal vibrations of the molecules to come into equilibrium with translational and rotational motion. Boltzmann admits that he has offered nothing more than "a series of imperfectly proved hypotheses" but claims that at least he has shown that the problem is not insoluble.[20]

The final resolution of these difficulties with the equipartition theorem had to wait for the development of the quantum theory. I will leave that story for another time, and turn now to a particularly interesting "imperfectly proved hypothesis" which Boltzmann had first proposed in 1871: "The different molecules of the gas will... pass through all possible states of motion."[21] Out of this statement has grown not only a new approach to the foundations of statistical mechanics, but also a lively specialty of modern pure mathematics.

Notes for §10.9

1. B. Stewart and P. G. Tait, *The Unseen Universe; or, Physical Speculations on a Future State* (London, 1875; 1st ed. pub. anonymously). P. M. Heimann, *Brit. J. Hist. Sci.* **6**, 73 (1972).

2. P. G. Tait, *Trans. R. S. Edinburgh* **33**, 65 (1886), reprinted in Tait's *Scientific Papers* (Cambridge University Press, 1890–1900), **2**, 124.
3. S. H. Burbury, *Phil. Mag.* [5] **21**, 481 (1886).
4. L. Boltzmann, *Wien. Ber.* **94**, 613 (1887); English trans. in *Phil. Mag.* [5] **23**, 305 (1887).
5. L. Boltzmann, *Wien. Ber.* **72**, 427 (1875).
6. P. G. Tait, *Phil. Mag.* [5] **23**, 433 (1887); *Proc. R.S. Edinburgh* **15**, 140 (1888). A. Cayley, *Proc. R.S. Edinburgh* **14**, 149 (1887).
7. W. Burnside, *Trans. R.S. Edinburgh* **33**, 501 (1887).
8. L. Boltzmann, *Berlin. Ber.* 1395 (1888).
9. S. H. Burbury, *Nature* **45**, 533 (1892). H. W. Watson, *Nature* **45**, 512 (1892). W. Burnside, *Nature* **45**, 533 (1892).
10. Kelvin, *Proc. R.S. London* **50**, 79 (1891); *Nature* **44**, 355 (1891).
11. Kelvin, *Baltimore Lectures on Molecular Dynamics and the Wave Theory of Light*, lectures delivered in 1884 with appendices added later (London: Clay, 1904).
12. Kelvin, *Phil. Mag.* [5] **33**, 466 (1892).
13. H. W. Watson and S. H. Burbury, *Nature* **46**, 100 (1892). L. Boltzmann, *Mün. Ber.* **22** (3), 329 (1892); English trans. in *Phil. Mag.* [5] **35**, 153 (1893). Rayleigh, *Phil. Mag.* [5] **49**, 98 (1900). H. Poincaré, *Rev. Gen. Sci.* **5**, 513 (1894). G. H. Bryan, *B.A. Rep.* **64**, 64 (1894). Kelvin, *Baltimore Lectures*, Appendix B (added in 1900). R. C. Bowden, *An Analysis of Lord Kelvin's Scientific Career* (Dissertation, University of North Carolina, Chapel Hill, 1972), pp. 209–18.
14. See note 3, §10.5.
15. S. H. Burbury, *Phil. Mag.* [5] **50**, 584 (1900); see also *A Treatise on the Kinetic Theory of Gases* (Cambridge, 1899); *Proc. London Math. Soc.* **26**, 431 (1895), **28**, 331 (1897), **29**, 225 (1898); *Phil. Trans.* **A187**, 1 (1896). R. Hesketh, *Times Higher Education Supplement*, p. 11 (20 April 1973).
16. G. H. Bryan, *Nature* **51**, 31 (1894).
17. E. P. Culverwell, *Nature* **51**, 78 (1894).
18. G. H. Bryan, *Nature* **51**, 152 (1894).
19. G. H. Bryan, *Nature* **51**, 319 (1895).
20. L. Boltzmann, *Nature* **51**, 413, 581 (1895).
21. L. Boltzmann, *Wien. Ber.* **63**, 397 (1871).

10.10 The ergodic hypothesis of Boltzmann and Maxwell

It is frequently stated that Maxwell and Boltzmann believed in the so-called ergodic hypothesis: a mechanical system, if left undisturbed, will pass through every point of the phase space lying on a certain energy surface. In other words, the positions and velocities of all the mass points (representing the atoms) will eventually take every possible value consistent with the given total energy of the system. If this hypothesis is true, then in computing the physical properties of such a system one can replace the average over a long trajectory (or history) of a single system by an average over all points on the energy surface.

There are several questions of physical, mathematical, and historical interest associated with the ergodic hypothesis, such as: (1) Can it ever be true? (2) Is it a necessary and/or sufficient condition for the validity of the equipartition theorem? (3) Did Maxwell and Boltzmann really believe it, and if so, why? (4) If it is not true, can a weaker hypothesis, for example the "quasi-ergodic hypothesis," be used instead? Questions (1), (2), and (3) can then be asked again of the quasi-ergodic hypothesis: the assumption that the system will pass within an infinitesimal distance of every point on the energy surface.

At the present time, only the first question can be given an unequivocal answer, *viz. no*. (The case of one-dimensional motion is a trivial exception.) In view of this, the fact that the validity of the ergodic hypothesis would have been a sufficient condition for the validity of the equipartition theorem is not much help; however, the quasi-ergodic hypothesis, which is probably true for some systems of interest in classical statistical mechanics, would also be a sufficient condition for equipartition "almost always" – except for an exceptional class of initial states.

The version of the ergodic hypothesis usually attributed to Maxwell and Boltzmann is somewhat stronger than most of the statements they actually made in this connection, and certainly stronger than anything they actually used in their proofs. It will therefore be necessary to disentangle the original statements of Maxwell and Boltzmann from the misconceptions that have subsequently grown up about them.

The Ehrenfests' article in the *Encyklopädie der mathematischen Wissenschaften*[1] is usually cited as the standard authority on this subject. While this article was itself an important contribution to the development of ergodic theory in statistical mechanics, it leaves much to be desired as an historical account of the work of Maxwell and Boltzmann, and most of the errors of later commentators in describing the origin of the ergodic hypothesis are probably the result of an uncritical acceptance of some careless statements made by the Ehrenfests. The first error is connected with the meaning of the word *ergodic*. Boltzmann introduced his *Ergoden* in 1884,[2] meaning by this term *not* a single system that goes through every point on the energy surface but rather an ensemble of systems with a certain distribution in phase space which he calls ergodic. While the Ehrenfests recognize this distinction, they greatly exaggerate the importance which Boltzmann assigned to the former property of a single system, and the word "ergodic" has consequently acquired a meaning completely different from its original one. (See further discussion below.) Another term, "continuity of path," which the Ehrenfests attribute to Maxwell, was apparently first used by J. H. Jeans in 1902 (23 years after Maxwell's death).[3] In view of this confusion of terminology and opinion, it will be useful to examine what Maxwell and Boltzmann themselves wrote on the subject. We shall use the word ergodic in its modern sense except in direct quotations; but we shall not follow the practice of some modern writers who have assigned the word ergodic to quasi-ergodic systems (on the grounds that the originally postulated ergodic systems cannot exist anyway).

Boltzmann's first detailed discussion of ergodic systems appears in a paper published in 1871.[4] He treats the motion of a point-mass in a plane under the influence of an attractive force described by a potential function $\frac{1}{2}(ax^2 + by^2)$, *i.e.* the compound harmonic motion which results in the so-called Lissajous figures. (These figures were first studied by Nathaniel Bowditch in 1815, and were more completely investigated by Lissajous in 1857.[5]) If the ratio of the periods of the two motions, a/b, is a rational number, then the point-mass will undergo a finite number of oscillations and then start over again and repeat the same motion. However, if a/b is irrational (*i.e.* the periods are incommensurable) then there is never an exact recurrence, and each oscillation is different from all the previous ones. The complete phase space is of course 4-dimensional, but may be visualized as a torus. If u and v are the velocities in the x and y directions, respectively, then all points on a trajectory must satisfy the conditions

$$\tfrac{1}{2}mu^2 + \tfrac{1}{2}ax^2 = c_1, \qquad \tfrac{1}{2}mv^2 + \tfrac{1}{2}by^2 = c_2$$

where c_1 and c_2 are the energies of the two components of motion, which are fixed initially. According to Boltzmann, the point-mass goes through the entire surface of a rectangle ("die ganze Fläche eines Rechtecks durchwandern"). Similarly in the case of a point moving in a force field described by a potential function $(a/r) + (b/r^2)$, the orbit is similar to an ellipse, but if the angles of two successive apside lines are rational multiples of π, then the point goes through all of a certain section of the plane ("durchwandert allmälig das ganze Stuck der Ebene") without ever returning exactly to the same point. Of course the crucial point—what did Boltzmann actually mean—depends on how one translates the phrases I have quoted. Did he mean that the trajectory goes through *every point* in the plane, or that the curve covers the area "densely"—*i.e.* goes as close as one likes to every point? We will be in a position to answer this question after giving a few more examples.

Boltzmann then asserted that the same property (whatever it is) should be true of a system of n points. In the next section of the same paper he treats the thermal equilibrium of gas molecules on the basis of the hypothesis that

> The great irregularity of the thermal motion, and the multiplicity of forces that act on the body from outside, make it probable that the atoms themselves, by virtue of the motion that we call heat, pass through all possible positions and velocities consistent with the equation of kinetic energy, and that we can therefore apply the equations previously developed to the coordinates and velocities of the atoms of warm bodies.

Thus the ergodic hypothesis for real physical systems is justified by the argument that the system will always be subjected to random external forces; Boltzmann did not believe at this point that the hypothesis had sufficiently strong support from the mathematical properties of unperturbed trajectories in phase space to be able to dispense with such physical arguments.

Now let us turn to Maxwell's most explicit statement on the ergodic hypothesis, which occurs in a paper written just before he died:[6]

> The only assumption which is necessary for the direct proof [of the equipartition theorem] is that the system, if left to itself in its actual state of motion, will, sooner or later, pass through every phase which is consistent with the equation of energy.
>
> Now it is manifest that there are cases in which this does not take place. The motion of a system not acted on by external forces possesses six equations besides the equation of energy, so that the system cannot pass through those phases which, though they satisfy the equation of energy, do not also satisfy these six equations.
>
> Again, there may be particular laws of force, as for instance that according to which the stress between two particles is proportional to the distance between them, for which the whole motion repeats itself after a finite time. In such cases a particular value of one variable corresponds to a particular value of each of the other variables, so that phases formed by sets of values of the variables which do not correspond cannot occur, though they may satisfy the several general equations.

But if we suppose that the material particles, or some of them, occasionally encounter a fixed obstacle such as the sides of the vessel containing the particles, then, except for special forms of the surface of this obstacle, each encounter will introduce a disturbance into the motion of the system, so that it will pass from one undisturbed path into another. The two paths must both satisfy equations of energy, and they must intersect each other in the phase for which the conditions of the encounter with the fixed obstacle are satisfied, but they are not subject to the equations of momentum. It is difficult in a case of such extreme complexity to arrive at a thoroughly satisfactory conclusion, but we may with considerable confidence assert that except for particular forms of the surface of the fixed obstacle, the system will sooner or later, after a sufficient number of encounters, pass through every phase consistent with the equation of energy.

Thus Maxwell proposed the ergodic hypothesis only for mechanical systems interacting with their surroundings, rather than for unperturbed systems. However, he did not require this interaction to be of a random nature on the microscopic level; if the walls of the container are rigid, their action on the molecules is completely deterministic, having the effect of knocking the system from one orbit to another on the energy surface. It is not obvious whether the resulting system is ergodic; probably it is not, but this model is rarely discussed by later writers.

Boltzmann first introduced the words *Ergoden* and *ergodische* in a paper on monocyclic systems in 1884.[2] Helmholtz[7] had defined a polycyclic system as "a dynamical system containing one or more periodic or circulating motions." If there is only one such motion, or if, owing to the existence of certain relations between the velocities of the different parts of the system, the circulating motions can all be defined by a single coordinate, the system is called *monocyclic*.[8] Moreover, it is assumed that the kinetic and potential energies do not depend on the actual values of the coordinates of the molecules, but only on their rates of change. These molecular coordinates are distinguished from the "controllable" coordinates of the system. When the state of the system is changed very slowly (corresponding to the "reversible" changes considered in thermodynamics) it is assumed that the changes in the controllable coordinates are negligible, and likewise the accelerations of the molecular coordinates are negligible. The purpose of constructing such systems is to find mechanical analogies for the Second Law of Thermodynamics—that is, to express the increment of energy added to a system in such a way that the kinetic energy (with which the temperature is then identified) is an integrating divisor for dQ. In this way the dynamical equivalent of the thermodynamic entropy could be found. Furthermore, by coupling together two monocyclic systems in such a way that their molecular coordinates but not their controllable coordinates are connected, and then showing that the coupled system forms a single monocyclic system, it would be possible to find a mechanical analogue of temperature equilibrium (the integrating divisors for the two systems must be equal).

The extent to which monocyclic systems actually provide a satisfactory mechanical analogy for thermodynamic systems will not be discussed here; at present we are concerned with them only insofar as they provided the stimulus for Boltzmann's introduction of *Ergoden*. We quote from Boltzmann's paper of 1884:

> A mass point moves, according to Newton's law of gravitation, around a fixed central body O, in an elliptical path. The motion is here clearly not monocyclic; however, by means of an artifice that I first used in the first section of my paper [4] and which Maxwell later used [6], we can transform it into a monocyclic motion. We imagine that the entire elliptic path is covered with masses, which should have at each point a density (number of masses per unit length) such that, although in the course of time the masses continuously flow through each cross-section of the path, the density at each point of the path remains constant. If one of Saturn's rings consisted of a homogeneous fluid or of a homogeneous swarm of solid bodies, then with an appropriate choice of ring cross-section at different places, it would provide an example for the motion under discussion.

Maxwell's first studies of the kinetic theory of gases had been preceded by his investigation of the stability of Saturn's rings according to different hypotheses as to their nature; here the example crops up again in Boltzmann's theory of statistical mechanics.

Boltzmann goes on to discuss various similar models, paying particular attention to the expressions for the entropy and kinetic energy, in order to discover under what conditions the latter is an integrating divisor for the former. In the course of this discussion he introduced what we now call *ensembles* of various kinds (Boltzmann's term was *Inbegriff*). An ensemble of mechanical systems all having the same energy but different values of the coordinates, determined by a certain distribution function, was called by Boltzmann an *Ergode*. The modern term for this concept, following Gibbs,[9] is "microcanonical ensemble." A single system that passes through all states compatible with a given fixed energy was called "isodic" by Boltzmann, but is now (following the Ehrenfests) called "ergodic."

Boltzmann returned to this subject in 1887, in his memoir on mechanical analogies of the Second Law of Thermodynamics.[10] He again discussed the Lissajous figures, and stated that in the cases where the periods are incommensurable,

> then the material point, in the course of a very long time, will traverse all the interior of a surface situated on a rectangle, and as soon as x is given, y is merely enclosed between two limits.
>
> We shall say in this case that one of the integral equations will be infinitely many valued....
>
> The first two constants of integration are therefore only what I have called the parameters of the path; no matter what the phase difference of the motions in the two coordinate directions may be, in the course of an infinitely long time all possible phase differences will always appear. All paths, for which the values of the first two constants of integration are the same, go into each other after a finite or infinite time, and all other quantities determine only the time at which the path will be traversed. We can thus say also: if the path is closed, all pairs of values of x and y, which correspond to a path, form a manifold of only one dimension.

Boltzmann then discusses the idea of inserting a small cylindrical obstacle in the system, with which the moving point can collide elastically, so that while its energy remains constant, momentum does not, and energy can be transferred from one degree of freedom to another:

> If the cylinder is placed so that it will be struck by all paths consistent with the equation of kinetic energy–thus in the case of central motion, very close to the circular path, and in the case of Lissajous motion very close to the origin of coordinates–then in the course of time, all combinations of values of x, y, dx/dt, and dy/dt consistent with the single equation of kinetic energy will be traversed.
>
> We shall consider at once the most general case, in that we assume that K integrals of the equations of motion are infinitely many-valued. After eliminating $t - \tau$, there remain only $2b - K - 1$ integral equations, which are not infinitely many-valued, and the

variables can in the course of time pass through all possible values consistent with the $2b - K - 1$ equations. We can imagine in this way a system in which $K = 2b - 2$, in which therefore all possible values of the variables consistent with the equation of kinetic energy will be traversed. An example of this is given by the above-mentioned Lissajous motion, or by central motion. An even simpler one is offered by all motions, as soon as $b = 1$. Such a system will have the same properties which warm bodies show empirically: their state is completely determined if one knows, in addition to the external and internal forces, the total energy. The probability of different states as well as all the properties of such a system can now be easily calculated.... Warm bodies possess indeed a property of still greater generality, in that the different phases that their state of motion assumes in the course of time are not experimentally observable; rather, because of the great number of atoms present, as soon as some atom passes to another state-phase, a neighboring one will assume the state-phase which the first one formerly possessed. Whence it certainly follows that different initial conditions will produce only completely accidental differences in the state of the warm body, while all observable properties will depend, aside from the internal and external forces, only on the total energy....

Boltzmann then proposes (as he and Maxwell had already done earlier) to consider an infinite number of systems with the same energy and all possible initial states. He defines a stationary distribution for which the number of systems, whose coordinates and momenta are within certain limits, remains always the same, and calls these systems *Ergoden*. This paper of Boltzmann's, published in 1887, is sometimes cited as his first formulation of the ergodic hypothesis, although as we have seen the word *Ergoden*, and the separate concept of a system that passes through all possible states on the energy surface, had been introduced earlier. Nevertheless the paper does represent a significant stage in the evolution of Boltzmann's own thinking, which might be characterized as his own formulation of Maxwell's approach. For Boltzmann had earlier[11] said that:

There is a difference in method between Maxwell and Boltzmann, inasmuch as Boltzmann measures the probability of a condition by the time during which the system possesses this condition on the average, whereas Maxwell considers innumerable similarly constituted systems with all possible initial conditions.

In a discussion of Kelvin's test-cases against the equipartition theorem in 1892, Boltzmann made the statement that "all possible sets of values of x, y, and θ which are consistent with the equation of *vis viva* are obtained with any required degree of approximation [*mit beliebiger Annaherung erreicht werden*] provided the motion continues for a sufficiently long time T."[12] Two paragraphs later he restated this assumption as follows: "In the course of an interval of time T the condition-point occupies all positions in a finite cylinder (the condition-cylinder)." Thus practically in the same breath, Boltzmann made a quasi-ergodic hypothesis and then restated it as an ergodic hypothesis—dropping the qualification "approximately" with no explanation.

This curious inconsistency raises the suspicion that when Boltzmann says he assumes a system will go through all possible states with the given energy, he may not mean that the trajectory goes through *every point* on the energy surface in the same sense that we would now give to such a statement. It must be pointed out here that Boltzmann had his own rather conservative concept of the infinitesimal and the infinite, which has been discussed recently by Dugas[13] and Klein.[14] As Dugas says, "Boltzmann is resolutely *finitist*: more precisely, he rejects all considerations of an *actual* infinity and considers an infinite set only as the limit of a collection of individuals whose number is very large but finite."

In particular, this means that when Boltzmann discusses the distribution of energy among a large number of molecules, he does not consider his proof to be completely satisfactory until he has carried it through with a finite number of molecules, each of which can have energy only in amounts that are integer multiples of some fixed energy. Concerning this procedure, Boltzmann remarks:

> This fiction does not, to be sure, correspond to any realizable mechanical problem, but it is indeed a problem which is much easier to handle mathematically and which goes over directly into the problem to be solved, if one lets the appropriate quantities become infinite. If this method of treating the problem seems at first sight to be very abstract, it nevertheless is generally the quickest way of getting to one's goal in such problems, and if one considers that everything infinite in nature never has meaning except as a limiting process, one cannot understand the infinite manifold of possible energies for each molecule in any way other than as the limiting case which arises when each molecule can take on more and more possible velocities.[15]

Klein points out the similarity between Boltzmann's derivation based on an energy "quantum" ϵ which eventually goes to zero, and Planck's method which led eventually to the discovery that a better distribution law for radiation could be obtained by keeping ϵ finite.[14]

It seems natural to suppose that for Boltzmann the distinction between a path going through *every point* and one merely going *arbitrarily close to every point* was meaningless, since the former property could only be understood in terms of the latter.

A similar confusion may be found in an article by Poincaré,[16] who appears to attribute both the ergodic and the quasi-ergodic hypothesis to Maxwell without pointing out the distinction between them. When a mathematician as brilliant and lucid as Poincaré fails to make this distinction, it is not surprising to find that physicists of the period were not very clear about it.

The ergodic hypothesis, sometimes expressed as a quasi-ergodic hypothesis, was discussed by Kelvin, Stoney, Rayleigh, Burbury, Einstein, Ornstein, and others in the period between 1890 and 1910.[17] On the other hand, Boltzmann himself seems to have abandoned the hypothesis during this period; he does not even mention it in his definitive work *Lectures on Gas Theory* (1896–98). Instead, he simply starts with the assumption that molecular motions are random (§14.6).

The first recognition of the mathematical distinction between ergodic and quasi-ergodic systems appears in a paper by Paul Hertz in 1910.[18] By this time the concurrent developments in mathematics (see next section) had begun to penetrate into the world of physics, as is shown by the following remark in Hertz's paper:

> On account of the great complexity of the system, it is plausible to assume that the path curve in its manifold entanglement describes almost completely the entire surface, and exhibits a property similar to that of the well-known Peano curve. The Peano curve goes exactly through every point of the surface; that considered here comes arbitrarily close to each point.

We now come to the famous Encyclopedia article of Paul and Tatiana Ehrenfest (1911). We have already suggested that many historical misconceptions about the ergodic hypothesis originated with this article, although to be fair we must also say that most of the misconceptions would vanish if one read the footnotes of the article carefully. Here is the relevant section:

The gas model as an ergodic system

10a. Ergodic mechanical systems.[88]

Note 88: One should be careful to distinguish between the following two concepts: (a) an ergodic system, (b) ergodic density distribution in the Γ-space. For the relationship of (a) to (b), see note 101.

One can define mechanical systems by the following property: The G-path describing their motion is an open curve[89] that covers every part of a multidimensional region densely.[90]

Note 89: Boltzmann (1871, 1887) gives among others the example of the motion of a point mass in a plane under the influence of an attractive force given by the potential $\frac{1}{2}(ax^2 + by^2)$, where the ratio a/b is irrational. This gives the open Lissajous figures when the ratio of the periods is irrational. The point mass during its motion passes arbitrarily closely to each point in certain rectangle.

Note 90: The Lissajous figure in the above example is in a way the projection of the corresponding G-path in the (x, y, u, v) space onto the (x, y) plane. For this reason it crosses itself; something which the corresponding G-paths never do. The G-path can be visualized topologically by an open geodetic line on the surface of a torus. This, without crossing itself, approaches arbitrarily closely all the ∞^2 points (x, y, u, v) which satisfy the equations

$$\frac{m}{2}u^2 + \frac{ax^2}{2} = c_1,$$

$$\frac{m}{2}v^2 + \frac{by^2}{2} = c_2,$$

(c_1 and c_2 are the energies given to the two components of the oscillation at time t_0). It is not difficult either to define purely geometrically sets of curves so that each single curve of the set approaches arbitrarily closely each of the ∞^3 points inside the torus.

Prompted by the existence of such systems, Boltzmann[91] and Maxwell[92] defined a class of mechanical systems in the following way:

The single, undisturbed motion of the system, if pursued without limit in time, will finally traverse "every phase point" which is compatible with its given total energy. A mechanical system satisfying this condition is called by Boltzmann an "ergodic system."[93]

[Note 91 refers to the papers listed in my Bibliography as Boltzmann 1868 and 1871c.]
[Note 92 refers to the paper listed in my Bibliography as Maxwell 1879a.]

Note 93: ἔργον = energy; ὁδός = the path: the G-path traverses all points of the energy surface. This terminology was first used by Boltzmann in 1887. Maxwell and with him the other British authors use in this connection the phrase, "assumption of the continuity of path" (meaning by "path" the G-path).

Boltzmann and Maxwell infer from this definition the following corollaries:

1. For an ergodic system all motions with the same total energy take place on the same G-path.[94]

Note 94: The reasoning is about as follows: The single G-path traverses each point in phase space which is on the energy surface. On the other hand, each phase point Γ lies on only one G-path (note 74), which then leads us to statement 1. Compare esp. Boltzmann (1887).

2. This means that all these motions differ only in the value of the constant c_{2rN}, which appears as a constant additive to the time (cf. Integral 23c).[95]

Note 95: In this respect see the statements of Boltzmann (1887) about the phase difference between the two components of the Lissajous motion when the ratio of the periods is irrational.

3. All of these motions yield the same value for the time average of any function $\phi(q,p)$ of the phase variables.[96]

Note 96: This time average is defined by

$$\lim_{\substack{T_1=-\infty\\T_2=+\infty}} \frac{1}{T_2-T_1}\int_{T_1}^{T_2} \phi(q,p)\,dt$$

§10.10] THE ERGODIC HYPOTHESIS

It is on account of this last property that the definition of ergodic systems and the assumption that the gas models are ergodic appear in Boltzmann's investigations (cf. §11).

However, the existence of ergodic systems (*i.e.* the consistency of their definition) is doubtful. So far, not even one example is known of a mechanical system for which the single G-path approaches arbitrarily closely every point of the corresponding energy surface.[97]

Moreover, no example is known where the single G-path actually traverses all points of the corresponding energy surface.[98]

Note 97: In order to make the G-path of the example in note 89 approach arbitrarily closely each of the ∞^3 Γ points, which satisfy the *one* equation

$$\frac{m}{2}(u^2 + v^2) + \frac{ax^2 + by^2}{2} = E_0,$$

Boltzmann introduces in the (x, y) plane an infinitesimal small elastic obstacle, with which the oscillating particle collides again and again. Similar procedures are used in other examples. In this respect see also Lord Rayleigh [see Rayleigh 1900, listed in Bibliography–S.G.B.]. It is on account of the complexity of the collisions of the molecules with each other that Boltzmann and Maxwell feel justified to assume that gas models are ergodic.

Note 98: To elucidate the difference between the following two requirements: (I) "to approach arbitrarily closely each point of the energy surface" and (II) "to traverse each point of the energy surface," let us consider again a geodesic line of a torus for which the ratio of the two numbers of turnings in the two directions is irrational, *e.g.* a bit bigger than $\frac{1}{2}$. Such a geodesic will intersect the meridian at infinitely many points P_n, which are densely distributed everywhere over the circumference. On the other hand, one can state that no matter how often

Nevertheless, not only is the latter the wording of the Boltzmann–Maxwell definition, but precisely this feature of the definition serves as the basis for the statement of the two authors that in the gas model, as an ergodic system, all motions with the same total energy traverse the same ϕ-path and hence give the same time average for every $\phi(q, p)$.[99]

one turns around the torus going along the geodesic line, one will never get from a point P_h to the diametrically opposite point Q on the meridian. Because if we did, then twice as many revolutions would bring us back to P_h, contrary to our original assumption. From this one can easily see that the set of those points P_h which can be reached by a given geodesic line form a denumerable subset in the continuum of all those points on the circumference which the geodesic line approaches arbitrarily closely.

Note 99: If, for the time being, we call systems satisfying requirement (I) of note 98 "quasi-ergodic," then, instead of statements 1), 2), and 3) in the text, we must say that for a "quasi-ergodic" system on each surface $E(q, p) = E_0$ there will be a continuum of $\infty^{(2rN-2)}$ different G-paths with different values of the constants $c_2, \ldots, c_{2rN-1}$. Hence one cannot extend the Boltzmann–Maxwell justification of statement 3) to quasi-ergodic systems.

We have quoted this long text and even longer set of notes from the Ehrenfests' article because any shorter extract or paraphrase would give an inadequate picture of their position. Mathematics, physics, and history are mixed together here in an extremely complicated way. I think they are trying to say that statistical mechanics, as conceived by Maxwell and Boltzmann, lacks a firm foundation because it relies on the ergodic hypothesis which is invalid for the only idealized mechanical system that Boltzmann had actually treated in sufficient detail for this purpose (the Lissajous figures). They dismiss without much further discussion the alternative quasi-ergodic hypothesis and the assumption that interactions with the surroundings of a system might make it effectively ergodic. But their most important contribution was to ask the mathematical question: Can any deterministic mechanical system be ergodic, as distinct from quasi-ergodic? For the first time the problem was clearly posed to mathematicians, who were now challenged to solve it. And they did.

Notes for §10.10

1. P. and T. Ehrenfest, *Enc. math. Wiss.* **4**, (32) (1911); English trans. by M. J. Moravcsik, *The Conceptual Foundations of the Statistical Approach in Mechanics* (Ithaca: Cornell University Press, 1959). M. J. Klein, *Paul Ehrenfest*, vol. **1**, *The Making of a Theoretical Physicist* (New York: American Elsevier Pub. Co., 1970), ch. 6.
2. L. Boltzmann, *Wien. Ber.* **90**, 231 (1884); *J. r. ang. Math.* **98**, 68 (1885).
3. J. H. Jeans, *Phil. Mag.* [6] **4**, 585 (1902). Bryan [*B. A. Rep.* **64**, 67 (1894)] had noticed that the assumption that a single system passes through every phase was not usually made.
4. L. Boltzmann, *Wien. Ber.* **63**, 679 (1871).
5. J. Lissajous, *Ann. Chem.* **51**, 147 (1857). N. Bowditch, *Mem. Am. Acad.* **3**, 413 (1815).
6. J. C. Maxwell, *Trans. Camb. Phil. Soc.* **12**, 547 (1879).
7. H. von Helmholtz, *Berlin. Ber.* **159**, 311, 755 (1884); *J. r. ang. Math.* **97**, 111, 317 (1884).
8. See also G. H. Bryan, *B. A. Rep.* **61**, 85 (1891).
9. J. W. Gibbs, *Elementary Principles in Statistical Mechanics* (New York: Scribner, 1902).
10. L. Boltzmann, *J. r. ang. Math.* **100**, 201 (1887).
11. L. Boltzmann, *Ann. Phys. Beibl.* **5**, 403 (1881); English trans. in *Phil. Mag.* [5] **14**, 299 (1882).
12. L. Boltzmann, *Mün. Ber.* **22** (3), 329 (1892), English trans. in *Phil. Mag.* [5] **35**, 153 (1893).
13. R. Dugas, *La Théorie Physique au sens de Boltzmann et ses prolognements modernes* (Neuchâtel-Suisse: Éditions de Griffon, 1959).
14. M. J. Klein, *Arch. Hist. Exact Sci.* **1**, 459 (1962). B. G. Baklaev, *Voprosy Istorii Estestvoznaniya i Tekhniki* **4**, 167 (1957).
15. L. Boltzmann, *Wien. Ber.* **76**, 373 (1877), trans. by Klein, *op. cit.* See also Boltzmann's remarks in his 1872 paper (English trans. in Brush, *Kinetic Theory* **2**, 118–19).
16. H. Poincaré, *Rev. Gen. Sci.* **5**, 513 (1894).
17. Kelvin, *Proc. R. S. London* **50**, 79 (1891); *Nature* **44**, 355 (1891); *Baltimore Lectures*, Appendix B. G. J. Stoney, *Phil. Mag.* [5] **40**, 362 (1895). Rayleigh, *Phil. Mag.* [5] **33**, 356 (1892), **49**, 98 (1900). S. H. Burbury, *Phil. Mag.* [5] **50**, 584 (1900). A. Einstein, *Ann. Phys.* [4] **9**, 417 (1902). L. S. Ornstein, *Amsterdam Verslagen* [4] **19**, 808, 947 (1911). G. H. Bryan, *B. A. Rep.* **64**, 67 (1894).
18. P. Hertz, *Ann. Phys.* [4] **33**, 225, 537 (1910).

10.11 Digression on the history of mathematics

Before we can properly understand the events that followed the publication of the Ehrenfests' article, it will be necessary to review some of the developments in pure mathematics during the second half of the 19th century. These developments touch in particular the notions of infinite numbers, and the measure and dimensionality of sets of points. The ergodic hypothesis asserts that a certain curve, which can on the one hand be placed in simple correspondence with a straight line (the time axis), can on the other hand be considered to pass through every point in a region of two or more dimensions (the phase space). Such a curve would appear to be one-dimensional from the former viewpoint, and multi-dimensional from the latter; that is, it would have a finite "area" or "volume" despite the fact that an ordinary line apparently has no area or volume. The possible existence of such a curve forces us to reconsider what we mean by the length, area, volume, or dimensionality of a set of points; it also requires us to adopt a consistent definition of orders of infinity which contradicts the "common-sense" ideas about infinity.

Since Newton's time, the concept of "infinitesimal" quantities has played an important role in applied mathematics, despite its dubious validity. In the 19th century, Cauchy began the project of putting calculus and the theory of functions on a more solid foundation. After Fourier had developed his method for expanding functions (including discontinuous functions) in trigonometric series, Riemann invented a function, defined by trigonometric series, which is continuous for irrational values of the variable and discontinuous for rational values. In 1834, Bolzano gave an example of a continuous function having no derivative, and the existence of such functions was also noted by Riemann in 1861 and Weierstrass in 1872.[1] Such examples showed the need for a better understanding of the continuum of real numbers, and for a formulation of calculus which did not depend on vaguely defined "infinitesimals" or "differentials." One result of the demand for rigor was Weierstrass's analysis employing the well-known "epsilon, delta" technique; I say "well-known" because even physicists come in contact with it in introductory calculus courses nowadays. Another result was the theory of sets of points, developed by George Cantor and others. Cantor published his work on set theory during the years 1871–97, so that his contributions were contemporary with those of Maxwell and Boltzmann. We shall be primarily concerned with Cantor's concept of different orders of infinity, which underlies the theory of "measure" of sets of points used in discussing the ergodic hypothesis.[2]

Galileo Galilei had observed, as long ago as 1638, that the usual method of counting the members of a class to determine its size does not work in the case of infinite classes, and that if one goes by one-to-one correspondence, he must conclude that there are no more integers than there are squares of integers.[3] However, Galileo failed to establish a definite method for comparing the sizes of infinite classes that cannot be put into one-to-one correspondence, and simply stated that the relations "equal," "greater," and "less" are not applicable to infinite classes.

According to Cantor, it is perfectly reasonable to assign magnitudes to infinite classes by using the criterion of one-to-one correspondence. Two classes whose members can be placed in one-to-one correspondence are said to be equivalent, or similar; for example, the class of all squares of integers is equivalent to the class of all positive integers. In this case, a class is equivalent to a part of itself, and this property furnishes a convenient definition for infinite classes. But it is not true that all infinite classes are equivalent, for there is another class—the set of all real numbers—whose members cannot be put into one-to-one correspondence with the integers. The first kind of infinite class, including integers, odd integers, squares, and so forth, is called "denumerably infinite" since the members of such classes can be "counted" or enumerated by assigning an integer $1, 2, \ldots$ to each member. The "cardinal number" of these classes is denoted by the Hebrew letter aleph, with a subscript zero: $\aleph_0$. Cantor showed that the class of all rational numbers—numbers that can be expressed as the ratio of two integers—is also denumerably infinite.[4] He then showed that the real numbers cannot be placed in one-to-one correspondence with the integers, and thus have a different cardinal number.

The consequence of Cantor's theory which most directly concerns us is the proof that any n-dimensional manifold can be put into one-to-one correspondence with any m-dimensional manifold, where m may or may not be equal to n. This possibility, first demonstrated by Cantor in 1878,[4] contradicts not only common sense but also Riemann's postulate for geometry than n quantities must be specified in order to determine position in an n-dimensional manifold.[5] Instead, position in n-dimensional space can be uniquely determined by giving the value of a single real continuous coordinate. The ergodic hypothesis for dynamical systems now looks somewhat more plausible than it might originally appear to the mathematically unsophisticated mind. However, the "mapping" from n dimensions to one dimension which Cantor uses in his proof lacks the property of continuity. Points that are close together in the n-dimensional manifold will not necessarily be mapped into points that are close together on the one-dimensional line. Cantor and other mathematicians immediately recognized that the possibility of a *continuous* one-to-one mapping might serve as the criterion for asserting that two manifolds have the same dimensionality, provided that one could prove that it is impossible to find a continuous one-to-one mapping between manifolds of different dimensionality. In this way the concept of dimension could be restored to mathematics. After much subsequent research by mathematicians, the proof was finally found by Brouwer; a general review of this subject is given in an article by Schoenflies.[6]

Peano's curve (discovered in 1890) goes through every point in a two-dimensional region in such a way that points on the latter region are continuous functions of points on the line.[7] However, the mapping is not one-to-one even though it is continuous; some points of the two-dimensional region correspond to only one point on the line, but others may correspond to two or four points. Such a curve could not therefore represent the path of the phase point of a dynamical system.

The remaining developments needed to establish the impossibility of ergodic systems were contributed by Borel in the 1890's[8] and by Lebesgue in 1902.[9] These investigations completed Cantor's method of "counting" infinite sets of points by providing a method for determining the length, area, volume, or more generally the "measure" of sets of points. The notion of measure is essential not only to modern probability theory but also to modern discussions of the foundations of statistical mechanics.[10] We mention here only a few results of measure theory: The measure of a point, or of a denumerable set of points, is zero. The measure of all the real numbers in a finite interval is the length of that interval. In other words, one may add a denumerable number of zeros and get the result zero, but one may not add a non-denumerable number. This property is sometimes expressed by saying that Lebesgue's measure is "countably additive." Statements which are true for all points for a set of points of measure zero are said to be true "almost everywhere."

Notes for §10.11

1. See C. B. Boyer, *The Concepts of the Calculus* (1949), reprinted as *The History of the Calculus* (New York: Dover Pubs., 1959), ch. VII; K. T. W. Weierstrass, *Mathematische Werke* (Berlin, 1895), **2**, 71–74.
2. See G. Cantor, *Contributions to the Founding of the Theory of Transfinite Numbers* (translated from German, with introduction and notes, by P. E. B. Jourdain) (Chicago: Open Court Publishing Company, 1915; reprinted by Dover Pubs.). See also J. W. Dauben, *Rete* **2**, 105 (1974).
3. Galileo Galilei, *Two New Sciences* (translated from Italian by S. Drake) (Madison: University of Wisconsin Press, 1974), pp. 40–41. According to N. L. Rabinovich [*Isis* **61**, 224 (1970)], Galileo was anticipated by Rabii Hasdai Crescas in the 14th century.
4. G. Cantor, *J. r. ang. Math.* **84**, 242 (1878); see also ref. 2. The usual way of proving this is as follows: write all the fractions with numerator 1 in a row, then all those with numerator 2 in another row, and so forth

$$\begin{array}{cccccc} \frac{1}{1} & \frac{1}{2} \to \frac{1}{3} & \frac{1}{4} & \frac{1}{5} & \cdots \\ \downarrow \nearrow \swarrow & & & & \\ \frac{2}{1} & \frac{2}{2} & \frac{2}{3} & \frac{2}{4} & \cdots \\ \swarrow & & & & \\ \frac{3}{1} & \frac{3}{2} & \frac{3}{3} & \cdots & \\ \downarrow & & & & \end{array}$$

The rational fractions are then counted by starting with $\frac{1}{1}$ and following the arrows (eliminating duplications); thus one establishes the following correspondence between the integers and the fractions

$$\begin{array}{ccccccc} 1 & 2 & 3 & 4 & 5 & 6 & 7 & \cdots \\ \frac{1}{1} & \frac{2}{1} & \frac{1}{2} & \frac{1}{3} & \frac{3}{1} & \frac{4}{1} & \frac{3}{2} & \cdots \end{array}$$

To prove that the real numbers cannot be placed in one-to-one correspondence with the integers, consider the real numbers between 0 and 1 and represent each real number by an infinite sequence of digits (including zeros) following a decimal point. Then suppose it is asserted that some sequence of real numbers has been constructed which enumerates all of them in a particular order, say for example

0.2356...
0.84557...
0.34343...

The proof consists in showing that one can always find a real number which is not included in any such enumeration. This number can be constructed by taking as its first digit a digit differing from the first digit of the decimal representation of the first real number in the sequence; as its second digit, one differing from the second digit of the second real number in the sequence, and so forth: thus 0.415.... This number is different from every real number in the sequence; hence it has not been included in the enumeration.

To prove that an n-dimensional manifold can be mapped onto a line, use the fact that any real number can be represented by an infinite sequence of integers (as in the decimal representation mentioned above); if there are n coordinates, we can write them in the form

$$e_1 = (\alpha_{1,1}, \alpha_{1,2}, \ldots, \alpha_{1,\nu}, \ldots)$$
$$\cdots\cdots\cdots$$
$$e_\mu = (\alpha_{\mu,1}, \alpha_{\mu,2}, \ldots, \alpha_{\mu,\nu}, \ldots)$$
$$\cdots\cdots\cdots$$
$$e_n = (\alpha_{n,1}, \alpha_{n,2}, \ldots, \alpha_{n,\nu}, \ldots)$$

where $\alpha_{\mu,\nu}$ is the νth integer used in the representation of the μth coordinate. But these n real numbers determine uniquely another number,

$$d = (\beta_1, \beta_2, \ldots, \beta_\nu, \ldots)$$

where

$$\beta_{(\nu-1)n+\mu} = \alpha_{\mu,\nu} \quad \begin{cases} u = 1, 2, \ldots, n \\ \nu = 1, 2, \ldots, \infty \end{cases}$$

Hence every point in the n-dimensional manifold can be labeled either by the n coordinates $e_1, \ldots, e_n$ or by the single coordinate d.

5. B. Riemann, "Ueber die Hypothesen, welche der Geometrie zu Grund legen," (Habilitationsvortrag, 1854), in *Gesammelte Mathematischer Werke und Wissenschaftlicher Nachlass* (Leipzig, 2nd ed. 1892, reprinted by Dover Pubs., New York, 1953), p. 272.

6. A. Schoenflies, *Jahresber. D. Math. Ver.*, Ergänz.-Bd. **2** (1908). To summarize the history briefly: proofs had been attempted by Lüroth (1878), Thomas (1878), Jürgens (1878), and Netto (1879) but because of various errors these proofs were not valid for the general case (as was pointed out by Jürgens in 1898) and the problem had to be reconsidered. After further work by Lüroth (1906), Baire (1905, 1907), and Fréchet (1908), a general proof was finally found by L. E. J. Brouwer, *Math. Ann.* **70**, 161 (1910), **71**, 97, 305, **72**, 55 (1912).

We shall not discuss the proof for the general case, but the proof that an n-dimensional manifold ($n \geq 2$) cannot be mapped continuously, one-to-one, onto a line, is fairly simple. First of all, it is clear that a manifold of one or more dimensions cannot be mapped one-to-one onto a single point. Now suppose that a two-dimensional manifold, M_2, is mapped one-to-one onto a line M_1. Imagine a single closed line U drawn in M_2, and consider its image A in M_1 (i.e. the points in M_1 onto which the points of U are mapped). Since A cannot be a single point, it must be a set containing a finite or infinite number of points of M_1 (actually an infinite number). However, A is different from M_1 since U is not the same as M_2. Now suppose one wishes to go from some point a, belonging to A, to another point m, belonging to M_1 but not to A. There will always be some "boundary point" α which must be traversed on the way; α is the "last" point of that part of the set A which lies between a and m. Now if the mapping of M_2 onto M_1 were continuous, then one could likewise go from the point on U corresponding to a to the point in M_2 but not on U corresponding to m, only by passing through the point in M_2 corresponding to α. But this is obviously untrue, since one can go from one point to another in two-dimensional space without passing through any specified third point. Hence the mapping cannot be continuous. The same obviously holds for any mapping of $M_n (n > 2)$ onto M_1.

7. G. Peano, *Math. Ann.* **36**, 157 (1890); see also E. H. Moore, *Transactions of the American Mathematical Society* **1**, 72 (1900).

8. E. Borel, *Leçons sur la théorie des fonctions* (Paris, 1898).
9. H. Lebesgue, *Annali Mat. Pura e appl.* [3] **7**, 231 (1902).
10. R. M. Lewis, *Arch. Rat. Mech. Anal.* **5**, 355 (1960). C. Truesdell, in *Ergodic Theories*, ed. P. Caldirola (New York: Academic Press, 1961), p. 21.
    According to M. Kac [*Probability and related topics in physical sciences* (New York: Interscience, 1959), p. 259] "The idea that probability theory can be formalized on the basis of measure theory occurs first in the classical paper of E. Borel, 'Sur les probabilités dénombrables et leurs applications aritmetiques,' *Rend. Circ. Mat. Palermo* **47**, 247–71 (1909)."

10.12 Proof of the impossibility of ergodic systems

In 1913 two papers appeared in the *Annalen der Physik*, containing independent proofs of the impossibility of ergodic systems. The proof of M. Plancherel, though published in full slightly later, was apparently worked out slightly before that of Artur Rosenthal.[1] Plancherel's proof appears to be much simpler, but in fact Plancherel refers to Rosenthal's paper for a justification of some of his assertions. He uses the Lebesgue theory of measure, whereas Rosenthal uses Brouwer's results on the conservation of dimensionality of a manifold under a one-to-one continuous mapping, together with Baire's notion of sets of first and second categories.

Without going into the technical details of these proofs, we can summarize the essential arguments of both as follows:

(1) It is possible to find a small region of the energy surface in which all the partial derivatives of the energy with respect to the coordinates and momenta are continuous functions of the coordinates and momenta, and such that at least one of these derivatives is always different from zero everywhere in the region. (If there are singular short-range interactions between the particles, it will be necessary to choose a part of the phase space in which no two particles are close together, but this can always be done.)

(2) Within this region it is possible to define a Lebesgue measure (Plancherel), or, equivalently, it is possible to map this region onto a $2\gamma N - 1$ dimensional cube (Rosenthal). (Each of the N particles is assumed to have γ degrees of freedom, so the entire phase space has $2\gamma N$ dimensions.)

(3) According to the ergodic hypothesis, the representative point of the system must pass through every point of this small region (call it G). This could happen in two ways. First, in a finite time interval, the representative point enters G, goes through all its points, and comes

out again. This means that the $2\gamma N - 1$ dimensional region G can be mapped onto a line of finite length, continuously and one-to-one. But this is impossible, according to Brouwer's proof of the invariance of dimensionality under continuous one-to-one mappings. It is also impossible according to Lebesgue's theory of measure, since a line is a set of measure zero with respect to a region of two or more dimensions.

The second alternative is that the representative point passes into and out of G, as it must according to Poincaré's recurrence theorem, an infinite number of times if it passes through once. If we add together all the points traversed in this infinite number of passages, will we get a set of points that contains all the points of G?

(4) The second alternative is also impossible, because each separate segment of the trajectory (corresponding to a single passage through G) is itself a set of measure zero. According to Cantor's theory, a line can contain only a denumerable set of non-overlapping intervals. Since Lebesgue measure is countably additive, a denumerable set of intervals, each having measure zero, will have total measure zero. Brouwer's theorem also applies to this case: a line is still one-dimensional even if it is infinitely long.

Hence it is impossible for the representative point to go through every point on the energy surface; a mechanical system cannot be ergodic.

As post mortem on the death of the ergodic hypothesis, we quote the remarks of Borel in his supplement to the French translation of the Ehrenfests' Encyclopedia article:[2]

> It is sufficient to be acquainted with the modern theory of sets to have the certainty that even the definition of ergodic systems is contradictory. Following the publication of the article of P. and T. Ehrenfest, in which the problem was posed in a precise way, this contradiction has been explicitly shown ... by A. Rosenthal and M. Plancherel .... Moreover, it seems that considerations of this kind [referring to the use of the results of Baire, Lebesgue, and Brouwer on the invariance of dimensionality and the measure of sets of points] are not necessary in the cases where the notion of ergodic systems has actually been applied, and particularly in the kinetic theory of gases; if one adopts the hypotheses of Boltzmann on the collision of molecules considered as elastic spheres, and if one represents the state of a gas composed of n molecules by a point in $3n$-dimensional space (of which the coordinates are the

centers of the n molecules) the trajectory of this point will be a polygon; if it is closed, this polygon will never have more than a finite number of sides, and consequently it will cut an arbitrarily chosen linear manifold of $3n$ dimensions in only a finite number of points; this fact absolutely contradicts the ergodic hypothesis. There would be no difficulties in extending this result to any hypotheses which might be made about the molecules, as soon as the trajectories conserve the character of continuity of trajectory of classical dynamics. The hypothesis of the rigorous abstract existence of ergodic systems is then contradictory; to tell the truth, it does not appear that this abstract hypothesis was ever really envisaged by physicists; they probably never believed in rigorously ergodic systems, in the mathematical sense–that is to say, remaining in ordinary geometry, in the trajectory that passes rigorously through all the interior points of a square, or through all the interior points of a cube; curves such as that of G. Peano were known only to those interested in set theory, and no one ever thought that such curves could be dynamical trajectories.

Notes for §10.12

1. M. Plancherel, *Arch. Sci. Phys.* [4] **33**, 254 (1912); *Ann. Phys.* [4] **42**, 1061 (1913). A. Rosenthal, *Ann. Phys.* [4] **42**, 796 (1913). English translations of both papers are in S. G. Brush, *Transport Theory and Statistical Physics* **1**, 287 (1971), along with some biographical information. Rosenthal inserted the following note in his paper: "The present work had already been edited when I learned from a notice in *Fortschritte der Physik* that the same matter was the subject of a report of Herr Prof. M. Plancherel (Freiburg, Switzerland) to the Berne meeting of the Swiss Physical Society (9 March 1912). The report of the meeting... contains no indication of the method of proof. It appears that his ideas move in part in the same direction as mine, but they are different in some essential points."
2. E. Borel, *Enc. Sci. Math.* **4** (1.1), 188 (1915).

CHAPTER 11

Interatomic Forces and the Equation of State

11.1 The Newtonian program*

Once the basic principles of the kinetic theory of gases had been accepted (though without rigorous mathematical proof) efforts could be directed toward more elaborate calculations of gas properties. Such calculations required initial assumptions about the nature of intermolecular forces. On the other hand, comparison of the results of these calculations with experiments (or in some cases the mathematical structure of the theory itself) led to conclusions about what kinds of forces might operate. Indeed, the possibility of *determining* the nature of intermolecular forces has been at various times one of the motivations for research on gas theory.

It was Newton who urged that a research program on intermolecular forces and the properties of matter be undertaken, following the successful example of his work on gravitational forces and planetary motion. As a prelude to our discussion of the application of kinetic theory to equilibrium properties (especially the equation of state) of gases in this chapter, and transport properties (especially viscosity and heat conduction) in the next two chapters, we examine Newton's

---
* Most of this section is reprinted from *Archive for Rational Mechanics and Analysis* **39**, 1 (1970), by permission of Springer-Verlag.

program as a proposal for a hypothetico-deductive approach to physics.

In the preface to the first edition of *Principia*, Newton wrote:

> We... treat mainly those [powers] that pertain to heaviness, lightness, elastic force, the resistance of fluids, and attractive or repulsive forces of that kind: And it is because of them that we propose these our own (so to speak) mathematical principles of philosophy. Indeed, all the difficulty of Philosophy seems to lie in this, that from the phenomena of motions we should investigate the forces of nature, and thereupon from these forces we should demonstrate the rest of the phenomena. And to this point pertain the general propositions which we have treated at length in the first and second books. In the third book, moreover, we have proposed an example of this business in the unfolding of the System of the World. There, indeed, from the heavenly phenomena, through propositions demonstrated mathematically in the first two books, are derived the forces of gravity by means of which bodies strive toward the sun and the several planets. Thereupon, from these forces through propositions likewise mathematical are deduced the motions of the planets, comets, moon, and ocean. Would it were possible to derive the rest of the phenomena of nature from mechanical principles by the same kind of reasoning! For many things move me to suspect somewhat that they all may depend upon certain forces, by which the little parts of bodies through causes not yet known are either driven against one another and crowd together according to regular figures, or are driven away from each other and drawn apart....[1]

For the historian it should be important to discern what Newton himself meant by these words. Thus, one should note that he does not refer specifically here to *atoms* and indeed very little of the solid substance of Newton's work has any relation to atomic theory. Moreover, as Truesdell has pointed out, there is nothing here to imply that the forces must be central or even pairwise, or that optical and electromagnetic properties are excluded by the qualification "mechanical." In Newton's time, the English word "mechanical" was used much more generally to mean "working like a machine" or "practical" as opposed to something involving thought, spirit, or speculation.[2] What Newton does emphasize in the preface where this program is proposed is that rigorous mathematical methods should be employed in working out the consequences of any assumption about forces, whatever that assumption may be. While Newton did indeed write out some speculations about the relations between interatomic forces and the chemical and physical properties of matter, which may be found in the famous Queries in his *Opticks*[3] and in an unpublished "conclusion" of the *Principia*,[4] he confined himself to mathematical demonstrations in the main text of the *Principia*.

But the historian must also take account of the meaning that was read into Newton's words by other scientists who lacked the ability or desire to master the contents of his book in detail. They assume he was simply restating the "mechanical philosophy" of René Descartes and Robert Boyle.[5] According to this philosophy, the purpose of theoretical physics is to construct explanations of phenomena in terms of atoms, assuming that the atom can have only "primary" mechanical properties such as mass, size, shape, and motion. The large amount of 18th-century research and writing inspired by this supposedly Newtonian program has recently been surveyed by Robert Schofield,[6] and Russell McCormmach has argued that it provided a unifying theme for the work of Henry Cavendish.[7]

Now it has been a hotly debated question whether "force" can be a legitimate mechanical property of an atom. According to the Cartesian tradition, the only admissible atomic force is a contact repulsion that prevents two atoms from occupying the same space; any kind of long-range attraction or repulsion is inconceivable as an inherent property of atoms, and must ultimately be explained in terms of contact actions (perhaps propagated through an "ether").[8] Whether you are willing to agree with this as a philosophical position or not, you have to recognize that it has been quite congenial to many scientists throughout history; it helps to account for the popularity of such speculations as the kinetic theory of gravity[9] and the vortex atom.[10]

At the other extreme is the theory of Roger Boscovich (1711–87), which eliminates mass, size, and shape as inherent properties of atoms and replaces them by a rather complicated force law for the interaction of point atoms.[11] Boscovich's ideas had some influence on later scientific theories, and a few of the more metaphysical thinkers of the early 19th century were quite content to reduce matter entirely to force, in some cases even eliminating the atom altogether.[12]

From a metaphysical viewpoint it might appear that the Cartesian and Boscovichean theories are mutually exclusive: the first denies the existence of force (except in a trivial sense as a label for that which prevents pieces of matter from interpenetrating), while the second denies the existence of matter (except in a trivial sense as the locus of a strongly repulsive force). But the most successful scientists pay little attention to such metaphysical issues, and so we must take seriously a model that seems to be a compromise between the two extremes: an atom with finite size and mass and a definite shape (usually spherical) which can also exert both long and short-range attractive and/or repulsive forces on other atoms. Given such a model, how can we use it to explain the properties of matter, and what do we gain by doing so?

Let us sketch a procedure that corresponds to a common conception of scientific method, as it ought to be applied in this case. Certain features of this procedure have been popularized in recent years by the philosopher of science Karl Popper,[13] but the basic ideas are quite old.

We could expect to start by knowing property P_1 of a substance; we then guess a plausible atomic model A_1, and by carrying out mathematical operation M_1 try to calculate P_1. If the theoretical and experimental values of P_1 agree sufficiently well–there is unfortunately no objective or generally accepted criterion for "sufficiently well," as Kuhn has pointed out[14]–we declare that a *prima facie* case has been made for the model A_1. We then use another operation M_2 on the same A_1 to calculate another property P_2. If the agreement is still satisfactory we continue the process, and our confidence in the validity of the model increases, though if we are good Popperians, we keep telling ourselves all the while that we have not *proved* that the model represents reality, and indeed can never hope to do so; it is only a conjecture that has so far survived all attempts at refutation.[15] As soon as one property P_i calculated from the model fails to agree with experiment we must at once reject A_1, pick another model A_2 and start all over again.

To apply this procedure to gases, we recall that the first quantitative properties of air to be studied in the 17th century were its pressure (or weight) and its volume. Following the earlier work of Torricelli, Pascal, and Guericke, Robert Boyle (1627–91) provided us with the simple relation: pressure × volume = constant. (I am here speaking of Boyle's significance retrospectively, as the modern scientist sees it; at the time, the *quantitative* "Boyle's law," suggested by Henry Power, Richard Townley, and others, was less important than the *qualitative* fact that the suction phenomena previously attributed to nature's abhorrence of a vacuum could be explained mechanically in terms of air pressure. This point has been discussed in §1.2.)

Boyle's law, which I will call property P_1, was published in 1662. Twenty-five years later Isaac Newton proposed a simple atomic model, A_1: gases are composed of particles that repel their neighbors with a force inversely proportional to the distance. Newton does not mention here any motion of the atoms or any thermal property of matter; his analysis presumes tacitly that only adjacent atoms exert any force upon each other. He states and proves the following theorem (*Principia*, Book II, §V, Prop. XXIII):

> If a fluid be composed of particles fleeing from each other, and the density be as the compression, the centrifugal forces of the particles will be inversely proportional to the distances of their centres. And, conversely, particles fleeing from each other, with forces that are inversely proportional to the distances of their centres, compose an elastic fluid, whose density is as the compression.

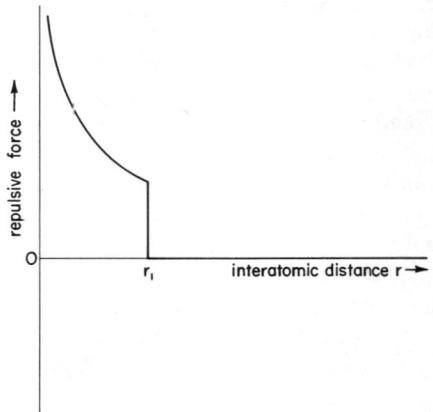

Fig. 11.1-1. Repulsive force $F(r) \propto 1/r$ for $r \leq r_1$; $F(r) = 0$ for $r > r_1$, where r_1 is the distance of the nearest neighbor. Model A_1 (Newton, 1687).

From the viewpoint of the hypothetico-deductive procedure outlined above, Newton has only accomplished the mathematical demonstration that $A_1 \rightarrow P_1$; the converse implication $P_1 \rightarrow A_1$ is valid if and only if one accepts the hypothesis that gases *are* composed of repelling atoms, and this had not yet been established. In other words, if we introduce the more general hypothesis A^*_1: "a fluid is composed of particles fleeing from (*i.e.* repelling) each other" with the understanding that the repulsive forces act only on nearest neighbors and that no account is to be taken of the effects of atomic motion, then the first part of Newton's theorem reads: $(P_1 + A^*_1) \rightarrow A_1$, where A_1 is a special case of A^*_1 in which the forces are inversely as the distance.

But Newton, unlike some of his followers, was careful to point out that $A_1^{\dagger}$ should not be taken for granted:

> But whether elastic fluids truly consist in little parts fleeing each other mutually, is a physical question. We have demonstrated mathematically a property of fluids consisting in little parts of this kind, so as to furnish philosophers a handle for taking up that question.[15]

Nor did Newton consistently adopt the model A_1 in his other writings; later, in his *Opticks*, he leaned toward the hypothesis that atoms are hard and impenetrable but have short-range *attractive* forces diminishing with distance.[3]

The model A_1 acquired considerable historical importance as the first quantitative interatomic force law to be inferred from the properties of gases, in spite of the fact that Newton himself did not take it very seriously. Judged (unfairly, of course) by modern standards, it must be regarded as a complete failure. Not only is it wrong in a quantitative sense; it is implausible since it implies that the force between two atoms depends on whether a third is in between them or they are neighbors. (If one does not make that stipulation the theory no longer gives the correct P_1; the pressure-volume relation depends on the size and shape of the container, as Newton himself pointed out, if it is assumed that $1/r$ forces act between all pairs of atoms.) Newton could justify this feature of this model only by suggesting an analogy with magnetic forces that can be reduced by interposing an iron plate between two magnetic bodies.

The reason for the failure of Newton's model is that the motion of the atoms has been ignored; we now realize that the pressure of a gas at ordinary density is almost entirely due to impacts of atoms against the walls rather than to continuously acting interatomic repulsive forces. But the amazing historical fact is that it took almost two centuries for the majority of physicists to accept that conclusion. With a few significant exceptions such as Leonhard Euler, Daniel Bernoulli, and John Herapath, most of them were so subservient to Newton's authority that they adhered firmly to what Newton had suggested only as a tentative hypothesis. One might say they rejected his hypothetico-deductive *method* while adopting what appeared to be its *conclusion*.

Of course there is more to it than that: during the 18th century, the idea that *heat* is associated with the interatomic repulsion force was added to Newton's theory, and accounted for much of its popularity. Yet it is rather disturbing to find not only most of the textbooks but also some first-rate scientists such as John Dalton and Thomas Young stating at the beginning of the 19th century that the repulsive force law for gas particles suggested by Newton has been definitely proved beyond any possible doubt.[16]

The first attempt to relate interatomic forces to gas properties was a disaster. What about the second? This was the 1758 theory of Boscovich, already mentioned. Atoms are point centers of force, but since they have no other properties except motion, the force law must do a lot more explanatory work. It must change from repulsive to attractive and back again several times, in order to account for the following properties:

	(properties)	(features of model)
P_2,	impenetrability of particles	repulsive force increases without limit as $r \to 0$ (r = distance between atoms)
P_3,	solid has an equilibrium volume at low pressures	force = 0 at distance r_1; force is repulsive for slightly smaller distances and attractive for slightly larger, so this is a point of stable equilibrium.
P_4,	solid changes to liquid which has larger equilibrium volume	force = 0 at r_3 (note that r_2 is point of *unstable* equilibrium
P_5,	at large distances any two pieces of matter have inverse square gravitational attraction	force $\propto 1/r^2$ as $r \to \infty$

There are many further ramifications of Boscovich's theory which I will not go into. Instead there are three things to be said about it: (1) The basic idea was attractive to some scientists, including major ones like Lord Kelvin.[17] (2) From a modern viewpoint it is definitely wrong as a *fundamental* postulate—we have now rejected the idea of explaining all these properties of matter simply by attributing a complicated system of forces to *point* atoms. Instead we prefer to start with a much simpler force law, say an electrostatic force between subatomic particles, and with the aid of more complicated assumptions about atomic structure and motions we may eventually *derive* an "effective" interatomic force law that has some resemblance to Boscovich's curve.

§11.1] THE NEWTONIAN PROGRAM 393

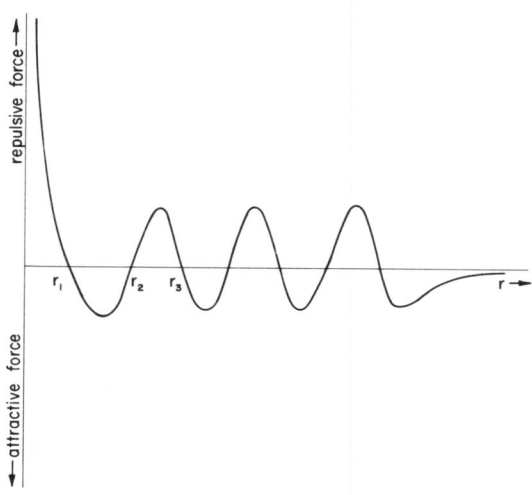

Fig. 11.1-2. Repulsive force for $r < r_1$; attractive force between r_1 and r_2; repulsive again between r_2 and r_3, etc.; attractive force $\propto 1/r^2$ for large r. Model A_2 (Boscovich, 1758).

(3) No quantitative calculations or predictions (with the possible exception of those of Kelvin) could be based on it during the period when it was entertained as a possible model of matter. The process of testing it against experiment never really got off the ground; one either accepted or rejected it for metaphysical reasons. So by the strict criteria of modern scientific methodology this was disaster #2. (Whether those criteria are really applicable to the behavior of scientists is another question.)

On the other hand we must at least credit Boscovich with extending the range of material properties that an atomic theory should try to explain: he has proposed the rather interesting idea that the same model might account for the existence of a substance in more than one physical state. If you look at some of the other speculations published in the 18th and early 19th centuries you will realize that this is not as trivial as it seems today. Many scientists seemed to think that the nature of the atom itself would have to be different in the gaseous, liquid, and solid states, if only because of the change in the amount of heat-fluid condensed around the atomic nucleus. The modern notion that one should try to explain all three states of matter (except metals and plasmas) with the same atomic model did not really take hold until after J. D. van der Waals published his famous theory in 1873. His theory predicted not only the liquid–gas transition but also the disappearance of that transition at the critical point.

I will come back to van der Waals' theory later in its proper chronological sequence but I mentioned the critical point because I wanted to introduce the third in the sequence of atomic force models. This was the one proposed by Laplace in 1822[18] to explain the phenomenon discovered in the same year by Cagniard de la Tour: if you seal a liquid with its own vapor in a strong container and heat it up, you find that eventually the meniscus dividing liquid and vapor disappears. This was interpreted at the time to mean that the liquid changes to gas, even though it has remained liquid far above its normal boiling temperature because of the high pressure. (The modern explanation that gaseous and liquid states are continuous above the critical point was first hinted at by John Herschel in 1830[19] but definitely established only with the work of Thomas Andrews in the 1860's; see §7.3).

Laplace had already developed a modification of Newton's gas theory, in which he explained the repulsion of gas atoms (still generally accepted) in terms of continual radiation and emission of caloric (heat) particles.[20] But he also postulated short-range attractive forces between atoms to account for phenomena such as surface tension, capillarity, and the cohesion of solids and liquids. The repulsive forces were supposed to operate at greater interatomic distances, in spite of Newton's attempt to restrict them to nearest neighbors only. With this model, A_3, Laplace could give a qualitative explanation of Cagniard de la Tour's experiment: the attractive forces dominate when atoms are close together (as in a solid or liquid) while the repulsive forces dominate when they are far apart (as in a gas); so by squeezing the atoms of a gas together one can shift the balance toward attraction and keep the substance in the liquid state at higher temperatures. So the boiling temperature increases with pressure, a property which we will call P_6. On the other hand, the amount of caloric in the substance increases with temperature, and eventually provides enough repulsive force to change the liquid to a gas even at high pressures: this is (apparent) property P_7.

The figures presented here are *not* taken from Laplace's paper, but represent my own guess as to what he had in mind.

Now Laplace's model is clearly wrong; in fact it is just the opposite of the one now generally accepted (short-range repulsive forces, long-range attractive forces). Even worse, it was not formulated quantitatively in such a way that it could be properly tested

§11.1] THE NEWTONIAN PROGRAM 395

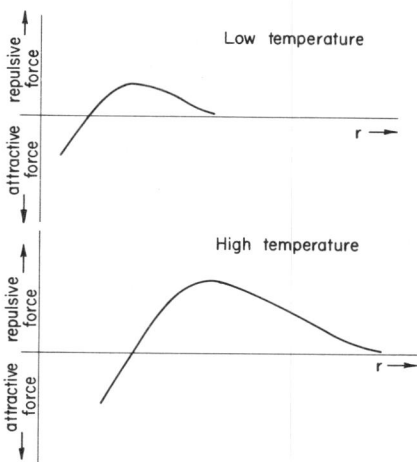

Fig. 11.1-3. The force law depends on temperature. As the temperature rises, the atom acquires more caloric, thereby increasing its size and repulsive force. At larger distances the repulsive force is proportional to $1/r$. Model A_3 (Laplace, 1822).

against experiments. Although Laplace did present a derivation of the ideal gas law based on his radiation-exchange theory of the repulsive force, he did not apparently realize that (as Newton had pointed out) if there is a long-range $1/r$ repulsive force between all pairs of atoms, the model no longer explains Boyle's law.

The caloric theory of heat was eventually rejected of course, and Laplace's model along with it, but *not* because A_3 led to an incorrect value of any particular property P_1. On the contrary, one might argue that the Laplace–Poisson explanation of the speed of sound, based on the concept of adiabatic compression and using empirical values for the ratio of specific heats, is somewhat more satisfactory than the 19th-century kinetic-theory explanation of this property (P_8); yet Laplace's theory was replaced by the kinetic theory for other reasons. (One reason, which might seem to be completely irrelevant, was that the wave theory of light won out over the particle theory around 1820–25; see §9.2.) All in all, it looks like defeat #3 for the program of relating interatomic forces to gas properties.

Of course I have exaggerated the failure of Laplace's theory: it did provide the best explanation of capillarity for many years, and thereby stimulated J. D. van der Waals in his own attempts to construct a more satisfactory atomic explanation of such phenomena; thus Laplace's theory did make some positive contributions to the progress of science.

Notes for §11.1

1. Isaac Newton, *Philosophiae Naturalis Principia Mathematica* (London, 1687, same in the 3rd ed., 1726), passage translated by Prof. C. Truesdell. The Motte–Cajori translation of this passage is somewhat misleading, especially in the last sentence where the phrase "secundum figuras regulares" is translated "in regular figures." [See *Sir Isaac Newton's Mathematical Principles of Natural Philosophy and his System of the World* (Berkeley: University of California Press, 1934), pp. xvii–xviii, reprinted in Brush, *Kinetic Theory* 1, 52–53.] According to Prof. Truesdell, Newton's use of the word "figura" is deliberately vague; there is no correct English translation, since Newton might have meant either "geometrical figure" or "according to regular qualities" or "according to qualities specified by rules." The last of these employs the only possible sense of "regularis" in classical Latin. (Letter from C. Truesdell to S. Brush, January 30, 1970; see also footnote 7 of Truesdell's paper cited in our note 2.)
2. C. Truesdell, *Texas Quarterly* 10, 238 (1967), reprinted in his *Essays in the History of Mechanics* (New York: Springer-Verlag, 1968), esp. pp. 177–182.
3. Isaac Newton, *Opticks* (London, 4th ed., 1730, reprinted by Dover, New York, 1952), pp. 375–402. Relations between Newton's speculations and contemporary experimental work have been discussed by W. B. Hardy, *Nature* 109, 375 (1922).
4. See A. Rupert Hall and Marie Boas Hall, *Unpublished Scientific Papers of Isaac Newton* (Cambridge University Press, 1962), pp. 334–344; Marie Boas Hall (ed.), *Nature and Nature's Laws* (New York: Harper & Row, 1970), pp. 323–333.
5. See, for example, the article by Marie Boas, *Osiris* 10, 412 (1952), or the more extended earlier treatment by E. A. Burtt, *The Metaphysical Foundations of Modern Physical Science* (New York: Humanities Press, 1952, reprint of the second edition of 1932). Historians still disagree as to how much Newton himself accepted this philosophy; at least he made it clear in the Leibniz–Clarke debate that he rejected the clockwork-universe concept of Boyle and Leibniz (cf. §14.1). See H. G. Alexander (ed.), *The Leibniz–Clarke Correspondence with extracts from Newton's Principia and Opticks* (New York: Philosophical Library, 1956). A comprehensive analysis of the relations between Newton's and Descartes' metaphysical positions may be found in Alexandre Koyré, *Newtonian Studies* (Cambridge, Mass.: Harvard University Press, 1965).
6. Robert E. Schofield, *Mechanism and materialism: British Natural Philosophy in an age of Reason* (Princeton: Princeton University Press, 1970).
7. Russell McCormmach, *Isis* 60, 293 (1969).
8. For general surveys see Max Jammer, *Concepts of Force* (Cambridge, Mass., Harvard University Press, 1957); Mary B. Hesse, *Forces and Fields* (London: Nelson, 1961); A. Koyré, *Newtonian Studies*.
9. This is the theory that gravity results from the bombardment of matter by streams of invisible particles; adjacent pieces of matter "shield" each other so the net effect is the same as an attractive force between them. The theory was popularized by G. L. LeSage in Geneva at the end of the 18th century; a large number of periodical articles published during the 19th century are listed in the *Royal Society Catalogue of Scientific Literature* (see index).
10. The theory that atoms are stable vortex motions in a continuous fluid was proposed by William Thomson (Lord Kelvin) and was popular among British physicists in the last quarter of the 19th century. See above, pp. 206–7.

11. R. J. Boscovich, *Philosophiae Naturalis Theoria redacta ad unicam legem virium in natura existentium* (Vienna, 1758); English trans., *A Theory of Natural Philosophy* (Chicago: Open Court, 1922, reprinted by MIT Press, Cambridge, 1966).
12. Such ideas may be found in the writings of the German *Naturphilosophen*, Schelling, Goethe, etc., and rather explicitly in a paper by I. H. Fichte, *Z. Philos. und Philos. Kritik* **24**, 24 (1854). I have discussed the relation between this tradition and the later attacks on atomism by Mach and others in §§1.7 and 8.1. The case for Boscovich's influence on later scientists is presented in a book of essays edited by L. L Whyte, *Roger Joseph Boscovich* (New York: Fordham University Press, 1961). The specific influence of Boscovich on Faraday, claimed by L. Pearce Williams in his biography *Michael Faraday* (New York: Basic Books, 1965) has recently been challenged by J. Brookes Spencer, *Arch. Hist. Exact Sci.* **4**, 184 (1967).
13. Karl R. Popper, *Logik der Forschung* (Vienna, 1934), English trans., *The Logic of Scientific Discovery* (London: Hutchinson, 1959); *Conjectures and Refutations* (New York: Basic Books, 1962).
14. T. S. Kuhn, *Isis* **52**, 161 (1961).
15. Isaac Newton, *Principia*, Book II, Prop. XXIII, Theorem XVIII, reprinted in S. G. Brush, *Kinetic Theory* **1**, 52–56 (from the Motte–Cajori translation); the last passage has been retranslated by Professor Truesdell.
16. J. Dalton, *Mem. Manchester Lit. Phil. Soc.* **5**, 540 (1802); Thomas Young, article "Cohesion" written in 1816 for the Britannica, reprinted in *Miscellaneous Works of the Late Thomas Young* (London 1855), vol. I, p. 454. R. S. Fleming, *Ann. Sci.* **31**, 561 (1974).
17. Kelvin, *B.A. Rep.* **59**, 494 (1889) and other references given on p. 225 of Whyte's book (*op. cit.*); R. Olson, *Isis* **60**, 91 (1969); see also recent issues of the *Isis* Critical Bibliography, entries indexed under Boscovich.
18. P. S. de Laplace, *Ann. Chim. Phys.* **21**, 22 (1822). The historical background of similar assumptions about short-range forces has been reviewed by W. B. Hardy, *Nature* **109**, 375 (1922). See also M. J. Klein, *Physica* **73**, 28 (1974); R. Fox, *Hist. Stud. Phys. Sci.* **4**, 89 (1974).
19. J. Herschel, *Preliminary discourse on the study of natural philosophy* (London, 1830), p. 234.
20. See S. G. Brush, *Kinetic Theory* **1**, 12–13.

11.2 The equation of state in the caloric theory*

We now examine the first serious attempt to use the Laplacian ideas about interatomic forces for quantitative calculations of gas pressure, that of the Swiss physicist Elie Ritter (1801–62), in 1846.[1] As Ritter's work is not generally known, and is not very accessible, it seems worthwhile to give a fairly complete account of it here.

Ritter did not discuss the caloric theory in any detail, but merely

---

* Most of this section is reprinted from *Am. J. Phys.* **29**, 593 (1961), by permission of the American Institute of Physics.

accepted the idea that caloric behaves like an ideal gas. He considered the pressure of a real gas to be the result of two causes: the repulsion due to caloric, and the attraction due to interatomic forces. He assumed in particular that the two effects could be treated independently and then added together; while this procedure had no theoretical justification, it permitted him to calculate the contribution of the forces in a manner independent of the caloric theory (aside from the assumption that the atoms are at rest). His result has the form

$$p = a\rho(1+\alpha t) - b\rho^2 \tag{1}$$

where ρ is the density, t the temperature (degrees centigrade), α the coefficient of expansion, and the constant b is essentially the virial of the interatomic forces (in modern terminology).

Ritter borrowed from Poisson[2] the mathematical apparatus necessary for this calculation. The pressure of a gas is the force per unit area exerted on a surface in the gas; Poisson and Ritter assumed that the atoms are fixed in a cubic array in space, neighboring atoms being a distance, Δ apart. Poisson showed that the pressure is

$$p = (1/6\Delta^3) \sum rf(r) \tag{2}$$

where $f(r)$ denotes the force exerted by an atom at a distance r; the sum is to be computed by choosing some central atom and summing the contributions from all the others.[3] Assuming that a spherical shell at a distance r from the central atom contains $4\pi r^2/\Delta^2$ atoms, Poisson wrote eq. (2) in the form

$$p = (2\pi/e\Delta^5) \sum_r r^3 f(r) \tag{3}$$

where now the sum is simply over possible values of the magnitude of r: $\Delta, 2\Delta, 3\Delta, \ldots$. In order to evaluate this sum he used the formula[4]

$$\sum_{r=\Delta}^{\infty} F(r) = (1/\Delta) \int_0^{\infty} F(r)\, dr - (1/2)F(0)$$
$$+ (2/\Delta) \int_0^{\infty} \left[\sum_{i=1}^{\infty} \cos(2\pi i r/\Delta)\right] F(r)\, dr$$

By successive integrations by parts the last term in this equation can be reduced to a power series in Δ, the coefficients being the derivatives of $F(r)$ at the origin. In this case $F(r) = r^3 f(r)$; so $F(0)$, and $F'(0)$, and $F''(0)$ were all set equal to zero. (Neither Poisson nor Ritter considered the possibility that $f(r)$ might become infinite at $r=0$, and their formulae are not very useful for calculations with singular forces.) The final result, obtained by Poisson and given again by Ritter, is that the contribution to the pressure is

$$p = (2\pi/3\Delta^6) \int_0^\infty r^3 f(r)\, dr + (\pi/180\Delta^2) f(0) - (\pi/756) f''(0) + \ldots \quad (4)$$

Since the density is proportional to $1/\Delta^3$, this could also be written as a series in descending powers of $\rho^{2/3}$

$$p = b\rho^2 + c\rho^{2/3} + d + e\rho^{-2/3} + \ldots \quad (5)$$

Poisson thought this formula might represent the total pressure of a gas, though he found it difficult to reconcile this with the empirical fact that the pressure is approximately proportional to the density, and he did not make much use of the formula.[5] Ritter, on the other hand, treated this contribution as a small correction to the ideal gas law, obtaining eq. (1) as a first approximation for attractive forces. It should be noted that b is a constant independent of temperature and density according to Ritter's theory.

In order to compare theory with experiment, Ritter derived expressions for the coefficients of expansion at constant pressure and volume. If the density at $t = 0°C$ is ρ, and one raises the temperature to $100°C$, keeping the pressure constant, the density will then be $\rho(1 + 100\alpha')$. (This is Ritter's definition of α'.) Eliminating the pressure, he obtained

$$1 + 100\alpha' = \frac{1 + 100\alpha + [(1 + 100\alpha)^2 + (1 - 2\rho/a)^2 - 1]^{1/2}}{2(1 - b\rho/a)} \quad (6)$$

Instead the volume is kept constant, the pressure will increase from p to $p(1 + 100\alpha'')$; eliminating p, the coefficient of expansion at constant volume was found to be

$$\alpha'' = a\alpha/(a - b\rho) \quad (7)$$

Both α' and α'' reduce to α for ideal gases ($b = 0$). Regnault[6] had found that hydrogen obeys the ideal gas law at pressures from 1 to 3.5 atmospheres, and that α has the value 0.0036613.

Regnault's experiments on air[6] showed deviations from the ideal gas laws, though the deviations were so small that Regnault himself attributed them to errors of observation. Ritter calculated the values of a and b in eq. (7) for three observations and obtained the results: (1) $a = 761.4714$, $b = 1.4714$; (2) $a = 760.2864$, $b = 0.2864$; (3) $a = 760.1865$, $b = 0.1865$, where the pressure is measured in millimetres of mercury. Taking $a = 761$ and $b = 1$, he then calculated α' and α'' for several pressures with the following results:

Pressure (atmospheres)	α' theory	α' expt.	α'' theory	α'' expt.
1	.0036716	.0036706	.0036661	.0036650
3	.0036865	.0036944	.0036758	.0036894
5	.0037038		.0036856	

Regnault's observations on carbon dioxide could be represented, according to Ritter, by the formula

$$p = 500.5522\rho^2 - 2.3230\rho^2,$$

and the coefficients of expansion were as follows:

Pressure (mm mercury)	α'' theory	α'' expt.	Pressure (mm mercury)	α' theory	α' expt.
758.47	.0036874	.0036856	760	.00370685	.0037099
901.09	.0036924	.0036943	2520	.00381886	.0038455
1742.73	.0037225	.0037523			
3589.07	.0037920	.0038598			

Ritter concluded that his theory gave the correct explanation for deviations from the ideal gas laws, although further observations at higher pressures would be desirable.

As in other parts of atomic theory, the basic physical principles of gas theory were stated long before they were proved; one must resist the temptation to glorify the isolated pioneer merely because subsequent work showed that he had guessed right. Thus a critical examination of Ritter's theory shows that by making several unjustified assumptions he was fortunate enough to arrive at an expression for the second virial coefficient very similar to the correct one; but that his theory could not have survived comparison with more accurate experimental data, and did not provide a good theoretical basis for further developments. It must therefore be regarded as an interesting curiosity, but not an anticipation of the more solid achievements of the later kinetic theorists.

It is interesting to compare Ritter's theory with Herapath's theory as expounded in his *Mathematical Physics* about the same time.

Herapath was still treating his atoms as billiard balls with no attractive forces, and he was able to derive from this model an equation of state which predicted that the pressure should be greater than the ideal gas pressure at high densities. Unfortunately most of the experimental data then available showed deviations in the opposite direction, which caused Herapath to have doubts about the validity of his theory. He was inclined to attribute such deviations to "imperfections" of the gas, a concept for which he did not give an explicit molecular interpretation, though one gathers that he had in mind the same cohesive forces on the pressure of a gas, and was delighted when in 1846 Regnault announced his discovery that hydrogen shows positive deviations from the ideal gas law at high pressures, in accordance with Herapath's theory.

Notes for §11.2

1. E. Ritter, *Mem. Soc. Phys. Geneve* **11**, 99 (1846).
2. S. D. Poisson, *J. École Polyt.* **13** (20), 33 (1831).
3. Note that in eq. (9) (§11.4) the quantity $rf(r)$ is summed over all pairs of atoms. In eq. (2) one chooses a single atom at $r = 0$ and sums over all the others; the volume is $v = N\Delta^3$, and the sum must be multiplied by $\tfrac{1}{2}N$ to get the correct number of pairs.
4. S. D. Poisson, *Mem. Acad. Sci. Inst. de France* **6**, 591 (1827).
5. S. D. Poisson, *J. École Polyt.* **13** (20), 157 (1831).
6. H. V. Regnault, *Ann. Chim. Phys.* [4] **4**, 5 (1842); *C.R. Paris* **20**, 975 (1845). See Wyatt and Randall, *The Expansion of Gases by Heat* (New York: American Book Co., 1902) for English translations of some of Regnault's work.

11.3 Interatomic forces in early kinetic theories*

To follow the line of argument developed in §11.1 we may regard the "hard sphere" or "billiard-ball" gas model going back to Daniel Bernoulli (1738), as model A_4. In the hands of John Herapath, this model led, for example, to a successful calculation of the speed of sound (property P_8), on the assumption that this speed must be about the same as that of an air molecule.[1] The first papers of Clausius and Maxwell, though not committed completely to this model, obtained results based primarily on it; there was some suggestion that the results might hold for other force laws.

---

* Most of this section is reprinted from *Arch. Rat. Mech. Anal.* **19**, 1 (1970), by permission of Springer-Verlag.

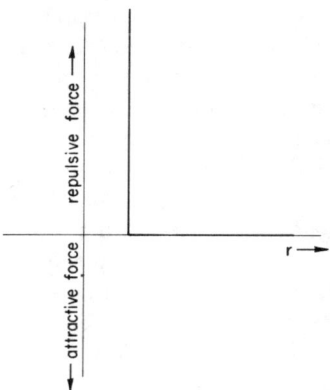

Fig. 11.3-1. The atom is a billiard ball: no force except for an infinite repulsion when two atoms come in contact. Model A_4 (Bernoulli, 1738).

Maxwell stated quite clearly that the kinetic hypothesis led to results in disagreement with two kinds of experiment: the ratio of specific heats of gases such as air, oxygen, hydrogen, and nitrogen; and the pressure-dependence of the viscosity coefficient (§5.3). The first disagreement was never satisfactorily resolved in the context of classical kinetic theory. The second turned out to be a victory for the theory, when further experiments inspired by Maxwell's theory indicated that the viscosity coefficient (and also the thermal conductivity coefficient) is independent of pressure, as the model predicts.[2] Although Maxwell had said at one point that a *single* failure could overthrow an hypothesis,[3] he seems to have been sufficiently impressed by the success of the kinetic theory in most other respects that he continued to develop it.

Maxwell's confidence in the kinetic theory, presumably based on his own experimental confirmation of his viscosity prediction, outweighed the skepticism which he continued to maintain in regard to specific heats. He proceeded to develop an elaborate version of the theory, taking into account the collision dynamics of particles interacting with a general r^{-n} force law, and introducing a general non-equilibrium distribution function. While he did not find a general solution of his equations, he did discover that in the special case of an r^{-5} repulsive force he could compute the transport coefficients without knowing the distribution function. According to this model, which we call A_5, the viscosity coefficient comes out to be directly proportional to the temperature, in agreement with Maxwell's own experiments.[4]

§11.3]　　　　　EARLY KINETIC THEORIES　　　　　403

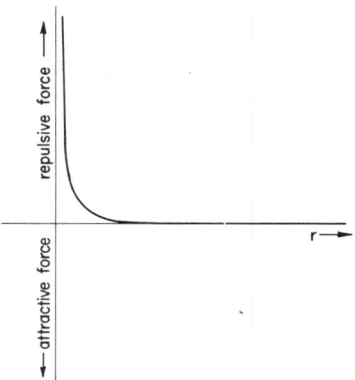

Fig. 11.3-2. Repulsive force proportional to r^{-5} at all distances. Model A_5 (Maxwell, 1867).

(The billiard-ball model, A_4, gives a viscosity proportional to the square root of the temperature.) Thus Maxwell proposed to adopt this model, and some of his British colleagues were carried along by his enthusiasm even after Maxwell himself realized that it could not be the true interatomic force law. German scientists such as O. E. Meyer did not think much of this new model for several reasons. First, the theory of attractive and repulsive interatomic forces was associated with a different philosophical viewpoint from the theory of billiard-ball atoms, as noted above. Second, the Joule–Thomson experiment had shown that long-range interatomic forces must be attractive rather than repulsive as Maxwell's model requires. Third, the experimental data of Meyer and others suggested that the exponent of the temperature in the viscosity law is definitely less than 1, although it was also definitely greater than $\frac{1}{2}$.[5] Of course if one rejected both A_4 and A_5 on the basis of the third argument there would be nothing left; it was considered preferable to go back to the billiard ball model but to assume that the diameter of the billiard-ball varies with temperature (*i.e.* with average collision speed).[6]

Notes for §11.3

1. This property can of course be derived from several other models, including the rotating vortex model as Euler showed in 1727 (see §3.3). We have also noted (§3.1) that J. J. Waterston, who adopted the billiard-ball model for most of his calculations,

actually thought Mossotti's model of attractive and repulsive forces was more likely to be correct.
2. See §§5.2 and 12.2 (viscosity) and §§13.4 and 13.5 (heat conduction).
3. J. C. Maxwell, *B.A. Rep.* **10**, 15 (1860), quoted in §5.3. Cf. A. F. Chalmers, *Brit. J. Phil. Sci.* **24**, 164 (1973).
4. J. C. Maxwell, *Phil. Trans.* **157**, 49 (1867); see Brush, *Kinetic Theory* **2**, 79.
5. See §§12.2 and 12.3.
6. O. E. Meyer, *Ann. Phys.* [2] **148**, 1, 203 (1873); a similar hypothesis was proposed by J. Stefan [*Wien. Ber.* **65** (2), 323 (1872)] in order to explain diffusion data.

11.4 The virial theorem*

Having decided on some particular force law, one then has to calculate the effect on the pressure. The starting point for most modern work on the equation of state is the virial theorem of Clausius.[1] The theorem states that the mean value of the kinetic energy in a system of material points is equal to the mean value of a quantity called the virial

$$\sum (m/2)\overline{V^2} = -(1/2) \sum \overline{(Xx + Yy + Zz)}, \qquad (8)$$

where x, y, and z represent the rectangular coordinates of the points, X, Y, and Z the respective components of the force acting on each point; the average value is taken over a time, in the case of a periodic motion, equal to a complete period, or, in the case of irregular motion, sufficiently long that the mean value becomes constant. It is further assumed that the system is in "stationary motion," *i.e.* that the points move within a limited space and the velocities do not change continuously in any particular direction.

The total virial may be divided into two parts: (1) the internal virial, which results simply from the forces which the points on each other; and (2) the external virial, resulting from external forces acting on the system. If we let $f(r)$ represent the force between two points at a distance r, the internal virial is the sum of $(-\frac{1}{2}rf(r))$ for each pair of points; and if the only external force is a pressure p confining the system to a volume v, the external virial is simply $3pv/2$. In this way Clausius arrived at the equation

$$E = (1/2) \sum rf(r) + (3/2)pv \qquad (9)$$

---

* Reprinted from *Am. J. Phys.* **29**, 593 (1961), by permission of the American Institute of Physics.

where E denotes the mean *vis viva* (kinetic energy) of the internal motions. This is essentially the same as the equation used by Ritter, except that Ritter used for E an empirical value derived from the ideal gas law, instead of setting it equal to the kinetic energy of atomic motion.[2]

Clausius gave the following proof of this theorem:

The equations of the motion of a material point are

$$m\left(\frac{d^2x}{dt^2}\right) = X; \quad m\left(\frac{d^2y}{dt^2}\right) = Y; \quad m\left(\frac{d^2z}{dt^2}\right) = Z$$

But we have

$$\frac{d^2(x^2)}{dt^2} = 2\frac{d}{dt}\left(x\frac{dx}{dt}\right) = 2\left(\frac{dx}{dt}\right)^2 + 2x\frac{d^2x}{dt^2}$$

or, differently arranged,

$$2(dx/dt)^2 = -2x\frac{d^2x}{dt^2} + \frac{d^2(x^2)}{dt^2} \tag{10}$$

Multiplying this equation by $(m/4)$, and putting the magnitude X for $m(d^2x/dt^2)$, we obtain

$$\frac{m}{2}\left(\frac{dx}{dt}\right)^2 = -\frac{Xx}{2} + \frac{m}{4}\frac{d^2(x^2)}{dt^2}$$

The terms of this equation may now be integrated for the time from 0 to t, and the integral divided by t; we thereby obtain

$$\frac{m}{2t}\int_0^t \left(\frac{dx}{dt}\right)^2 dt = -\frac{1}{2t}\int_0^t Xx\,dt + \frac{m}{4t}\left[\frac{d(x^2)}{dt} - \left(\frac{d(x^2)}{dt}\right)_0\right]$$

where $(d(x^2)/dt)_0$ denotes the initial value of $d(x^2)/dt$....

The last term of the equation, which has its factor included in the square brackets, becomes, when the motion is periodic, $= 0$ at the end of each period, as the end of the period $d(x^2)/dt$ resumes the initial value $(d(x^2)/dt)_0$. When the motion is not periodic, but irregularly varying, the factor in brackets does not so regularly become $= 0$; yet its value cannot continually increase with the time, but can only fluctuate within certain limits; and the divisor, t, by which that term is affected, must accordingly cause the term to become vanishingly small with very great values of t. Hence, omitting it, we may write

$$\frac{m}{2}\overline{\left(\frac{dx}{dt}\right)^2} = -\frac{\overline{Xx}}{2}$$

As the same equation is valid also for the remaining coordinates ... and for a system of any number of points we have ...

$$\sum \frac{m}{2}\overline{V^2} = -\frac{1}{2}\sum \overline{(Xx + Yy + Zz)}$$

It should be noted that the important contribution of Clausius was the elimination of the last term in eq. (10), since that equation itself was already well known in classical mechanics.[3]

Clausius himself did not use the virial theorem to calculate the equation of state, but only in connection with his more general discussions of the relations between thermodynamics and mechanics.

Notes for §11.4

1. R. Clausius, *Bonn Ber.* p. 114 (1870); English trans. reprinted in Brush, *Kinetic Theory* **1**, 172.
2. Cf. note 3, §11.2.
3. Yvon-Villarceau, *C.R. Paris* **75**, 232, 377, 990 (1872). C. G. J. Jacobi, *Vorlesungen über Dynamik*, lectures at Königsberg, 1842–43 (Berlin, 2nd. ed., 1884), p. 22. R. Lipschitz, *J. r. ang. Math.* **66**, 363 (1866).

11.5 The van der Waals equation*

The first attempt to apply the virial theorem was the famous equation of state of van der Waals (1837–1923), proposed in his Leiden thesis in 1873.[1] Van der Waals did not make any special assumptions about the form of the force law, but he tried to draw some general conclusions about the effects of such forces. Although his discussion of these forces pertained mainly to liquids, it was reasonable to expect the same

---

* Reprinted from *Am. J. Phys.* **29**, 593 (1961) by permission of the American Institute of Physics.

§11.5]   THE VAN DER WAALS EQUATION   407

equation to apply to gases where the correction to the ideal gas law would be small anyway.

> He assumed that such forces have a very short range, so that "... we need only take account (in considering the force on any given particle) those other particles which are within a sphere of very small radius having the particle as centre, and termed the 'sphere of action,' the forces themselves becoming insensible at distances greater than the radius of the sphere.[2]

If the density is constant throughout,

> ... it follows that all those points will be in equilibrium about which we can describe a sphere of action without encroaching on the boundary. By this of course is meant that the particles will be in equilibrium as far as attraction alone is concerned; not necessarily so when the molecular motion is also taken into account—though this will actually be the case for the mass taken as a whole. In other words, the forces X, Y, and Z are zero for all points within the mass. Consequently the expression $\Sigma(Xx + Yy + Zz)$ vanishes.... The particles for which the forces may be put equal to zero constitute *a priori* by far the greater part of the mass, leaving only a comparatively small number on which uncompensated forces act. These last lie on the boundary and form a layer whose thickness is the radius of the sphere of action; and the forces on these particles are directed inwards. If about one such particle we describe the sphere of action, part of this sphere will be external to the liquid, and this part will represent the space which would be occupied by the particles which would if present annul the forces. So the remaining force acting inwards is equal in magnitude to the attraction which the particle in question would have experienced from the action of the particles which are absent.[3]

The equation of state, including this additional pressure, may thus be written

$$(p + p')v = \sum mV^2/2 \tag{11}$$

where p is the total exerted by the boundary, and p' is the molecular pressure at the surface. Van der Waals believed that p would be much greater than p' in gases, but p' would be greater in liquids.

In order to determine how p' depends on density, van der Waals appealed to the hypothesis of molecular motion:

> ... consider an infinitely thin column in the boundary layer, and imagine a part of space below this layer, within the body, containing every molecule that could attract the column. If in this space there were a molecule at rest, we should require to know the law of force to be able to estimate its attraction on the column. But if this molecule is in motion, and can occupy any part of the space indifferently, the above difficulty for the most part disappears; and we can take the attraction exerted by the molecule to be the mean of the attractions which it would exert in its different possible positions in the space. The same consideration applies to a second molecule which may be within the space at the same time as the first. In short, the attraction exerted by the matter in the space mentioned is proportional to the quantity of matter, or to the density. The same holds for the molecules within the column, so that the attraction is proportional to the square of the density, or inversely proportional to the square of the volume.[4]

In order to take into account the influence of finite molecular extension, van der Waals used a different method; here he avoided a mistake which had been made by his predecessors.

> Of course the effect of the extension will be to make the volume within which the motion takes place smaller than it seems to be. At first I considered that the difference between the external volume and the volume taken up by the molecules was the space within which the motion takes place. But I trust to be able to prove, by further considerations, that up to a certain degree of condensation of matter the external volume must be diminished by four times the volume of the molecules, and for greater condensation that it must be diminished by a continuously diminishing multiple of this volume.[5]

The factor of four was deduced by considering the mean free path of an atom. Clausius had taken into account, in his derivation of the mean-free-path formula (see the third paper of this series), the extension of the molecules in a plane perpendicular to the direction of motion, but not in the direction of motion itself, and hence his result was too large:

> ... just as if the free path of a ball thrown against a wall were said to be the distance of the centre of the ball from the wall when the

§11.5] THE VAN DER WAALS EQUATION 409

motion began; whereas the free path is that distance minus the radius of the ball. Thus, considering the diameter of the molecule we get a shorter free path, and consequently a proportionately greater number of encounters. But then the opposing pressure must be greater in proportion.[6]

If λ is the mean distance of the molecules, supposed to be arranged in cubical order, and each molecule is regarded as a sphere of diameter σ, then according to Clausius the mean free path of a single moving molecule is

$$l = \lambda^3/\pi\sigma^2$$

when all the others are assumed at rest, or

$$l_1 = 3\lambda^3/4\pi\sigma^2$$

when the others are moving with the same velocity.

We must now find how much the diameter of the molecules diminishes the path. If all the impulses were in the motion joining the centres of the molecules considered as spheres, then l_1 would have to be diminished by the distance between the centres when impact occurs. For half the diameter of the molecules must be subtracted at the beginning as well as at the end of the free path. Thus

$$l_2 = l_1 - \sigma$$

or

$$l_2 = \frac{\lambda^3 - 4\pi\sigma^3/3}{4\pi\sigma^3/3}; \qquad \frac{l_2}{l_1} = \frac{\lambda^3 - 4\pi\sigma^3/3}{\lambda^3}$$

Considering that $(\sigma/2)$ is the radius of the molecule here regarded as a sphere, and that $n\lambda^3$ is equal to the unit volume here taken as v; also that $(4\pi n\sigma^3/3)$ is eight times the volume of the molecules themselves; we get

$$\frac{l_2}{l_1} = \frac{v - 8b_1}{v}$$

where b_1 is the volume of the molecules.

The encounters, however, are only central exceptionally; and therefore, in the mean, l_1 must be dimished by less than σ. From the following considerations we can find what fraction of σ is to be subtracted from l_1. At the instant of impact the centre of the moving molecule lies on a sphere of radius σ described about the centre of the second molecule. Consider this sphere bisected by a plane perpendicular to the direction of motion. For central impact the centre of the moving molecule has its greatest distance from this plane, and for intermediate cases the centre is at other points of the hemispherical surface. The diminution of the path is the distance at impact of the centre of the moving molecule from the plane, and hence the mean diminution of the mean path is the mean ordinate of the hemisphere. But since the centre is equally likely to fall on any point of the hemisphere, we must take the mean ordinate for equal elements of the hemispherical surface, and not, as might easily be thought, for equal elements of the plane. We have to find

$$\int Z\, d\omega \Big/ \int d\omega$$

where $d\omega$ is the element of surface. This is the ordinate of the centre of gravity of a hemispherical surface, and is known to be half the radius. Hence from l_1 we must substract $(\sigma/2)$ and not σ. Putting $l_1 - (\sigma/2) = l_2$ we get

$$\frac{l_3}{l_1} = \frac{v - 4b_1}{v}\ [7]$$

If one takes into account the fact that the molecules have a Maxwellian distribution of velocities, the factor $(\frac{4}{3})$ in the above formula for l_1 must be replaced[8] by $\sqrt{2}$, but this does not change the final result: the volume available to the molecules is to be written as $v - b$, where b is equal, as a first approximation, to four times the volume of the molecules themselves.[9] Thus van der Waals derived his equation of state

$$(p + a/v^2)(v - b) = \sum mV^2/3 = R(1 + \alpha t) \tag{12}$$

This equation has been very useful in correlating and interpreting the properties of gases, and its success may probably be attributed to the addition of the term (a/v^2) to represent the effect of interatomic forces. (The constant b was at this stage simply an adjustable parameter, chosen to fit experimental data, and therefore making it equal to *four* times the volume of the molecules has little practical significance.) However, the derivation of the equation is not very

§11.6] EQUATION OF STATE FOR HARD SPHERES

satisfactory from a theoretical point of view,[10] and since it is discussed quite thoroughly in most textbooks on the kinetic theory it does not seem necessary to go into the subject here. Later work, based on more direct application of the virial theorem, is of more interest than empirical modifications of the van der Waals equation, and has given more information about interatomic forces than can be obtained from such empirical equations of state.

From the viewpoint of the history of kinetic theory we must acknowledge the influence of van der Waals on later Dutch physicists who made important contributions to statistical mechanics in the 20th century (Ornstein, Debye, Kramers, Uhlenbeck, and many others).

Notes for §11.5

1. J. D. van der Waals, *Over de Continuiteit van den Gan en Vloeistoftoestand* (Leiden, 1873); English trans. by R. Threlfall and J. F. Adair, in *Physical Memoirs* (London, 1890), **1**, Part 3.
2. Threlfall–Adair translation, p. 342.
3. *Ibid.*, p. 343.
4. *Ibid.*, p. 388.
5. *Ibid.*, pp. 372–73.
6. *Ibid.*, p. 373.
7. *Ibid.*, pp. 374–75.
8. *Ibid.*, pp. 376–84; J. C. Maxwell, *Phil. Mag.* [4] **19**, 19 (1860); Brush, *Kinetic Theory* **1**, note on pp. 160–61.
9. J. S. Rowlinson [*Nature* **244**, 414 (1973)] notes that Clausius obtained a factor 8, and Maxwell a factor 16, in later attempts to "correct" the van der Waals equation. Rowlinson's article includes several other interesting details about the history of the van der Waals equation.
10. See, *e.g.* Maxwell, *Nature* **10**, 477 (1874); P. G. Tait, *Trans. R.S. Edinburgh* **36**, 257 (1891).

11.6 Later discussions of the equation of state for hard spheres*

A better deduction of the equation of state for hard spheres was given by H. A. Lorentz (1853–1928) in 1881.[1] Lorentz started by writing the contribution to the virial from collisions of spheres of diameter σ as

$$A = -(1/2) \sum (K\sigma)$$

---

* *Reprinted from Am. J. Phys.* **29**, 593 (1961), by permission of the American Institute of Physics.

where K is the repulsive force which the spheres exert on each other at a distance σ, and the sum is over all pairs of molecules which collide at a particular time. Actually K is zero except at the instants when collisions occur, and then it is infinite. To evaluate the sum it is more convenient to average over a time interval τ, and then interchange the order of summation and integration

$$A = -(1/2)\tau) \int \sum (K\sigma) \, d\tau = -(\sigma/2\tau) \int \sum (K) \, d\tau$$
$$= -(\sigma/2\tau) \sum \left[\int K \, d\tau \right]$$

The quantity $[\int K \, d\tau]$ is just the momentum change in a collision, so we have

$$A = -(m\sigma/2\tau) \sum U_n$$

where U_n is the relative velocity measured along the line of centers. This expression can then be written in terms of the velocity-distribution function $f(u)$: the number of collisions during the time τ between two molecules with velocities lying in the ranges $(u, u + du)$ and $(u', u' + du')$, such that the angle between u and u' is between ϕ and $\phi + d\phi$, and the angle between the relative velocity U and the line of centers is between χ and $\chi + d\chi$ (thus $U_n = U \cos \chi$), is

$$(\pi\sigma^2\tau/v)f(u)f(u')U \sin \phi \sin \chi \cos \chi \, du \, du' \, d\phi \, d\chi$$

Now substituting $U^2 = u^2 + u'^2 - 2uu' \cos \phi$ and integrating over all the variables, Lorentz obtained the result

$$A = -(\pi m\sigma^3/4v) \int_0^\infty \int_0^\infty \int_0^\pi \int_0^{\pi/2} f(u)f(u')(u^2 + u'^2 - 2uu' \cos \phi)$$
$$\times \sin \phi \sin \chi \cos^2 \chi \, du \, du' \, d\phi \, d\chi$$
$$= -(\pi m\sigma^3/6v) \int_0^\infty \int_0^\infty (u^2 + u'^2)f(u)f(u') du \, du'$$

It is not actually necessary to know what $f(u)$ is, since it is sufficient to express the answer in terms of N and the mean-square velocity

$$\int_0^\infty f(u) du = \int_0^\infty f(u') du' = N,$$
$$\int_0^\infty u^2 f(u) du = \int_0^\infty u'^2 f(u') du' = N\overline{u^2}$$

§11.6]   EQUATION OF STATE FOR HARD SPHERES

Hence

$$A = (\pi m \sigma^3 / 3v) N^2 \overline{u^2}$$

and using the virial theorem we find for the equation of state

$$pv = (mN/3)\overline{u^2}(1 - b/v)^{-1}, \qquad (13)$$

where b has the same meaning as in van der Waals' equation.

Should the "b" term (sometimes called the "co-volume") be on the right or left side of the equation of state? Equation (13) agrees with eq. (12) to first order in b/v, and neither is claimed to be accurate to second order. (In the following we ignore attractive forces, setting $a = 0$). Yet there is an important difference between them, which was the subject of a controversy between Tait, who favored (13), and Korteweg and Rayleigh, who preferred (12).[2] Tait criticized the original derivation of van der Waals on the grounds that p and v in the virial equation must be the actual pressure and volume of the system, and are not subject to correction.[3] Rayleigh defended van der Waals, and pointed out that the equation $p(v - b) = RT$ is exact for a one-dimensional gas.[4] Korteweg said that if one multiplies the collision rate by a factor $v/(v - b)$ to take account of the effect of finite molecular size, Lorentz's formula becomes

$$pv = RT\{1 + (b/v)[v/(v - b)]\} = RT[v/(v - b)]$$

which is identical with van der Waals' equation.[5]

To take account of deviations from this equation of state at higher densities, Korteweg proposed the general form

$$p(v - xb) = RT$$

where x varies smoothly between 1, at low densities, and $3\sqrt{2}/\pi \approx 1.35$ at high densities (The latter number is derived from the close packed volume of hard spheres.)[6]

Korteweg's form has the advantage that it expresses one's intuitive ideas about the behavior of the system, at least qualitatively, over a wide range of pressures, even if one cannot calculate x as a function of (b/v) at this stage of the development of the theory. Unfortunately these intuitive ideas are inaccurate; Korteweg's equation, assuming x to be a continuous function of density, fails to describe the fluid–solid phase transition which is now known to occur in small systems of hard spheres.[7]

The suggested method of correcting low-density formulas by

introducing factors which take account of the change in collision rate due to finite molecular size works well at moderate densities. As refined by Enskog,[8] it can be successful even at fairly high densities, but it involves an assumption which ultimately breaks down. The assumption is that when one increases the density, the same sequence of collisions will occur, the only difference being that the time scale is altered. The possibility of discovering any discontinuous change in properties of the system, such as a phase transition, is thus eliminated. Since the existence of the hard-sphere phase transition was not known at the time, Tait's objections appeared to be rather pedantic.

Van der Waals tried to calculate the next correction term by an extension of his original method, and obtained the result[9]

$$pv = RT[1 + (b/v) + (15b^2/32v^2) + \ldots]$$

Boltzmann[10] and Jäger[11] independently calculated the coefficient of b^2/v^2; both obtained $\frac{5}{8}$ instead of $\frac{15}{32}$. Boltzmann communicated to van der Waals his calculation of the next coefficient[12] which did not agree with that calculated by J. J. van Laar[13] using a method suggested by van der Waals,[14] and invited van der Waals to discuss the reasons for the discrepancy. Van der Waals avoided giving a direct answer,[15] but later his son showed that the Boltzmann–Jäger result is correct, and can be deduced by van der Waals' method, suitably corrected.[16] The father then accepted the Boltzmann–Jäger value as correct.[17] Kohnstamm traced the difficulty to an ambiguous use of the phrase "length of path" by van der Waals and Korteweg, which had led them to treat the effect of molecular volume on mean free path incorrectly; the treatment of Clausius, Jäger, and Boltzmann was considered satisfactory and worthy of further development.[18]

In a later paper, van der Waals adopted the position that the observed variation of the volume correction with density cannot be explained by computing these correction terms to higher approximations with fixed b, but is due rather to an actual compression of the molecule itself.[19] In fact the mathematical complexity involved in computing these corrections from the theory grows quite rapidly, and the fifth virial coefficient (coefficient of b^4/v^4) was not known even approximately until electronic computers were employed in the 1950's.[20] On the other hand it has been possible to obtain some upper and lower bounds on values of higher virial coefficients, and these results allow some conclusions to be drawn about the convergence of the series and the possibility of a phase transition.[21]

The encounter between Boltzmann's group and van der Waals'

group at the end of the 19th century marked the transition from "kinetic" to "statistical" methods in the treatment of equilibrium problems in gas theory. Boltzmann's techniques were seen to be more powerful and general than those of van der Waals, especially in dealing with molecular models which include attractive forces,[22] though the intuitive approach of van der Waals was needed to initiate this area of research and is still considered useful by many modern physicists.

Notes for §11.6

1. H. A. Lorentz, *Ann. Phys.* [3] **12**, 127, 660 (1881), reprinted in his *Collected Papers*, eds. P. Zeeman and A. D. Fokker (The Hague: M. Nijhoff, 1934–39), **6**, 40.
2. The Tait–Rayleigh–Korteweg controversy and the third virial coefficient were discussed by P. A. Kohnstamm, *Amsterdam Verslagen* [4] **12**, 948 (1904); *Amsterdam Proc.* **6**, 794 (1904); see also the papers cited below in note 18.
3. P. G. Tait, *Trans. R.S. Edinburgh* **36**, 257 (1891); *Nature* **44**, 546, 627 (1891). On p. 546 (*loc. cit.*) Tait says: "Partly on account of its unfamiliar language, but more especially on account of a very definite unfavorable opinion expressed by Clerk-Maxwell [*Nature* **10**, 477 (1874)] I did not attempt to read the pamphlet [of van der Waals] when it appeared; and it was not until 1888 that, in consequence of some hints from Dr. H. Du Bois, I hastily perused it in its German form."
4. Rayleigh, *Nature* **44**, 499, 597 (1891).
5. D. J. Korteweg, *Nature* **45**, 152 (1891). See Tait's reply, *Nature* **45**, 199 (1891).
6. D. J. Korteweg, *Nature* **45**, 277 (1891). L. Boltzmann, *Lectures on Gas Theory*, pp. 226–27.
7. B. J. Alder and T. Wainwright, *J. Chem. Phys.* **27**, 1208 (1957); **31**, 459 (1959); *Phys. Rev.* [2] **127**, 359 (1962). The transition was originally predicted by J. G. Kirkwood and Elizabeth Monroe, *J. Chem. Phys.* **9**, 514 (1941), see p. 522.
8. D. Enskog, *Kungliga Svenska Vetenskapsakademiens Handlingar* **63** (4) (1922), English trans. in Brush, *Kinetic Theory* **3**, 226. See discussion in §12.8.
9. J. D. van der Waals, *Amsterdam Verslagen* [4] **5**, 150 (1896).
10. L. Boltzmann, *Wien. Ber.* **105**, 695 (1896).
11. G. Jäger, *Wien. Ber.* **105**, 15 (1896).
12. L. Boltzmann, *Amsterdam Verslagen* [4] **7**, 477 (1899); *Amsterdam Proc.* **1**, 398 (1899). For further discussion of the fourth virial coefficient see P. Ehrenfest, *Wien. Ber.* **112**, 1107 (1903); H. Happel, *Ann. Phys.* [4] **21**, 342 (1906); R. Majumdar, *Bull. Calcutta Math. Soc.* **21**, 107 (1929).
13. J. J. van Laar, *Amsterdam Verslagen* [4] **7**, 350 (1899), English trans. in *Amsterdam Proc.* **1**, 273 (1899). The calculation is given in more detail in *Arch. Mus. Teyler* **6**, 237 (1900).
14. J. D. van der Waals, *Amsterdam Verslagen* [4] **7**, 160 (1899); *Amsterdam Proc.* **1**, 138 (1899).
15. J. D. van der Waals, *Amsterdam Verslagen* [4] **7**, 537 (1899), English trans. in *Amsterdam Proc.* **1**, 468 (1899), French trans. in *Arch. Néerl.* [2] **4**, 299 (1901).
16. J. D. van der Waals, Jr., *Amsterdam Verslagen* [4] **11**, 640 (1903); *Amsterdam Proc.* **5**, 487 (1903); *Arch. Néerl.* [2] **8**, 285 (1903).

17. J. D. van der Waals, *Amsterdam Verslagen* [4] **12**, 82 (1903); *Amsterdam Proc.* **6**, 123 (1903). "My son has ... pointed out that also according to the direct method a value of α equal to that calculated by Boltzmann is found, if we form another conception of the influence of the pressure than I had formed and since then I am inclined to adopt the coefficients calculated according to the method of Boltzmann as accurate." See also J. D. van der Waals, *Arch. Néerl.* [2] **4**, 299 (1901), **9**, 381 (1904); *Boltzmann-Festschrift* (1904), p. 305; J. J. van Laar, *Arch. Mus. Teyler* **7**, 185 (1901). A detailed analysis from the viewpoint of modern theory is given by J. H. Nairn and J. E. Kilpatrick, *Am. J. Phys.* **40**, 503 (1972).
18. P. A. Kohnstamm, *Amsterdam Verslagen* [4] **12**, 961 (1904); *Amsterdam Proc.* **6**, 787 (1904). J. D. van der Waals, *Amsterdam Verslagen* [2] **10**, 321 (1876); *Arch. Néerl.* **12**, 201 (1877). D. J. Korteweg, *Amsterdam Verslagen* [2] **10**, 349 (1876); *Arch. Néerl.* **12**, 241 (1877). R. Clausius, *Die Kinetische Theorie der Gase* (Braunschweig, 1889), p. 60. G. Jäger, *Wien. Ber.* **105**, 97 (1896). L. Boltzmann, *Vorlesungen über Gastheorie* **II**. Teil (Leipzig, 1898), p. 164; English trans., *Lectures on Gas Theory*, p. 364.
19. J. D. van der Waals, *Boltzmann-Festschrift* (1904), p. 305; *Amsterdam Verslagen* [4] **21**, 800, 1074 (1912-13).
20. M. N. and A. W. Rosenblueth, *J. Chem. Phys.* **22**, 881 (1954). F. H. Ree and W. G. Hoover, *J. Chem. Phys.* **40**, 939 (1964), **46**, 4181 (1967). H. L. Frisch, *Adv. Chem. Phys.* **6**, 229 (1964). J. S. Rowlinson, *Proc. R.S. London* **A279**, 147 (1964).
21. J. Groeneveld, *Physics Letters* **3**, 50 (1962).
22. Boltzmann (see next section); M. Reinganum, *Ann. Phys.* [4] **6**, 533 (1901); W. H. Keesom, *Amsterdam Verslagen* [4] **20**, 1390, 1406, **21**, 492 (1912), English trans. in *Amsterdam Proc.* **15**, 240, 256, 417 (1912).

11.7 Second virial coefficient for continuous force law*

In order to calculate the contribution to the pressure of forces which are continuous functions of the distance, it is necessary to know something about the spatial distribution of molecules in the gas; Poisson's static model is clearly inadequate if the forces become strong when two molecules get close to each other. The required formula is of course the Maxwell–Boltzmann distribution law,[1] which states that the relative probability of a molecular configuration with potential energy V is $e^{-V/kT}$ (T = absolute temperature in °K and k = Boltzmann's constant). For the calculation of the second virial coefficient it is sufficiently accurate to assume that this formula can be applied to a pair of molecules, neglecting their interactions with other molecules in the gas.

This method was used by Boltzmann[2] to calculate the second virial coefficient for molecules interacting with a force law $f(r) = Kr^{-5}$

---

* The first part of this section is reprinted from *Am. J. Phys.* **29**, 593 (1961), by permission of the American Institute of Physics.

(Maxwellian molecules). The potential energy $V(r)$ is then $\int_r^\infty f(r)\,dr = K/4r^4$, and the contribution to the virial is

$$\sum rf(r) = 3NB = (2\pi N^2/v)\int_0^\infty r^3 f(r)\exp(-V(r)/kT)\,dr$$
$$= (2\pi N^2 K/v)\int_0^\infty r^{-2}\exp(-K/4r^4 kT)\,dr \qquad (14)$$

Evaluating the integral, Boltzmann obtained the result

$$pv = RT(1 + B/v) = RT(1 + a(N/v)(K/kT)^{3/4})$$

where a is a constant which has the value (not given explicitly by Boltzmann) of $(\pi/3\sqrt{2})\Gamma(\frac{1}{4})$.[3]

Equation (14) is in principle a solution of the problem considered in this chapter, in the sense that it permits one to calculate the first-order correction to the ideal gas law whenever the force law is given—provided the integral converges. (The method does not work, for example, when the force is a Coulomb attraction, $V(r) = e^2/r$, and therefore ionized gases cannot be described in this way.) There is also a clearly defined procedure for calculating higher virial coefficients, even though the calculation may not be practical for most force laws. The main practical difference between Boltzmann's method and Ritter's method is that the former gives a temperature-dependent virial coefficient, so that by comparing theoretical formulae with observations one can obtain information about the force law which operates.

With Boltzmann's theory at hand, it is now possible to go back to the van der Waals equation and ask: What precise force law must be assumed in order to derive such an equation of state? There is no doubt that it involves a hard sphere of definite diameter d, since the atoms are excluded from a space $2d^3/3$. It also includes an attractive force whose effect on the pressure is represented by the term a/V^2, i.e. the pressure is decreased by the force. But there has been considerable confusion about the nature of the attractive force implied by the van der Waals equations, beginning with van der Waals himself. He simply asserted that the results of the Joule–Thomson experiment showed that there are attractive forces, but that these forces must vanish beyond some distance, whereas a repulsive force must dominate at much shorter distances. He did not propose any definite law for the attractive force, so that his result a/V^2 must be regarded as a lucky guess based on physical "intuition" rather than the consequence of a mathematical derivation.[4] He was supported on this point by Maxwell, but again only on an intuitive basis.[5]

Whereas van der Waals stated that the attractive force would decrease rapidly to zero at large distances, Boltzmann complained in 1898 that this had never been proven satisfactorily, and claimed instead that the van der Waals equation could be derived only on the assumption that the force has a small but *constant* value at large distances. When he told this to van der Waals, the latter replied, first, that he had never made any such assumption himself, and second, that he thought it was physically improbable.[6]

Some writers have stated that the a/V^2 term implies an inverse fourth power attractive force. Others, misled perhaps by the nomenclature "Van der Waals forces" for the quantum-mechanical dispersion forces calculated by Wang, Slater, Eisenschitz, and London in the late 1920's, have talked about inverse seventh power forces. While I hesitate to contradict some of the eminent scientists who have written on this subject, I am reliably informed by my colleagues in statistical mechanics that the latest work confirms Boltzmann's conclusion (at least for the cases that can be solved exactly): the van der Waals equation of state is compatible with a very long range attractive force that does not vary with distance but has a very small magnitude.[7] More precisely, the force law for model A_6 is the limit of a function $F(r) = -c$ for $r \leq R$ as $R \to \infty$ and $c \to 0$, keeping the product cR fixed; together with a billiard-ball repulsive core mentioned above. The van der Waals equation can also be derived from other models with very weak but very long-range attractive forces, combined with a repulsive core.[8]

Notes for §11.7

1. See §§6.1 and 10.5; L. Boltzmann, *Wien. Ber.* **72**, 427 (1875).
2. L. Boltzmann, *Wien. Ber.* **105**, 695 (1896).
3. J. H. Jeans, *The Dynamical Theory of Gases* (Cambridge University Press, 4th ed., 1925, reprinted by Dover Pubs., New York, 1954), p. 134.
4. J. D. van der Waals, *Over de Continuiteit*... (Leiden, 1873), see *e.g.* the remarks on p. 437 of the Threlfall–Adair translation. Some historical remarks on "van der Waals forces" may be found in the article by H. Margenau, *Reviews of Modern Physics* **11**, 1 (1939); see also B. Hardy, *Nature* **109**, 375 (1922).
5. J. C. Maxwell, *Nature* **10**, 477 (1874).
6. Boltzmann, *Lectures on Gas Theory*, pp. 219, 220, 375.
7. M. Kac, G. E. Uhlenbeck, and P. C. Hemmer, *J. Math. Phys.* **4**, 216 (1963).
8. I thank Dr. W. J. Camp of Sandia Laboratories for a letter giving information on this point.

11.8 Gibbs' statistical mechanics

To one who has followed the late 19th-century Dutch work on the theory of the equation of state mentioned above, it will not be surprising that Dutch scientists were among the first in Europe to adopt the Statistical Mechanics of Gibbs. J. Willard Gibbs (1839–1903), the greatest American theoretical physicist of the 19th (and perhaps any) century, does not receive as much attention in this book as he really deserves, mainly because his influence on the course of science was not strongly felt until the end of the period with which we are concerned. Gibbs took the "mechanical theory of heat," a collection of ill-stated empirical generalizations still stinking of the factory, encumbered by crude models and analogies whose naiveté attracted only the scorn of meticulous thinkers like Mach and Duhem, and transformed it into the elegant system of "thermodynamics" which we know today. In the course of this work he also produced many new and useful results, pertaining especially to the equilibrium of heterogeneous systems. Having made Clausius' "general" theory of heat really general, he then turned to the special theory, otherwise known as the kinetic theory of gases, and raised it to a new level of abstraction in his "statistical mechanics," providing at the same time with his "ensembles" a powerful calculational technique superior even to that of Boltzmann.[1]

The adoption of the ensemble method for calculating equilibrium properties completed the transition from "kinetic" to "combinatorial" methods which was initiated by Boltzmann. Since it permits one to ignore the complicated details of collision processes, so that one simply needs to enumerate the possible configurations and compute their energies, the technique of equilibrium statistical mechanics offers great practical advantages; these advantages are especially evident in quantum statistics.

Boltzmann's original combinatorial method, based on his relation between entropy and probability (introduced in 1877), involved enumerating all the ways that a fixed total amount of energy could be divided among a large number of atoms. The energy of an individual atom was thus not strictly independent of the energies of the other atoms, although it could be shown that it became independent, and was given by the usual Maxwell–Boltzmann distribution law, in the limit of an infinite number of atoms with infinite total energy, the energy per atom remaining fixed. Boltzmann later introduced ensembles consisting of large numbers of systems (see §10.10) but he considered these mainly as technical devices for illustrating fundamental principles and not as aids to practical computation. For that matter, Gibbs himself did not make much use of his ensembles, but he did describe their properties in such a way that later workers could easily make use of them.

The advantage of an ensemble is that the energies of individual atoms can be treated as independent variables for the purpose of statistical averaging; this is accomplished by considering systems with all possible energies from zero to infinity, the probability of occurrence of an energy E being proportional to the Boltzmann factor $e^{-\beta E}$, $\beta = 1/kT$. The collection of all these systems is known as a *canonical ensemble*. Gibbs showed that the value of any thermodynamic property calculated with a canonical ensemble is precisely the same as that for a suitably chosen *microcanonical ensemble* in which all systems have the same energy. The reason is that only a small fraction of systems in the ensemble have properties differing significantly from the average.

A further generalization is the *grand canonical ensemble*, in which not only the energy but also the number of particles is allowed to vary; the probability of occurrence of a system with N particles is proportional to $e^{-\mu N}$. Just as in the canonical ensemble one specifies the temperature instead of the total energy, so in the grand canonical ensemble one specifies μ instead of N. In order to obtain the properties of the system with a specified average number of particles, one has to calculate the average value of N as a function of μ and then solve the resulting equation to get μ as a function of N, so that μ can be eliminated. (In practice the technique is often equivalent to finding the maximum of a function of several variables subject to certain conditions by the method of Lagrange multipliers.) The grand canonical ensemble is especially useful in problems involving chemical equilibrium among several components; there is then a μ for each component, which turns out to be just the "chemical potential" (partial free energy per particle) for that component. In quantum statistics, moreover, one often wants to assign the particles to various quantum states and then average over all possible numbers of particles in each state; such averages are much easier to compute if one can avoid the restriction that the total number of particles is fixed.

Note for §11.8

1. The published works of Gibbs are easily available in his *Collected Works* (London and New York: Longmans, Green and Co., 1928, reprinted by Yale University Press,

1948, and by Dover Pubs., 1960). Secondary sources include the two-volume collection of papers, *A Commentary on the Scientific Writings of J. Willard Gibbs*, eds. F. G. Donnan and A. Haas (New Haven: Yale University Press, 1936); M. Fierz, in *Theoretical Physics in the Twentieth Century*, eds. M. Fierz and V. F. Weisskopf (New York: Interscience, 1960), p. 161; M. J. Klein, *Hist. Stud. Phys. Sci.* **1**, 127 (1969); L. Rosenfeld, *Acta Phys. Polonica* **14**, 3 (1955); L. P. Wheeler, *Josiah Willard Gibbs* (New Haven: Yale University Press, 1951); F. W. Stevens, *Science* **66**, 159 (1927); M. J. Klein, *DSB* **5**, 386 (1972). E. B. Wilson, *Bull. Am. Math. Soc.* **37**, 401 (1931).

CHAPTER 12

Viscosity and the Maxwell–Boltzmann Transport Theory*

12.1 Introduction

Although Herapath and Waterson had made some attempt to deal with aspects of phenomena such as sound propagation and diffusion, most of their efforts had been directed toward explaining equilibrium properties of gases. Clausius, Maxwell, and Boltzmann realized that a satisfactory kinetic theory must explain non-equilibrium processes as well, though the state of empirical knowledge of such processes was still quite unsatisfactory. Accurate experimental data on viscosity, heat conduction, and diffusion were not available until much later (with a few exceptions), and the kinetic theorists had to do their calculations without the advantage of knowing the answer beforehand. This circumstance made it all the more convincing when experiments confirmed the theoretical predictions; we have already noted that Maxwell's discovery that the viscosity of a gas is independent of its density helped to establish the validity of the kinetic theory.

One may distinguish two types of non-equilibrium processes: in the first, there is a continuous flow of mass, momentum, or energy resulting from variations in concentration, velocity, and temperature imposed externally; in the second, a gas that is initially in a non-

---

* Most of this chapter (except for §12.5) is reprinted from *Am. J. Phys.* **30**, 269 (1962), by permission of the American Institute of Physics.

equilibrium state spontaneously moves toward equilibrium in the absence of external interference. The two types are of course intimately related, as shown by Maxwell's theory of relaxation processes[1] and by modern work on the "fluctuation-dissipation theorem" – transport coefficients can be expressed in terms of the relaxation of fluctuations from an equilibrium state, and conversely.[2]

Phenomena of the second type, such as the relaxation of fluctuations in the local temperature or average velocity, are of considerable interest from a theoretical viewpoint, since they provide a justification for the application of the equilibrium theory whenever the external conditions do not lead one to expect any nonuniformity. More generally, a microscopic theory of these "irreversible" phenomena helps to explain the Second Law of Thermodynamics (principle of dissipation of energy) and even, some would say, the existence of a preferred "direction of time." A detailed discussion of that subject will be presented in ch. 14.

The most important example of the first type of irreversible process is viscosity – the internal friction between different layers of a gas forced to move at different velocities. Since it is possible to obtain by simple approximate methods a direct relationship between the viscosity coefficient and the molecular diameter (or the exponent "n" if a repulsive force varying as r^{-n} is assumed), measurements of the former give information about the latter, which, combined with similar information obtained from deviations from the ideal gas law, permit one to estimate the size of a molecule. Other transport coefficients such as heat conductivity and diffusion are not as useful for this purpose, since the theory is more complicated and the experiments more difficult. On the other hand, the mere existence of certain "cross-phenomena" such as thermal diffusion was not suspected until they were predicted by the kinetic theory, and thus their measurement provided another confirmation of the theory.

In this chapter we shall mainly be concerned with the general theory of transport phenomena in dilute gases and its application to the calculation of the viscosity coefficient in particular. We use the expression "coefficient of viscosity" when referring to the precise concept embodied in the definition based on Newton's hypothesis. The word "viscosity" has been derived from the Latin *viscum, viscus* meaning "mistletoe" or "birdlime." The latter indicates the confusion between the adjectives "viscous" and "sticky" which was formerly prevalent. The Oxford English Dictionary traces the words "viscosite" and "viscouse" back to the 15th century; in Jonson's *Alchemist*

(1612) there is the passage[3]

> Sub. And what's your mercury?
> Face. A very fugitive; he will be gone, sir.
> Sub. How know you him?
> Face. By his viscosity, his oleosity, and his suscitability.

Newton[4] assumed that if two parallel laminae having an area of contact A move with constant velocities v_1 and v_2, the force required to maintain the constant difference of velocity is proportional to A and to $(v_1 - v_2)$; the viscosity coefficient η is now usually defined as the force required per unit area to maintain unit gradient of velocity,

$$F = \eta A \frac{\Delta v}{\Delta z},$$

where the z axis is perpendicular to the laminae. It is found experimentally that η is independent of Δv over a fairly wide range up to a critical velocity depending on the ratio of the viscosity to the density; the viscous behavior of liquids in this range is called "Newtonian flow." The ratio η/ρ is known as the "kinematic viscosity." The viscosity coefficient η is usually expressed in dynes, centimeters, and seconds, the corresponding unit being called the "poise" in honor of Poiseuille who conducted the first extensive experimental investigations of liquid viscosities. The coefficient of viscosity of water at 20°C is approximately 0.01 poise or one centipoise.

The viscosity of a liquid is most easily measured by determining the time required for a given amount to run through a capillary tube. (This method does not give absolute viscosities very accurately because of the difficulty of measuring the width of the tube precisely and making various other corrections, but it can be used to determine relative viscosities of different liquids.) Conversely, the amount of liquid which has run through the tube provides a measure of time, as in the ancient Egyptian water clocks, and Höppler[5] has pointed out that the Egyptians were aware of the variation of viscosity with temperature.

When Maxwell turned his attention to gas viscosity, he could draw on the hydrodynamical theory of viscous fluids developed by Navier, Stokes, and others.[6] This theory did not depend on any assumption about the mechanism of viscosity, though many scientists imagined that viscosity was a result of molecular interactions. Little was known about the viscosity of gases; Maxwell's theoretical explanation was apparently completely original, owing nothing to experiment.[7] Indeed, most of the later experimental work on gas transport properties was stimulated by the kinetic theory. The main results of the kinetic theory of viscosity do not require any very accurate knowledge of interatomic forces, since they depend only on the molecular velocity and the average time between collisions. Interatomic forces affect the temperature dependence of viscosity by making the collision probability vary with velocity, but it takes quite accurate measurements to establish this effect.[8] Maxwell's mean-free-path treatment became so firmly enshrined in textbooks that earlier ideas about the mechanism of liquid viscosity were almost forgotten. Consequently, it is now customary to begin any discussion of liquid viscosity by emphasizing that momentum is transferred mainly by direct interactions between neighboring molecules, rather than by the motion of molecules from one region to another as in a gas.

We begin by discussing various attempts to improve the mean-free-path method; but the main part of the chapter deals with the more general methods proposed by Maxwell (1866) and Boltzmann (1872), which take account of the change of the velocity distribution in a nonuniform gas. Aside from a few special cases, the calculation of transport coefficients by these methods was not accomplished until about 1916 (Chapman and Enskog). Jäger (1899) and Enskog (1921) proposed modifications in the dilute-gas theory to obtain formulae for the viscosity of dense (hard-sphere) gases.

Notes for §12.1

1. Maxwell, *Phil. Trans.* **157**, 49 (1867); see pp. 29–32 in Brush, *Kinetic Theory* 2.
2. For further discussion and references see Brush, *Chemical Reviews* **62**, 513 (1962); *Kinetic Theory*, 3, 69–72.
3. Ben Jonson, *The Alchemist* (London, 1612), Act 2, Scene 5.
4. *Principia*, Book II, §IX, Hypothesis: "The resistance arising from the want of lubricity in the parts of a fluid, is, other things being equal, proportional to the velocity with which the parts of the fluid are separated from one another." Motte–Cajori trans., p. 385.
5. F. Höppler, *Kolloid-Zeitschrift* **98**, 1 (1942).
6. C. Truesdell, *J. Rat. Mech. Anal.* **1**, 125 (1952), **2**, 593 (1953). H. Rouse and S. Ince, *History of Hydraulics* (Iowa City: State University of Iowa Institute of Hydraulics Research, 1957, reprinted by Dover Pubs., New York, 1963).
7. As noted in §5.2, the only known experiments seemed to contradict his conclusions.
8. On the attempts of Lennard-Jones and others to determine force laws from viscosity measurements see S. G. Brush, *Arch. Rat. Mech. Anal.* **39**, 1 (1970).

12.2 Mean-free-path formulae for viscosity

In his first paper of 1860, Maxwell calculated the viscosity of a gas of N hard spheres of diameter σ by considering the momentum transferred by collisions between molecules coming from parts of the gas in which the mean velocity is different.[1] His proof depends on the assumptions that the frequency of collisions experienced by a molecule is determined by a "mean free path" L which depends only on σ and N, and that after each collision the molecule can be considered to have the mean velocity characteristic of the place where the collision occurred. He thus derived the expression

$$\eta = \rho L \bar{v}/3 \tag{1}$$

for the viscosity η of a portion of gas whose density is ρ and whose mean velocity is $\bar{v}$.

If we write

$$L = c/N\pi\sigma^2, \tag{2}$$

the problem is then reduced to the calculation of the numerical constant c.

Clausius[2] showed that when only one molecule is moving and the rest are stationary, $c = 1$; when they are all moving with the same speed in random directions, he found $c = \frac{3}{4}$. Maxwell set c equal to v/V, where v is the actual velocity of the molecule considered and V is its velocity relative to the others; he then simply stated that if u is the actual velocity of the others, $V = (v^2 + u^2)^{\frac{1}{2}}$, and so if u is equal to v, then $V = v\sqrt{2}$ and hence[3] $c = 1/\sqrt{2}$. Clausius considered this argument incorrect, and therefore published the details of his own calculation[4] in which he pointed out that the relative velocity is really $(u^2 + v^2 - 2uv\cos\theta)^{\frac{1}{2}}$, where θ is the angle between the directions of u and v. If this is averaged over all values of θ, the result is

$$V = v + u^2/3v \quad \text{when } u < v, \quad \text{and} \quad V = u + v^2/3u \quad \text{when } u > v \tag{3}$$

If u is equal to v, both expressions are equal and one gets the result $c = \frac{3}{4}$, as stated. Actually, Maxwell's result is correct if one calculates V by averaging over all values of u and v, using the Maxwell velocity-distribution law, although it is not evident from Maxwell's argument mentioned above that he actually did this.[5] Clausius eventually agreed that the factor should be $1/\sqrt{2}$.[6]

Tait later criticized this method of determining the mean free path;

§12.2] MEAN-FREE-PATH FORMULAE FOR VISCOSITY

he contended that instead of defining the mean path as the quotient of the average speed by the average number of collisions per particle per second, one should calculate the mean path corresponding to each value of the velocity and then average over velocities.[7] Tait's mean free path is thus

$$L = \frac{1}{N\pi\sigma^2} \int_0^\infty \frac{v}{V(v)} f(v) \, dv$$
$$= \frac{1}{N\pi\sigma^2} \int_0^\infty \frac{f(v) v \, dv}{\int_0^v f(u)\left(v + \frac{u^2}{3v}\right) du + \int_v^\infty f(u)\left(u + \frac{v^2}{3u}\right) du} \quad (4)$$

Substituting for $f(u)$ the Maxwell distribution,

$$f(u) = (4/a^3\sqrt{\pi})u^2 e^{-u^2/a^2}$$

(and similarly for $f(v)$), and evaluating the integral numerically, Tait obtained the result

$$L = 0.677/N\pi\sigma^2$$

Tait also derived a more accurate expression for the viscosity by taking account of the fact that the faster molecules have a longer free path and thus give a greater contribution to the viscosity than is indicated by eq. (1). The error in the earlier theory is essentially that of replacing the average of a product by the product of the averages of two quantities: *i.e.* $(\overline{Lv})$ by $(\overline{L})(\bar{v})$. The correct value of $(\overline{Lv})$, according to Tait, is[8]

$$(\overline{Lv}) = \frac{1}{N\pi\sigma^2} \int_0^\infty \frac{v^2 f(v) \, dv}{V(v)} \cong 0.838 a/N\pi\sigma^2 \quad (5)$$

Recalling that the mean velocity is $\bar{v} = 2a/\sqrt{\pi}$, we may write Tait's expression for the viscosity as

$$\eta = (1.051)(\rho\bar{v}/3)(1/\sqrt{2}N\pi\sigma^2), \quad (6)$$

which shows that the viscosity is about 5% greater than that predicted by Maxwell.

The other assumption made by Maxwell is also not quite correct, because of the phenomenon of "persistence of velocities"–after a collision, a molecule is still likely to have a component of velocity in its original direction. This effect was studied by Jeans,[9] who concluded that the expression (6) should be multiplied by another factor of approximately 1.382. It is at this point that the whole concept of "mean free path" begins to break down, since it is very difficult to account exactly for this persistence effect in the method of mean free paths. A correct treatment must also include a determination of the velocity-distribution function, because Maxwell's distribution is only approximately valid if the mean velocity varies throughout the gas. We will return to this point in §12.4.

Regardless of the exact numerical value adopted for the constant c, eq. (1) predicts that the viscosity of a gas should be proportional to the mean velocity of the molecules, *i.e.* to the square root of the absolute temperature. Maxwell's second theory, based on the assumption of a fifth power repulsive force, leads to an expression directly proportional to the temperature (see the next section). Maxwell's own experiments on the viscosity of air[10] gave the same result, and this encouraged him to adopt the hypothesis of a fifth power repulsive force.

Later and more accurate experiments showed that the viscosity of most gases does not in general follow this law, but varies as some power of the temperature, between $\frac{1}{2}$ and 1, the exponent itself not being the same for different gases, and sometimes not even for the same gas under different conditions.[11] Stefan and Meyer suggested that these observations might be explained by assuming that the diameter of a molecule varies with temperature.[12] Since the hard-sphere model was recognized to be only an approximation, representing the effect of short-range repulsive forces, it was reasonable to expect that two colliding molecules would be able to come closer together at higher temperatures when their mean velocities are greater, and thus the diameter would appear to change. From a theoretical point of view, however, it was clearly not very satisfactory to regard the diameter as an empirical parameter which had to be chosen differently at every temperature.

An important advance was made by Sutherland in 1893.[13] He pointed out that one could assume that the molecule itself had a fixed diameter, but exerted *attractive* forces on other molecules at larger distances. The effect of such forces would be to cause collisions between molecules that would otherwise pass near to each other without hitting, and thus increase the effective diameter. The effect of these forces on the probability of collision would depend on the relative velocity of the two molecules, and hence on the temperature.

Sutherland's argument is very simple: suppose b is the perpendicular distance from one molecule (considered at rest) to the asymptote of the path of the other (*i.e.* the path it would follow in the absence of forces). Denoting by a the radius of the molecules, V the relative velocity, r the distance, $u = 1/r$, $h = bV$, and $m^2 F(u)$ the force function and $m^2 f(u)$ the potential function, then for the usual differential equation of the orbit we have

$$\frac{d^2 u}{d\theta^2} + u - \frac{mF(u)}{h^2 u^2} = 0, \tag{7}$$

with its first integral the equation of energy,

$$(h^2/2)[(du/d\theta)^2 + u^2] = v^2/2 = mf(u) + \tfrac{1}{2}V^2, \tag{8}$$

v^2 being the velocity at any reciprocal-distance u.

Now when this orbit is such that there is no collision, we can determine the nearest distance to which the molecules approach one another (an apsidal distance) by the condition $du/d\theta = 0$; denote the reciprocal of this distance by w, it is then given by

$$\frac{h^2 w^2}{2} = mf(w) + \frac{V^2}{2}, \tag{9}$$

or

$$mf(w) - \frac{b^2 V^2 w^2}{2} + \frac{V^2}{2} = 0. \tag{10}$$

Now there will be a collision if $1/w$ is greater than $2a$, that is, if w is greater than $1/2a$; hence the greatest value of b for which a collision is possible is given by

$$mf(1/2a) - \frac{b^2 V^2}{2(2a)^2} + \frac{V^2}{2} = 0, \tag{11}$$

or

$$b^2 = (2a)^2 \left[1 + \frac{2mf(1/2a)}{V^2}\right], \tag{12}$$

and there is a collision for every value of b from 0 up to that given by the last equation.... Hence, molecular force causes the spheres to behave as regards collisions as if they were larger spheres devoid of force, the diameter-squared $(2a)^2$ being enlarged in the proportion $1 + 2mf(1/2a)/V^2 : 1$.

Hence, in the theory of viscosity as worked out for forceless molecules, we need only increase the square of the molecular sphere-diameter in this proportion to take account of molecular force. As the expression diminishes with increasing V^2, that is with increasing temperature, we see at once why the apparent result of increasing temperature was to make the molecules shrink: increase of temperature does not make the real molecules shrink (at least to the extent imagined), but produces shrinkage of the imaginary enlarged forceless spheres which could exhibit the same viscosity as the real molecules.[14]

It is clear from Sutherland's formula that the viscosity should behave like

$$\eta \sim \frac{\sqrt{T}}{1 + C/T},\tag{13}$$

and Sutherland showed that all the available data on viscosities of gases could be fitted within experimental error by choosing a single value of C for each gas.[15]

In 1900 Rayleigh[16] gave a clever dimensional analysis which showed that the viscosity of a gas composed of points repelling each other as the nth power of the distance must vary as $\bar{v}^{(n+3)/(n-1)}$, where $\bar{v}$ is the root-mean-square velocity and is thus proportional to $\sqrt{T}$.[17] This formula includes as special cases Maxwell's two results for hard spheres ($n = \infty$) and fifth power repulsion. Rayleigh assumed that the viscosity is independent of density, and thus

> the only quantities (besides the density) on which η can depend are m the mass of a particle, v the velocity of mean square, and k the repulsive force at unit distance. The dimensions of these quantities are as follows:
>
> $\eta = (\text{mass})^1(\text{length})^{-1}(\text{time})^{-1}$,
> $m = (\text{mass})^1$,
> $v = (\text{length})^1(\text{time})^{-1}$,
> $k = (\text{mass})^1(\text{length})^{n+1}(\text{time})^{-2}$.

Thus, if we assume $\eta = m^x v^y k^z$, we have

$$1 = x + z, \qquad -1 = y + (n+1)z, \qquad -1 = -y - 2z,$$

whence

$$x = (n+1)/(n-1), \qquad y = (n+3)/(n-1), \qquad z = 2/(n-1).$$

Accordingly,

$$\eta = a m^{(n+1)/(n-1)} v^{(n+3)/(n-1)} k^{-2/(n-1)},$$

where a is a purely numerical coefficient. For a given kind of molecule, m and k are constant. Thus

$$\eta \propto v^{(n+3)/(n-1)} \propto T^{(n+3)/(2n-2)}.\tag{14}$$

According to Rayleigh,

the best experiments on air show that, so far as a formula of this kind can represent the facts, $\eta \propto T^{0.77}$. It may be observed that $n = 8$ corresponds to $\eta \propto T^{0.79}$.

Rayleigh also realized that, for the purposes of testing the kinetic theory which must assume spherically symmetric molecules for simplicity, experiments on monatomic gases like argon were more important than the more common experiments on oxygen, hydrogen, and nitrogen.

Notes for §12.2

1. *Phil. Mag.* [4] **19**, 31 (1860); see *Kinetic Theory* **1**, 164–66.
2. R. Clausius, *Ann. Phys.* [2] **105**, 239 (1859); see *Kinetic Theory* **1**, 140.
3. Maxwell, *Phil. Mag.* [4] **19**, 19 (1860); see *Kinetic Theory* **1**, 160.
4. Clausius, *Phil. Mag.* [4] **19**, 434 (1860).
5. See the footnote by W. D. Niven, editor of Maxwell's *Papers*, reprinted in *Kinetic Theory* **1**, 160–61, for details of the calculation using the Maxwell velocity distribution.
6. In a letter to William Thomson in 1871, Maxwell made the following remark on this discrepancy in numerical factors: "Clausius made objection No. 1 to an integration founded on his theory of uniform velocity of molecules. (This is the first commitment of Clausius to such a theory.) As he was sure to be converted and I was lazy, I said 0. Objection No 2 and c. to theory of diffusion and conduction were well founded ...." (See H. T. Bernstein, *Isis* **54**, 212, 214 (1963).) As it turned out, Clausius was indeed converted without any further effort on Maxwell's part: see *Ann. Phys.* [3] **10**, 92 (1880); *Lumière Electrique* **17**, 241 (1885).
7. P. G. Tait, *Trans. R.S. Edinburgh* **33**, 73 (1886). See also L. B. Loeb, *Kinetic Theory of Gases* (New York: McGraw-Hill Book Co., Inc., 1927), p. 186; E. H. Kennard, *Kinetic Theory of Gases* (New York: McGraw-Hill Book Co., Inc., 1938), pp. 142–45.
8. P. G. Tait, *Trans. R.S. Edinburgh* **33**, 257, 260, 277 (1886). Boltzmann criticized Tait's definition of mean free path, and said that he had previously (and more accurately) evaluated the same integral; see *Wien. Ber.* **96**, 891 (1888); *Phil. Mag.* [5] **25**, 81 (1888). Tait replied by defending his own work but admitting that he had not read Boltzmann's work "because the methods employed seemed to me altogether unnecessarily intricate." *Proc. R.S. Edinburgh* **15**, 140 (1888); *Phil. Mag.* [5] **25**, 172 (1888).
9. J. H. Jeans, *Phil. Mag.* [6] **8**, 700 (1904); *The Dynamical Theory of Gases* (Cambridge: Cambridge University Press, 1921), pp. 260, 275. See also M. v. Smoluchowski, *Rozprawy Krakow* **A46**, 129 (1906), French trans. in *Bull. Acad. Sci. Cracovie*, p. 202 (1906); G. Jäger, *Wien. Ber.* **127**, 849 (1918).
10. Maxwell, *Phil. Trans.* **157**, 62 (1867); *Phil. Mag.* [4] **35**, 145 (1868); *Scientific Papers*, **2**, 42.
11. O. E. Meyer, *Ann. Phys.* [2] **148**, 1, 203 (1873). J. Puluj, *Wien. Ber.* **69**, 287 (1874), **70**,

243 (1875), **73**, 589 (1876). A. v. Obermayer, *Wien. Ber.* **71**, 281 (1875), **73**, 433 (1876). J. R. Partington, *An Advanced Treatise on Physical Chemistry*, **1**, §VII.f.
12. J. Stefan, *Wien. Ber.* **46**, 8, 495 (1861), **65**, 339 (1872). O. E. Meyer, *Ann. Phys.* [2] **148**, 233 (1873).
13. W. Sutherland, *Phil. Mag.* [5] **36**, 507 (1893).
14. *Ibid.*, pp. 511–12.
15. See also P. Breitenbach, *Ann. Phys.* [3] **67**, 803 (1899), [4] **5**, 166 (1901). More precise calculations based on this model were made by M. Reinganum, *Phys. Z.* **2**, 241 (1901), *Ann. Phys.* [4] **10**, 334 (1902), Chapman and Enskog (papers cited below, §§12.6 and 12.7); C. G. F. James, *Proc. Cambridge Phil. Soc.* **20**, 447 (1921); H. R. Hasse and W. R. Cook, *Phil. Mag.* [7] **3**, 977 (1927). See also S. Chapman and T. G. Cowling, *The Mathematical Theory of Non-Uniform Gases* (Cambridge: Cambridge University Press, 1939, 2nd ed., 1952), pp. 182–84.
16. Rayleigh, *Proc. R.S. London* **66**, 68 (1900).
17. For proof using modern techniques see C. Truesdell, *Z. Phys.* **131**, 273 (1952) and ch. IV of his forthcoming book on kinetic theory.

12.3 Maxwell's transport theory

In the absence of a more fundamental approach, the attempts of Tait (1886–87) and Sutherland (1893) to patch up Maxwell's approximate theory would have been of considerable value, especially for applications such as determination of atomic parameters. Yet one might question why Tait and Sutherland should have spent their time refining a theory which could never be more than approximately correct, when the basic principles of the modern theory had already been established by Maxwell (1866) and Boltzmann (1872). The explanation seems to be that Boltzmann himself, who was regarded as the foremost authority on the theory at that time, had given up his attempts to solve his integro-differential equation determining the distribution function in a nonuniform gas. This failure deterred less ambitious mathematicians, and no further progress was made until Chapman and Enskog attacked the problem again (1912–17).

In his important paper of 1866, Maxwell adopted the procedure of deriving general equations for the variation of physical properties of a medium (local density, velocity, and energy) which could be compared with the macroscopic equations known in hydrodynamics; the phenomenological transport coefficients appearing in the latter could then be identified with the kinetic-theory expressions in the former. Thus for viscosity, the relevant hydrodynamic equation is that of Navier[1] and Stokes[2]

§12.3]   MAXWELL'S TRANSPORT THEORY   433

$$\rho\frac{\partial u}{\partial t}+\frac{dp}{dx}-\eta\left\{\frac{d^2u}{dx^2}+\frac{d^2u}{dy^2}+\frac{d^2u}{dz^2}\right\}-\frac{1}{3}\eta\frac{d}{dx}\left(\frac{du}{dx}+\frac{dv}{dy}+\frac{dw}{dz}\right)=X\rho,$$

(15)

where ρ is the density of the fluid, p the pressure, (u, v, w) the velocity of an element of the fluid, and X the x component of the external force acting on the element. Two other equations may be written down by permuting x, y, and z; the set of three equations is now usually written in vector notation,

$$\rho\frac{\partial \mathbf{V}}{\partial t}+\nabla p-\eta\nabla^2\mathbf{V}+\frac{\eta}{3}\nabla\nabla\cdot\mathbf{V}=\mathbf{F}\rho$$

It should be noted that this equation refers to an incompressible fluid, terms involving bulk viscosity having been omitted. Maxwell did not prove, as has often been claimed, that the bulk viscosity is zero; in effect he assumed it.[3]

In order to calculate the viscosity coefficient η it is necessary to investigate the rate of change of the local average velocity due to collisions; this requires a detailed study of the dynamics of collisions. Maxwell characterized molecular collisions by three parameters: b, the distance between the asymptotes of the orbits of the two molecules; 2θ, the angle by which the direction of relative velocity is changed by the encounter; and ϕ, the angle between the plane of the orbits and a plane containing V (the relative velocity) and parallel to the x axis. The two molecules have masses M_1, M_2, and the force between them is assumed to be of the form Kr^{-n}; their velocity components are ξ_1, η_1, ζ_1 and ξ_2, η_2, ζ_2 before the collision, and ξ_1', η_1', etc. afterwards. According to Maxwell,[4]

$$\xi_1' = \xi_1 + \frac{M_2}{M_1+M_2}\left\{(\xi_2-\xi_1)(2\sin^2\theta)+\left[(\eta_2-\eta_1)^2 + (\zeta_2-\zeta_1)^2\right]^{\frac{1}{2}}\sin 2\theta \cos \phi\right\}$$

(16)

The angle of deflection, 2θ, may be expressed in terms of b

$$\frac{\pi}{2}-\theta=\int_0^{x'}\frac{dx}{\left[1-x^2-\frac{2}{n-1}\left(\frac{x}{a}\right)^{n-1}\right]^{\frac{1}{2}}},$$

(17)

where

$$x=b/r, \quad a=b\left[\frac{V^2M_1M_2}{K(M_1+M_2)}\right]^{1/(n-1)},$$

and x' is the least positive root of the equation

$$1 - x^2 - \left(\frac{2}{n-1}\right)(x/a)^{n-1} = 0$$

Maxwell now considers two systems of molecules moving with relative velocity V; if there are dN_1 molecules of one system and dN_2 of the other in a volume element having the specified velocities, then the number of encounters with specified values of V, b, and ϕ will be $Vb\,db\,d\phi\,\delta t\,dN_1\,dN_2$ in a time interval δt. If Q is any molecular property which changes to Q' after such an encounter, then it is required to calculate

$$\delta Q\,dN_1/\delta t = (Q' - Q)Vb\,db\,d\phi\,dN_1\,dN_2 \tag{18}$$

integrated over all values of ϕ, b, N_1, and N_2. The first integration is trivial, and one is left with terms depending on $\sin^2\theta$ or $\sin^2 2\theta$. The integration with respect to b may be converted to an integration with respect to a, using the relation

$$b\,db = [K(M_1 + M_2)/V^2 M_1 M_2]^{2/(n-1)}\,a\,da, \tag{19}$$

and one must then evaluate the two integrals

$$A_1 = \int_0^\infty 4\pi a\,da\,\sin^2\theta, \qquad A_2 = \int_0^\infty \pi a\,da\,\sin^2 2\theta$$

in which θ becomes a known function of a as soon as n is specified. It is clear from (19) that the relative velocity will appear in the result in the form $V^{-4/(n-1)}$ multiplied by one more factor of V from (18), i.e. $V^{(n-5)/(n-1)}$

It has been necessary to give Maxwell's derivation up to this point in considerable detail, so that it will be clear where the special properties of the r^{-5} repulsive force come from. If one sets $n = 5$, then V disappears from all the integrals, which then become simply average values of Q calculated with whatever the correct distribution function may be.[5] Even though one does not know what this correct function is, one may still identify the average values of ξ and ξ^2 with macroscopic quantities.

After some further calculation, Maxwell arrived at the result[6]

$$\eta = p/3kA_2\rho, \quad \text{where} \quad k = (K/2M^3)^{\frac{1}{2}}, \tag{20}$$

and A_2 was found to have the value 1.3682. The viscosity is thus proportional to the absolute temperature if the gas obeys the ideal gas laws; it is also independent of density, as Maxwell had shown in his previous paper.[7]

* * *

Maxwell's calculation of transport coefficients from the kinetic theory gave a strong impetus to experimental work in the 1860's and 1870's. As a result, some new properties of gases were established, and methods for estimating atomic parameters were developed.

One of the most active workers in this field was Oscar Emil Meyer (1834–1915), a German physicist who contributed to the early acceptance of kinetic theory in Germany both by his experimental work and by publishing an elementary textbook.[8] Meyer had begun his first experiments on viscosity, published in 1861, before the appearance of Maxwell's first paper on kinetic theory. He reported the result $\mu = 0.000360$ for air at 18°C (cgs units) and found it remarkable that this is only 37 times smaller than the viscosity of water, despite the fact that water is 770 times more dense. He stated that the high viscosity of air is difficult to explain with the "usual hypothesis of molecular constitution of bodies, according to which the molecules are in fixed positions which can be changed only by external forces." Such a theory would attribute viscosity to forces between neighboring particles, but it is hard to believe that the forces between air molecules are so much stronger than those between water molecules. On the other hand, he noted that the results are compatible with the Bernoulli–Clausius view of the constitution of gases. Viscosity would be due, in this theory, to the equalization of speeds of different layers of gas by individual molecules moving from one layer to another. Thus the viscosity would be greater if the speed of the molecules is greater, and indeed Clausius had estimated that the speed of gas molecules is very high.[9]

Since his own early work had already made him favorably inclined toward the kinetic-theory explanation of gas viscosity, Meyer could immediately recognize the importance of Maxwell's new theory. In a paper first read to the Deutsche Naturforscher Versammlung at Stettin, on 23 September 1863, and published in 1865, he began by reviewing the two current theories of forces; those that attribute attractive and repulsive forces to individual atoms, and those that attribute only impenetrability to them. The first, he says, has dominated most realms of mathematical physics during the past century, while the second is important primarily in connection with the mechanical theory of heat and the gas laws.[10] This distinction has to be remembered in reading Meyer's further remarks; when he refers to the theory of molecular collisions he assumes implicitly that the molecules act like billiard balls with no possibility of action at a distance:

> On the other hand the further development of the hypothesis of molecular collisions has led to consequences which are not yet confirmed by observation, and which in part seem so improbable that experimental proof is indispensable.
>
> Maxwell has founded a theory on the hypothesis of molecular collisions, which leads to the striking result [11] that the viscosity coefficient of a gas, as well as its thermal conductivity, is independent of the density or pressure of the gas. Stefan [12] has pointed out that, if the hypothesis of molecular collisions is correct, the viscosity coefficient of a gas must increase with temperature, whereas observations indicate that for liquids it decreases with temperature. Clausius [13] developed from the hypothesis the law that the thermal conductivity of a gas will be independent of density and increase with temperature.

(As we mentioned earlier, this "law" should be credited to Maxwell except that he failed to state it explicitly.) Meyer continues:

> These results of the theory, insofar as they concerned viscosity, were not unknown to me when I did my first work. However, I did not indicate at that time my views on them since my observations were not sufficient to test the validity of the hypothesis.
>
> This gap in my observations I have now filled through new experimental investigations, in which I set myself the task of *establishing through measurements how the viscosity coefficient of air varies with pressure and temperature.* . . . I cannot assume that the subject has great intrinsic interest, but I believe it does have value in relation to Maxwell's theory as a means of testing the admissibility of the hypothesis of molecular collisions. For such a test the study of viscosity appears especially appropriate, since from the viewpoint of the hypothesis it is one of the simplest processes involving molecular motions and collisions.
>
> The results of my observations definitely favor Maxwell's theory and consequently also the validity of the hypothesis of molecular collisions. This is confirmed by observations which Graham has made on the flow of gases through capillary tubes. [14] I do not claim in this confirmation a final proof of the hypothesis; I think that further tests of a different kind, namely thermal observations, are needed.

§12.3]   MAXWELL'S TRANSPORT THEORY   437

Meyer then presented a detailed analysis of the earlier experiments from which values of the viscosity coefficient have been deduced; it appears that these are misleading with respect to Maxwell's theory, because in reducing the data it had been assumed that the viscosity of air decreases as the pressure is lowered. For example, in Stokes' calculation of the viscosity coefficient from Baily's experiments on the oscillations of a pendulum in air, which gives a value much smaller than Meyer's, there is involved a correction factor which is based on the assumption that rarefied air has no perceptible effect on the pendulum motion.[15] Meyer asserts that there is on the contrary a significant effect, and that Bessel's correction factor is preferable.

Meyer then stated that his own earlier results, using Coulomb's vibrating-disk method, are too large, whereas Stokes' are too small. His new results are, from the first series of observation (all at about 22°C)

pressure	viscosity (cgs units)
757.5 mm Hg	0.000332
500.6	0.000307
251.1	0.000236

Although these results do not completely confirm the prediction that viscosity is independent of pressure, at least they show that it decreases much more slowly than does the pressure. Another series of observations gave the results

temperature	pressure	viscosity
21°C	749.1	0.000288
20.6	499.7	0.000385
19.6	250.5	0.000318

If these are all averaged together it would appear that Maxwell's law (pressure-independence of viscosity) is approximately valid, but more precise measurements are obviously needed.

Meyer also attempted to test the predicted temperature-dependence of the viscosity coefficient. However, in his method one observes directly only the quantity $\sqrt{\mu\rho}$ (ρ = density) and the temperature-dependence of ρ would be in the opposite direction to that of ρ. (According to Maxwell's theory for elastic spheres $\mu \propto \sqrt{T}$.) The results showed a slight increase in viscosity with temperature but the data were insufficient for a satisfactory test.

In his theoretical discussion, following Clausius, Meyer used expansions in powers of the mean free path and of the velocity gradients. One advantage of that procedure is that Meyer, like Clausius, recognized that the results of the theory will not be valid at very low densities when the mean free path becomes large; there was no such qualification in Maxwell's writings until 1879, though he presumably would have admitted it if pressed.

In another paper (1866), Meyer used data on the flow of gases through tubes to make an independent determination of the viscosity coefficient.[16] He proposed to use the results already published by Graham and pointed out that these results give an indirect confirmation of Maxwell's law, although Maxwell did not seem to be aware of this in 1860 when he stated that the only experiment with which he was familiar contradicted his prediction.

The viscosity coefficient can be calculated from data on flow through tubes provided it can first be shown that Poiseuille's law is obeyed for gases as well as liquids, i.e. that the rate of flow is proportional to the fourth power of the radius of the tube, and inversely proportional to its length. Unfortunately Graham has not given enough data to check this. If one uses the results he does report, assuming Poiseuille's law holds, then according to Meyer one finds a viscosity coefficient whose value is not much more than half that which is obtained by the Coulomb vibrating-disk method. The reason for the discrepancy, Meyer thinks, is that external friction (between the gas and the sides of the tube) has a different effect in the two methods. In the disk method it is usually assumed that such friction is effectively infinite, i.e. the fluid sticks to the solid surface without any slipping at all; but this may not be true for flow through a tube. By assuming that the results from the disk method calculated in this way are correct, Meyer could estimate the slip coefficient from the data on flow through a tube. For wide tubes, where the effect of slip should be relatively much less, Meyer thought Maxwell's theory is valid, as is indicated by the fact that the computed viscosities do increase with temperature. Graham himself had noted that cold air transpires faster than warm, even though it is denser, and this fact by itself could be considered a qualitative confirmation of Maxwell's theory.[17]

According to Meyer, the slip coefficient ζ, which is used as a correction factor in the form $(1+4\zeta/R)^{-1}$ to the viscosity coefficient for flow through a tube of radius R, is proportional to the viscosity coefficient for different gases. Hence the "external friction constant," defined as $E = \mu/\zeta$, should be about the same for all gases. Using this

§12.3]   MAXWELL'S TRANSPORT THEORY   439

assumption and taking $\mu = 0.000275$ for the viscosity coefficient of air, he computed values of μ from Graham's transpiration coefficients for 18 other gases. There seem to be no simple regularities in these numbers, except that pairs of gases having the same molecular weight also have the same viscosity; nitrogen and carbon monoxide are similar, and carbon dioxide and nitrous oxide are similar. Yet olefiant gas has the same molecular weight as the first pair, but not the same viscosity. From the theory, one expects (Meyer noted) that the viscosity should depend not only on molecular weight but also on the size and speed of the particles.

In a postscript added in December 1865 to his experimental paper on gas viscosity,[18] Maxwell stated that he had learned of Meyer's work from Stokes after the main part of his own paper had been communicated to the Royal Society. He noted the discrepancy between Meyer's results and his own, and attributed it to a defect in Meyer's experimental arrangement:

> In November 1863 I made a series of experiments with an arrangement of three brass disks placed on a vertical axis exactly as in M. Meyer's experiments, except that I had then no air-tight apparatus, and the disks were protected from currents of air by a wooden box only.
>
> I attempted to determine the viscosity of air by means of the observed mutual action between the disks at various distances. I obtained the values of this mutual action for distances under 2 inches, but I found that the results were so much involved with the unknown motion of the air near the edge of the disks, that I could place no dependence on the results unless I had a complete mathematical theory of the motion near the edge.
>
> In M. Meyer's experiments the time of vibration is much shorter than in most of mine. This will diminish the effect of the edge in comparison with the total effect, but in rarefied air both the mutual action and the effect of the edge are much increased. In this calculation, however, the effect of the three edges of the disks is supposed to be the same, whether they are in contact or separated. This, I think, will account for the large value which he has obtained for the viscosity, and for the fact that with the brass disks which vibrate in 14 seconds, he finds the apparent viscosity diminish as the pressure diminishes, while with the glass disks which vibrate in 8 seconds it first increases and then diminishes.

A further disadvantage of Meyer's method, according to Maxwell, is that it gives only the value of $\sqrt{\mu\rho}$, from which μ must be determined. "For these reasons I prefer the results deduced from experiments with fixed disks interposed between the moving ones," *i.e.* his method.[19]

Shortly after the publication of Maxwell's experiments on viscosity, Meyer reconsidered his own work and launched a new series of experiments.[20] Although his previous result, $\mu = 0.000275$ for the viscosity coefficient of air, was significantly higher than Maxwell's result [$\mu = 0.0001878(1 + \alpha t)$, where t = centigrade temperature], he had also deduced from Graham's observations a lower value, $\mu = 0.000178$. Since Maxwell's formula gives values within the limits of his own two results "their correctness can scarcely be questioned." But he felt it necessary to repeat Maxwell's version of the observations with his own apparatus, using round disks; there seemed to be some irregularities in the first set of results so he made some minor changes in the setup.

Meyer's new data, taken in 1868, yielded the results, for about 18°C

$\mu = 0.000197$ (from combining one series of observations)

and

$\mu = 0.000190$ (from a second series)

Maxwell's formula would give 0.000200 for this temperature, which Meyer considered to be sufficiently good agreement. He also computed a revised table of transpiration coefficients and viscosities for other gases.

Meyer turned next to the problem of the temperature-dependence of the viscosity coefficient. Maxwell's original mean-free-path theory based on the elastic sphere model had predicted that the viscosity coefficient should be proportional to the square root of the absolute temperature, and Meyer had accepted this result, not necessarily as an accurate quantitative law but simply as a striking qualitative result contradicting common-sense knowledge of the properties of fluids. Maxwell's second theory, based on the model of point atoms with repulsive forces inversely as the fifth power of the distance, had led to the prediction $\mu \propto T$, and Maxwell had claimed that his own data verified this prediction. In 1873, Meyer published some new results for a rather limited temperature range (0°C to 21°C) which he said agreed with the formula

$$\mu = \mu_0(1 + \alpha t)^{\frac{3}{2}} = \mu_0(T/273)^{\frac{3}{2}}$$

In spite of his previous concession that Maxwell's experimental results had been superior to his own, Meyer now suggested that Maxwell's temperature-dependence was incorrect because of a failure to control the temperature of the apparatus precisely enough.[21]

Maxwell's new viscosity theory was also disturbing on theoretical grounds. Having viewed Maxwell's first theory as being solidly based on the elastic-sphere model, Meyer could not easily follow Maxwell's switch to what he saw as a completely incompatible hypothesis, the model of the atom as a point center of force. Moreover, Meyer pointed out that the Joule–Thomson experiment indicated that the long-range forces between gas molecules must be attractive, not repulsive as Maxwell was now assuming. (Of course such an attraction was equally inconsistent with this philosophical commitment to elastic spheres, but Meyer did not seem to be able to resolve that problem.)

Although Meyer would clearly have preferred to retain the elastic sphere model, he recognized that the $\sqrt{T}$ behavior predicted by this model does not agree with anyone's experimental data. He attributed this discrepancy to failure to take account of the internal motions of molecules. If we try to force the data to fit the elastic-sphere formula, we find that the diameter of the spheres must decrease with increasing temperature. This might seem surprising at first, Meyer said, since if the internal parts of the molecule acquire more energy one would expect the size of the molecule to *increase* with temperature. But the contradiction can be resolved by noting that when two molecules collide they can penetrate further into each other because there will be more "open space" between their parts (at high temperatures). Meyer is aware that Josef Stefan has already proposed a temperature-dependent diameter (see below) but he does not accept Stefan's rationalization of this model in terms of aether envelopes.

Further experiments by Meyer, Puluj,[22] von Obermayer,[23] Warburg, and Barus gave exponents in the range from 0.65 to 0.96 for various gases. For a time there was a tendency for experimenters to fit their data with simple fractional exponents, presumably in the expectation that this would ultimately reveal some profound fact about the forces between gas molecules.

In his paper on rarefied gases (1879), Maxwell cited some of these experiments and stated that they "shew that the viscosity of air varies according to a lower power of the absolute temperature than the first, probably the 0.77 power."[24] Maxwell ignored the fact that it was already evident that this exponent would have significant different values for some other gases. Indeed, he was not really very much concerned about using a "realistic" model at this point; he preferred to continue using his inverse fifth power model, which implied a viscosity coefficient directly proportional to absolute temperature, in order to be able to carry out his mathematical calculations.

Notes for §12.3

1. C.-L.-M.-H. Navier, *Mem. Acad. Sci. Inst. France* (2) **6**, 389 (1822).
2. G. G. Stokes, *Trans. Camb. Phil. Soc.* **8**, 287 (1849) [read 1845], reprinted in his *Mathematical and Physical Papers*, with a new preface by C. Truesdell (New York: Johnson Corp., 1966), **1**, 75.
3. C. Truesdell, *J. Rat. Mech. Anal.* **1**, 125 (1952), see p. 229.
4. J. C. Maxwell, *Phil. Trans.* **157**, 57 (1867), see Brush, *Kinetic Theory* **2**, 36.
5. Boltzmann explained this special property in another way: the decrease of the "effective" cross section with increasing velocity is exactly compensated by the increase in the number of collisions caused by the increasing relative velocity [*Wien Ber.* **81**, 117 (1880)].
6. Four slightly different values of A_2 have subsequently been reported: (a) 1.3740, by K. Aichi and T. Tanukadate, in H. Nagaoka, *Nature* **69**, 79 (1903); (b) 1.3700, by S. Chapman, *Manchester Mem.* **66** (1) (1922); (c) 1.3703, by H. Hancock, in E. Ikenberry and C. Truesdell, *J. Rat. Mech. Anal.* **5**, 1 (1956); (d) 1.3706, by L. D. Higgins and F. J. Smith, *Molecular Physics* **14**, 399 (1968). It should be noted that A_2 as defined by Chapman and Cowling (*Mathematical Theory of Non-Uniform Gases*) differs from A_2 as defined by Maxwell by a factor π.
7. For further discussion of Maxwell's transport theory see G. R. Kirchhoff, *Vorlesungen über die Theorie der Wärme* (Leipzig, 1894), p. 134ff.; J. H. Jeans, *The Dynamical Theory of Gases*, ch. VIII; Ikenberry and Truesdell, *op. cit.*; C. Truesdell, *J. Rat. Mech. Anal.* **5**, 55 (1956).
8. O. E. Meyer, *Die Kinetische Theorie der Gase* (Breslau, 1877, 2nd ed., 1899).
9. O. E. Meyer, *Ann. Phys.* [2], **113**, 383, (1861).
10. O. E. Meyer, *Ann. Phys.* [2] **125**, 177, 401, 564 (1965). See §11.1 for a review of these two traditions.
11. He quotes Maxwell's phrase "*a very startling consequence*–*Phil. Mag.*, vol. **19**, p. 32."
12. J. Stefan, *Wien. Ber.* **46**, 12 (1863).
13. R. Clausius, *Ann. Phys.* [2] **115**, 49, 56 (1862).
14. T. Graham, *Phil. Trans.* **136**, 573 (1846), **139**, 349 (1849).
15. Cf. §5.2. Stokes admitted that his reduction of Baily's observations did incorrectly assume that there was no viscous effect due to the air (at about one inch of mercury pressure) but did not refer to Meyer.
16. O. E. Meyer, *Ann. Phys.* [2] **127**, 253, 353 (1866).
17. Graham, *Phil. Trans.* **139**, 387 (1849).
18. J. C. Maxwell, *Phil. Trans.* **156**, 249 (1866); *Papers* **2**, 1.
19. *Papers* **2**, 25.
20. O. E. Meyer, *Ann. Phys.* [2] **143**, 14 (1871).
21. O. E. Meyer, *Ann. Phys.* [2] **148**, 203 (1873), see p. 227.

22. J. Puluj, *Wien. Ber.* **69**, 287 (1874), **70**, 243 (1875), **73**, 589 (1876); *Ann. Phys.* [3] **1**, 296 (1877).
23. A. von Obermayer, *Wien. Ber.* **71**, 281 (1875), **73**, 433 (1876).
24. J. C. Maxwell, *Phil. Trans.* **170**, 231 (1880); quotation from *Papers* **2**, 692.

12.4 Boltzmann's equation and the H-theorem

A rigorous kinetic theory of irreversible processes depends on a knowledge of the distribution of velocities throughout the gas under various conditions. Maxwell's distribution law provides this information in the special case of a uniform gas at equilibrium, although Maxwell's proof of the law was criticized on various grounds by later writers. In 1872, Boltzmann attempted to solve the problem by establishing an equation describing the changes in the distribution resulting from collisions between molecules, and he later gave a more general equation applicable when there are external forces. If this equation can be solved for the distribution function, then one can use Maxwell's theory to calculate all the transport coefficients. Boltzmann showed further that, even if the equation could not be solved, two important results could be extracted from it: (1) once Maxwell's equilibrium distribution is established in the gas, it is not changed by molecular collisions in the absence of external forces; (2) a quantity H, depending on the distribution function, can never increase in the course of time. By identifying H with minus the entropy of the system, Boltzmann thus found a microscopic proof of the Second Law of Thermodynamics (the "H-theorem").[1]

To the reader who has gone through the derivation of Boltzmann's equation given in modern textbooks, Boltzmann's original proof would probably seem unnecessarily long and complicated—particularly since it applies only to the case when there are no external forces and the gas is uniform.[2] We shall therefore mention only the most important points in the derivation.

The first of these is Boltzmann's assumption[3] that the number of collisions in which the two colliding atoms have kinetic energies in the range $(x, x + dx)$ and $(x', x' + dx')$ before the collision is proportional to $f(x, t)f(x', t)\, dx\, dx'$, where $f(x, t)$ is the distribution function for the kinetic energy of single atoms at time t. It implies that there is no correlation between the variables x and x'. After the collision, one atom has energy ξ and the other must have energy $x + x' - \xi$; the probability distribution of ξ is determined by the force law and by x

and x', and was simply written $\psi(x, x', \xi)$ by Boltzmann. (The two results mentioned above do not depend on its specific form.)

The next step was to write the change in the function $f(x, t)$, during a small time interval, caused by collisions. Two kinds of collisions must be considered: those in which one molecule had energy x before the collision, and those in which one had energy x afterwards. The first type reduces $f(x, t)$ and the second increases it. Thus

$$f(x, t + \tau)\, dx = f(x, t)\, dx - \int dn + \int dv, \qquad (21)$$

where

$$\int dn = \tau\, dx \int_0^\infty \int_0^{x+x'} f(x, t) f(x', t) \psi(x, x', \xi)\, dx'\, d\xi$$

and

$$\int dv = \tau\, dx \int_0^x \int_{x-u}^\infty f(u, t) f(x, t) \psi(u, v, x)\, du\, dv$$
$$+ \tau\, dx \int_x^\infty \int_0^\infty f(u, t) f(v, t) \psi(u, v, x)\, du\, dv,$$

where the limits on the integrals are determined by the conservation of energy; u and v represent the energies of two atoms before a collision which results in one of them acquiring an energy x afterwards. Boltzmann then showed that by choosing new variables, interchanging the order of integration, and invoking some necessary symmetry properties of the function $\psi(x, x', \xi)$, the integrals over the variables (u, v) could be combined with those over (x', ξ). If τ represents a very short time, $f(x, t + \tau) - f(x, t)$ may be replaced by $\tau \partial f(x, t)/\partial t$, and thus Boltzmann obtained finally[4]

$$\frac{\partial f(x, t)}{\partial t} = \int_0^\infty \int_0^{x+x'} \left[\frac{f(\xi, t)}{\sqrt{\xi}} \frac{f(x + x' - \xi', t)}{(x + x' - \xi')^{\frac{1}{2}}} \right.$$
$$\left. - \frac{f(x, t)}{\sqrt{x}} \frac{f(x', t)}{\sqrt{x'}} \right] (xx')^{\frac{1}{2}} \psi(x, x', \xi)\, dx\, d\xi \qquad (22)$$

The first result now follows immediately if one substitutes Maxwell's distribution

$$f(x, t) = C\sqrt{x}\, e^{-hx},$$

where C and h are constants; the factor in square brackets vanishes identically, and therefore $\partial(x, t)/\partial t = 0$.

Boltzmann then introduced the quantity

$$E = \int_0^\infty f(x,t)\{\log[f(x,t)/\sqrt{x}] - 1\}\,dx \tag{23}$$

The second result follows on calculating the time derivative of E; it can be shown that $dE/dt \leq 0$ provided that f satisfies eq. (22), while $dE/dt = 0$ if and only if f is the "equilibrium" Maxwell distribution.[5] Moreover, for a system in equilibrium, the quantity $(-E)$ has the same properties as the entropy, and it is reasonable to extend the definition of entropy to non-equilibrium systems by using eq. (23). Boltzmann's theorem shows that a gas in any arbitrary initial state will, as a result of collisions, tend to approach to Maxwellian state, and this state, once reached, is permanent; at the same time the entropy increases to its maximum possible value.

Boltzmann proved too much here, for his conclusions are apparently incompatible with the reversibility of the original laws of motion. This problem is discussed in detail in ch. 14.

It should also be noted that Boltzmann's equation is really a special case of Maxwell's transport equation, the quantity Q being identified with the number of molecules having specified velocities; the function which Boltzmann calls $\psi(x, x', \xi)$ is essentially the same as the collision integral discussed by Maxwell.

It is convenient to write the final form of the Boltzmann equation in the notation used by Enskog, since Enskog's solution of it is the only one we shall discuss (see §12.7 and table 12.7-1). In a nonuniform gas one must include the "streaming terms," which express the time variation of $f(x, y, z, r_x, r_y, r_z, t)$ arising from the time dependence of $x, y, z, r_x, r_y,$ and r_z as well as from the direct dependence on t, and the collision term (on the right-hand side of eq. (b), note 5) can be written explicitly in terms of the impact parameter b and the relative velocity g:[6]

$$\frac{\partial f}{\partial t} + r_x \frac{\partial f}{\partial x} + r_y \frac{\partial f}{\partial y} + r_z \frac{\partial f}{\partial z} + X \frac{\partial f}{\partial r_x} + Y \frac{\partial f}{\partial r_y} + Z \frac{\partial f}{\partial r_z}$$
$$= \int \int_0^\infty \int_0^{2\pi} (f'f_1' - ff_1)gb\,do_1\,db\,d\epsilon \tag{24}$$

Notes for §12.4

1. L. Boltzmann, *Wien. Ber.* **66**, 275 (1972); English trans. in Brush, *Kinetic Theory* **2**, 88.
2. "By the study of Boltzmann I have been unable to understand him. He could not

understand me on account of my shortness, and his length was and is an equal stumbling-block to me. Hence I am very much inclined to join the glorious company of supplanters and to put the whole business in about six lines."–J. C. Maxwell, letter to P. G. Tait, August 1873, quoted in C. G. Knott's *Life and Scientific Work of Peter Guthrie Tait* (Cambridge University Press, 1911), p. 114. Compare Boltzmann's description of Maxwell's style, which he compared to that of composer who is no writer of program-music, obliged to set the explanation above the score: "... wie der Musiker bei den ersten Takten Mozart, Beethoven, Schubert erkennt, so würde der Mathematiker nach wenigen Seiten seiner Cauchy, Gauss, Jacobi, Helmholtz unterscheiden. Höchste äussere Eleganz, mitunter etwas schwaches Knochengerüst der Schlüsse charakterisiert die Franzosen, die grösste dramatische Wucht die Engländer, vor allen Maxwell. Wer kennt nicht seine dynamische Gastheorie? Zuerst entwickeln sich majestätische die Variationen der Geschwindigkeiten, dann setzen von der einen Seite die Zustandsgleichungen, von der andern die Gleichungen der Zentralbewegung ein, immer höher wogt das Chaos der Formeln; plötzlich ertönen die vier Worte: 'Put $n = 5$.' Der böse Dämon V [V = relative velocity, see paragraph before eq. (20)] verschwindet, wie in der Musik eine wilde, bisher alles unterwühlende Figur der Bässe plötzlich verstummt; wie mit einem Zauberschlage ordner sich, was früher unbezwingbar schien. Da ist keine Zeit, zu sagen, warum diese oder jene Substitution gemach wird; war das nicht fühlt, lege das Buch weg; Maxwell ist kein Programm-Musiker, der über die Noten deren Erklärung setzen muss. Gefügig speien nun die Formeln Resultat auf Resultat aus, bis überraschend als Schlusseffekt nich das Wärmegewicht eines schweren. Gases gewonnen wird und der Vorhang sinkt...." This originally appeared in the volume *Gustav Robert Kirchhoff, Festrede zur Feier des 301. Grundungstages der Karl-Franzens-Universitat zu Graz* (Leipzig, 1888), pp. 29–30, and has since been reprinted and translated in several publications. It is interesting to note that Boltzmann himself had once studied with Anton Bruckner.
3. The same assumption was made by Clausius, *Ann. Phys.* [2] **105**, 239 (1859), and Maxwell, eq. (18) in the text.
4. This differs from Boltzmann's equation as usually given in modern texts because Boltzmann used energy as the variable and hence the denominators $\sqrt{\xi}$ etc. are needed. Boltzmann points out that the positive sign for these roots must always be used because of the meaning of f and ψ.
5. The time derivative of E is

$$\frac{dE}{dt} = \int_0^\infty \log[f(x,t)/\sqrt{x}\,]\,\frac{\partial f(x,t)}{\partial t}\,dx$$

The theorem states that dE/dt can never be positive if $f(x, t)$ satisfies eq. (22). The proof consists in deriving from eq. (25) four expressions for dE/dt, each representing a different choice of independent variables. (The energies before and after the collision are related by the equation $x + x' = \xi + \xi'$, any one of these four variables may be chosen as dependent on the other three.) The process is simple in principle, but some care must be exercised in fixing the limits of integration. Each expression for dE/dt has the form

$$\frac{dE}{dt} = \int_0^\infty \int_0^\infty \int_0^{x+x'} (\log s)(\sigma\sigma' - ss')r\,dx\,dx'\,d\xi, \qquad \text{(a)}$$

where

$$s = f(x, t)/\sqrt{x}, \qquad s' = f(x', t)/\sqrt{x'}, \qquad \sigma = f(\xi, t)/\sqrt{\xi}$$

and
$$\sigma' = f(x + x' - \xi, t)/(x + x' - \xi)^{\frac{1}{2}}$$
and
$$r = (xx')^{\frac{1}{2}}\psi(x, x', \xi)$$

The three other expressions are derived from eq. (a) by the permutations $(s, s')(\sigma, \sigma')$, $(s, \sigma)(s', \sigma')$, and $(s, \sigma')(s', \sigma)$. Adding together the four equations and dividing by four, Boltzmann obtained the result

$$\frac{dE}{dt} = \frac{1}{4}\int_0^\infty \int_0^\infty \int_0^{x+x'} \log(ss'/\sigma\sigma')(\sigma\sigma' - ss')r\,dx\,dx'\,d\xi \tag{b}$$

He then pointed out that the product $\log(ss'/\sigma\sigma')(\sigma\sigma' - ss')$ is always negative or zero, and r is necessarily positive. Hence $dE/dt \leq 0$. Note that $dE/dt = 0$ when $\sigma\sigma' = ss'$, as will be the case when f is the Maxwell distribution.

6. *Wien. Ber.* **66**, 324 (1872); *Kinetic Theory* **2**, 133.

12.5 Hilbert's work on the Boltzmann equation*

According to Boltzmann's formulation (1872), a quantitative treatment of transport processes depends on the solution of a certain integro-differential equation. One might therefore expect some interaction between the development of kinetic theory and the development of the mathematical theory of integral equations; in fact there was hardly any interaction until Hilbert's work 40 years later. It is "well known" that "the first complete theory for a particular type of integral equation was given by Abel (1802–29) in 1823" according to a recent article by Bernkopf.[1] Further systematic work was done later in the 19th century by Volterra, Poincaré, and Fredholm. David Hilbert (1862–1943) was already recognized as one of the leading mathematicians of the 20th century when he began to publish a series of papers on the subject in 1904.[2] He approached the theory of integral equations by considering them as limits of infinite system of linear equations, which could be solved by using infinite determinants. He was especially interested in equations of the form

$$f(s) = \varphi(s) + \int_a^b K(s, t)\varphi(t)\,dt,$$

---

* Reprinted from *Kinetic Theory* 3, 6–7, by permission of Pergamon Press.

where $f(s)$ and $K(s, t)$ are given functions, and $\varphi(t)$ is the unknown function to be determined. If $K(s, t)$, called the "kernel" of the integral equation, is symmetric in its two arguments s and t, the analysis of the equation is greatly simplified and a number of useful theorems can be proved.[3]

When Hilbert decided to include a chapter on kinetic theory in his treatise on integral equations,[4] it does not appear that he had any particular interest in the physical problems associated with gases. He did not try to make any detailed calculations of gas properties, and did not discuss the basic issues such as the nature of irreversibility and the validity of mechanical explanations which had exercised the mathematician Ernst Zermelo in his debate with Boltzmann in 1896–97.[5] A few years later, when Hilbert presented his views on the contemporary problems of physics, he did not even mention kinetic theory.[6] We must therefore conclude that he was simply looking for another possible application of his mathematical theories, and when he had succeeded in finding and characterizing a special class of solutions (later called "normal") which are determined by a sequence of linear integral equations with symmetric kernels, his interest in the Boltzmann equation and in kinetic theory was exhausted. However, it should be mentioned that he did encourage some of his students to work on mathematical problems arising in kinetic theory,[7] and he is said to have lectured on this subject at Göttingen for several years.[8]

Chapman criticized Hilbert's approach partly on the grounds that Boguslawski obtained incorrect results by using it.[9] However, this criticism does not seem to apply to the essential features of Hilbert's work, since the work of Enskog and later mathematical studies have started from a similar approach. The results of Hilbert, Chapman, and Enskog are all similar insofar as they do not attempt to follow the detailed evolution of the distribution function starting from an arbitrary initial distribution; instead they select a very special, so-called "normal" solution, for which the linear transport coefficients are well defined. It is assumed that after a short time (corresponding to a few collisions for each molecule) any solution of the Maxwell–Boltzmann equations must approach asymptotically any appropriately chosen normal solution. The validity of this assumption has been studied in some recent works.[10] In particular, Grad (1958, 1963) has discussed in detail the differences between the Hilbert method and the Chapman–Enskog method and shows that they should both ultimately lead to identical results.[11]

Notes for §12.5

1. M. Bernkopf, *Arch. Hist. Exact Sci.* **3**, 1 (1966); E. Hellinger and O. Toeplitz, *Enc. math. Wiss.*, eds. Zweiter Band, Dritter Teil, and Zweite Halfte (Leipzig: Teubner, 1923–27), p. 1335.
2. D. Hilbert, *Nachr. K. Ges. Wiss. Göttingen* (1904–10), reprinted in *Grundzüge einer allgemeinen Theorie der linearen Integralgleichungen*, (Leipzig, Teubner, 1912); see summary by E. Hellinger in Hilbert's *Gesammelte Abhandlungen* (Berlin, Springer, 1932–35), Bd. III, p. 94.
3. Hilbert, *Grundzüge*, ch. XIV; see also W. V. Lovitt, *Linear Integral Equations*, (New York, McGraw-Hill, 1924), ch. V.
4. *Grundzüge*, ch. XXII, reprinted in *Math. Ann.* **72**, 562 (1912); English trans. in Brush, *Kinetic Theory* **3**, 89.
5. See Zermelo's papers translated in Brush, *Kinetic Theory* vol. 2.
6. D. Hilbert, *Math. Ann.* **92**, 1 (1924). See Blumenthal's remarks on p. 416 of Hilbert's *Gesammelte Abhandlungen*, Bd. III.
7. H. Bolza, M. Born, and T. v. Kármán, *Nachr. K. Ges. Wiss. Göttingen*, p. 221 (1913); E. Hecke, *Math. Ann.* **78**, 398 (1918), *Math. Z.* **12**, 274 (1922); Dissertations 56, 58, and 59 listed at the end of Hilbert's *Gesammelte Abhandlungen*, 3.
8. E. Guth (private communication).
9. S. Chapman and T. G. Cowling, *Mathematical Theory of Non-Uniform Gases* (1952 ed.), p. 384; S. Boguslawski, *Math. Ann.* **76**, 431 (1915). This paper has nothing to do with the Boltzmann equation but simply suggests that if there is an expansion of some quantity in powers of the mean free path λ, one can start by taking $\lambda = 0$, corresponding to zero viscosity, solve the hydrodynamic equations, and use this result to get the next approximation. Boguslawski did use a Maxwell–Boltzmann distribution in a later paper on the temperature-dependence of dielectric constants: *Zhur. Russ. Fiz.-Khim. Obsh.* **46** (Fiz. Otd.), 81 (1914). See also V. V. Struminskii, *Soviet Physics Doklady* **9**, 733 (1965).
10. See *Kinetic Theory* **3**, ch. V.
11. H. Grad, *Handbuch der Physik* **12**, 205 (1958); *Phys. Fluids* **6**, 147 (1963); *Proc. 3rd Int. Symposium on Rarefied Gas Dynamics, Paris, 1962* (New York: Academic Press, 1963), I, 26.

12.6 Chapman's transport theory

Boltzmann's equation was published in 1872; during the next four decades, Boltzmann and others devoted a prodigious amount of effort to attempts to solve it, yet it was not until 1916–17 that Sydney Chapman and David Enskog independently published solutions good enough for the calculation of transport coefficients in a dilute gas. Maxwell continued his work in transport theory especially in connection with the radiometer problem (§5.5) but did not escape the restriction to molecules interacting with inverse fifth power repulsive forces. Boltzmann[1] published several long papers on the theory of viscosity and diffusion, but never succeeded in calculating numerical values for the coefficients, although he managed to derive differential equations whose numerical solution would have given the desired results.[2] Other work by Stefan,[3] Langevin,[4] Brillouin,[5] Enskog,[6] and Chapman[7] involved approximations whose accuracy could not be estimated except by comparison with later exact solutions.[8] Some important special cases were treated exactly or discussed mathematically by Lorentz,[9] Hilbert (§12.5), Lunn,[10] and Pidduck,[11] but these did not give practical results for the general case.

The modern theory of transport processes in gases was developed independently by Chapman and Enskog. Chapman's work[12] follows the method of Maxwell, being based on the use of transfer equations, while Enskog's (§12.7) depends on the solution of Boltzmann's equation for the velocity-distribution function. Both had announced some of their results in earlier papers, based on incomplete or approximate calculations: Enskog[6] predicted in 1911 that a temperature gradient in a mixture would produce diffusion,[13] and Chapman[7] gave approximate formulae for the transport coefficients which turned out to be fairly accurate.

Chapmann expressed the velocity-distribution function in a slightly non-uniform gas in the form (see table 12.7-1)

$$f(U, V, W) = f_0(U, V, W)[1 + F(U, V, W)]$$
$$= (hm/\pi)^{\frac{3}{2}} \exp[-hm(U^2 + V^2 + W^2][1 + F(U, V, W)], \quad (25)$$

where $h = 1/2RT$, and (U, V, W) are the components of the molecular velocity relative to the local average velocity, $U = u - u_0$, etc.; F depends on the variations of temperature and local velocity in the gas. Since Chapman's theory attempted to deal only with first-order effects, no second derivatives or products of derivatives were included in F; and since F was supposed to vanish in a uniform gas, it could contain no terms independent of the space derivatives of T and the local velocity. Because F must be invariant with respect to any orthogonal transformation of the coordinate axes, it can depend only on the invariant quantities

$$C^2 \equiv U^2 + V^2 + W^2$$

$$S \equiv \frac{\partial u_0}{\partial x} + \frac{\partial v_0}{\partial y} + \frac{\partial w_0}{\partial z}$$

$$DT \equiv \left(U\frac{\partial}{\partial x} + V\frac{\partial}{\partial y} + W\frac{\partial}{\partial z}\right)T$$

$$S' \equiv U^2 \frac{\partial u_0}{\partial x} + V^2 \frac{\partial v_0}{\partial y} + W^2 \frac{\partial w_0}{\partial z} + VW\left(\frac{\partial w_0}{\partial y} + \frac{\partial v_0}{\partial z}\right)$$
$$+ WU\left(\frac{\partial u_0}{\partial z} + \frac{\partial w_0}{\partial x}\right) + UV\left(\frac{\partial v_0}{\partial x} + \frac{\partial u_0}{\partial y}\right)$$

Chapman thus assumed that F could be written

$$F = (DT)P_1(C^2) + SP_2(C^2) + S'P_3(C^2), \tag{26}$$

where P_1, P_2, and P_3 are functions of C^2 to be determined later.

The next step was to write down transfer equations, following the method of Maxwell:

The rate of change of $\nu\bar{Q}$ (ν = local number density), the aggregate value of $Q(u, v, w)$ per unit volume, may be analyzed into three parts, viz. that due to molecular encounters (which we denote by ΔQ), that due to the passage of molecules in or out of the volume element considered, and that due to the action of the external forces. The equation expressing this analysis may readily be shown [15] to be

$$\frac{\partial}{\partial t}(\nu\bar{Q}) = \Delta Q - \sum_{x,y,z}\left[\frac{\partial}{\partial x}(\nu\bar{u}\bar{Q}) - \frac{\nu}{m}X\left(\frac{\partial Q}{\partial u}\right)\right] \tag{27}$$

We may define ΔQ by the statement that $(\Delta Q)\,dx\,dy\,dz$ is the change produced by molecular encounters during time dt in the sum ΣQ taken over all the molecules in the volume element $dx\,dy\,dz$: evidently $\Sigma Q = \nu\bar{Q}\,dx\,dy\,dz$.[16]

In calculating the mean values required in eq. (27), Chapman neglected F, except in the case of the collision term Q where the contribution from the unity term [eq. (25)] cancels out. The results were

(I) $Q = u(u^2 + v^2 + w^2)^s$,

$$\Delta UC^{2s} = \frac{1}{T}\frac{\partial T}{\partial x}\left[\frac{(2s+3)!!}{(2hm)^{s+1}}\frac{s\nu}{3}\right], \tag{28}$$

(II) $Q = u^2(u^2 + v^2 + W^2)^s$,

$$\Delta U^2 C^{2s} = \left(2\frac{\partial u_0}{\partial x} - \frac{\partial v_0}{\partial y} - \frac{\partial w_0}{\partial z}\right)\left[\frac{(2s+5)!!}{(2hm)^{s+1}}\frac{2\nu}{45}\right] \tag{29}$$

Other similar results can be obtained by permutation of the x, y, and z axes.

The analysis of the dynamics of an encounter leads to the expression

$$\Delta Q_1 = 2\nu_1\nu_2 \int\int\int\int\int\int\int (Q_1' - Q_1) \times (f_1 f_2 C_R) p\, dp\, d\epsilon\, dU_1\, dV_1\, dW_1\, dU_2\, dV_2\, dW_2, \quad (30)$$

which is essentially the same as that used by Maxwell. Substituting from eq. (25) for f_1 and f_2, one finds that the contribution from the unity term is the difference between two integrals having exactly the same form but different labels for the variables, and is therefore zero.

It is consistent with the degree of approximation being used to omit the term $F_1 F_2$ which arises from $f_1 f_2$, and hence one is left with integrands of the form $Q(F_1 + F_2)$. By assumption [eq. (26)] F consists of a sum of two parts, one odd in the velocities and one even, which Chapman now denotes by

$$O(U, V, W) = \frac{1}{T}\left(U\frac{\partial T}{\partial x} + V\frac{\partial T}{\partial y} + W\frac{\partial T}{\partial z}\right) P_1(C^2),$$

$$E(U, V, W) = (c_{11} U^2 + c_{22} V^2 + c_{33} W^2 + c_{23} VW + c_{31} WU + c_{12} UV) P(C^2), \quad (31)$$

where

$$c_{11} = 2\frac{\partial u_0}{\partial x} - \frac{\partial v_0}{\partial y} - \frac{\partial w_0}{\partial z},$$

$$c_{12} = 3\left(\frac{\partial u_0}{\partial y} + \frac{\partial v_0}{\partial x}\right), \quad \text{etc.}$$

The dependence of F on the derivatives $\partial T/\partial x$, $\partial u_0/\partial x$, etc. is assumed to be the same as that of ΔQ [cf. eqs. (28), (29)]. Chapman further assumes that P_1 and P can be expanded as power series in C^2

$$P_1(C^2) = -B_0 \sum_{r=0}^{\infty} \frac{(2hm)^r}{(2r+3)!!} \frac{\beta_{r-1}}{r} C^{2r},$$

$$P(C^2) = -C_0(2hm) \sum_{r=0}^{\infty} \frac{(2hm)^r}{(2r+5)!!} \gamma_r C^{2r},$$

(the factor r in the denominator in the first expression is to be omitted when $r = 0$).

Because of the form of the integral in eq. (31), only terms in QF which are even in U, V, and W separately contribute anything to ΔQ; hence if Q is odd in U, only the part of F which is likewise odd in U need be considered, and conversely. It turns out that the viscosity coefficient thus depends only on $P(C^2)$, and the thermal conductivity only on $P_1(C^2)$.

After several pages of involved calculations, Chapman derived the following equation giving the change in Q due to molecular encounters

$$\Delta UC^{2(s+1)} = \frac{1}{T}\frac{\partial T}{\partial x}\left(\sum_{r=0}^{\infty}\beta_r b_{rs}\right)\frac{(2s+5)!!}{(2hm)^{s+2}}\frac{(s+1)\nu}{3}, \tag{32}$$

$$\Delta U^2 C^{2s} = c_{11}\left(\sum_{r=0}^{\infty}\gamma_r c_{rs}\right)\frac{(2s+5)!!}{(2hm)^{s+1}}\frac{2\nu}{45}, \tag{33}$$

where the coefficients b_{rs} and c_{rs} can be determined if the law of force is given (the c_{rs} are not to be confused with the coefficient appearing in eq. (31)). By comparing eqs. (32) and (28), and (33) and (29), one may determine the coefficients β_r and γ_r

$$\sum_{r=0}^{\infty}\beta_r b_{rs} = 1, \tag{34}$$

$$\sum_{r=0}^{\infty}\gamma_r c_{rs} = 1 \tag{35}$$

The solution of these equations is

$$\beta_r = \nabla_r(b_{rs})/\nabla(b_{rs}), \tag{36}$$

$$\gamma_r = \nabla_r(c_{rs})/\nabla(c_{rs}), \tag{37}$$

where $\nabla(b_{rs})$ and $\nabla(c_{rs})$ denote the infinite determinants formed from the arrays (b_{rs}) and (c_{rs}), and $\nabla_r(b_{rs})$ and $\nabla_r(c_{rs})$ denote the determinants obtained by replacing such element of column r in $\nabla(b_{rs})$ and $\nabla(c_{rs})$, respectively, by unity.

While the determination of all the coefficients β_r and γ_r, which is required for the complete specification of the velocity-distribution function, is rather difficult, the coefficients of viscosity and thermal conductivity depend only on the sums $\Sigma_r\gamma_r$ and $\Sigma_r\beta_r$. In the case of Maxwellian molecules, Chapman showed that $\beta_0 = \gamma_0 = 1$, $\beta_r = \gamma_r = 0$ ($r > 0$), so that the velocity-distribution function is simply

$$f(U, V, W) = (hm/\pi)^{\frac{3}{2}} \exp\left[-hm(U^2 + V^2 + W^2)\right]$$

$$\times\left\{1 - (\mu_5 A_1)^{-1}\left[\frac{75}{4T}\left(U\frac{\partial T}{\partial x} + V\frac{\partial T}{\partial y} + W\frac{\partial T}{\partial z}\right)\left(-1 + \frac{2}{5}hmC^2\right)\right.\right.$$

$$\left.\left. + \frac{10hm}{3}(c_{11}U^2 + c_{22}V^2 + c_{33}W^2 + 2c_{13}UW + 2c_{23}VW + 2c_{12}UV)\right]\right\}.$$

and where[17]

$$_5A_1 = (15/2)(Km)^{\frac{1}{2}} \int_0^\infty (\sin^2 \chi)\alpha \, d\alpha = (15/2\pi)(1.3682)(Km)^{\frac{1}{2}}$$

In the case of rigid elastic spheres, all the β_r and γ_r must be included, and the determinants and sums must be evaluated numerically. Chapman found

$$\sum_{r=0}^\infty \beta_r = 1.026, \qquad \sum_{r=0}^\infty \gamma_r = 1.016$$

Approximate values of these sums were also given for molecules repelling with an inverse-power force, and for rigid elastic attracting spheres (Sutherland's model). In all cases the sums were slightly greater than unity, the largest value being 1.038 for $\Sigma\beta_r$ when the ratio C/T [cf. eq. (13)] becomes infinite.

By comparing the pressure system in the gas as calculated from the distribution function,

$$P_{xx} = \rho\overline{U^2}, \qquad P_{xy} = \rho\overline{UV}, \qquad \text{etc.}$$

with the corresponding hydrodynamic equations,

$$P_{xx} = p - (2/3)\eta\left(2\frac{\partial u_0}{\partial x} - \frac{\partial v_0}{\partial y} - \frac{\partial w_0}{\partial z}\right)$$

$$P_{xy} = -\eta\left(\frac{\partial v_0}{\partial x} + \frac{\partial u_0}{\partial y}\right),$$

Chapman was able to identify the coefficient of viscosity as

$$\eta = \frac{\rho}{10hm} C_0 \sum_0^\infty \gamma_r \tag{38}$$

The constant C_0 depends on the force law through the angle of deflection χ. When the appropriate values of C_0 are substituted, the results are

$$\eta = (\Sigma\gamma_r)\frac{5m}{16\pi^{1/2}\sigma^2}\left(\frac{kT}{m}\right)^{1/2}\left(\frac{1}{1+C/T}\right) \tag{39}$$

for rigid attracting spheres, which reduces in the limit of simple rigid spheres $(C \to 0)$ to

$$\eta = (1.016)\frac{5m}{16\pi^{1/2}\sigma^2}\left(\frac{kT}{m}\right)^{1/2}; \tag{40}$$

for centers of force r^{-n},

$$\eta = (\Sigma \gamma_r) \frac{75m}{8\pi^{\frac{1}{2}}{}_nA_1\Gamma\left(4 - \frac{2}{n-1}\right)} (kT/m)^{(n+3)/2(n-1)} \qquad (41)$$

The constant ${}_nA_1$ is related to the angle of deflection by the formula

$${}_nA_1 = 18 \left(\frac{Km}{n-1}\right)^{2/(n-1)} \int_0^\infty [1 - P_2(\cos \chi)]\alpha \, d\alpha$$

and is independent of temperature and density.

Comparison of eq. (40) with the formula of Jeans,[18] which corrected eq. (6) for the effect of persistence of velocities, showed that while the dependence on R, T, m, and σ is correct, the numerical constant given by Jeans is 12% too small. Comparison of eq. (41) with eq. (14) shows that the temperature dependence predicted by Rayleigh is correct.

Notes for §12.6

1. L. Boltzmann, *Wien. Ber.* **81**, 117 (1880), **84**, 40, 1230 (1881), **86**, 63 (1882), **88**, 835 (1883), *Jahresber. D. Math. Ver.* **6**, 130 (1899).
2. C. L. Pekeris, *Proc. Nat. Acad. Sci. U.S.A.* **41**, 661 (1955).
3. J. Stefan, *Wien. Ber.* **65**, 323 (1872).
4. P. Langevin, *Ann. Chim. Phys.* [8] **5**, 245 (1905).
5. M. Brillouin, *Ann. Chim. Phys.* [7] **20**, 440 (1900).
6. D. Enskog, *Phys. Z.* **12**, 56, 533 (1911).
7. S. Chapman, *Phil. Trans.* **A211**, 433 (1912).
8. See Chapman and Cowling, *Mathematical Theory of Non-Uniform Gases* (1952 ed.), 382–85.
9. H. A. Lorentz, *Amsterdam Verslagen* [4] **13**, 493, 565, 710 (1904–5), English trans. in *Amsterdam Proc.* **7**, 438, 585, 684 (1905).
10. A. C. Lunn, *Bull. Am. Math. Soc.* **19**, 455 (1913).
11. F. B. Pidduck, *Proc. London Math. Soc.* **15**, 89 (1916).
12. S. Chapman, *Phil. Trans.* **A216**, 279 (1916), **A217**, 115 (1917). A long summary was published in *Proc. R.S. London* **A93**, 1 (1916–17) and is reprinted in Brush, *Kinetic Theory* **3**, 102. In a letter to the author, 11 July 1961, Chapman recalled the origin of his work in kinetic theory: "In the summer of 1910, after taking the last of my mathematical examinations at Cambridge University, I asked Larmor to suggest a subject of research for me in applied mathematics. I had already started writing papers on pure mathematics. Larmor drew my attention to the work of Knudsen and Smoluchowski, and I made some extensions of their results, but never published them. But I found Maxwell's [*Phil. Trans.*] papers and did not know how many mathematicians had tried to generalize his work. Thus with the ignorant hardihood of youth I attempted a problem that Larmor would certainly have thought unfit to suggest to such a novice. It seems it is sometimes good not to know too much."

Further recollections by Chapman were contained in his lecture "The Kinetic Theory of Gases Fifty Years Ago" at the Boulder Theoretical Physics Institute (1966), reprinted in Brush, *Kinetic Theory* 3, 260. For some additional biographical information see *Sydney Chapman, Eighty, from his Friends*, eds. S.-I. Akasofu, B. Fogle, and B. Haurwitz (Boulder: University of Colorado Press, 1968); T. G. Cowling, *Biog. Mem. Fell. Roy. Soc.* **17**, 53 (1971); W. O. Roberts, *Icarus* **13**, 354 (1970); J. Lear, *Sat. Rev.*, July 6, 1968, p. 43; Transcript of 1966 interview with Chapman, at Amer. Inst. Phys. Center for History of Physics; and Appendix to this chapter.

13. See Brush, *Kinetic Theory* 3, 8, 18–20.
14. S. Chapman, *Phil. Trans.* **A216**, 284 (1916).
15. J. H. Jeans, *Dynamical Theory of Gases*, p. 313.
16. Chapman, *op. cit.*, p. 285.
17. This is the same integral which Maxwell calls A_2; Chapman's theory reduces to that of Maxwell for "Maxwellian" molecules (inverse fifth power force).
18. See note 9, §12.2.

12.7 Enskog's solution of the Boltzmann equation

During the period when Chapman was working out his transport theory in Cambridge, the Swedish physicist David Enskog was attacking the same problem along different lines, leading ultimately to the same results.[1] Since Chapman and Enskog were each working independently without knowledge of the other's results, there seems to be no point in arguing about which one deserves priority; Chapman actually published expressions for the transport coefficients first, but comparison of these with Enskog's results "revealed a few algebraic and arithmetical errors which affected some of Chapman's formulae for the coefficients for a mixture."[2]

Enskog's theory was based on a series solution of the Boltzmann equation, Eq. (26), which he wrote in the form (see table 12.7-1 for notation)

$$\frac{\partial f}{\partial t} + D(f) = J(f, f_1),$$

$$D(f) = r_x \frac{\partial f}{\partial x} + r_y \frac{\partial f}{\partial y} + r_z \frac{\partial f}{\partial z} + X \frac{\partial f}{\partial r_x} + Y \frac{\partial f}{\partial r_y} + Z \frac{\partial f}{\partial r_z},$$

$$J(f, f_1) = \int \int_0^\infty \int_0^{2\pi} (f' f_1' - f f_1) g b \, do_1 \, db \, d\epsilon$$

The nonlinear term $J(f, f_1)$ creates considerable difficulties and makes most ordinary methods for solving differential equations useless; we have already noted that Boltzmann gave up after several years of work on this problem, and Hilbert had given only a formal prescription for the solution which he himself did not carry out.

Table 12.7-1. A guide to the notation used by Maxwell, Boltzmann, Chapman, and Enskog in describing molecular encounters.[a]

Quantity	M	B	C	E
Velocity components	ξ, η, ζ	ξ, η, ζ	U, V, W	r_x, r_y, r_z
Relative velocity	V	V	C_R	
Mean square velocity: i.e. $\frac{3}{2}$ times the constant dividing the square of the velocity in the Maxwell distribution function	$\dfrac{3\alpha^2}{2}$	$\dfrac{3}{2h}$	$\dfrac{3}{2hm}$	$\left(\dfrac{3}{2hm} = \dfrac{3}{2c}\right)$
Velocity-distribution function	N	f	f	f
Element of velocity space	$d\xi\, d\eta\, d\zeta$	$d\omega$	$dU\, dV\, dW$	do
Impact parameter	b	b	p	b
Angle between the plane containing the orbits of the colliding molecules and a plane containing the relative velocity	ϕ	ϕ	ϵ	ϵ

[a] All four used the prime (') to indicate values of quantities after the collision and subscripts 1 and 2 to distinguish the two molecules.

Enskog and Hilbert both started by writing[3]

$$f = f^{(0)}/\lambda + f^{(1)} + \lambda f^{(2)} + \ldots, \tag{42}$$

where λ is a parameter ultimately to be set equal to 1, which helps to keep track of the order of the various terms (the power of λ^{-1} appearing in any term is the same as that of the density appearing in the same term). Enskog also assumed that each term $\partial f^{(s)}/\partial t$ could similarly be expanded in powers of λ,

$$\frac{\partial f^{(s)}}{\partial t} = \frac{\partial_0 f^{(s)}}{\partial t} + \lambda \frac{\partial_1 f^{(s)}}{\partial t} + \lambda^2 \frac{\partial_2 f^{(s)}}{\partial t} + \ldots, \tag{43}$$

so that by substituting eqs. (42) and (43) into the Boltzmann equation and requiring that the resulting equation be valid for all values of λ – i.e. by equating coefficients of corresponding powers of λ – he obtained the set of relations

$$J(f^{(0)}, f_1^{(0)}) = 0, \tag{44}$$

$$\frac{\partial_0 f^{(0)}}{\partial t} + D(f^{(0)}) = J(f^{(0)} f_1^{(1)} - f^{(1)} f_1^{(0)}), \tag{45}$$

$$\frac{\partial_1 f^{(0)}}{\partial t} + \frac{\partial_0 f^{(1)}}{\partial t} + D(f^{(1)}) = J(f^{(0)} f_1^{(2)} + f^{(1)} f_1^{(1)} + f^{(2)} f_1^{(0)}), \quad \text{etc.}$$

Thus $f^{(s)}$ can be determined if $f^{(0)}$, $f^{(1)}$, up to $f^{(s-1)}$, are known. Equation (44) has the general solution [cf. eq. (28)]

$$f^{(0)} = \rho T^{-\frac{3}{2}} \left(\frac{m}{8\pi^3 k^3}\right)^{\frac{1}{2}} \exp\{-(m/2kT) \\ \times [(r_x - u)^2 + (r_y - v)^2 + (r_z - w)^2]\},$$

as pointed out by Boltzmann;[4] the identification of the constants ρ, T, u, v, and w with the local density, temperature, and velocity components is somewhat arbitrary from a mathematical viewpoint, but leads to a physically reasonable solution. Enskog then proved that each of the terms $f^{(1)}$, $f^{(2)}$, etc., and hence the function f, can depend only on these five parameters and their space and time derivatives at the point (x, y, z) and the time t. These parameters are derived from the molecular quantities

$$\psi^{(1)} = 1$$
$$\psi^{(2)} = r_x$$
$$\psi^{(3)} = r_y \qquad (46)$$
$$\psi^{(4)} = r_z$$
$$\psi^{(5)} = r_x^2 + r_y^2 + r_z^2 = r^2,$$

which comprise the summational invariants of a collision; any distribution function must satisfy the requirements

$$\int f\psi^{(i)} \, do = 0, \qquad (i = 1 \text{ to } 5) \qquad (47)$$

but Enskog's method requires that each term in the series (42) must also satisfy conditions of the type (47). The next step is to specify the definition of the terms on the right-hand side of eq. (46); this was done by writing down expressions of the form

$$\int \psi^{(i)} \left(\frac{\partial_0 f^{(0)}}{\partial t} + Df^{(0)}\right) do = \int \psi^{(i)} J(f^{(0)} f_1^{(1)} - f^{(1)} f_1^{(0)}) \, do \qquad (48)$$

The right-hand side of eq. (48) must be zero since $\psi^{(i)}$ is unchanged by the collision, and hence a necessary condition for eq. (45) to be soluble is that the left-hand side of eq. (48) be zero. Since (integrating by parts) for any function ϕ

$$\int \phi \frac{\partial f}{\partial t} \, do = \frac{\partial}{\partial t} \int \phi f \, do - \int f \frac{\partial \phi}{\partial t} \, do = \frac{\partial (n\bar{\phi})}{\partial t} - n \overline{\frac{\partial \phi}{\partial t}},$$

§12.7] ENSKOG'S SOLUTION 459

where

$$\bar{\phi} = \int \phi f \, do \bigg/ \int f \, do \quad \text{and} \quad n = \int f \, do,$$

the identification of $\partial_0 f^{(0)}/\partial t$, $\partial_1 f^{(0)}/\partial t$, etc. may be accomplished by calculating

$$\bar{\phi} = (1/n) \int f\phi \, do = (1/n) \int \sum_{r=0}^{\infty} f^{(r)} \phi \, do = \sum \bar{\phi}^{(r)},$$

where

$$\bar{\phi}^{(r)} = (1/n) \int \phi f^{(r)} \, dr$$

Similarly, $\partial \bar{\phi}/\partial t$ may be divided into parts $\partial_r \bar{\phi}/\partial t$ depending on $f^{(r)}$ only, and then $\partial f/\partial t$ may be expressed as

$$\frac{\partial f}{\partial t} = \sum_{i=1}^{5} \frac{\partial f}{\partial \psi^{(i)}} \frac{\partial \psi^{(i)}}{\partial t} = \sum_{i=1}^{5} \sum_{r=0}^{\infty} \frac{\partial f}{\partial \psi^{(i)}} \frac{\partial_r \psi^{(i)}}{\partial t} = \sum_{r=0}^{\infty} \frac{\partial_r f}{\partial t},$$

where

$$\frac{\partial_r f}{\partial t} = \sum_{i=1}^{5} \frac{\partial f}{\partial \psi^{(i)}} \frac{\partial_r \psi^{(i)}}{\partial t}$$

When applied to $\partial_0 f^{(0)}/\partial t$, this process leads to the result

$$\frac{\partial_0 f^{(0)}}{\partial t} + D(f^{(0)}) = f^{(0)} \left[\frac{1}{T(hm)^{1/2}} \frac{\partial T}{\partial x} \xi \left(\tau^2 - \frac{5}{2} \right) + \cdots \right.$$
$$\left. + 2 \left(\xi^2 - \frac{\tau^2}{3} \right) \frac{\partial u}{\partial x} + \cdots + 2 \xi \eta \left(\frac{\partial v}{\partial x} + \frac{\partial u}{\partial y} \right) + \cdots \right], \qquad (49)$$

where

$$\xi = (r_x - u)\sqrt{c}, \qquad \eta = (r_y - v)\sqrt{c}, \qquad \zeta = (r_z - w)\sqrt{c},$$
$$c = hm = m/2kT, \qquad \tau^2 = \xi^2 + \eta^2 + \zeta^2,$$
$$f^{(0)} = A \, e^{-\tau^2} = \rho T^{-\frac{3}{2}} \left(\frac{m}{8\pi^3 k^3} \right)^{\frac{1}{2}} e^{-\tau^2}$$

Equation (49) may now be compared with eq. (45), and $f^{(1)}$ may thus be divided into parts proportional to the temperature and velocity gradients. The rest of Enskog's calculation is similar to that of Chapman, and leads to the same expressions for the coefficients of viscosity and heat conductivity.

Notes for §12.7

1. D. Enskog, *Kinetische Theorie der Vorgänge in mässig verdünnten Gasen* (Uppsala: Almqvist & Wiksells Boktrycheri-A.B., 1917); English trans. in Brush, *Kinetic Theory* **3**, 125. For a sketch of Enskog's career see *ibid*, pp. 7–9, and Appendix to this chapter. Enskog attended Hilbert's lectures on kinetic theory in 1912–13 (J. Mehra, *The Physicist's Conception of Nature* (Boston: Reidel, 1973), p. 178.)
2. Chapman and Cowling, *Mathematical Theory of Non-Uniform Gases* (1952 ed.), p. 385; Enskog, *Arkiv Mat. Astr. Fysik* (Stockholm) **16** (16) (1922).
3. Enskog, *Kinetische Theorie*, p. 21; on p. 32 he says: "Wollen wir das Verhältnis dieser Approximationsmethode zur Hilbertschen angeben, so können wir sagen, dass die Glieder von Hilberts Entwicklung einer Kombination von sukzessiver Annäherung in bezug auf die Geschwindigkeitskomponenten mit einer ebensolchen in bezug auf die Zeit entsprechen, während diejenigen der Reihe (47) [same as eq. (42) in the text with $\lambda = 1$] eine von der Zeit nicht unmittelbar abhängige in bezug auf die Geschwindigkeitskomponenten darstellen. Da die Differentialgleichungen für ρ, T, u, v, w bei Berechnung einer endlichen Zahl von Gliedern nur angenähert sind, so stellt freilich auch unsere Methode nur eine Annäherung an die zeitlichen Verhältnisse dar, aber das oben gesagte gilt doch mit dem Grad von Genauigkeit, der überhapt erreicht wird. Berechnet man f in der n-ten Annäherung, so wird nach unserer Berechnungsweise $f^{(0)}$ in derselben Annäherung als Funktion der Dichte, Temperatur usw. des Gases mit der Zeit invariant, nach Hilberts Methode aber nur in erster Annäherung. Nennen wir die Konstanten in dem nach Hilberts Methode berechneten $f^{(0)}$: ρ', T', u', v', w', so ist in einem Volumenelement, das sich mit den Geschwindigkeitskomponenten u', v', w' bewegt $\rho' T'^{-\frac{3}{2}}$ konstant, $\rho T^{-\frac{3}{2}}$ kann sich wiederum durch Wärmeleitung oder Reibung sehr wohl auf das hundert-, tausendfache usw. vergrössern. Es ist dann klar, dass dieses $f^{(0)}$ im allgemeinen gar keine Annäherung darstellen kann. Will man ein gutes Resultat haben, so muss man jedenfalls viel mehr Glieder berechnen alsnach der obigen Methode." For English trans. of this passage see Brush, *Kinetic Theory* **3**, 143–44.
4. L. Boltzmann, *Lectures on Gas Theory*, p. 139.

12.8 Viscosity of dense gases

While the preceding results of Chapman and Enskog depend on the molecular diameter, they are actually valid only at low densities, and correspond to the ideal gas approximation in which the second and higher virial coefficients are ignored. As early as 1899, Gustav Jäger proposed a method for expressing the viscosity of a dense gas in terms of the viscosity at low densities.[1] He proceeded by analogy, arguing that just as the ideal gas formula,

$$pv = RT$$

becomes

$$pv = RT(1 + 4b/v + 10b^2/v^2 + \ldots)$$

when the effect of finite molecular size is taken into account, the dilute gas viscosity formula

$$\eta = k\rho cL$$

(c = mean velocity, k = a numerical constant equal to $\frac{1}{3}$ in the elementary theory, but which Jäger considered should be $\frac{5}{12}$) could be modified in the same way,

$$\eta = k\rho cL(1 + 4b/v + 10b^2/v^2 + \ldots)$$

In addition to this change, Jäger believed that the mean free path should also be corrected because of the effect of excluded volumes on the collision rate; this corresponds to taking account of the third virial coefficient in the equation of state. Thus

$$L = \frac{1}{N\pi d^2(1 + 5b/2v + \ldots)} \frac{c}{r} = \frac{L_0}{A}$$

where c/r is the ratio of the mean absolute velocity to the mean relative velocity of two colliding molecules, denoted by c in eq. (2)

$$L_0 = \frac{c}{N\pi d^2 r}$$

and

$$A = 1 + 5b/2v + \ldots = \frac{(pv/RT) - 1}{4b/v}$$

The viscosity is therefore, neglecting second-order terms in b/v,

$$\eta = \eta_0(1 + 3b/2v)$$

In another paper,[2] Jäger argued that the factor L in the formula $\eta = k\rho cL$ should be replaced by $L - \alpha d$, where α is some fraction of the diameter of a molecule which he calculated as $\frac{1}{2}$. Writing

$$\eta = k\rho c(L + \alpha d)(1 + 4bA/v),$$

he then obtained

$$\eta = \eta_0 \frac{1}{A} + \frac{4b}{v} + \frac{6br\alpha}{cv} + \frac{24b^2 A\alpha r}{cv^2}$$

Using Clausius' value $\frac{4}{3}$ for the ratio r/c, and $\alpha = \frac{1}{2}$, he found

$$\eta = \eta_0 \frac{1}{A} + \frac{8b}{v} + \frac{16Ab^2}{v^2} \tag{50}$$

This formula is remarkable chiefly in its resemblance to one derived in 1921 by Enskog, using more rigorous methods[3]

$$\eta = \eta_0 \frac{1}{A} + \frac{16b}{5v} + (0.7614)\frac{16Ab^2}{v^2} \tag{51}$$

Enskog's derivation of eq. (51) was an extension of his method discussed in §12.6; like Jager, he assumed that the collision rate in a dense gas would be changed by a factor A which could be related to the equation of state. He modified Boltzmann's equation by replacing the product ff_1 in the collision integral, in which both f and f_1 are evaluated at the same point, by $f(x, y, z, \ldots)f_1(x_1, y_1, z_1, \ldots)$, where the points (x, y, z) and (x_1, y_1, x_1) must be separated by the distance d. In this way he could take account of the transport of momentum across finite distances in collisions, which is important in dense gases and liquids, as well as the transport of momentum by the motion of molecules from one part of the gas to another.

With the development of modern methods in transport theory after 1945, several attempts were made to "improve" Enskog's theory of dense gases by systematic calculation of correction terms analogous to virial coefficients in the equation of state. These attempts were generally unsuccessful until 1965 when it was discovered that no such "density expansion" of transport coefficients in simple powers of density actually exists. Intricate and laborious calculations done since then have shown that Enskog's result, eq. (51), is accurate to within a few percent even at high densities.[4]

Notes for §12.8

1. G. Jäger, *Wien. Ber.* **108**, 447 (1899); see also his earlier papers, *Wien. Ber.* **102**, 253 (1893), **105**, 15, 97 (1896).
2. *Wien. Ber.* **109**, 74 (1900); see also *Wien. Ber.* **111**, 697 (1902) and M. Brillouin, *Leçons sur la Viscosité des Liquides et des Gaz* (Paris: Gauthier-Villars, 1907), Seconde Partie, p. 119.
3. D. Enskog, *Kungliga Svenska Vetenskapsakademiens Handlingar*, Ny Följd, **63** (4) (1922), English trans. in Brush, *Kinetic Theory* 3, 226. Chapman and Cowling, *Mathematical Theory of Non-Uniform Gases*, ch. 16.
4. Brush, *Kinetic Theory* **3**, 59–80.

Appendix 12.A: Biographies of Chapman and Enskog

While book-length biographies of Maxwell and Boltzmann are available, information about Chapman and Enskog is not as accessible. We therefore include here a profile of Chapman which was published in the London *Observer*, 7 July 1957; and a biography of Enskog by Hilding Faxén translated from the *Svenskt Biografiskt Lexikon* 13, 765 (Stockholm, Albert Bonnier, 1950).

Profile of Sydney Chapman

Scientific research is increasingly done by teams. The days of towering individuals who dominate whole tracts of scientific territory on their own, seem to be passing.

At the same time, scientists have been becoming respectable, even *bourgeois*. They sport discreet dark suits, are skilful and confident on committees; they have secretaries and efficient filing systems. There are still absent-minded professors with egg on their waistcoats, but they are more usually classicists than physicists.

This is why Professor Sydney Chapman, President of the Commission for the International Geophysical Year, and one of the most senior and distinguished of geophysicists, is something of a rarity.

He is an individualist, who has never been associated with scientific empires or famous schools. But he has taken large and far-reaching scientific problems and made them his own. His books remain standard works for many years and will always be landmarks in their field.

He is also undoubtedly an eccentric–in a wholly charming way. He arrives at Royal Society soirées in full evening dress, riding a bicycle. His greatest books have been written on the backs of old examination papers. He is intensely shy, and ill at ease with formality. This perhaps explains why he often seems more relaxed and forthcoming in the company of Americans.

Physically, he reminds one slightly of Popeye the Sailor. He is wiry, tough, with a determined jaw and considerable physical stamina. (At one geophysical congress in Rome, over which he was presiding, he would walk each day through the broiling streets from his hotel up to the congress center five miles away.)

But, unlike Popeye, who displays no symptoms of cerebral activity, Chapman possesses remarkable mathematical abilities, and matches his physical endurance with astonishing powers of concentration and mental stamina. He can be an exhausting person to work with, and thinks nothing of discussing a scientific problem for five or six hours at a stretch.

The pattern of his career has been conventional, providing no obvious clue to his odd personality. Born at Eccles in 1888, son of a clerical worker, he went with scholarships to Salford Technical School, Manchester University, and Cambridge, where he became a wrangler in his second year. There, returning one wet Sunday afternoon to his rooms, he found a visitor who said, "I'm the Astronomer Royal," and offered him a job at Greenwich Observatory.

His interest in geophysical problems was encouraged at Greenwich where his task was to build up a new magnetic observatory and bring the work on the earth's magnetic field up to date. From Greenwich, he went to Manchester, and then moved to Imperial College as head of the mathematics department, where he remained for 22 years until 1946. Then he became a professor at Oxford until he reached retiring age in 1953.

During the First World War he was a member of the I.L.P. and a conscientious objector, and spent some time working in a boys' club in Bermondsey. Hitler, he says, converted him from pacifism, and for part of the last war he was a scientific adviser to the Army Council.

One of Chapman's principal scientific achievements was to show that the moon causes regular tides in the earth's atmosphere which can be detected by a minute analysis of changes in barometric pressure. Among his current interests are the similar tides which the moon produces in the earth's crust in the form of minute deformations. Measurements of these crustal tides will be made during the I.G.Y.

He is a world authority on geomagnetism, and has done important work on the upper atmosphere and aurorae. One of his major works turned out to be of vital practical significance: a massive study on the behavior of gases culminated in 1939 in a book published jointly with T. G. Cowling, called *The Mathematical Theory of Non-Uniform Gases*. It is a standard work on the subject, and very heavy going— "like chewing glass," says Chapman. Yet the "gaseous diffusion" process used during the war to separate fissile uranium-235 for atomic bombs from natural uranium was based on this work.

But his restless, energetic and original personality seems to have been given full rein only after his retirement from Oxford. As a professor, his scientific distinction was not matched by his distinction as a teacher. Students found that his shyness made him difficult to approach, and his lectures were not always remarkable for their audibility. He might have retired, like many professors, into a dim limbo of sporadic committees and gentle bumbledom.

Instead, he burst forth from his academic cocoon as a sort of scientific nomad of considerable splendour. Together with his wife – the four children of the marriage are now grown up – he divides his time between a house in Alaska, where he is advisory director of the Geophysical Institute, a house 5000 feet up in Boulder, Colorado, where he is associated with the high altitude observatory, and Europe.

As President of the I.G.Y. Commission he has had to travel considerably. He was recently in Brussels, is now in Colorado, and plans to visit Teheran later this year. He appears to be having the time of his life. He still bicycles when he can – he reckons to have bicycled in every continent of the world – and he walks, swims, and enjoys music. He is much appreciated in America for his simplicity and enthusiasm, his interest in the work of other people as much as his own, and, one may suspect, for his mild eccentricities. It was during some informal conversations with a group of young American scientists that the idea of the International Geophysical Year was born.

He is still very active scientifically, pursuing his own research and guiding that of others. But he has achieved an unfettered freedom which suits him. Brown as a nut, and still shy, he will never become a formidable scientific Grand Old Man. His scientific work will always be a landmark in geophysics. But his charm and odd personality will be remembered by the layman in association with the I.G.Y.

This huge international scientific effort means a lot to him, not only for what it will achieve scientifically, but because he cares deeply for the international nature of science, and the cooperation of scientists. Because he has not a trace of pomposity, and his own particular brand of charm, he has got on excellently with scientists of all nations. He was a good – indeed the obvious – person to choose for the job.

[Reprinted by permission of The Observer, London.]

Biography of David Enskog

ENSKOG, David, born 22 April 1884, in Västra Ämtervik, Värmland, Sweden. Died 1 June 1947 in Sankt Eriks Hospital, Stockholm. Parents: Nils Olsson, a preacher, and Karolina Jonasdotter. Entered the Karlstads läroverk [high school] in 1900, where he took his "maturity examination" 8 June 1903. Matriculation examination at

Uppsala University the same year; "philosophy candidate" [B.A.] 31 January 1907; "philosophy licentiate" [M.A.] 15 September 1911; doctoral dissertation presented 14 April 1917, received Ph.D. 31 May 1917. Deputy secondary school teacher in Karlstad, 1 September to 19 October 1907; in Stockholm, 1911–12; extraordinary lectureship in Skövde starting in 1913; lectureship in Gävle, starting 26 April 1918 (in mathematics and physics). He held a traveling fellowship at Göttingen and Munich (Letterstedtsk stipend) in 1922–23; appointed to a lectureship at Norrmalm, 13 September 1929; Professor in mathematics and mechanics at the Royal Institute of Technology, Stockholm, 12 December 1930; censor at the matriculation examinations from 1931; teacher of mathematics and mechanics at the Navy school, 1933–39; chairman of the Swedish national committee for physics, 1940. Served on various other committees. Received the Wallmarks prize in 1928, and the Svante Arrhenius gold medal in 1946.

He was married to Anna Aurora Jönsson, 27 December 1913, in Karlstad; she was born 12 January 1886, in Fryksande (Värmland), the daughter of Johan Jönsson, a farmer, and Karin Pettersson.

Enskog was the son of a well-liked preacher in Värmland. After grade school and three years at the Karlstads läroverk, he passed his matriculation examination with high marks. At Uppsala he studied very hard, and took the examinations in mathematics, mechanics, astronomy, chemistry, and philosophy, with further studies in physics and geography. The year after his doctoral dissertation (1917) he was a lecturer in mathematics and physics in Gävle. After a short time as lecturer at Norrmalm, he was nominated to the professorship of mathematics and mechanics at the Royal Institute of Technology in Stockholm (1930).

Enskog's first great scientific discovery was thermal diffusion (1912). If a mixture of gases is placed in a vessel, of which one part is colder than the rest, then one of the gases will collect in the colder part and the other in the warmer. This theoretical result was tested a few years later. Independently of Enskog's work, the same phenomenon was predicted by Chapman in England, and confirmed experimentally by Dootson in 1916. It became one of the industrially useful methods for separating gases which cannot be separated by chemical methods, such as different isotopes of the same element.

A major part of Enskog's scientific work concerned viscosity, heat conduction, and diffusion in gases and liquids. By the beginning of this century it had been proved that gases and liquids consist of rapidly moving small particles, called molecules. With the help of this knowledge one should be able to explain viscosity, heat conductivity, and diffusion, and by comparing theory with experiment one could find the forces which act between molecules when they collide. This problem had been formulated mathematically during the period 1860–80 by J. C. Maxwell, at Cambridge, and L. Boltzmann, who taught at Austrian and German universities. One could write down an integro-differential equation which could only be solved with unrealistic assumptions about the intermolecular forces. Many scientists were trying to find the general solution of this equation, but it was first solved by S. Chapman and Enskog, who simultaneously and independently found a physically valid solution. Though they used different methods, their results were the same. The methods were in both cases ones of successive approximation. Thus Enskog solved the gas theoretical problem. At the same time, however, it was difficult to judge the value of his series expansion because he had not proved, any more than his predecessors, the convergence of the series. One had to be content with establishing that the results agreed with experiment in certain important special cases, until Burnett in 1939 proved the convergence of the series. Enskog himself turned his attention to the theory of liquids and compressed gases, and used his method to explain how the viscosity decreases with temperature in a liquid, in contrast to the behavior of dilute gases. Chapman found Enskog's method preferable to his own and used it as the basis for his comprehensive book (1939, see "sources") which he presented to Enskog.

Enskog also developed his methods in a more mathematical direction to study the solution of Fredholm integral equations. By Fredholm's own method, one obtains the solution as a quotient of two infinite determinants. Thus it is possible to get a mathematical proof, but one can seldom succeed in obtaining a practical solution of the equation. Apart from this procedure, developed by Schmidt, Enskog's method was the first to lead to practically useful results, and Enskog showed how his general theory of the integral equation can also yield Schmidt's method.

Before the introduction of modern quantum mechanics, Enskog's contributions were of great interest, but then this part of science rapidly developed in a different direction. Enskog also improved G. G. Stokes's formula for sound absorption in gases and liquids.

Enskog was very thoroughgoing as a teacher and examiner. In his scientific work he was primarily a theorist, but he was also interested to some extent in experimental work. He was an honest and unassuming person, quiet and considerate in manner.

Letters from Enskog may be found in the library of the Vetenskapsakademiens (Academy of Science).

Published works: see the "List of publications of the staff of the Royal Institute of Technology, Stockholm" [Kgl. Tekniska högskolans handl. **1**, 64–650 (1947)].

Sources: Västra Ämterviks town birth record-book for 1884; Fryksande town birth record-book, 1886; death book of Bromma, Stockholm, for 1947. S. Chapman and T. G. Cowling, *The Mathematical Theory of Non-Uniform Gases* (Cambridge, 1939), historical summary, pp. 380–88; K. Kärre and N. Svartholm, "David Enskog" [Kosmos, Fysiska uppsatser utg. av Sv. fysikersamf., **26** (1948)]; Lärarmatrikeln 1922 and 1934; personal communications.

[Translated by permission of Svenskt Biografiskt Lexikon and Albert Bonniers Förlag, Stockholm.]

CHAPTER 13

Heat Conduction and the Stefan–Boltzmann Law*

13.1 Introduction

At the end of the 18th century not many physicists would have thought it worthwhile to investigate the question: Why do you often burn your mouth when you eat a piece of hot apple pie? For that matter, not many physicists were earning their living by designing winter clothing and developing nutritious soups for soldiers and paupers. Only the notorious American traitor Benjamin Thompson, Count Rumford of the Holy Roman Empire, was persistent enough in studying and publicizing the various modes of heat transfer to make this subject one of the major scientific concerns of the next century.[1]

My own interest in 19th-century research on heat transfer is restricted to two main themes: the development and testing of a molecular explanation of heat conduction in gases, and the origin of the fourth power relation between radiation and temperature. The former is part of the history of the kinetic theory of gases, discussed in earlier chapters; the latter provides the basis for Planck's work on black-body radiation, which led to his quantum hypothesis.[3] I hope to demonstrate that these two themes are much more closely related than has usually been recognized in discussion of the history of modern physics.

---
* Reprinted with minor revisions from *Archive for History of Exact Sciences* 11, 38–96 (1973) by permission of Springer-Verlag.

Just as it turns out to be necessary to coordinate our discussions of heat conduction and radiation, we will also have to mention convection (though without making any attempt at a comprehensive review of that topic). In both cases the reason is that the conceptual distinction between the three modes of heat transfer, accepted in elementary texts for more than a century, covers up the fact that in the real world it is exceptionally difficult to separate them. Whatever success has been achieved in doing so owes much to a complicated interplay between theory and experiment, occurring mainly in the second half of the 19th century; this interplay is central to our subject. Another misconception fostered by some modern physics textbooks is that the black-body radiation law belongs to "equilibrium statistical mechanics" whereas heat conduction is dealt with under the heading of "transport theory" – the "transfer" aspect of radiation being thus subordinated to its role in the development of quantum theory.

* * *

For 19th-century scientists the subject of heat transfer, like all other branches of physics, had to begin with Newton. Unfortunately Newton's writings on this subject form such a tiny part of his entire output that historians have done very little to illuminate their origins and development, so that while "Newton's law of cooling" seems to be very important in the history of heat transfer, it pales into insignificance besides his work on mechanics, gravity, optics, and the calculus. An exception is the recent paper of Ruffner, which does contain a careful analysis of the genesis of the law of cooling, emphasizing Newton's attempts to extend existing temperature scales to higher temperatures, and noting the relation of this work to Newton's discussion of comets in the *Principia*.[3]

Newton's paper "Scala graduum Caloris," published in 1701,[4] presented a scale of temperatures ("degrees of heat") which he had determined by using a linseed-oil thermometer and a piece of hot iron. He estimated the temperature of the iron by assuming that:

> the heat which hot iron, in a determinate time, communicates to cold bodies near it, that is, the heat which the iron loses in a certain time, is as the whole heat of the iron; and therefore, if equal times of cooling be taken, the degrees of heat will be in geometrical proportion, and therefore easily found by the tables of logarithms.

That is *not* Newton's Law of Cooling; in fact it is inconsistent with another statement found in the same paper, that the values of the "excess of the degrees of heat of the iron...above the heat of the atmosphere...were in geometrical progression, when the times are in an arithmetical progression." This second statement appears to mean that the rate of cooling is proportional to the *difference* between the heat (temperature) of a body and that of its surroundings, and indeed that is the interpretation adopted by later scientists. Ruffner says that this discrepancy is "a minor point" but I think it is worth suggesting that Newton might well have intended a distinction between the total heat emitted (which would depend only on the temperature of the body itself) and the net heat lost (after subtracting the heat *received* from surrounding bodies at a rate proportional to *their* temperature). This interpretation would eliminate the inconsistency between the two statements and make it a little more plausible why the difference of temperatures appears in the Law of Cooling; but Newton's published paper does not make this distinction clear.

It should also be noted that Newton did not distinguish between different modes of heat transfer. He did specify that the hot iron "was laid not in a calm air but in a wind that blew uniformly on it, that the air heated by the iron might be always carried off by the wind, and the cold air succeed it alternately." Under such conditions, if the temperature was not very high, we can guess that convection would be more important than radiation, and conduction would be negligible. It is hardly fair for later investigators to claim that they have "disproved Newton's law of cooling" unless they have conducted experiments, as he did, with a steady breeze.[5]

Finally, one might almost say that Newton used his Law of Cooling as an axiom to construct his temperature scale, rather than claiming it to be an experimental result; though he did state that the results are consistent with those obtained from the ordinary linseed-oil thermometer.

In the 18th century several experimenters announced that they had found deviations from Newton's law of cooling.[6] The question acquired further interest in connection with speculations on geological history, the earth being regarded as a hot gaseous mass which has been gradually cooling down, condensing to a liquid and partially solidifying. Buffon, inspired perhaps by some remarks on the subject in Newton's *Principia*,[7] conducted a series of experiments on the cooling of iron spheres of various sizes, extrapolating to a sphere the size of the earth.[8] This geophysical problem in turn stimulated the researches of J. B. J. Fourier on heat conduction in solids, with the accompanying question of heat transfer through the atmosphere being subsidiary but not forgotten.[9]

* * *

Another of Newton's experiments on the cooling of bodies raised the question of whether the transfer of heat is due to the surrounding air or could take place at least to some extent in the absence of air. In Query 18 of the *Opticks* he wrote:

> If in two large tall cylindrical Vessels of Glass inverted, two little Thermometers be suspended so as not to touch the Vessels, and the Air be drawn out of one of these Vessels, and these Vessels thus prepared be carried out of a cold place into a warm one; the Thermometer *in vacuo* will grow cold almost as soon as the other Thermometer. Is not the Heat of the warm Room convey'd through the *Vacuum* by the Vibrations of a much subtiler medium than Air, which after the Air was drawn out remained in the *Vacuum*? And is not this Medium the same with that Medium by which Light is refracted and reflected, and by whose Vibrations Light communicates Heat to Bodies, and is put into Fits of easy Reflexion and easy Transmission? And do not the Vibrations of this Medium in hot Bodies contribute to the intenseness and duration of their Heat? And do not hot Bodies communicate their Heat to contiguous cold ones, by the Vibrations of this Medium propagated from them into the cold ones? And is not this Medium exceedingly more rare and subtile than the Air, and exceedingly more elastick and active? And doth it not readily pervade all Bodies? And is it not (by its elastick force) expanded through all the Heavens?

It was hard to say "no" to Sir Isaac.

Late in the 18th century, and continuing through the 19th, there was considerable interest in the phenomena of radiant heat.[10] Most of the early investigators believed that radiation could be transmitted through a vacuum in the absence of any matter. In investigating the cooling of bodies, it was therefore necessary to distinguish between that part of the cooling which was due purely to radiation, and that part which was due to conduction or convection by the surrounding gas. The simplest assumption would be that these parts can somehow be determined separately and then added together to get the total heat transfer.

As Newton had suggested (in the above quotation) one might try to measure the heat transfer due to radiation by evacuating the air from the system, thus eliminating conduction and convection. As we will see later on, this procedure involves two assumptions whose inaccuracy was not clearly recognized by the early investigators: first, that a sufficiently low pressure has been attained so that conduction and convection really are negligible; second, that the radiation part of the heat transfer is unaffected by the presence of a gas (*i.e.* that there is no appreciable absorption of radiation by the gas).

Convection as a mode of heat transfer associated with observable motion of gas was discovered by Carlo Rinaldini in 1657, according to W. E. K. Middleton.[11] It was discussed in the meteorological context in the 18th century by Franklin, Lambert, Ducarla-Bonifas, and de Saussure,[11] and a mathematical theory of convection in fluids was formulated by Euler in 1764.[12] Rumford discovered convection currents in liquids in his "hot apple pie" paper published in 1797; he was primarily interested in the negative aspects of heat transfer, *i.e.* what *prevents* heat from flowing through a coat containing many small cavities, or through a concoction of stewed apples or soup? His argument was that when the fluid–air or water–was obstructed by a network of solid barriers in the fabric or the pastry, then there is no other way for the heat to travel, hence it stays where it is for quite a long time.[13]

At this point we see that the problem of distinguishing radiation from conduction and convection has been obscured by the problem of distinguishing conduction from convection: the thrust of Rumford's argument was understood to be that fluids do not *conduct* heat; they can only transfer it by convection, or perhaps by radiation. But this distinction depends on theoretical ideas prevalent at the end of the 18th century: Rumford, though an adherent of the doctrine that heat is molecular motion, rather than a substance, did not advocate a *kinetic* theory of gases in the modern sense. He did not think of the air molecules as moving sizable distances from one place to another, but rather as vibrating around more or less fixed equilibrium positions. On the other hand, any transfer of heat from one molecule to its neighbor would be defined, in his scheme, as "conduction" rather than "radiation." Here is how Rumford expresses himself:

> though the particles of air individually, or each for itself, are capable of receiving and transporting heat, yet air in a quiescent state, or as a fluid whose parts are at rest with respect to each other, is not capable of conducting it, or giving it a passage; in

short, that heat is not capable of *passing through a mass of air*, penetrating from one particle of it to another, and that it is to this circumstance that its non-conducting power is principally owing.[14]

In a later paper he wrote:

> ... air is a *non-conductor* of Heat.... Heat cannot pass through it without being transported by its particles, which, in this process, act individually or independently of each other.[15]

Rumford's distinction between heat transfer by conduction and by particle transport depends on his static theory of gas structure, and becomes almost meaningless as soon as one adopts the kinetic theory, since conduction of heat in gases is then attributed to a process in which particle transport rather than interparticle radiation plays the essential role.[16]

The first phase of the attempt to establish the concept of thermal conductivity in gases as distinct from radiation and convection has been described in a valuable article by A. C. Burr.[17] But I do not think Burr has fully appreciated the extent to which this distinction is affected by the changing theoretical ideas about the nature of gases and of heat in the first half of the 19th century. Thus he refuses to consider experiments in which the "law of cooling" was studied with a thermometer surrounded by a gas, with no attempt to eliminate convection, since he does not think these experiments were relevant to the establishment of thermal conductivity as a distinct property of gases. His account is inadequate insofar as the subject cannot really be so narrowly defined—such experiments did play an important part in the development of the subject.

The quotations from Rumford's early papers given by Burr convey the impression that Rumford concluded that the amount of heat transferred through a vacuum by radiation is negligible compared to that transferred through air by convection. However, in an experiment conducted in 1786, Rumford mounted a thermometer inside a globe which was then evacuated by draining mercury from it; he measured the time it took for the thermometer to register a change in external temperature, and found that it was not much longer than that for a thermometer in a similar air-filled globe.[18] Further evidence of the quantitative importance of radiation was presented in his "Experiments on cooling bodies," published in 1804. Here Rumford reported that when two bottles were filled with boiling water, one bottle having thick glass sides and the other thin tin sides, the glass bottle cooled down almost twice as fast. Conclusion:

If we admit the hypothesis that hot bodies are cooled, not by losing or acquiring some material substance, but by the action of colder surrounding bodies, communicated by undulations or radiations excited in an etherial fluid, the results of this experiment may be easily explained.[19]

The tin would presumably retard the cooling, by acting as a reflector for radiation.

The last quotation hints at another aspect of Rumford's conception of heat: he believed in "frigorific" (cold) radiation as well as heat radiation.[20] But that idea was not very popular in the 19th century, and Rumford's use of it may have further confused later readers of his papers.

According to Burr, Rumford's conclusion that fluids do not conduct heat was generally accepted until the work of Magnus in 1860. As evidence Burr offers little else than a quotation from Robison.[21] While it is true that Humphry Davy and Thomas Young agreed with Rumford that heat transfer is effected "principally" or "almost entirely" by the motion of particles,[22] Rumford's ideas on heat transfer were disputed by a number of distinguished scientists in the first decades of the 19th century.[23] Of these, John Leslie is of special importance to our subject because he conducted a series of original investigations himself; the relation between Rumford and Leslie has recently been discussed in some detail by Richard Olson.[24] We give a summary of Leslie's work in Appendix 13A.

In opposition to Rumford's view that heat could be transmitted through a vacuum by ether waves, Leslie rejected the ether and claimed that heat could never pass through a completely empty space. He suggested that Rumford and others had not achieved a perfect vacuum, and performed a series of experiments to show that the rate of cooling of a thermometer decreases as the pressure is reduced; he asserted that if the air surrounding a body were completely removed, no heat could be lost at all.[25] Leslie argued further, against Rumford (and William Herschel), that heat radiation involves some combination of light with particles of ordinary matter, and thus, for example, cannot pass through solids except perhaps by some process of absorption and reemission. But he distinguished between pulsating motions of air—periodic expansions and contractions—as a means of transmitting radiant heat from one solid body to another, and an overall mass motion of air carrying heat by the process later known as convection.

Later writers rejected almost unanimously Leslie's view that heat cannot pass through a vacuum; indeed, the theory that radiant heat is transmitted, like light, by ether waves, played a significant role is assisting the transition from the caloric theory to thermodynamics between 1830 and 1845.[26] Yet the type of experiment on which Leslie had based his conclusion was repeated 60 years later by William Crookes, using presumably much better vacuum apparatus, with similar results: the rate of heat transfer was found to decrease sharply at very low pressures.[27]

In reading what scientists said about radiant heat transfer in the first half of the 19th century (and earlier), we must remember that what they considered to be experiments *in vacuo* were usually done at pressures large enough that the effects of residual air were not at all negligible. It was not until the publication of Maxwell's kinetic theory in 1860 that it was generally realized that gases might conduct heat at low densities as well as they do at ordinary ones. This will be one of the main points advanced in this chapter: that Maxwell's theory forced a revision of the conventional interpretation of heat-transfer experiments, thereby assisting the discovery of the Stefan–Boltzmann radiation law.

Notes for §13.1

1. S. C. Brown, *Benjamin Thompson – Count Rumford, Count Rumford on the Nature of Heat* (New York and Oxford: Pergamon Press, 1967); see especially the paper reprinted on pp. 28–51. Brown, *Count Rumford: Physicist Extraordinary* (Garden City, N.Y.: Doubleday, 1962).
2. See §13.7, note 32.
3. J. A. Ruffner, *Arch. Hist. Exact. Sci.* **2**, 138 (1963). Other historical discussions may be found in E. Mach, *Principien der Wärmelehre* (Leipzig, 1896, 4th ed., 1923); D. S. L. Cardwell, *From Watt to Clausius* (Ithaca: Cornell University Press, 1971), p. 18.
4. *Phil. Trans.* **22**, 824 (1701); English trans. in abr. ed. of *Phil. Trans.* (London, 1809), **4**, 572; both reprinted in ed. I. B. Cohen, *Isaac Newton Papers & Letters on Natural Philosophy* (Cambridge, Mass.: Harvard University Press, 1958), pp. 259–68. On knowledge of Newton's authorship of this paper see W. E. K. Middleton, *A History of the Thermometer* (Baltimore: Johns Hopkins Press, 1966), 57–58.
5. J. Black, *Lectures on the Elements of Chemistry* (Philadelphia, 1807), p. 84; A. C. Mitchell, *Trans. R. S. Edinburgh* **40**, 39 (1899); J. R. Partington, *An Advanced Treatise on Physical Chemistry* (London: Longmans, 1952), **3**, 268.
6. G. Martine, "Essay on the heating and cooling of bodies" (St. Andrews, 1739), reprinted in his *Essays and Observations on the construction and graduation of thermometers, and on the Heating and Cooling of Bodies* (Edinburgh, 4th ed., 1787), pp. 51–87; G. W. Richmann, *Novi Commentarii Acad. Sci. Imper. Petropol.* **1**, 174 (1948); J. C. P. Erxleben, *Novi Commentarii Societatis Regiae Scientiarum Gottingensis* **8**, 74 (1777).
7. *Sir Isaac Newton's Mathematical Principles of Natural Philosophy and his System*

§13.1]   INTRODUCTION   477

       of the World, A. Motte's translation revised by F. Cajori (Berkeley: University of California Press, 1934), p. 522.
8. Buffon, *Introduction à l'Histoire des Minéraux* (1774); see *Oeuvres Complètes de Buffon*, ed. M. Flourens, nouv. ed. (Paris: Garnier Frères, n.d.), **9**, 82, 89, 307–8, 348–453.
9. See §2 of the following chapter. Fourier extended his theory of heat conduction to incompressible fluids in a brief paper read to the Académie des Sciences in 1820, published posthumously in *Mém. Acad. Sci.* **12**, 507 (1833), reprinted in *Oeuvres de Fourier*, ed. G. Darboux (Paris, 1888–90), **2**, 595.
10. See the surveys by E. S. Barr, *Infrared Physics* **1**, 1 (1961), **2**, 67 (1962), **3**, 195 (1963); *Am. J. Phys.* **28**, 42 (1960); E. S. Cornell, *Ann. Sci.* **1**, 217 (1936); S. P. Langley, *Pop. Sci. Mon.* **34**, 212, 385 (1889); A. Wolf, *A History of Science, Technology, and Philosophy in the 18th century*, 2nd ed. (London and New York: Macmillan, 1952), **1**, 206–12. For an early (1777) definition of radiant heat as distinct from convection, see *The Collected Papers of Carl Wilhelm Scheele* (London: Bell, 1931), p. 124.
11. W. E. Knowles Middleton, *Physis* **10**, 299 (1968).
12. C. Truesdell, *Rational Fluid Mechanics 1687–1765*, Editor's Introduction to vol. II.12 of Euler's *Works* (Zürich: Orell Füssli, 1954), CVII.
13. Rumford, "The Propagation of Heat in Fluids" (1797), reprinted in Brown, *Benjamin Thompson* (*op. cit.*, footnote 1).
14. *Phil. Trans.* **76**, 273 (1786), **82**, 48 (1792); quotation from Rumford's *Collected Works*, ed. S. C. Brown (Cambridge, Mass.: Harvard University Press, 1968), **1**, 109.
15. Rumford, "The Propagation of Heat in Fluids," in Brown, *Benjamin Thompson*, p. 29, or Rumford's *Works* **1**, 121.
16. The conflict between Rumford's ideas about heat transfer and the kinetic theory (which he is popularly supposed to have accepted) has been discussed by D. S. L. Cardwell, *Mem. Manchester Lit. Phil. Soc.* **106**, 108 (1963–64); see also his *From Watt to Clausius* (Ithaca, N.Y.: Cornell University Press, 1971), p. 100.
17. A. C. Burr, *Isis* **21**, 169 (1934); see also his earlier paper on the concept of thermal conductivity, *Isis*, **20**, 246 (1933). On heat conduction in liquids see C. Chree, *Phil. Mag.* [5] **24**, 1 (1887).
18. Rumford, *Phil. Trans.* **76**, 273 (1788); Brown, *Benjamin Thompson*, p. 124.
19. Rumford, *Nicholson's J. Nat. Phil.* **12**, 70 (1805); *Collected Works*, ed. Brown, **2**, 1.
20. See Brown, *Benjamin Thompson*, pp. 202–5.
21. Burr, *op. cit.*, p. 175; J. Robison, note on p. 339 of his ed. of J. Black's *Lectures on the Elements of Chemistry* (Philadelphia, 1807).
22. H. Davy, *Phil. Trans.* **107**, pt. I, 44 (1817), esp. p. 61; T. Young, *A Course of Lectures on Natural Philosophy and the Mechanical Arts* (London, 1807; New York: Johnson Reprint, 1971), **1**, 635.
23. J. Dalton, *Mem. Manchester Lit. Phil. Soc.* **5**, 373 (1802, read 1799); Berthollet, *Nicholson's J. Nat. Phil.* **8**, 198 (1804); T. Thomson, *Nicholson's J. Nat. Phil.* **4**, 529 (1801); J. Murray, *Nicholson's J. Nat. Phil.* **1**, 165, 241 (1802); J. B. Biot, *Bull. Soc. Philomath. Paris* **3**, 36 (1801); J. A. Deluc, *Ann. Phys.* **1**, 464 (1799); W. Henry, *The Elements of Experimental Chemistry* (London, 11th ed., 1829), p. 112; T. S. Traill, *Encyclopedia Britannica* (7th ed.) **11**, 180 (1842), art. I.2.
24. R. Olson, *Ann. Sci.* **26**, 273 (1970). Leslie's major work was *An Experimental Inquiry into the Nature and Propagation of Heat* (London, 1804).
25. J. Leslie, *Ann. Phil.* **14**, 6 (1819); see Olson, *op. cit.*
26. See ch. 9.
27. W. Crookes, *Proc. R. S. London* **31**, 239 (1881).

13.2 The Dulong–Petit law of cooling

In his *New System of Chemical Philosophy* (1808), John Dalton proposed a new temperature scale, based on the assumption that the expansion of water or mercury is proportional to the square of the temperature measured to its freezing point.[1] According to this new scale, he claimed, the discrepancies from Newton's law of cooling discovered by Leslie and others would disappear. While Dalton was not very successful as a constructor of temperature scales,[2] he did have the happy inspiration to try to measure the quantitative variation of cooling rate with gas density. From his own experiments he concluded that the cooling effect of air is proportional to the cube of its density; he proposed to represent the time of cooling by the formula

$$t = \frac{1}{0.004 + 0.006 d^{\frac{1}{3}}}$$

where the time of cooling in air at atmospheric pressure is taken as 100.[2]

While Dalton's investigation was not as extensive as Leslie's (see Appendix 13A), it probably attracted more attention. But the validity of Newton's law of cooling remained in doubt for another decade.[3] Finally – at least for the next generation of scientists – the question was settled by an elaborate series of experiments conducted by P. L. Dulong and A. T. Petit.[4] Their memoir was awarded a prize by the Académie des Sciences in 1818, and the new law of cooling which they proposed was generally accepted throughout most of the scientific world for the next half-century. A full century later a respected textbook asserted that the Dulong–Petit law "seems to apply with considerable accuracy through a much wider range of temperature differences than that of Newton" and is based on "one of the most elaborate series of experiments ever conducted."[5] Nevertheless it has now been so thoroughly cast into obscurity that the phrase "law of Dulong and Petit" means something entirely different to the modern scientist, namely an approximate relation between specific heats and molecular weights.[6]

Dulong and Petit presented in the first part of their memoir a detailed study of the temperature scale, in which they strongly criticized Dalton's experiments and interpretations. Turning to the problem of the law of cooling, they pointed out that no "empirical" formula could be completely satisfactory since it would probably be inaccurate outside the range of conditions under which data were originally taken in order to construct the formula. Instead, it was necessary to use a reliable *theory* of radiation to obtain the general form of the law, and then to fix the numerical constants by experiment. Using the notion that radiation is a dynamic process in which each body is in a state of equilibrium, absorbing and emitting at the same time (Prevost's law of exchanges[7]), they argued that the rate of cooling of a body at temperature $t + \theta$ in a vacuum surrounded by a container at temperature θ must be determined by the general formula

$$v = \text{rate of cooling} = F(t + \theta) - F(\theta)$$

Their own experiments indicated that the dependence on t and θ was of the form

$$v = \phi(t)a^\theta$$

Hence it followed that $F(\theta) = ma^\theta + \text{constant}$, and

$$v = ma^\theta(a^t - 1)$$

where the coefficients were found to have the numerical values

$$a = 1.0077 \quad \text{and} \quad m = 2.037$$

One consequence of the exponential form is that the heat radiated by a body will never become zero for finite values of its temperature. According to Dulong and Petit, this means that the hypothetical "absolute zero" of temperature must be an infinite number of degrees below ordinary temperatures. That does not imply, however, that a body contains an infinite amount of heat, since the heat capacity must decrease at low temperatures in such a way that the total heat at any finite temperature T is

$$\int_{-\infty}^{T} c(T)\, dT = \text{a finite quantity}$$

The above formula for v refers to the rate of cooling in a "vacuum" which contains somewhat less than 2 mm (mercury) pressure.[8] Dulong and Petit also investigated the rate of cooling in air at various temperatures and pressures. They reported that the cooling due to air (having subtracted the "radiation" cooling determined from the vacuum experiments) depends on the temperature-difference t but not on the temperature of the air itself; it does depend on the pressure. Their data taken at pressures from 0.045 to 0.72 meters Hg can be fitted by the formula

$$v = mp^c t^b$$

where $b = 1.233$ and $c = 0.45$ for any body cooling in air. (Leslie had found $c = 0.2$ and Dalton claimed that $c = \frac{1}{3}$.)

For other gases studied by Dulong and Petit, the data on cooling could be represented by the same formula with the same value of b but different values of c (0.38 for hydrogen, 0.517 for carbonic acid, 0.501 for olefiant gas [ethylene]). The third constant, m, depends on the dimensions of the cooling body as well as on the gas; typical values were 0.00919 for air, 0.0318 for hydrogen, 0.00887 for carbonic acid, and 0.01227 for olefiant gas.

While Dulong and Petit refer briefly to Leslie's suggested distinction [see Appendix] between the abduction [conduction] of heat by air (by a process similar to that in solids) and heat transfer by the flow of air currents, they do not seem to think that the former process is of much importance. In their formula for the total rate of cooling,

$$v = ma^\theta(a^t - 1) + mp^c t^b,$$

the first term is generally supposed to refer only to radiation and the second term primarily to convection, with conduction being ignored.

The work of Dulong and Petit was frequently mentioned and often quoted at some length in writings on heat transfer during the next few decades.[9] A new investigation, conducted for the purpose of checking and extending their results, was announced in 1846 by F. de la Provostaye and P. Desains, who stated that the Dulong–Petit formula did not accurately represent heat transfer under conditions beyond those originally studied. In particular, it was found that the exponents b and c decrease and then increase as the air pressure is reduced, if the container is small; but for a large container they tend to the fixed values 1.233 and 0.45 respectively as reported by Dulong and Petit.[10]

In 1847 the American scientist John William Draper, Professor of Chemistry at the University of New York, reported some experiments motivated by the question: At what temperature do bodies become self-luminous? Citing previous estimates of 635° (Newton), 812° (Davy), 947° (Wedgwood), and 980° (Daniell), he described his own experiments on platinum from which he concluded that the temperature is 977°F. (His temperature scale was based on the assumption that the coefficient of expansion of platinum as measured at lower temperatures by Dulong and Petit is the same at all temperatures; the actual

§13.2]   DULONG–PETIT LAW OF COOLING   481

expansion observed was 1/222 of the original length of the platinum bar.) He asserted that the same result applied to all heated solids and even to lead, which is liquid at that temperature. He accepted the Dulong–Petit radiation law, except that he noted that it does not take account of the distribution of different wavelengths in the radiant heat; Draper announced the general principle that "as the temperature of an incandescent body rises, it emits rays of light of an increasing refrangibility." He stressed that similar properties pertain to both the heat and light emitted; both increase more rapidly than does the temperature of the body. Near the end of his paper, Draper presented a table of intensities of radiant heat emitted by platinum at different temperatures; he made no attempt to compare these values with those predicted by the Dulong–Petit law. Draper's paper was reprinted in his *Scientific Memoirs* in 1878, just in time for his data to be used by Josef Stefan in his attempt to establish a new radiation law in place of the Dulong–Petit formula.[11]

The first hint of how the new physical ideas of the 1840's might affect theories of heat transfer is found in a paper by Ludwig Wilhelmy, published in 1851.[12] Wilhelmy argued that since recent work had established the complete identity of heat and light, one can no longer use theories of the Fourier–Poisson type, since these imply that heat is a substance, and he seemed to think that the Dulong–Petit law was associated with such theories. He stated that the propagation of heat by radiation is really the same process as its propagation by conduction – a remark which he attributed to Fourier.[13] This would be true if one regarded heat as a substance transmitted through space from one particle to another, or if one regarded it as composed of ether-vibrations, according to the wave theory of heat; presumably Wilhelmy took the latter position. But he drew from this statement the consequence that there can be no heat conduction in the usual sense in a gas, since the molecules can freely change their positions. Thus Wilhelmy accepted the idea that molecules are in motion in a gas, and even the further idea that the amount of motion increases as the temperature rises, but has not yet been able to disentangle thermal molecular motion from the motion of the ether. He considered it natural that heat propagation would be accelerated by increased thermal vibrations of molecules, and this effect should help to explain the increased rate of cooling at high temperatures found by Dulong and Petit, apart from the possible effect of temperature on the emission coefficient of the body itself. His own formula was based on the explicit assumption that the rate of heat transfer by radiation depends on the pressure of the gas surrounding a cooling body.

Notes for §13.2

1. J. Dalton, *A New System of Chemical Philosophy*, **1** (London, 1808; 2nd ed., 1842); see p. 8 in the reprint by Citadel Press, New York, 1964.
2. *Ibid.*, p. 95.
3. See F. de la Roche, *J. de Physique* **75**, 201 (1812); J. B. Biot, *Bull. Soc. Philomath. Paris*, p. 21 (1815); C. Despretz, *Ann. Chim. Phys.* **6**, 184 (1817).
4. P. L. Dulong and A. T. Petit, *Ann. Chim. Phys.* **7**, 113, 225, 337 (1817). On the background of this prize competition see R. Fox, *The Caloric Theory of Gases* (New York: Oxford University Press, 1971), p. 236.
5. T. Preston, *Theory of Heat* (London, 1894, 3rd ed., 1919), 504–12. In 1913, Irving Langmuir wrote that "Dulong and Petit's formulas, as supplemented by Peclet's work... have been and are still taken, in engineering practice, as an accurate expression for heat losses from a surface..."– *Trans. Am. Electrochem. Soc.* **23**, 299 (1913).
6. P. L. Dulong and A. L. Petit, *Ann. Chim. Phys.* [2] **10**, 395 (1819). Their determination of specific heats is based on measurements of cooling times and thus depends indirectly on the radiation law.

    To my knowledge Fox (*op. cit.*) is the only historian who has paid serious attention to the Dulong–Petit *radiation* law in recent years. There is a brief account in D. S. L. Cardwell, *From Watt to Clausius* (Ithaca: Cornell University Press, 1971).
7. P. Prevost, *J. de Physique* **18**, 314 (1791); see D. B. Brace, *The Laws of Radiation and Absorption* (New York: American Book Co., 1901).
8. Dulong and Petit, *op. cit.*, p. 245. The uncertainty in the actual value of the pressure was discussed in 1878 by Stefan (see below, §13.7).
9. A. Ure, *A Dictionary of Chemistry and Mineralogy* (London, 4th ed., 1831), pp. 257–58. S. D. Poisson, *Théorie mathématique de la chaleur* (Paris, 1835), ch. II. T. Webster, *The Theory of the Equilibrium and Motion of Fluids* (Cambridge, 1836), p. 128. P. Kelland, *Rept. 11th Meeting Brit. Assn.*, Part 1 (1841), p. 1. Muncke, *Gehler's Physikalisches Wörterbuch*, **10** (1), 440 (Leipzig, 2nd ed., 1841). F. Lunn, *Encyclopedia Metropolitana* **IV**, 272 (London, 1845). J. D. Forbes, *The progress of mathematical and physical science principally from 1775 to 1850* (Dissertation Sixth for the 8th ed. of the *Encyclopedia Britannica*, pub. separately in 1858), pp. 952–53. J. Herschel, *Meteorology* (Edinburgh, 1861, reprinted from the *Encyclopedia Britannica*), p. 43. W. Hopkins, *Phil. Trans.* **150**, 379 (1860). E. Peclet, *Traité de la Chaleur* (Paris, 3rd ed., 1861), **3**, 439. E. Becquerel, *La Lumière* (Paris, 1867), **1**, 82–86. G. J. Stoney, *Proc. R.S. London* **17**, 40 (1869).
10. F. de la Provostaye and P. Desains, *Ann. Chim. Phys.* [3] **16**, 337 (1846).
11. J. W. Draper, *Am. J. Sci.* [2] **4**, 388 (1847); *Phil. Mag.* [3] **30**, 345 (1847); *Scientific Memoirs* (New York, 1878), pp. 23–45. The table on p. 44 of this book is reproduced by J. Stefan in his 1879 memoir discussed below, §13.7. Draper's estimates of the temperatures corresponding to various degrees of brightness of a metal apparently formed the basis for the values 525° and 1200° attached to Tyndall's experiments by Adolph Wüllner which then gave Stefan the first hint of his T^4 law; see A. Wüllner, *Die Lehre von der Wärme* (Leipzig, 1871), pp. 215–16. On Draper's career in general see D. Fleming, *John William Draper and the Religion of Science* (Philadelphia: University of Pennsylvania Press, 1950).
12. L. Wilhelmy, *Ann. Phys.* [2] **84**, 119 (1851). Ludwig Ferdinand Wilhelmy (1812–64) is

known mainly for his contributions to chemical kinetics; see E. Farber, *Chymia* 7, 135 (1961). The substance of his 1851 paper can also be found in a larger work, *Versuch einer mathematisch-physikalischen Wärmetheorie* (Heidelberg, 1851), p. 88f.
13. J. B. J. Fourier, *Ann. Chim. Phys.* 6, 295 (1817); see *Oeuvres de Fourier*, ed. G. Darboux (Paris, 1890), 2, 351–86.

13.3 Heat conduction in gases, before Maxwell

In the 1840's several authorities expressed the opinion that while heat conduction might actually occur in gases, there seemed to be no reliable way to distinguish it from radiation and convection. (At least by this time the modern term "convection" had been introduced though not universally adopted.[1]) Some of these opinions give an indication of the background of Maxwell's early ideas.

In the *Encyclopedia Britannica* we find the statement:

> The conducting power of gases is not so easily ascertained, as in liquids because it is difficult to separate their conducting power from the effect of radiation of heat through them. The experiments of Sir John Leslie and of Dr. Dalton, however, decidedly show a difference in the conducting power of gases, which is also more nearly in the direct ratio of their specific gravity than of their other properties; hydrogen having the lowest conducting power, atmospheric air one considerably higher, and carbonic acid the greatest of all the gases subjected to this examination.[2]

Similarly, in the *Encyclopedia Metropolitana*, we read:

> Experiments toward ascertaining the conducting power of gases have been made by Mr. Dalton, Sir H. Davy, and perhaps by others. They ascertained the times of cooling required for thermometer bulbs placed in different airs. But a little consideration will teach us, that this mode of cooling does not apply to our present question, because many other incidental causes must have interfered with the results, or, in truth, the whole investigation refers to radiation. How can we separate the effects due to that action? or how much are we to allow for the circulating currents in the gases? These questions it seems impossible to answer, and without an absolute correction could be obtained for these disturbing causes, we feel that the real question at issue remains undecided.

> Although perhaps the effect of currents might be guarded

against, as in the case of liquids, yet we do not see how the process of radiation is to be prevented in media of such tenuity; and no experiments had until very recently been made, enabling us to ascertain how much of effect might be ascribed to conducting power, after subtracting the effect due to radiation under similar circumstances in vacuo.[3]

The solution was not simply to do more and better experiments, but also to replace some of the theoretical ideas that had been used in analyzing the earlier experiments. One example of such replacement is found in Wilhelmy's 1851 paper, mentioned above, but Wilhelmy was still conflating conduction and radiation. Another paper published in the same year, by G. G. Stokes, is somewhat more relevant in establishing the context of the kinetic theory of heat conduction, since we know that Maxwell discussed some of his early ideas with Stokes. Stokes proposed to develop a theory of heat conduction in crystals which would be independent of the hypothesis of "molecular radiation" employed by Duhamel and others—"a hypothesis," Stokes stated, "which for my own part I regard as very questionable." Elaborating his views later in the same paper, he wrote:

> For my own part I believe conduction to be quite distinct from internal radiation, although the theory which makes conduction to be nothing more than molecular radiation and absorption seems to be received by many philosophers with the most implicit reliance. No doubt internal radiation may, and I believe generally if not always does, accompany conduction; and when the distance which a ray of heat can travel before it is absorbed is insensible, we may include internal radiation in the mathematical theory of conduction, and even, if we please, in our definition of the word *conduction*. Of course the distance which we may regard as insensible will depend partly on the dimensions of the body, partly upon certain lengths relating to the state of temperature in the interior, and depending upon the problem with which we have to deal.[4]

* * *

One important clue in the complex mystery of heat transfer was the difference in conducting powers of different gases. According to Burr, the first systematic study of this aspect of the subject was made by Joseph Priestley around 1780.[5] He filled a vessel, in which was mounted a thermometer, with different kinds of gases, then plunged it into hot or cold water and timed the interval needed for the mercury in the thermometer to reach some particular temperature. He reported that hydrogen "conducted heat much better than any kind of air, the mercury ascending the same space in about half the time it took up in common air." Carbon dioxide conducted heat considerably worse than common air.[6] Further studies were made by Leslie, Dalton, and Davy, but aside from confirming the exceptionally high conductivity of hydrogen their results did not agree quantitatively, and in some cases not even qualitatively. (It is somewhat puzzling that the *Britannica* article quoted above reversed the conclusions about relative conducting powers of gases.)

A new type of experiment was conducted by Thomas Andrews in 1841, and later became quite popular, perhaps in part because of its technological interest. The method was based on observing the color of a platinum wire, sealed in a glass tube which could be filled with various gases, as the wire was heated continuously by an electric current. The wire is bright red in air, duller in other gases, and not red at all in hydrogen, even when observed in total darkness; this was taken as an indication that the hydrogen conducts the heat much more rapidly away from the wire.[7] A quantitative measure of cooling power could also be obtained from the intensity of the electric current, and such numbers were reported for 13 gases; but Andrews was not able to convert his results directly into thermal units that could be compared with data from other experiments.

Similar results were reported by W. R. Grove in 1845 in a paper titled "On the application of voltaic ignition to lighting mines."[8] Although he did not attempt a systematic study comparable to that of Andrews, Grove's publication and a subsequent one in 1849[9] attracted some attention and served to revive the question of heat conduction in gases. It seems rather curious that the hot-wire type of experiment should have helped to establish the property of conductivity, since as Grove himself pointed out in 1845, one might suspect that the unusual effect of hydrogen is due to "a specific action on the ignited wire, somewhat in the nature of catalysis" rather than to conduction through the gas. In fact, Grove attempted to test the hypothesis that the phenomenon he observed is due to the "rapid cooling effect of the hydrogen" by using a very narrow tube so as to "impede the circulation of the heated currents" but found that this modification had no effect on the heat emitted by the wire.[10] He concluded that the greater cooling effectiveness of hydrogen is due partly to "the mobile or vibratory character of the particles by which heat is more rapidly abstracted" but mainly "to a molecular action at the surfaces of the ignited body and of the gas."[11]

It was Grove's paper, published in German translation with a note by Poggendorff,[12] that later stimulated the important work of Gustav Magnus. It also aroused the interest of Rudolf Clausius as early as 1852, but at this time Clausius seemed to be more concerned with the relation between temperature, electric conductivity of the wire, and its heat loss, than with the problem of heat conduction through gases.[13] Even in his papers on kinetic theory published in 1857 and 1858, Clausius did not mention heat conduction in gases.

* * *

The investigations of Gustav Magnus, which according to Burr "finally established the fact that gases conduct heat,"[14] were completed about the same time as the publication of Maxwell's first paper on kinetic theory (which included a calculation of the coefficient of thermal conductivity). In evaluating the claim that Magnus "discovered" heat conduction in gases, we must of course examine closely what Magnus himself meant by conduction. He argued that the high conductivity of hydrogen compared to other gases cannot be explained, as others had suggested, by attributing it to greater "mobility" of individual hydrogen particles; on the contrary, "it is the difference of specific gravity" (caused by the absorption of heat in the gas) "which produces currents." Thus if hydrogen and carbon dioxide absorbed the same amount of heat from a hot body—as seems to be indicated by the fact that their specific heats are the same—one would expect the cooling power of hydrogen to be *less* than that of carbon dioxide, if the heat transfer is due primarily to particle motion. Hence "there remains no other explanation for the more rapid cooling in hydrogen than that this gas can transmit heat from particle to particle, in other words can conduct it, and that it possesses this property in a higher degree than other gases."[15] In other words, Magnus is still thinking of molecular radiation as the mechanism for conduction, and is contrasting this with convection; he does not recognize the possibility of conduction by molecular motion in the sense that Maxwell used.

In his detailed report on his own experiments, Magnus realized that the problem is a little more complicated than the above quotations indicate. Even if the superior ability of hydrogen to transfer heat cannot be attributed to convection currents, it is still possible that this property may simply be due to its permitting a greater amount of radiation to penetrate it. In the earlier discussions of "radiation corrections" it had generally been assumed that one could simply determine the radiation cooling in a vacuum and then subtract that amount from the cooling observed in a gas to obtain the net cooling due to the gas itself. But that involves the assumption that the gas offers no resistance to the passage of radiation–it is perfectly "diathermanous."[16]

By the middle of the 19th century the possibility that gases may absorb radiant heat had to be taken seriously in some areas of physics. For example it was well known that the pattern of trade winds near the earth's surface is related to the convection currents produced by the sun's differential heating of the atmosphere in the equatorial regions, though the effect of the earth's rotation on this pattern had only recently been clarified.[17] Macedonio Melloni had studied local variations in the amount of solar radiation reaching the earth, and had attributed these to absorption by water vapor in the atmosphere.[18] But before 1860 it was not believed that any of the "permanent" gases–oxygen, nitrogen, hydrogen, carbon dioxide, etc.–absorb appreciable amounts of radiant heat. One of the first attempts to investigate this question was made by R. Franz, who studied the absorption of radiation passing through different gases in a tube 90 cm long. He reported that the absorption amounts to 3% for air, 2.6% for hydrogen, and 2.86% for carbon dioxide.[19]

Magnus mentioned the results of Franz as evidence that radiation does not penetrate hydrogen more readily than it does air, so that the much larger heat transfer by hydrogen still requires an explanation. However, he also recognized that Franz's results should be verified using the same heat source as in his own conductivity experiments (boiling water), since diathermancy may depend on the nature of the source. (In other words, he is suggesting that the "spectral distribution" of the radiation may influence the absorption–surely not a very original thought for 1860.) In further experiments Magnus satisfied himself that hydrogen does not have any anomalous absorption properties that might account for its high apparent conductivity, but on the other hand it seemed to be impossible to determine the precise amount of radiation transfer through a gas independently of the perturbing influences of the walls of the apparatus.[20] His final conclusions were that hydrogen transfers heat better than a vacuum, while all other gases transfer it not as well; that the heat transfer in hydrogen is greater, the greater its density; that all gases resist the passage of radiation to some extent; and that hydrogen *conducts* heat like a metal.[21]

The effect of imperfect diathermancy on the measurement of thermal conductivity was not to be satisfactorily established for some time, though Tyndall's experiments on this subject had already begun in 1859. But before continuing our discussion of that line of research we must introduce the main topic of this chapter: the kinetic theory of heat conduction in gases.

Notes for §13.3

1. W. Prout defined convection as that mode of propagation of heat which causes the temperature of a thermometer to rise when it is "away from the direct influence" of the heat source, for example in a dark chimney above a fire: "a portion of the air passing through and near the fire, has become heated, and has carried up the chimney the temperature acquired from the fire"–*Chemistry, Meteorology, and the Functions of Digestion, considered with reference to Natural Theology* (the 8th Bridgwater Treatise, London, 1834), p. 65; 4th ed., 1855, p. 43. According to S. Brown, the word "convection" did not come into general use for another 20 years (see *Benjamin Thompson – Count Rumford*, p. 28). W. H. Brock [*Notes and Rec. Roy. Soc. London* **24**, 281 (1970)] suggests that it was popularized by J. F. Daniell. Prout's definition does not attempt to specify a means for distinguishing convection from conduction.
   A brief sketch of the history of theories of convection has been published by E. U. Schlünder, *Chemie Ingenieur Technik* **42**, 905 (1970).
2. [T. S. Traill] *Encyclopedia Britannica*, 7th ed., **9**, 180 (Edinburgh, 1842), reprinted with additions in the 8th ed., 1856; quotation from p. 182.
3. [F. Lunn] *Encyclopedia Metropolitana*, **4**, 302–3 (London, 1845).
4. G. G. Stokes, *Cambridge and Dublin Math. J.* **6**, 215 (1851); reprinted in *Mathematical and Physical Papers by the late Sir George Gabriel Stokes*, reprint of the 2nd ed., **3**, 203 (New York: Johnson Reprint Corp., 1966); quotation from p. 207. One respected authority who treated radiation and conduction as essentially the same process was J. Herschel: *A Preliminary Discourse on the Study of Natural Philosophy* (London, 1830; New York: Johnson Reprint, 1966), 206–7.
5. A. C. Burr, *Isis* **21**, 169 (1934).
6. J. Priestley, *Experiments and observations on different kinds of air, and other branches of natural philosophy, connected with the subject* (Birmingham, England, 1790; New York: Kraus Reprint, 1970), **2**, Bk. VII, §IX; quotation from p. 459.
7. T. Andrews, *Proc. R. Irish Acad.* **1**, 465 (1841); *Scientific Papers* (London, 1889), p. 66.
8. W. R. Grove, *Phil. Mag.* [3] **27**, 442 (1845).
9. W. R. Grove, *Phil. Mag.* [3] **35**, 114 (1849); Burr gives extensive quotations from this paper.
10. This experiment (and its interpretation) is reminiscent of R. Brown's experimental proof that the observed motion of small particles in fluids cannot be caused by the invisible molecular motion of the fluid that is responsible for its evaporation; see below, §15.1.
11. W. R. Grove, *Phil. Trans.* **139**, 49 (1849), quotation from p. 58. At the end of this

paper Grove stated that he has just received from T. Andrews a copy of the latter's paper published in 1840 describing a similar experiment.
12. "Das Erkalten eines galvanisch glühenden Drahts in verschiedenen Gasen geschieht wohl *mutatis mutandis* nach denselben Gesetzen, wlech Dulong und Petit für das Erkalten eines auf gewöhnliche Weise erhitzen Körpers festgestellt haben. Auf ein so erhitzten Körper wirkt auch das Wasserstoffgas am schnellsten erkaltend, dann folgt ölbindendes Gas, Kohlensäure, etc."–J. C. Poggendorff, *Ann. Physik* [2] **71**, 197 (1847). This note is cited by Magnus in his paper of 1861 (see below).
13. R. Clausius, *Ann. Phys.* [2] **87**, 501 (1852).
14. Burr, *op. cit.*, p. 179. The priority for this discovery has also been claimed for E. Peclet [*Traité de la chaleur*, Paris, 1828, 2nd ed., 1843, 3rd ed., 1844] by A. Witz, *Ann. Chim.* [5] **15**, 433 (1878). As far as I can determine, Peclet's work is much less substantial than that of Leslie.
15. G. Magnus, *Monatsber. K. Akad. Wiss. Berlin*, p. 485 (1860); *Phil. Mag.* [4] **20**, 510 (1860).
16. The standard terminology derives from a paper by M. Melloni, *Compt. Rend. Acad. Sci. Paris* **13**, 808 (1841); see E. S. Barr, *Infrared Physics* **2**, 67 (1962).
17. See H. L. Burstyn, *Bull. Am. Meteorol. Soc.* **47**, 890 (1966).
18. Barr, *op. cit.*
19. R. Franz, *Ann. Phys.* [2] **94**, 337 (1855). The refutation of this result by Tyndall and others is discussed below in §13.6, footnote 5.
20. Magnus, *Ann. Phys.* [2] **112**, 543 (1861).
21. The logic of this inference from the experimental results has been criticized by Preston, who says: "The only certain inference we seem able to make is that the flow of heat to the thermometer, or the heat-carrying power of the space, is increased by hydrogen and diminished by other gases, and there are no *a priori* grounds for the supposition that hydrogen possesses a conducting power similar to metals more than any other gas"–*Theory of Heat*, 3rd ed. (London, 1919), p. 664. The criticism that hydrogen simply absorbs less radiation than air, rather than having a large "metallic" conductivity, was vigorously pressed against Magnus' conclusions by H. Buff, *Ann. Phys.* [2] **158**, 177 (1876); *Phil. Mag.* [5] **4**, 401 (1877).

13.4 Maxwell's kinetic theory of heat conduction

As we have seen in ch. 4, the kinetic theory of gases was revived in the 1850's by Krönig and Clausius, the older writings of Daniel Bernoulli, Herapath, and Waterston having by that time fallen into obscurity. Since Bernoulli, Herapath, and Waterston had almost nothing to say about heat conduction,[1] and Krönig and Clausius did not mention it at all in their early papers, we can begin with Maxwell's theory published in 1860. We need only recall that Maxwell borrowed from Clausius the concept of *mean free path* traveled by a gas molecule from one collision to the next.[2]

Although Maxwell was the first to propose a kinetic theory of heat conduction, and indeed introduced the modern view that heat conduction is a special case of a general "transport process" of which diffusion and viscosity are other, parallel, examples, he nevertheless treated the subject very casually, one might even say sloppily. Perhaps this was because he was under the impression that no experimental test was yet possible:

> It would be almost impossible to establish the value of the conductivity of a gas by direct experiment, as the heat radiated from the sides of the vessel would be far greater than the heat conducted through the air, even if currents could be entirely prevented.[3]

But the real difficulty before 1860 was not in such technical aspects of the experiment, but in the *theoretical* conceptions of the nature of heat conduction in gases: it was thought to be a process of *radiation* from one molecule to another, rather than a consequence of molecular motion that was nevertheless distinct from convection. And, insofar as conduction involved the *presence* of a gas, it was naturally assumed that its effectiveness depends on the *amount* of gas present. Before gaseous heat conduction could be discovered in the laboratory, it first had to be properly defined in theory.

According to Maxwell, the mechanism of heat conduction in gases is the motion of fast molecules from hotter regions to cooler regions, where they collide with slower molecules and give up some of their kinetic energy. The process is different from convection since it is not driven by density differences between hot and cold regions, which set up macroscopic currents in the gas, but by the random wanderings of individual molecules throughout the entire available space. A surprising feature of this mechanism is that its effectiveness is very nearly independent of density: a rarefied (but not *extremely* rarefied) gas conducts heat as well as a moderately dense one. A clear understanding of this feature turns out to be essential to the correct interpretation of experiments on heat transfer.

Maxwell assumed that molecules are emitted randomly in all directions from any infinitesimal region of the gas, at a rate that depends on the average speed and density of the molecules in that region. The fraction of these molecules that reach another region a certain distance away will then be determined by the mean free path, which in turn depends on the molecular diameter and on the number of molecules in unit volume (mean free path $L \propto 1/nd^2$). The rate of transfer of matter across a specified (imaginary) plane in the gas is then determined by integrating these quantities over all emitting regions,

§13.4]   MAXWELL'S KINETIC THEORY   491

with a factor that decreases exponentially with distance from the plane in accordance with the mean-free-path theory. The result is that "the quantity of matter transferred across unit of area by the motion of agitation in unit of time" is

$$q = -\frac{1}{3}\frac{d}{dx}(\rho v L)$$

where $\rho = MN$ is the density, $M =$ mass of each molecule, $N =$ number of molecules, and $v =$ "mean velocity of agitation."

To compute the rate of transfer of heat across the same plane, Maxwell assumed that he could simply replace M by E in this equation: E is the total energy of a single molecule, including both the kinetic energy of motion of its center of gravity and the energy of internal motions of the molecule. At this point Maxwell was not insisting on complete equipartition of internal energy, but rather he assumed with Clausius that there is simply a constant ratio between the two kinds of energy; thus he put

$$E = \tfrac{1}{2}\beta M v^2$$

"where β is a coefficient, the experimental value of which is 1.634."[4] When the indicated substitution is made, and the gradient dv/dx is replaced by the temperature gradient dT/dx, the result is

$$J_q = -\frac{3}{4}\beta \rho L V(1/T)\frac{dT}{dx}$$

where J is the mechanical equivalent of heat. Maxwell then put in some numerical values and came out with the conclusion that "the resistance of a stratum of air to the conduction of heat is about 10 000 000 times greater than that of a stratum of copper of equal thickness."

Although Maxwell did not say so explicitly, it can be seen from his formula that the coefficient of thermal conductivity (the factor multiplying dT/dx) should be independent of density and proportional to the square root of the absolute temperature, just like the viscosity coefficient,[5] and that lighter gases such as hydrogen should have the greatest conductivity at a given temperature and pressure.

* * *

Maxwell's subsequent work on kinetic theory and especially on heat conduction was greatly influenced by that of Clausius.[6] Clausius had of course studied Maxwell's original paper of 1860 with considerable care; in fact, his own translation of the paper into German, with several notes on the derivation of various formulae, is still preserved at the Deutsches Museum in Munich, along with other manuscripts.

In 1862, Clausius published a long memoir on heat conduction in gases, which is of interest for several reasons even though the theory he presents is rather cumbersome and mostly obsolete.[7] First, he feels it necessary to reply to the objections of Emil Jochmann, who had argued intuitively that if the kinetic theory were a valid description of gases, then temperature-differences would be immediately eliminated by molecular collisions, resulting in extremely large thermal conductivity in all gases, contrary to experience.[8] Clausius replied that the short mean free path, which he had previously invoked to deal with a similar objection by Buys-Ballot concerning the mixing of gases, would limit the rate of heat conduction; and he dismissed the one-dimensional analogy of colliding elastic spheres suggested by Jochmann as being irrelevant to the real situation in gases. Clausius also provided a long historical footnote, enumerating the various precursors of the kinetic theory of whom he had learned since publishing his earlier papers: G. L. LeSage, P. Prevost, Lucretius, Gassendi, Boyle, Parent, Hermann, Daniel and Jean Bernoulli.

Proceeding then to Maxwell's theory, Clausius considered a thin stratum in the gas, parallel to the two boundaries which are maintained at different temperatures. If the x-axis runs perpendicular to these boundaries, and the highest temperature is found at the smallest value of x, then there will be a flow of heat in the direction of increasing x. Moreover, any two individual molecules that collide in the stratum will have, on the average, a net center-of-mass motion in the $+x$ direction, in addition to their accidental motions distributed at random in all directions. Hence the molecules emitted from the stratum after colliding will have on the average a small velocity component in the $+x$ direction. Clausius noted that Maxwell had disregarded this effect.

But Clausius himself assumed that, while the accidental motion of the molecules is equally likely to be in any direction, the *speeds* are all the same. From his remarks further on in the paper, it appears that while Clausius does accept Maxwell's argument that the speeds will *not* be all the same, he does not choose to use the particular velocity distribution suggested by Maxwell; he prefers instead to ignore the "accidental variations" on the grounds that they "cannot in any case contribute to cause more *vis viva* to traverse a given plane in one direction than in another, since, whatever may be their individual effects, their influence must be the same in both directions." He does admit that the velocity distribution must be taken into account in computing numerical values that depend on average values of functions of the velocity.[9]

Clausius then pointed out that if the temperature varies from one place to another in the gas, the density and mean free path must also vary. But he wanted to avoid this complication and therefore postulated that only a single mean free path need be considered, the one corresponding to the normal state of the gas (1 atm pressure, $T = 0°C$).

Clausius accepted and formalized Maxwell's proposal that the three transport processes—diffusion, viscosity, and heat conduction—should be treated from a single consistent viewpoint, relating them to the transport by molecular motion of mass, momentum, and energy across an imaginary surface in the gas. He used an elaborate but somewhat arbitrary set of series expansions (in powers of the mean free path), arriving finally at a formula for the thermal conductivity coefficient almost identical to Maxwell's—when Maxwell's result was translated into the same variables there was a factor of $\frac{1}{2}$ in place of the $\frac{5}{12}$ found by Clausius.[10] Clausius took some pains to point out that this near-agreement was only apparent, and that Maxwell had arrived at his result by erroneous reasoning (and a fortuitous cancellation of two errors), for his formulae refer to the situation in which there is a progressive motion of mass as well as energy in one direction.

In commenting on his final result for heat conduction, Clausius noted that the coefficient increases in proportion to the square root of the absolute temperature, but is independent of pressure. The first behavior is similar to that of the velocity of sound, and therefore requires no particular explanation. As for the second,

> This is explained by the circumstance that, although the number of molecules which can convey the heat is greater in a gas which is rendered more dense by increased pressure, the distances traversed by the individual molecules are smaller. This latter conclusion might lead to absurdity, if it were assumed to be applicable to the gas under every conceivable condition of compression or expansion. It must, however, be borne in mind that there are obvious limits to the application of it to conditions of the gas which depart very much from the mean condition: on the one hand, the gas must not be so much compressed as to produce a too great departure from the laws of permanent gases which have been taken as the foundation for the whole course of reasoning;

and on the other hand, it must be not be so much expanded that the mean length of excursion of the molecules becomes so great that its higher powers cannot be disregarded.[11]

Of course the same remarks apply equally well to Maxwell's earlier results for viscosity, diffusion, and heat conduction, but Maxwell himself did not bother to make them. All he said was, in respect to the result that viscosity does not depend on density, "Such a consequence of a mathematical theory is very startling, and the only experiment I have met with on the subject does not seem to confirm it."[12] It was Clausius who, being much more strongly committed to the kinetic theory at this point, felt a need to justify its results and point out their limits of validity. (He also had to deal with a few skeptics—Buys-Ballot, Hoppe, Jochmann, and others—who would raise objections if he did not anticipate and answer them in advance.) So it was Clausius rather than Maxwell who first published a qualitative explanation of the fact that gas transport coefficients are independent of density (in the paragraph quoted above) even though the fact itself is clearly Maxwell's discovery.[13] As it turned out, the theoretical prediction that heat conduction does not decrease with decreasing density (except at extremely low density) was to be of crucial importance in disentangling the three modes of heat transfer.

In estimating numerical values of heat conduction coefficients, Clausius assumed that the mean value of the molecular speed is simply the one that gives the correct *vis viva* or quantity of heat contained in the gas. He had no way to estimate the mean free path except by taking Maxwell's values, derived from comparison of his theory with viscosity and diffusion data. These were of the order of 1/400 000 English inch, or 1/16 000 000 meter. On this basis, Clausius estimated that the conducting power of air should be 1/1400 of that of lead. The considerable discrepancy between this value and Maxwell's estimate (ten million times smaller than copper) is due primarily, as Clausius pointed out in a footnote, to Maxwell's errors in converting from one system of units to another.

Unfortunately there were hardly any experimental data with which to compare these numbers. Only a qualitative test could be made: the theory predicts that the thermal conductivity should be inversely proportional to the specific gravity for different gases with the same specific heats and mean free paths. This is in accord with the well-known result that the lightest gas, hydrogen, conducts heat much better than the others.[14]

* * *

Maxwell's immediate response to this paper of Clausius is found in his unpublished manuscript, "On the conduction of heat in gases."[15] Maxwell adopted, at least temporarily, Clausius' argument that the mean free path should be assumed to vary from one place to another in the gas. He also accepted the conditions "that the transfer shall be of heat and not of matter, and that every intermediate slice of air shall be in equilibrium," admitting that "in my former paper I paid little attention to this subject as I had no experimental data to compare with the theory."

Maxwell attempted to interpret the awkward constant β, which represents the ratio of total molecular energy to translational kinetic energy, by suggesting that the medium may consist "partly of perfect spheres and partly of other bodies" so that one can write

$$\beta = 1 + q$$

where $q = 0$ if the particles are all spherical and $q = 1$ if none of them are. Substitution of experimental specific heat data then leads to the conclusion that "we must suppose 0.634 of the weight of air to consist of non spherical particles and 0.366 of its weight to consist of perfectly spherical particles." But the same reasoning leads immediately to a negative value for the quantity of spherical particles in the case of steam, so the hypothesis does not seem very plausible.[16]

Maxwell did not complete or publish his revised mean-free-path theory of heat conduction, presumably because he recognized that the mean-free-path method itself is unsatisfactory. The problem was taken up again in his long memoir "On the Dynamical Theory of Gases" published in 1867, in which he said:

> It is to Professor Clausius, of Zurich, that we owe the most complete dynamical theory of gases.... there were several errors in my theory of the conduction of heat in gases which M. Clausius has pointed out in an elaborate memoir on that subject.[17]

But Maxwell did not discuss any of the technical points raised by Clausius; as in his 1860 paper, he simply treated heat as another quantity that might be transferred by molecular motion and collisions, with the proviso that the transfer is to be computed with respect to a plane that is moving with the mean velocity of the gas itself. But now, instead of treating molecules as elastic spheres that travel in straight lines until they hit another sphere, transferring to it an amount of momentum and energy depending on the mean values for the region of this collision as compared to the previous one, Maxwell gives a more accurate calculation based on the collision dynamics for point-centers of force. Although this part of the calculation is carried through to completion for only one special force law, the basic equations for stress and heat flux are formulated in a manner that can in principle be applied to any reasonable molecular model. Moreover, stress and heat flux are defined in a sufficiently general way that both may depend on gradients of temperature and hydrodynamic velocity and (in mixtures) on gradients of component concentrations. While Maxwell restricted his initial calculations to situations in which heat flux is proportional to temperature gradient, the more general formulation is needed in dealing with rarefied gases, or problems involving high-speed flow, where a single "thermal conductivity coefficient" is insufficient to specify the thermal properties of the system, and the distinction between conduction and convection may break down.[18]

Maxwell expressed his final result for the thermal conductivity coefficient in the form

$$C = \frac{5}{3(\gamma - 1)} \frac{p_0}{\rho_0 \theta_0} \frac{\mu}{s}$$

where γ = ratio of specific heats, μ = viscosity, s = specific gravity, and p_0, ρ_0, and θ_0 are the pressure, density, and temperature of a standard gas. As was later discovered by Boltzmann (see below) the factor $\frac{5}{3}$ should be replaced by $\frac{5}{2}$, but otherwise the formula is now considered correct for the special case of point atoms interacting with repulsive forces that vary inversely as the fifth power of their distance, and very nearly correct for monatomic gases in general.[19]

Maxwell's estimation of numerical values for particular gases is still rather casual—he is not aware of the existence of any experimental data and does not even cite the papers of Magnus. He estimates that "iron at 25°C conducts heat 3525 times better than air at 16.6°C" and, rather than give a direct comparison with Clausius' theoretical formula, he says:

> M. Clausius, from a different form of the theory, and from a different value of μ, found that lead should conduct heat 1400 times better than air. Now iron is twice as good a conductor of heat as lead, so that this estimate is not far different from that of M. Clausius in actual value.

Since it is assumed that the ideal gas law holds, the ratio $p_0/\rho_0\theta_0$ is a constant, and the thermal conductivity coefficient, like the viscosity coefficient, is independent of density but directly proportional to the temperature.[20] On the other hand Clausius' theory, like Maxwell's 1860 theory, was based on the elastic sphere model, and predicts a $\sqrt{T}$ law. So far there was no experimental data pertaining to this point. Maxwell noted at the end of his paper that oxygen, nitrogen, and carbonic acid should all have roughly the same conductivity as air; curiously, he did not mention the experiments indicating that hydrogen has a much greater conductivity, though this fact had been noted by Clausius in 1862 (see above), and was still the only confirmation of the theory available in 1867.

* * *

In 1872, Ludwig Boltzmann published his own version of transport theory, basically equivalent to Maxwell's second theory but based on solving an integro-differential equation for the distribution function f instead of on equations for the transfer of mass, momentum, and energy.[21] Boltzmann's result for a gas of point particles with inverse fifth power repulsive forces would have been identical with Maxwell's except that he discovered a mistake in Maxwell's calculation: the factor $\frac{5}{3}$ in the expression for C quoted above should have been $\frac{5}{2}$.[22] In Boltzmann's paper the relation between heat conduction, viscosity, and specific heat is expressed in almost exactly the standard form that was to be adopted later

$$k = \tfrac{5}{2}\mu c_v$$

(k = coefficient of thermal conductivity, c_v = specific heat at constant volume).

Boltzmann remarked that "if one includes the intramolecular motion following Maxwell's method," *i.e.* by introducing the parameter β to represent the ratio of total molecular energy to translational energy, then the same formula should also apply to polyatomic gases. But Boltzmann did not accept this conclusion:

> ... this seems very arbitrary to me, and if one includes intramolecular motion in some other way, he can easily obtain significantly different values for the heat conduction constant. It appears that an exact calculation of this constant from the theory is impossible until we know more about the intramolecular motion.[23]

But by this time Josef Stefan had published an experimental determination of the conductivity coefficients for several gases, so Boltzmann concluded that "our experimental knowledge is here much better than our theoretical knowledge." That assessment suggests that Boltzmann is overly concerned with determining numerical values of coefficients, and underrates the substantial advance in physical understanding achieved by Maxwell's theory.

Notes for §13.4

1. There are some brief remarks in Herapath's *Mathematical Physics*, **1**, 186f. (London, 1847; New York: Johnson Reprint Corp., 1972).
2. See §4.4 and §5.2; an English trans. of Clausius' paper of 1858 on the mean free path may be found in Brush, *Kinetic Theory*, 2 (New York: Pergamon Press, 1965); the original German is reprinted in Brush, *Kinetische Theorie* 2 (Berlin: Akademie Verlag/Braunschweig: Vieweg & Sohn, 1970).
3. J. C. Maxwell, *Phil. Mag.* [4] **19**, 19, **20**, 37 (1860), reprinted in *The Scientific Papers of James Clerk Maxwell* (Cambridge, 1890; New York: Dover Pubs., 1952, 1965); quotation from *Papers* **1**, 405.
4. Maxwell, *Papers* **1**, 403.
5. These results can now be obtained more directly, with the benefit of hindsight and dimensional analysis, by reflecting on the nature of the physical parameters involved in the kinetic-theory model but avoiding the mean-free-path approximation. Cf. Rayleigh, *Proc. R. S. London* **66**, 68 (1900); C. Truesdell, *Z. Phys.* **131**, 273 (1952) and more recent work (see ch. IV of his forthcoming book on the kinetic theory).
6. On the relation between Clausius and Maxwell see E. Garber, *Hist. Stud. Phys. Sci.* **2**, 299 (1970).
7. R. Clausius, *Ann. Phys* [2] **115**, 1 (1862); *Phil. Mag.* [4] **23**, 417, 512 (1862).
8. E. C. G. G. Jochmann, *Ann. Phys.* [2] **108**, 153 (1859).
9. See the translation in *Phil. Mag.* [4] **23**, 420, 425, 428. The problem of a drift of molecules from hot to cold regions in Clausius's theory was discussed later by O. Reynolds, *Phil. Mag.* [5] **43**, 142 (1897) and W. Sutherland, *Phil. Mag.* [5] **44**, 52 (1897), in connection with the theory of the radiometer. On Clausius' attitude toward Maxwell's distribution function see Garber, *op. cit.*
10. In a simplified derivation based on the assumption that all molecules travel a distance between collisions just equal to the average mean free path, Victor von Lang obtained a formula similar to that of Clausius except that the coefficient $\frac{5}{12}$ was replaced by $\frac{1}{6}$. This was soon afterwards adjusted to $\frac{1}{2}$ after Stefan obtained experimental data suggesting this value; see V. v. Lang, *Sitzungsber. Akad. Wiss. Wien* **64** (2), 485 (1871), **65** (2), 415 (1872). This example of the "flexibility" of mean-free-path theories was later noted by Boltzmann [*Verh. 69. Vers. D. Naturf. u. Ärzte*, p. 19 (1897)] and Enskog [*Kinetische Theorie der Vorgänge in mäßig verdünnten Gasen* (Uppsala, 1917)]; S. G. Brush, *Kinetic Theory* (New York: Pergamon Press, 1972), **3**, 125 and probably helped to encourage the view that a more fundamental approach was needed.
11. Quoted from the translation in *Phil. Mag.* [4] **23**, 529.
12. Maxwell, *Papers* **1** 391. For further discussion of this point see S. G. Brush, *Am. J. Phys.* **39**, 631 (1971).

13. In his letter to Stokes in 1859, Maxwell did say that the viscosity coefficient is independent of density because "in a rare gas the mean path is greater, so that the frictional action extends to greater distances" [Brush, *Kinetic Theory* 1, 26–27] but he did not put even this much explanation into the final version of the 1860 paper.

H. A. Lorentz mentioned the density-independence of heat conduction as a notable example of the prediction of new phenomena by scientific theories; see *Rev. Univ. Bruxelles* 26, 445 (1921); *Collected Papers* (The Hague: Nijhoff, 1935), 8, 393.

14. This explanation for the unequal conductivities of different gases was rediscovered by Friedrich Mohr, though Clausius quickly claimed his priority. Mohr, *Ber. Deutsch. Chem. Ges.* 4, 85 (1871); Clausius, *Ber. D. Chem. Ges.* 4, 269 (1871).

15. To be published together with other papers of Maxwell on kinetic theory, in a book edited by S. G. Brush, C. W. F. Everitt, and E. Garber. The contents of the manuscript are summarized by Garber, *op. cit.* (footnote 6).

16. Cf. Brush, *op. cit.* (footnote 12), p. 638

17. J. C. Maxwell, *Phil. Trans.* 157, 49 (1867); *Phil. Mag.* [4] 32, 390 (1866), 35, 129, 185 (1868); *Papers* 2, 26; Brush, *Kinetic Theory* 2, 23 (quotation from pp. 27–28).

18. J. C. Maxwell, *Phil. Trans.* 170, 231 (1880); *Papers* 2, 681. C. Truesdell, *J. Rational Mech. Anal.* 1, 125 (1952); E. Ikenberry and C. Truesdell, *J. Rational Mech. Anal.* 5, 1 (1956); C. Truesdell, *J. Rational Mech. Anal.* 5, 55 (1956).

19. S. Chapman, *Phil. Trans.* A216, 279 (1917); D. Enskog, *op. cit.* (footnote 10); S. Chapman and T. G. Cowling, *Mathematical Theory of Non-Uniform Gases* (Cambridge University Press, 1939, 3rd ed., 1970); Brush, *Kinetic Theory* 3, 16, 114, 182.

20. This result can also be obtained by dimensional analysis; see papers cited in footnote 5.

21. L. Boltzmann, *Sitzungsber. Akad. Wiss. Wien* 66 (2), 275 (1872); reprinted in his *Wissenschaftliche Abhandlungen* (Leipzig: Barth, 1909; New York: Chelsea, 1968), 1, 316 and in Brush, *Kinetische Theorie* (Berlin: Akademie-Verlag/Braunschweig: Vieweg & Sohn, 1970), 2, 115; English trans. in Brush, *Kinetic Theory* (New York: Pergamon Press, 1966), 2, 88.

22. See p. 141 of the translation in Brush, *op. cit.*; the mistake was rediscovered by H. Poincaré, *Compt. Rend. Acad. Sci. Paris* 116, 1020 (1893).

23. Boltzmann, *ibid.*

13.5 Experimental tests of Maxwell's theory

The first study of heat conduction in gases designed specifically to test the kinetic theory predictions was published in 1871 by Friedrich Narr.[1] He cited Clausius' formula for heat conduction, in the form

$$\frac{\text{velocity of cooling}}{\text{temperature gradient}} = m \propto \frac{\gamma}{\sqrt{\sigma}} \epsilon$$

where γ = ratio of specific heats, σ = specific gravity, and ϵ = mean free path. Assuming that ϵ is the same for all gases, and taking $m = 100$

for air, he computed the following table:

	m (observed by Narr)	m (calculated from Clausius's formula)
Hydrogen	551	380
Air	100	100
Nitrogen	98	101
CO_2	81	79

He suggested that the discrepancy for hydrogen might be attributed to the fact that the mean free path is really longer there than in the other gases, since the hydrogen molecule is smaller. Narr gave only relative values for the different gases, so his data cannot be directly related to the mean free paths found by other methods.

A more thorough investigation of heat conduction in gases was undertaken a little later by Josef Stefan. Stefan, one of Boltzmann's teachers at Vienna, had previously been concerned with various problems in gas theory and hydrodynamics, though before 1870 he had not participated in the development of the kinetic theory.[2]

Stefan reviewed the various difficulties encountered by previous experimenters and claimed to have developed a satisfactory method himself.[3] For the coefficient of thermal conductivity of air, he found a numerical value almost exactly the same as that deduced theoretically by Maxwell in 1867:

> My experiment gives the number 0.0000558, taking the centimeter, gram, and second as units, so that the conductivity of air is nearly 20 000 times smaller than that of copper, and about 3400 times smaller than that of iron. Maxwell states that air must conduct about 3500 times worse than iron, and, expressing the results of his formula in the units given above, I find the value 0.000054 for air.[4]
>
> Few physical theories have produced such strikingly confirmed predictions, and one must indeed regard the dynamical theory of gases as one of the best founded physical theories.
>
> Also another law, which is given by this theory, namely the independence of the conductivity from density, has been proved correct in a completely indisputable way by this experiment.
>
> Similarly the relative values of the conductivities of different gases agree with the formulae of Maxwell.

This idyllic state of perfect agreement between theory and experiment did not last very long. By the time Stefan's paper appeared Boltzmann, as noted above, had already discovered an arithmetic error in Maxwell's calculations, and when this error was corrected the theoretical value was 50% larger. Stefan suggested that the discrepancy might be explained by assuming that intramolecular motion does not contribute to heat conduction, at least to the extent supposed in Maxwell's theory.[5]

In 1876, Adolph Kundt and Emil Warburg performed a series of experiments on heat conduction at low pressures.[6] A primary reason for going to low pressures was the belief that convection currents should be diminished whereas, if Maxwell's theory is valid, the conductivity should maintain a constant value. Their results indicated that the conductivity is indeed independent of pressure down to about 1 mm Hg for air, and down to about 9 mm Hg for hydrogen. They also suggested that at lower pressures there should be a temperature discontinuity at the interface between a hot solid and a cooler gas, analogous to the "slipping" (velocity discontinuity) which they found in their viscosity experiments; but they were not able to establish this phenomenon experimentally.[7]

Kundt and Warburg pointed out that the fact that air maintains its thermal conductivity down to such low pressures casts doubt on the validity of the Dulong–Petit cooling law. The experimental proof of this law, by Dulong and Petit and subsequently by de la Provostaye and Desains, depended on the assumption that the heat loss from a thermometer into a "vacuum" depends entirely on radiation, aside from a small calculable correction. However, Dulong and Petit did not attain pressures lower than 0.5 or 1 mm Hg, and a recalculation using recent data showed that the heat loss by conduction in their experiment would have been from 20% to 50% of that by radiation, contrary to their assumptions.

In 1875, Winkelmann attempted to test some of the more detailed predictions of the kinetic theory of heat conduction. His first paper presented a comparison of his experimental values for different gases, which showed rough agreement with the theoretical formulae for elastic spheres and inverse fifth power repulsive forces.[8] In a second paper (1876), he investigated the temperature-dependence of the coefficient, and found that it is directly proportional to the absolute temperature, as predicted by Maxwell for particles interacting with inverse fifth power forces.[9] However, Winkelmann argued that this theoretical model was untenable since, according to the results of the Joule–Thomson experiment, the long-range forces between molecules must be attractive, not repulsive. He therefore regarded his experimental data as a confirmation of Maxwell's temperature-law but not of his atomic model. Winkelmann also suggested that a modification of the Dulong–Petit cooling law is needed.

In a third paper (1876), Winkelmann had to retract his previous conclusion that the heat conduction coefficient is directly proportional to the temperature. He had not allowed for the fact that the specific heat of mercury *decreases* with temperature, and when this was taken into account it turned out that the heat conduction coefficient varies with temperature in the same way as does the viscosity, namely, approximately as the $\frac{3}{4}$ power of the absolute temperature.[10]

Boltzmann examined the new experimental data and proposed a modified theory for polyatomic gases in a paper published in 1876.[11] His main concern was to establish the value of the ratio of the heat conduction coefficient to the product of viscosity coefficient and specific heat at constant volume, $k/c_v\mu$. According to Maxwell's kinetic theory this ratio, later denoted by f, should be a constant characteristic of a given gas, which is now usually called the Eucken factor. Boltzmann stated that if only the translational motion of the molecule could be transferred in collisions, one would expect $f = 15(\gamma - 1)/4$, where γ = ratio of specific heats, whereas if the total energy of the molecule is involved, it should have the $v = 2.5$. Comparison with the experimental data of Stefan, Kundt, and Warburg, and Winkelmann suggested that the actual value lies somewhere in between but is closer to the first one.[12]

In 1877, Oscar Emil Meyer published his textbook, *Die Kinetische Theorie der Gase*, in which he presented a new way of estimating the heat conduction coefficient for a gas of elastic spheres. He obtained the result $f = 1.537$. One of the integrals was later recalculated, leading to a modified result $f = 1.6027$, which appeared in the second edition of the book (1899).[13] While Meyer's value was in better agreement with some experimental data than the value of 2.5 that follows from Maxwell's theory for inverse fifth power forces, the subsequent work of Chapman and Enskog showed that it did not have as good a theoretical basis.[14]

From the viewpoint of modern theorists, experimental results for monatomic gases are of much greater interest than those for polyatomic gases or mixtures such as air; it was already realized by 1876 that the numerical values for polyatomic gases differed from the theoretical predictions for reasons involving the transfer of internal molecular energy in collisions, even though the theory might be valid for point particles or elastic spheres.[15] The first gas known to be monatomic was mercury vapor; Schleiermacher measured its thermal conductivity in 1889, and reported the value $f = 3.15$, using an earlier viscosity measurement by Koch.[16] In 1903, Schwarze measured thermal conductivities of the monatomic gases argon and helium, and gave the values $f = 2.501$ and 2.507 respectively.[17] Bannawitz reported $f = 2.501$ for neon in 1915.[18] After that there was apparently never much doubt that the theoretical value $f = 2.5$ for monatomic gases is correct.[19]

For polyatomic gases the situation was more complicated; until fairly recently, kinetic theorists have not been able to improve on Eucken's semi-empirical formula for f, proposed in 1913.[20] It must probably be admitted that the classical kinetic theory of gases cannot give a completely satisfactory quantitative treatment of heat conduction without invoking some arbitrary postulate about the extent to which internal molecular energy is transferred in collisions. But aside from this defect, which affects only the numerical values of the coefficients, the kinetic theory of heat conduction has been reasonably successful in dealing with a property that has always presented rather severe experimental difficulties.[21]

Extensive work has been done on the temperature-dependence of the coefficient. Originally it was hoped that experimental measure-

Table 13.5-1. Heat conduction coefficient for hydrogen, in cal/cm sec °C (experimental data represented by the formula $k = k_0(1 + \alpha t)$)

Date	Investigator	Reference	k_0	α
1872	Stefan	3	0.00045	
1875	Winkelmann	8	0.0003324	
1876	Winkelmann	10		0.00277
1877	Winkelmann	26		0.002855
1881	Graetz	27	0.0003190	0.0016
1883	Winkelmann	28		0.00208
1886	Winkelmann	29		0.00206
1888	Schleiermacher	30		0.00275
1890	Eichhorn	31		0.002045
1891	Winkelmann	32	0.0003829	0.00175
1900	Eckerlein	33	0.0003186	0.00422
1911	Eucken	34	0.0003340	
1917	Weber	35	0.0004165	
1927	Hercus and Laby	37	0.000406	
1934	Dickins	38	0.000417	0.0028
1934	Kannuluik and Martin	39	0.000413	
Modern value		40	0.000402	0.0031

Table 13.5-2. Heat conduction coefficient for air
(experimental data represented by the formula $k = k_0(1 + \alpha t)$)

Date	Investigator	Reference	k_0	α
1872	Stefan	3	0.000056	
1875	Winkelmann	8	0.0000525	
1876	Winkelmann	9		0.003648
1876	Winkelmann	10		0.00277
1877	Winkelmann	26		0.00268
1881	Graetz	27	0.00004838	0.00185
1881	Christiansen	41		0.001504
1883	Winkelmann	28		0.00208
1886	Winkelmann	29		0.00206
1888	Schleiermacher	30	0.0000562	0.00281
1890	Eichhorn	31		0.002045
1891	Winkelmann	32	0.0000555	0.00190
1892	Graetz	42	0.0000500	
1893	Winkelmann	43	0.0000568	
1895	Kutta	44	0.00005715	
1900	Eckerlein	33	0.00004677	0.00362
1901	Müller	45	0.0000569	0.00196
1903	Schwarze	46	0.00005719	0.00253
1907	Pauli	47		0.00197
1917	Weber	35	0.0000568	
1918	Hercus and Laby	48	0.0000540	
1926	Gregory and Archer	36	0.00005914	
1927	Hercus and Laby	37	0.0000585	
1931	Mann and Dickins	49	0.0000579	0.0028
1933	Gregory and Archer	50	0.0000585	
1934	Hercus and Sutherland	51	0.0000572	0.00298
1934	Dickins	38	0.0000584	0.0029
1934	Milverton	52	0.0000581	
1934	Kannuluik and Martin	39	0.0000576	
Modern value		53	0.0000576	0.00315

ments could help to choose between the original elastic-sphere model (which predicts $k \propto T^{\frac{1}{2}}$) and the inverse fifth power force model (which predicts $k \propto T$). Despite a few early attempts to interpret the experimental data as confirming one or the other of these hypotheses, it soon became clear that for real gases the exponents of T are somewhere between $\frac{1}{2}$ and 1;[22] and that the difficulty of getting accurate thermal conductivity data makes this approach less effective as a means of determining the intermolecular force law than the use of viscosity and equation of state data.[23]

The experimentalists adopted the practice of reporting their data in terms of the temperature coefficient α defined by the equation

$$k = k_0(1 + \alpha t)$$

where k_0 is the conductivity coefficient at 0°C and t = temperature in °C. The value $\alpha = 1/273 = 0.00366$ would then correspond to the relation $k = k_0(T/273)$ predicted by the inverse fifth power force model, whereas the value $\alpha = 1/546 = 0.00183$ would correspond approximately to the relation $k = k_0(T_0/273)^{\frac{1}{2}}$ predicted by the elastic-sphere model, for small values of t. (Presumably the theorists would have preferred that the data be used to determine an *exponent* of the absolute temperature, especially after Rayleigh showed how this exponent could be directly related to the exponent in the interatomic force law.[24]) In tables 13.5-1 and 13.5-2 we give the values of α reported for air and hydrogen, up to 1935. By the time quantitative agreement had been reached on the correct value of α, it was too late to use such data to determine the exponent in the interatomic force law, since the point-center of repulsive force was no longer accepted as an atomic model, and more reliable data were available from other kinds of experiments.[25]

Notes for §13.5

1. F. Narr, *Ann. Phys.* [2] **142**, 123 (1871).
2. J. Stefan, *Sitzungsber. Akad. Wiss. Wien* **27**, 375 (1857), **36**, 85 (1859), **53** (2), 529 (1866). In 1863, Stefan noted that the Fourier heat conduction equation predicts an infinite velocity of propagation of heat during an infinitesimal time interval, and suggested how this implausible result might be avoided by treating heat flow in terms of the collisions of elastic spheres, as in the kinetic theory: *Sitzungsber. Akad. Wiss. Wien* **47** (2), 326 (1863).
3. J. Stefan, *Sitzungsber. Akad. Wiss. Wien* **65** (2), 45 (1872). Burr [*op. cit.*, §13.3, footnote 5] gave the date 1873 for this paper, citing only the French summary published in *Journal de Physique* **2**, 148 (1873). Stefan used the method of observing the cooling rate of a thermometer immersed in the gas.
4. These numbers, like those given below, are in calories/centimeter · second · degree.
5. J. Stefan, *Sitzungsber. Akad. Wiss. Wien* **72** (2), 69 (1876). Burr (*op. cit.*) notes the "remarkable agreement" between Stefan's experimental result and Maxwell's theoretical prediction but fails to state that the latter was erroneous.
6. A. Kundt and E. Warburg, *Ann. Phys.* [2] **155**, 337, 525, **156**, 177 (1875).
7. See below, §13.8, for Smoluchowski's work on this effect.
8. A. Winkelmann, *Ann. Phys.* [2] **156**, 497 (1875).
9. A. Winkelmann, *Ann. Phys.* [2] **157**, 497 (1876).

10. A. Winkelmann, *Ann. Phys.* [2] **159**, 177 (1876); *Tageblatt 52. Vers. Ges. D. Naturf. u. Ärzte*, p. 181 (1879). The decrease in specific heat of mercury with temperature was confirmed by many later studies [see J. R. Partington, *An Advanced Treatise on Physical Chemistry* (London: Longmans, Green and Co., 1951), **2**, 212], thus demolishing a presupposition that had probably been used by many earlier experimenters, on the basis of data suggesting an *increase* in specific heat with temperature reported by Dulong and Petit, *J. École Polyt.* **11**, 226 (1820) and Regnault, *Ann. Phys.* [2] **62**, 79 (1844).

    Results indicating a $T^{\frac{1}{2}}$ law for viscosity were reported by O. E. Meyer, *Ann. Phys.* [2] **148**, 1, 203 (1873), and by A. von Obermayer, *Sitzungsber. Akad. Wiss. Wien* **71** (2), 281 (1875), **73** (2), 433 (1876).
11. L. Boltzmann, *Sitzungsber. Akad. Wiss. Wien* **72** (2), 458 (1875); *Ann. Phys.* [2] **157**, 457 (1876).
12. See also J. Plank, *Sitzungsber. Akad. Wiss. Wien* **74** (2), 215 (1877). C. Truesdell has suggested that f should properly be called the "Maxwell number." Note that the "Prandtl number" is defined as $\mu c_p / k$, where c_p = specific heat at constant pressure.
13. O. E. Meyer, *Die Kinetische Theorie der Gase* (Breslau, 1877; 2nd ed., 1899); see the English trans. of the 2nd ed. by R. E. Baynes, *The Kinetic Theory of Gases* (London, 1899), pp. 465–66.
14. See §13.4, footnote 19.
15. J. Stefan, *Sitzungsber. Akad. Wiss. Wien* **72**, 69 (1876).
16. A. Schleiermacher, *Ann. Phys.* [3] **36**, 346 (1889). See the criticism of this determination by Meyer, *Kinetic Theory of Gases*, p. 296. According to Zaitseva, Schleiermacher used too low a value for the viscosity in computing f; when a better value for viscosity is used, it comes out 2.69, in good agreement with modern results. L. S. Zaitseva, *Zhur. tekhn. fiz.* **29**, 497 (1959), English trans. in *Soviet. Physics-Tech. Phys.* **4**, 444 (1959).
17. W. Schwarze, *Ann. Phys.* [4] **11**, 303 (1903). According to A. Eucken [*Phys. Z.* **14**, 324 (1913)] the value for helium is too high because Schwarze miscalculated the value of c_v.
18. E. Bannawitz, *Ann. Phys.* [4] **48**, 577 (1915).
19. See S. Chapman and T. G. Cowling, *The Mathematical Theory of Non-Uniform Gases* (Cambridge University Press, 2nd ed., 1952), p. 236; more recent values are given by W. G. Kannuluik and E. H. Carman, *Proc. Phys. Soc.* **B65**, 701 (1952).
20. A. Eucken, *Phys. Z.* **14**, 324 (1913); J. A. Pollock, *J. Roy. Soc. N.S. Wales* **49**, 249 (1915), *Phil. Mag.* [6] **31**, 52 (1916), *Proc. Roy. Soc. N.S. Wales* **53**, 116 (1919): K. Schäfer, *Z. phys. Chem.* **B53**, 149 (1943); Chapman and Cowling, *op. cit.*, 237–43; J. O. Hirschfelder, C. F. Curtiss and R. B. Bird, *Molecular Theory of Gases and Liquids* (New York: Wiley, 1954), pp. 573–74.
21. Edward Mason has pointed out to me that the defect may be more serious if one has to make f temperature-dependent, as seems necessary in some cases. On the other hand the same type of empirical correction should show up in the treatment of sound absorption data, which would give an additional check on the values of f used to fit heat conduction measurements (private communication). For a recent assessment of the theory see J. A. Barker, M. V. Bobetic, and A. Pompe, *Molecular Physics* **20**, 347 (1971). Values of thermal conductivity determined from viscosity data using the Chapman–Enskog theory are now considered more reliable than those found by direct experiments: see H. J. M. Hanley, R. D. McCarty & H. Intemann, *J. Research Nat. Bur. Stds.* **74A**, 331 (1970); J. Kestin, S. T. Ro and W. Wakeham, *Physica* **58**, 165 (1972).

22. See the papers of Winkelmann, Meyer, and von Obermayer cited above.
23. See S. G. Brush, *Arch. Rat. Mech. Anal.* **39**, 1 (1970); *Kinetic Theory* (New York: Pergamon Press, 1972), **3**, 21–31.
24. See §13.4, footnote 5.
25. Brush, *op. cit.*, footnote 23.
26. A. Winkelmann, *Ann. Phys.* [3] **1**, 63 (1877).
27. L. Graetz, *Ann. Phys.* [3] **14**, 232 (1881).
28. A. Winkelmann, *Ann. Phys.* [3] **19**, 649 (1883).
29. A. Winkelmann, *Ann. Phys.* [3] **29**, 68 (1886).
30. A. Schleiermacher, *Ann. Phys.* [3] **34**, 623 (1888), **36**, 346 (1889).
31. W. Eichhorn, *Ann. Phys.* [3] **40**, 697 (1890).
32. A. Winkelmann, *Ann. Phys.* [3] **44**, 177, 429 (1891).
33. P. A. Eckerlein, *Ann. Phys.* [4] **3**, 120 (1900).
34. A. Eucken, *Phys. Z.* **12**, 1101 (1911).
35. S. Weber, *Ann. Phys.* [4] **54**, 325, 437 (1917).
36. H. Gregory and C. T. Archer, *Phil. Mag.* [7] **1**, 593 (1926).
37. E. O. Hercus and T. H. Laby, *Phil. Mag.* [7] **3**, 1061 (1927).
38. B. G. Dickins, *Proc. R. S. London* **143A**, 517 (1934).
39. W. G. Kannuluik and L. H. Martin, *Proc. R. S. London* **144A**, 496 (1934).
40. C. Y. Ho, R. W. Powell and P. E. Liley, *J. Phys. Chem. Ref. Data* **1**, 279 (1972). They give $k = 0.001665$ W/cm °K for 270°K and $k = 0.001717$ W/cm °K for 280°K; I have interpolated and converted into cal/cm sec °C. The value of α comes from a private communication from P. E. Liley.
41. C. Christiansen, *Ann. Phys.* [3] **14**, 23 (1881).
42. L. Graetz, *Ann. Phys.* [3] **45**, 298 (1892).
43. A. Winkelmann, *Ann. Phys.* [3] **48**, 180 (1893).
44. W. Kutta, *Ann. Phys.* [3] **54**, 104 (1895) (correction of Winkelmann's value given above in footnote 32).
45. E. Müller, *Die Abhängigkeit des Wärmeleitungscoefficienten der Luft von der Temperatur* (Erlangen, 1901); *Phys. Z.* **2**, 161 (1901).
46. W. Schwarze, *Ann. Phys.* [4] **11**, 303 (1903).
47. E. Pauli, *Ann. Phys.* [4] **23**, 907 (1907).
48. E. O. Hercus and T. H. Laby, *Proc. R. S. London* **95A**, 190 (1918).
49. W. B. Mann and B. G. Dickens, *Proc. R. S. London* **134**, 77 (1931).
50. H. S. Gregory and C. T. Archer, *Phil. Mag.* [7] **15**, 301 (1933).
51. E. O. Hercus and D. M. Sutherland, *Proc. R. S. London* **145A**, 599 (1934).
52. S. W. Milverton, *Phil. Mag.* [7] **17**, 397 (1934).
53. K. Tödheide, F. Hensel and E. U. Franck, *Landolt-Börnstein Zahlenwerte und Funktionen*, 5. Teil, Bandteil b (New York: Springer-Verlag, Sechste Auflage 1968), p. 39. Their value, 241 μW/cm grd, has been converted into cal/cm sec °C. The value of α comes from numerical values furnished by P. E. Liley.

13.6 The temperature of the sun

The elaborate efforts devoted to calculating and measuring the heat conduction coefficients of gases have not completely dislodged Count Rumford's assertion that fluids do not conduct heat; even today we think of quiescent air as an "insulator," in designing houses and winter clothing. Thus heat conduction in gases would be of rather limited interest in the history of science if it were not associated with a cluster of problems involving radiation: What is the temperature of the sun? At what rate does it radiate energy to the earth? How much of this radiation is absorbed by the earth's atmosphere? What is the relation between the atomic structure of matter and its ability to radiate and absorb energy at different frequencies? The last of these questions is somewhat beyond the scope of this chapter but we can at least indicate how radiation problems were involved in the prehistory of modern atomic physics.

Before 1870 it was generally assumed that the Dulong–Petit cooling law, discussed in §13.2, gives the correct relation between the temperature of a body and the amount of heat radiation it emits, either in a vacuum or in the presence of air.[1] But could the formula be extrapolated to radiators as hot as the sun? In 1838, Pouillet used the Dulong–Petit formula to estimate the temperature of the sun, and arrived at a value between 1461° and 1761° (depending on what assumption was made about the emissivity of the sun's surface).[2] Although Pouillet claimed that he had himself verified the formula for temperatures above 1000°, other scientists eventually became skeptical about assigning such a low temperature to the sun. Platinum, for example, melts at about 1750° and can be vaporized by focusing the sun's rays on it through a lens, which would seem to indicate that the sun itself must be hotter than 1750°.[3] On the other hand, estimates based on Newton's law of cooling, such as that by J. J. Waterston in 1861, were enormously larger—around 13 million degrees.[4]

Any estimate of the sun's temperature which depends on measuring its rate of radiation emission must of course depend on some assumption about the fraction of the radiation absorbed by the earth's atmosphere. Such assumptions had no reliable experimental basis until around 1860 when John Tyndall succeeded in measuring the absorption of radiant heat by different gases. He found that carbon dioxide absorbs about 90 times as much as dry air or hydrogen, using a copper tube filled with boiling water as the heat source.[5] This result was soon recognized to be of great importance in meteorology, and it stimulated a number of further investigations in the 1880's and later; it was suggested, for example, that measurements of solar heat reaching the ground could be used to determine the proportion of carbon dioxide in the atmosphere overhead.[6] The suggestion of Svante Arrhenius, that variations in atmospheric carbon-dioxide content can trigger large-scale climatic changes such as glacial epochs, is frequently revived by environmentalists, despite theoretical calculations showing that large increases in carbon dioxide content would have little effect on the amount of radiation absorbed.[7]

Tyndall's experiments on the absorption of radiation by water vapor are of interest for similar reasons. His initial report that undried air, containing less than $\frac{1}{2}$% water vapor, produced an absorption 13 times as great as dried air,[8] led to a controversy with Magnus, who suggested that the London air used in Tyndall's experiments was probably polluted by small solid particles which account for the absorption.[9] To meet this objection Tyndall repeated the experiment with supposedly purer air from other locations outside London, and still found the same results.[10] Magnus, in his own experiments, found no difference in absorption between dry and moist air.[11] The controversy could not be resolved until experiments were done with radiation of a definite wavelength.[12]

Until the degree of "diathermancy" of different gases could be established, the results of conductivity measurements on gases and vapors (especially those composed of polyatomic molecules) were uncertain. Tyndall's dispute with Magnus led him to question whether Magnus had really eliminated convection in the experiments alleged to prove the existence of conduction; and he challenged the contention of Magnus that "inasmuch as hydrogen is more athermanous than atmospheric air, the greater heating of the thermometer in the first experiments could only be due to the greater conductivity of hydrogen."[13]

In the 1870's and 1880's there was considerable interest in problems of solar physics.[14] The Dulong–Petit cooling law was the subject of vigorous debate, especially in Paris following the publication of Angelo Secchi's book on the sun which contained a new estimate of its temperature.[15] Secchi and others argued that the Dulong–Petit law should be rejected because it yields an unreasonably low temperature for the sun.[16] John Ericsson, the Swedish-American inventor who built the *Monitor*, flatly asserted:

> Probably no doctrine in physics has occasioned such serious misconception as that propounded by MM. Dulong and Petit. The advance of every branch of knowledge connected with radiant heat has been retarded by the adoption of their doctrine regarding its transmission.[17]

At the same time it was hard to defend the validity of Newton's law of cooling at high temperatures, and the Dulong–Petit law still had some supporters.[18]

Before 1878, there had been few serious attempts to replace the Dulong–Petit law by an alternative other than that of Newton. One of the first of the new formulas was that of Francesco Rossetti,

$$y = aT^2(T - \theta) - b(T - \theta),$$

where y is the thermal effect measured by thermoelectric instruments, T = absolute temperature of the radiating body, and θ = temperature of the surrounding medium. The first term represents radiation *in vacuo*, the second the "influence of the surrounding air." Thus Newton's formula $y \propto (T - \theta)$ was modified by assuming that the emissive power of the body is not constant but increases as the square of the temperature. This assumption was said to be confirmed by the experiments of Tyndall.[19] Rossetti stated that his equation represents the available data better than the Dulong–Petit law for temperatures above 300°C, including some data up to 2670°. He estimated the sun's temperature to be between 10 000° and 20 000°, depending on how much radiation is assumed to be absorbed by the earth's atmosphere.[20]

In a more extensive investigation published at about the same time, Aimé Witz reviewed the data on aerial cooling and its dependence on the pressure of the gas surrounding the hot body. He found that the Dulong–Petit factor, $p^{0.45}$, is not valid for pressures greater than 1 atm; rather, the cooling velocity seemed to be almost independent of pressure, up to about 3.5 atm. He suggested that the "mobility" of molecules might vary inversely as density, and if the cooling power of a gas depends jointly on mobility and density the effect of these two factors might cancel out. But Witz, like Rossetti, made no reference to the Clausius–Maxwell papers on heat conduction, which might have provided him with a firmer basis for his arguments.[21] (Here is yet another example of the failure of French scientists to pay sufficient attention to work being done in Britain and Germany during this period.)

In this section I have suggested some features of the problem situation in the 1870's just before Josef Stefan did his work leading to the establishment of the T^4 law. While my emphasis has been on concerns related to solar physics, it should be mentioned that technological applications provided another reason for interest in radiation from hot bodies, as illustrated by S. P. Langley's comparison of the sun's radiation with that from the Bessemer converter at the Edgar Thompson steelworks near Pittsburgh.[22] The technological interest in radiation studies was responsible for substantial support for the German research in this area in the 1880's and 1890's.[23]

Notes for §13.6

1. See refs. in footnotes 9–11, §13.2. The validity of the law was questioned by Wilhelmy (*op. cit.*, §13.2, footnote 12), and T. Box, *A practical Treatise on Heat* (London, 1868), p. 163f.
2. [C. S. M.] Pouillet, *Compt. Rend. Acad. Sci. Paris* **7**, 24 (1838); *Ann. Phys.* [2] **45**, 25, 481 (1838); English trans. in *Taylor's Scientific Memoirs* **4**, 44 (1846).
3. A. J. Meadows, *Early Solar Physics* (New York: Pergamon Press, 1970), pp. 8–9.
4. J. J. Waterston, *Monthly Notices Roy. Astron. Soc.* **22**, 60 (1862), reprinted in *The Collected Scientific Papers of John James Waterston* (Edinburgh: Oliver & Boyd, 1928); see also *Monthly Notices Roy. Astron. Soc.* **20**, 196 (1860) (not reprinted in the *Papers*).
5. The experiments of Franz in 1855, mentioned in §13.3, were not considered reliable: "Franz... discovered a supposed absorption by dry air in a 3-foot tube of 3.54 per cent; but this was attributed by Tyndall to the fact that Franz employed glass plates to close the ends of the tube, and as glass largely absorbs the non-luminous radiation, the plates soon become warm and radiate heat to the pile. In this situation of affairs, when the cool gas is admitted into the tube it rapidly lowers the temperature of the radiating glass plates both by conduction and convection, so that the total radiation to the pile is reduced just as if the gas exercised a true absorption of the radiant heat." – T. Preston, *Theory of Heat* (3rd ed., 1919), p. 522. This is an elaboration of Tyndall's brief remarks which may be found on p. 38 of his *Contributions to Molecular Physics in the Domain of Radiant Heat* (London, 1872), reprinted from *Phil. Trans.*, 1861. Tyndall's experimental method has been discussed in some detail by D. E. Williamson, *Am. Sci.* **39**, 672 (1951). For the carbon dioxide experiment see Tyndall's *Contributions*, p. 80 (read to the Royal Society of London, 30 Jan. 1862).
6. E. Lecher, *Sitzungsber. Akad. Wiss. Wien, Math.-Naturwiss. Kl.*, **82** (2), 851 (1880), **86** (2), 52 (1883); *Ann. Phys.* [3] **12**, 466 (1881). E. Lecher and J. Pernter, *Ann. Phys.* [3] **12**, 180 (1881). H. Heine, *Ann. Phys.* [3] **16**, 441 (1882).
7. Arrhenius, *Verh. Deutsch. Naturf.*, Th. 2, Hälfte 1, p. 41 (1895); *Bihang till Kongl. Svenska Vetenskaps-Akademiens Handlingar* (Stockholm), **22** (Afd. 1, No. 1); *Phil. Mag.* [5] **41**, 237 (1896). T. C. Chamberlin, *J. Geol.* **7**, 545, 667, 751 (1899). Early critiques were published by K. Ångström, *Ann. Phys.* [4] **3**, 720 (1900); C. Schaefer, *Ann. Phys.* [4] **16**, 93 (1905); H. Rubens and R. Ladenburg, *Verh. Deutsch. Phys. Ges.* **7**, 170 (1905). The theory was revived by G. S. Callendar, *Q. J. Royal Meteorol. Soc.* **64**, 223 (1938) and later papers. A good elementary introduction to the subject is the article by G. N. Plass, *Am. J. Phys.* **24**, 376 (1956), but Plass's quantitative conclusions have been frequently criticized: see L. D. Kaplan, *Tellus* **12**, 204 (1960), **13**, 301 (1961); F. Möller, *J. Geophys. Res.* **68**, 3877 (1963); S. Manabe and R. T. Wetherald, *J. Atmos. Sci.* **24**, 241 (1967); H. Landsberg, *Science* **170**, 1265 (1970); S. I. Rasool and S. H. Schneider, *Science* **173**, 138 (1971); S. F. Singer, ed., *Global Effects of Environmental Pollution* (New York: Springer Verlag, 1970), pp. 25, 139 (papers by S. Manabe and J. M. Mitchell, Jr.).
8. Tyndall, *Contributions*, pp. 41, 109.

9. See Tyndall, *Contributions*, pp. 61, 124, 129.
10. *Ibid.*, pp. 129–30.
11. G. Magnus, *Ann. Phys.* [2] **112**, 531 (1861).
12. Abney and Festing, *Proc. R. S. London* **35**, 80 (1883); S. P. Langley, *Am. J. Sci.* [3] **28**, 163 (1884); *Phil. Mag.* [5] **18**, 289 (1884). Later work is summarized by F. E. Fowle, *Smiths. Misc. Coll.* **68**, no. 8 (1917).
13. Tyndall, *Contributions*, p. 382. Cf. Buff's criticisms mentioned above, §13.3, footnote 21.
14. Meadows, *op. cit.* (footnote 3); O. Chwolson, *Repert. f. Meteorologie* (St. Petersburg) **15**, 1 (1892).
15. A. Secchi, *Le Soleil* (Paris, 1870, 2nd ed., 1875–77), **2**, Livre VI.
16. A. Secchi, *Compt. Rend. Acad. Sci. Paris* **74**, 26, 301 (1872), **78**, 719 (1874). J. Homer Lane, *Am. J. Sci.* [2] **50**, 57 (1870), reprinted in Meadows, *op. cit.* (footnote 3). L. Soret, *Arch. Sci. Phys.* [2] **44**, 220, **45**, 252 (1872), **52**, 89 (1875), **55**, 217 (1876), [3] **1**, 86 (1878); *Ann. école norm. Paris* **3**, 435 (1874). S. P. Langley, *Proc. Am. Acad.* [2] **6**, 106 (1879).
17. J. Ericsson, *Nature* **5**, 505, **6**, 106 (1872); quotation from his *Contributions to the Centennial Exhibition* (New York, 1876), pp. 44–45. For accounts of Ericsson's life and his "caloric engine," which created a sensation in New York in 1853, see W. C. Church, *The Life of John Ericsson* (New York, 1891); E. S. Ferguson, *U.S. Nat. Mus. Bull.* **228**, 41 (1961).
18. A. Boutan, *J. de Physique* **1**, 154 (1872). Vicaire, *Compt. Rend. Acad. Sci. Paris* **74**, 31, 461 (1872), **78**, 1012 (1874). J. Violle, *Compt. Rend. Acad. Sci. Paris* **78**, 1425, 1816, **79**, 746 (1874); *Ann. école norm. Paris* **4**, 363 (1875); *Arch. Sci. Phys.* [2] **55**, 207 (1876); *Ann. Chim. Phys.* **10**, 289 (1877). The last-cited memoir won a prize of 2000 francs offered by the Académie des Sciences for a method of estimating the temperature of the sun, although the judges stated that none of the contestants had completely settled the problem; see the report by Fizeau, Jamin, Faye, Berthelot, and Desains, *Compt. Rend. Acad. Sci. Paris* **84**, 813 (1877). A smaller prize was awarded to A. Crova, who accepted "provisionally" the validity of the Dulong–Petit law as late as 1880: *Ann. Chim. Phys.* [5] **19**, 472 (1880).
19. F. Rossetti, *Mem. Accad. Lincei* **2**, 169 (1878); *Ann. Chim. Phys.* [5] **17**, 177 (1879); *Phil. Mag.* [4] **8**, 324, 438, 537 (1879). See also *Nuovo Cimento* **6**, 101 (1879); *J. Physique* **8**, 257 (1879).
20. The problem of atmospheric corrections was discussed by Violle (1877 memoir, cited in footnote 18) but was not generally considered satisfactorily resolved until the work of Langley, Abney, Paschen, Rubens, and others in the 1880's and 1890's.
21. A. Witz, *Ann. Chim. Phys.* [5] **15**, 433 (1878); see also *Compt. Rend. Acad. Sci. Paris* **89**, 228 (1879), **92**, 405 (1881); *Ann. Chim. Phys.* [5] **18**, 208 (1879), **23**, 131 (1881). Witz also published a useful review of the research on the radiation-temperature problem in the 1870's: *Rev. Quest. Sci.* **7**, 229 (1880).
22. Langley, *op. cit.* (footnote 16). Violle, in his 1877 memoir cited in footnote 18, also mentioned that he had compared the sun's radiation with that from an iron furnace at the Martin-Siemens factory. J. D. Bernal is apparently alluding to this type of observation when he asserts that the development of the radiation pyrometer as a means of assessing furnace temperatures led through the study of energy distribution in radiation to Planck's quantum theory: *Science and Industry in the Nineteenth Century* (London: Routledge & Kegan Paul, 1953; Bloomington: Indiana University Press, 1970), p. 36.

23. Hans Kangro, *Vorgeschichte des Planckschen Strahlungsgesetzes* (Wiesbaden: Steiner, 1970).

13.7 The Stefan–Boltzmann law

In view of the extensive (though somewhat conflicting) experimental data on the radiation-temperature relation available by 1879, it seems strange that the original stimulus for the modern radiation law was a single pair of rough observations by John Tyndall, made in 1864; and even stranger that the account of these observations had to be filtered through a textbook writer as well as a translator before reaching the mind that was to found a new radiation law on them.

In his paper "On luminous and obscure radiation," Tyndall remarked that in solid metals, "augmented temperature introduces waves of shorter periods into the radiation," whereas the period of vibration of radiation emitted by vapors is independent of temperature. He continued:

> It may be asked, "what becomes of the long obscure periods when we heighten the temperature? Are they broken up or changed into short ones, or do they maintain themselves side by side with the new vibrations?" The question is worth an experimental answer.
>
> A spiral of platinum wire, suitably supported, was placed within the camera of the electric lamp at the place usually occupied by the carbon points. This spiral was connected with a voltaic battery; and by varying the resistance it was possible to raise it gradually from a state of darkness to an intense white heat.... A thermo-pile was ... moved into the region of obscure rays beyond the red of the spectrum. Altering nothing but the strength of the current, the spiral was reduced to darkness, and lowered in temperature till the deflection of the galvanometer fell to 1°. Our question is, "What becomes of the waves which produce this deflection when new ones are introduced by augmenting the temperature of the spiral?"
>
> Causing the spiral to pass from this state of darkness through various degrees of incandescence, the deflections given in table 13.7-1 were obtained. The deflection of 60° here obtained is equivalent to 122 of the first degrees of the valvanometer. Hence the intensity of the obscure rays in the case of the full white heat is 122 times that of the rays of the same refrangibility emitted by the dark

Table 13.7-1

Appearance of Spiral	Deflection by Obscure rays	Appearance of Spiral	Deflection by Obscure rays
Dark	1°	Full red	27°
Dark	6	Bright red	44.4
Faint red	10.4	Nearly white	54.3
Dull red	12.5	Full white	60
Red	18		

spiral used at the commencement. Or, as the intensity is proportional to the square of the amplitude, this, in the case of the last deflection, was eleven times that of the waves which produced the first. The *wavelength*, of course, remained the same throughout.[1]

A German translation of Tyndall's paper was published in the *Annalen der Physik*, where it was read by Adolph Wüllner.[2] In his textbook published in 1871, Wüllner stated that the galvanometer reading of 60° in Tyndall's experiment corresponded to an amount of heat proportional to 122, whereas for "faint red" [*schwachroth*] the amount was 10.4°. On the basis of earlier work by Draper,[3] Wüllner assigned the temperatures 525° and 1200° to these appearances, and noted that the intensity of radiation had increased nearly twelvefold (ratio of 122 to 10.4).[4] The apparent error of failing to convert the number 10.4° from a galvanometer deflection to a corrected heat intensity is actually negligible in this case, since according to Tyndall the galvanometer response is nearly linear up to about 30°.[5]

In Stefan's 1879 paper the T^4 law was introduced in a rather mysterious way in the first section, and it was not until near the end that he revealed where it came from:

> ...I will note here the remark which Wüllner makes in his textbook on the report of Tyndall's experiment on radiation of a platinum wire made to glow by an electric current, since this remark first induced me to assume that heat radiation is proportional to the fourth power of the absolute temperature.
>
> From weak red heat (about 525°) to complete white heat (about 1200°) the intensity of radiation increases from 10.4 to 122, thus nearly twelvefold (more precisely 11.7). The ratio of the absolute temperatures $273 + 1200$ and $273 + 525$ gives when raised to the fourth power 11.6.[6]

Although the fortuitous character of the experimental origin of Stefan's radiation law has sometimes been remarked,[7] it seems to have been generally forgotten that Tyndall's observation referred specifically to the intensity of radiation emitted at a fixed wavelength rather than to the total emission, though it is only for the latter quantity that the T^4 law is now considered valid.

Stefan had begun the paper by presenting an extensive critical review of the Dulong–Petit memoir of 1817. He showed that the Dulong–Petit data can be represented equally well by a T^4 law or by the original exponential formula of Dulong and Petit, but then went on to argue that the numerical values reported by those authors do not really have the meaning usually ascribed to them. In the original experiment, a thermometer was placed inside a large copper sphere and the air was then pumped out; it was stated that in most of the experiments the pressure of the remaining air was no more than 2 mm. Even this number is uncertain since, when the paper was reprinted in the *Journal de l'école Polytechnique*, the pressure was given as 3 mm rather than 2 mm.[8] Dulong and Petit had assumed that the transfer of heat through the air would decrease uniformly as the air pressure decreased, and by taking measurements at 720, 360, 180, 90, and 45 min pressure they thought they had established well enough the law of this decrease. Since the rates of cooling observed at 2 mm differed only slightly from those obtained by extrapolating the data from higher pressure down to zero pressure, they assumed that the data for 2 mm represented the true rate of cooling by radiation into a vacuum, and subtracted this amount, as a "radiation correction," from the rate observed at higher pressures. What they did not realize was that the two modes of heat transfer through air—conduction and convection—behave quite differently as the pressure varies. It is just at this point that theoretical ideas play an essential role in the interpretation of experiments.

According to Stefan, most of the cooling which Dulong and Petit observed at high pressures was due to convection, and this process does indeed decrease in importance as the air pressure decreases. But conduction is independent of pressure over a large range, and is about as great at 2 mm pressure as at atmospheric pressure, according to the Maxwell–Clausius theory as confirmed experimentally by Stefan himself and by others. Thus what Dulong and Petit called the "radiation correction" was really the sum of radiation and conduction. Any extrapolation of their results to higher temperatures would be quite misleading if (as is indeed the case) the temperature-dependence of radiation is significantly different from that of conduction.

Stefan then attempted to reinterpret the Dulong–Petit data in the light of more recent knowledge about heat conduction in gases, in order to retrieve the actual radiation-temperature relation for a vacuum. Unfortunately Dulong and Petit did not publish enough details of their original experimental apparatus and uncorrected cooling rates to enable this to be done unambiguously, and the confusion between 2 and 3 mm pressure in the two published versions of their paper made matters even worse. It was not even certain how much convection was actually taking place, since this probably depended on how soon the temperatures were measured after the apparatus was pumped out. Nevertheless the Dulong–Petit data were still, after more than 60 years, the most comprehensive available, and Stefan thought it was worth some effort to salvage them in the absence of anything better. So he computed a theoretical heat-conduction correction, based on plausible assumptions about the dimensions of the Dulong–Petit apparatus, assuming that the conduction coefficient is a linear function of temperature, with a temperature coefficient of 0.0027,[9] and that its value for air at 0°C is 0.000054.

Stefan found that the law of cooling of a blackened thermometer was not significantly affected by the heat-conduction correction, but for the silvered thermometer the corrections were a sizable fraction of the total. Hence, "all conclusions which Dulong and Petit have drawn from their observations on the radiation of a silver surface lose their foundations." For example, the result that the ratio of cooling velocities for the bare and silvered thermometer is 5.7 at all temperatures has no absolute meaning, since it depends on the arrangement of the apparatus, and cannot imply anything about the ratio of emissivities of glass and silver.

The T^4 law was found to fit the newly-corrected Dulong–Petit data as well as the Dulong–Petit formula, or better in some cases, but the improvement was not very striking. Stefan had not yet given any plausible reason for selecting the fourth power from all other possible candidates that might be qualitatively similar, and which might fit the data about as well. The T^4 law did begin to gain an advantage over the Dulong–Petit law when, in later sections of Stefan's paper, the experiments of de la Provostaye and Desains, Desprez, Draper, and Ericsson were analysed. But it is hard to believe that the T^4 law would ever have been accepted on the basis of Stefan's paper alone.

* * *

The first direct attempt to test Stefan's T^4 law was undertaken by

L. Graetz in Strassburg, using an apparatus similar to that of Kundt and Warburg. A thermometer was placed in the middle of a glass sphere which was then evacuated with a Geissler pump to a pressure so low that (according to Graetz) there is no heat conduction by the remaining air. For temperatures from 0° to 250°C he found that Stefan's law is definitely preferable to that of Dulong and Petit.[10] It is scarcely a coincidence that Graetz soon afterward published a series of measurements of heat conduction coefficients of gases; the experimental methods for determining radiation and conduction through gases were inevitably linked closely together.[11]

In 1881, the Danish physicist L. Lorenz published a paper on the conductivity of metals for heat and electricity. He found it necessary to make a correction for the heat lost by the metal to the air, in order to interpret experimental data, and proposed a formula in which Stefan's T^4 radiation law was used in conjunction with an additional term proportional to $(T - T_0)^{\frac{5}{4}}$ for the heat loss through air.[12] (Lorenz did not make a clear distinction between conduction and convection in this paper.)

Further support for Stefan's law came from Lecher,[13] Christiansen,[14] and Schneebeli.[15] But Violle, while finally abandoning the Dulong–Petit law, stated that Stefan's law underestimated the radiation emitted at high temperatures,[16] while Young[17] and Riviere[18] ignored Stefan's law in their discussions of the need to modify the Dulong–Petit law. P. G. Tait was almost alone among physicists in retaining the Dulong–Petit law as late as 1884 without making any mention at all of Stefan's law.[19] Sir William Siemens, without mentioning either Dulong–Petit or Stefan, proposed yet another formula based on his own experiments with an electric arc. Inverting the usual argument that the Dulong–Petit law predicts too low a temperature for the sun, Siemens concluded that the sun's temperature must not be *more* than 3000°C because "the energy emitted from a source much exceeding this limit would no longer be luminous, but consist mainly of ultra-violet rays, rendering the sun invisible, but scorching and destructive of all life."[20]

Our (perhaps incomplete) survey of the literature in the five years following the publication of Stefan's radiation law shows that five scientists gave it at least qualified support, but four others appeared not to have heard of it, and one rejected it. Yet only one of the ten still accepted the Dulong–Petit law.

* * *

As we will see in reviewing the later experimental work, Stefan's radiation law could not be established as clearly superior to its competitors for at least a decade after its publication. It was therefore of great importance that Ludwig Boltzmann in 1884 provided an abstract argument which singled out the T^4 law as a strict consequence of the Second Law of Thermodynamics combined with Maxwell's electromagnetic theory. It was this theoretical derivation which encouraged physicists to give Stefan's law the benefit of the doubt, and goaded experimentalists to search for a "perfect black body" for which the T^4 formula would be valid, even if it did not apply accurately to the radiation of many substances under ordinary conditions.

Boltzmann's theoretical discovery emerged from a tradition of speculation and experiments on the *pressure* of radiation or light, going back to the 17th century if not earlier.[21] Shortly after the establishment of thermodynamics, William Thomson had pointed out that radiant heat or light must produce a mechanical effect "since the communication of heat to a body is merely the excitation of certain motions among its particles." He estimated that about 1 horsepower of ordinary mechanical effect might be produced by an engine exposing 1800 square feet of surface to receive solar heat during a warm summer day in the country. He also insisted that such a conversion of heat into mechanical effect must be subject to Carnot's principle, but did not say precisely what this entailed.[22]

The problem of radiation pressure was raised again by the Crookes Radiometer in the 1870's, together with the publication in 1873 of Maxwell's calculation of the pressure and energy of electromagnetic waves. Maxwell found that "in a medium in which waves are propagated there is a pressure in the direction normal to the waves, and numerically equal to the energy in unit of volume."[23] But the magnitude of the action of radiation on the radiometer vanes was soon found to be much greater than could be accounted for by Maxwell's theory, and while this fact at one point caused Maxwell to consider the possibility that his theoretical calculation might be wrong, the problem was soon resolved when the radiometer effect was attributed to the action of the residual gas.[24] Attempts to explain the radiometer effect thus helped to initiate theoretical work on rarefied gas dynamics which could later be used to clarify the mechanism of heat conduction at very low pressures.[25] But it also had another repercussion: it stimulated the thinking of Adolfo Bartoli, in Italy, about the general problem of radiation pressure in connection with thermodynamics. His monograph of 1876 was ignored by scientists in other countries,[26] but a paper of the American scientist H. T. Eddy on "Radiant Heat an exception to the

Second Law of Thermodynamics"[27] gave Bartoli another chance to put forth his ideas. He published an extract from his earlier book in a short article in *Il Nuovo Cimento*, which was later translated into German in Exner's *Repertorium*.[28] Even before this, E. Wiedemann had informed Boltzmann of Bartoli's book of 1876, and Boltzmann published a general analysis of Bartoli's contention that the existence of radiation pressure is required by the Second Law of Thermodynamics.[29] In a second paper, Boltzmann presented a derivation of the T^4 radiation law, which is now known as the Stefan–Boltzmann law.[30]

In the Bartoli–Boltzmann analysis, emphasis is placed on the quantity of radiation energy which must be present in a space maintained at a particular temperature, rather than on the rate of cooling of a hot body.[31] One result of this change in viewpoint has been to obscure the historical origin of the Stefan–Boltzmann law in a research tradition, going back to Newton, that was primarily concerned with the problem of heat transfer. In that tradition, radiation was only one mode of transfer, and it could be separated only with great difficulty from other modes such as conduction and convection by air. Thus the "Stefan" half of the law is not simply an experimental analysis of black-body radiation, as it is usually understood, but rather the last phase of an effort to extract the contribution of radiation from the total rate of cooling or rate of heat transfer from a hot body. The "Boltzmann" half involves the manipulation of a completely different set of theoretical concepts. (A sketch of of the derivation is given in Appendix 13B.) The beautiful coincidence of the two results was so striking that there never seemed to be any doubt that both expressed the same fundamental law of nature, even though one referred to the rate of cooling of a hot solid and the other to the energy density of heat radiation in a vacuum. Hence the "Stefan–Boltzmann law."

As is well known, the problem of determining the frequency-distribution of black-body radiation was already attracting considerable interest by 1884, and was to lead eventually to the invention of quantum theory by Max Planck in 1900. There seems to be no need to go into this aspect of the subject since several excellent accounts are already available.[32] But since the existence of an accepted law for the *total* radiation energy did play an important role in discussions of possible frequency-distribution laws in the 1880's and 1890's, it is useful to review the progress of gradual establishment of the Stefan–Boltzmann law in this period.

* * *

In 1885, an extensive study of radiation by August Schleiermacher gave results that seemed to contradict Stefan's formula when applied to platinum and copper oxide radiators. But Schleiermacher, an experimentalist, seems to have been so intimidated by Boltzmann's theoretical derivation of the T^4 law that he hesitated to state flatly that his own results contradicted it; instead he suggested that the absorption of radiation by the surface might change with temperature in such a way as to make the net emission vary erratically. He was willing to believe that Stefan's law would be absolutely correct for a perfectly black body, but he did not suggest how such an object was to be obtained. A more important consequence of his work was the development of an improved version of the Andrews–Grove "hot wire" method for measuring heat transfer. Schleiermacher pointed out the advantage of keeping the wire at a constant temperature (indicated by its electrical resistance) as compared to the determination of the rate of cooling of a thermometer, which could only give heat-transfer properties averaged over a finite temperature range. His method ultimately turned out to be more useful in determining the heat-conduction coefficients of gases than in testing the radiation law.[33]

P. G. Tait, writing on "Radiation and Convection" in the *Britannica*, stated the Dulong–Petit law without criticizing its validity, but also mentioned Boltzmann's "highly interesting" speculation leading to Stefan's T^4 law.[34] Violle maintained the validity of one of his own formulas, $mTb^{T^2}a^T$ where a depends on wavelength.[35] J. T. Bottomley, in England, reported experiments which "do not appear to give any support" to Stefan's law.[36] Schleiermacher, without retracting his earlier experimental results, assumed that Stefan's law was valid in a paper published in 1888.[37] H. F. Weber proposed a new law of the form $T\,e^{aT}$, which he said was in better agreement with the measurements of Graetz.[38] But Graetz himself disputed this, asserting that his data were more accurately represented by Stefan's formula than by Weber's.[39] Edler announced a new series of experimental results, which he compared with the formulas of Stefan and Weber; although Stefan's law came off somewhat better than Weber's in this case, Edler preferred to revive Wilhelmy's exponential law,[40] which he claimed to be a better representation of his data than either of the others.[41]

The American scientist William Ferrel found that the Dulong–Petit data, as corrected for heat conduction by Stefan, could be equally well fitted by a modified Dulong–Petit law of the form $[(1.0082)^{\Delta T} - 1]$ or a modified Stefan law of the form $T^{4.2}$. Thus Stefan's choice of the exact value 4 for the exponent had no experimental justification as compared with that of Dulong and Petit's formula, if one is willing to adjust the parameters in both of them. Ferrel rejected Rossetti's formula because it does not have the form of the difference of values of the same function for two different temperatures (as one would expect if Prevost's law of exchanges were valid). Ferrel stated that the Dulong–Petit law cannot be valid at extremely low temperatures since the amount of heat radiated does not go to zero at absolute zero temperature; on the other hand he did not think that any law of the form T^n can represent the true law if n is a constant.[42] In another paper Ferrel discussed Weber's formula and decided that it was not possible to fit all the data with the same value of the constant a, but he did not entirely reject the formula.[43]

The state of uncertainty suggested by the above review persisted as late as 1893 or 1894, with experimenters bemoaning their inability to get reliable results consistent with any one law; up to this time, it would appear that the best argument for the T^4 law was not its ability to fit experimental data significantly better than other formulae, but simply the fact that it possessed an appealing theoretical derivation.[44] Research continued in several countries, stimulated in part by the possible technological applications as well as by the desire to settle what seemed to be a fundamental problem in physics.[45] The solution came around 1895 when it was recognized that it was necessary to do experiments with perfectly "black" bodies simulated by a cavity (*Hohlraum*) maintained at a definite temperature, from which radiation was allowed to emerge through a small hole. The deviation from "blackness" increases with the size of the hole relative to the size of the cavity.[46] Experiments of this type, carried out by Lummer and Pringsheim, Paschen, Wilson, Mendenhall and Saunders, and others in the years 1895–99 finally did provide a definite proof of the validity of the Stefan–Boltzmann law.[47]

Of course, once the law had been established under these very restrictive conditions, it could be recognized as the preferred first approximation (often with little attention paid to further corrections) for vastly different sources of radiation such as the sun or even the earth's atmosphere.[48]

Notes for §13.7

1. J. Tyndall, *Phil. Mag.* [4] **28**, 329 (1864), reprinted in *Contributions to Molecular Physics in the Domain of Radiant Heat* (London, 1872), quotation from pp. 257–58.

A discussion of Tyndall's data translated into modern terms has been given by C. H. Davies, *Nature* **157**, 737, 879 (1946).
2. J. Tyndall, *Ann. Phys.* [2] **124**, 36 (1865).
3. See §13.2, footnote 11.
4. A. Wüllner, *Die Lehre von der Wärme vom Standpunkte der mechanischen Wärmetheorie* (Leipzig, 2nd ed., 1870–71, 3rd ed., 1874–75), pp. 215–16.
5. See Tyndall, *Contributions*, pp. 2, 57–58; *Heat A Mode of Motion* (New York, 4th ed., 1873), Appendix to ch. X.
6. J. Stefan, *Sitzungsber. Akad. Wiss. Wien* **79** (2), 391 (1879), quotation translated from p. 421.
7. E. Bauer, *Ann. Chim.* [8] **29**, 5 (1913), note 2 on p. 22; M. Jammer, *The Conceptual Development of Quantum Mechanics* (New York: McGraw-Hill, 1966), p. 6. According to C. E. Mendenhall, Stefan deduced his law "from a discussion of bad observations on imperfect radiators, for which it does not hold–a case in which two negatives have apparently been equivalent to an affirmative, so to speak" *Pyrometry*, The papers and discussion of a Symposium, Chicago, 1919 (New York: Amer. Inst. Mining & Metall. Engrs., 1920), p. 63. One should keep in mind that Stefan did use data from several other sources to justify his T^4 law.
8. Dulong and Petit, *J. École Poly.* **11**, 189 (1820).
9. This was approximately Winkelmann's value reported in 1877; cf. table 2 in §13.5.
10. L. Graetz, *Ann. Phys.* [3] **11**, 913 (1880).
11. See §13.5, tables 1 and 2, and footnote 27.
12. L. Lorenz, *Ann. Phys.* [3] **13**, 422, 582 (1881).
13. E. Lecher, *Ann. Phys.* [3] **17**, 477 (1882).
14. C. Christiansen, *Ann. Phys.* [3] **19**, 267 (1883).
15. H. Scheebeli, *Ann. Phys.* [3] **22**, 430 (1884).
16. J. Violle, *Compt. Rend. Acad. Sci. Paris* **92**, 866, 1204 (1881); see his review of Stefan's paper in *J. Physique* **10**, 317 (1881).
17. C. A. Young, *The Sun* (New York, 1881), p. 267.
18. Riviere, *Compt. Rend. Acad. Sci. Paris* **95**, 452 (1882).
19. P. G. Tait, *Heat* (London, 1904, "reprinted with corrections" from the 1st ed. of 1884), p. 278.
20. W. Siemens, *Proc. Roy. Inst.* **10**, 315 (1884); *Nature* **28**, 19 (1883); *Proc. R. S. London* **35**, 166 (1883); *The Royal Institution Library of Science, Astronomy* (Amsterdam: Elsevier, 1970), **1**, 207 (quotation from p. 209).
21. See E. T. Whittaker, *A History of the Theories of Aether and Electricity* (London: Nelson, rev. ed., 1951), **1**, 273–76; M. L. Schagrin, *Am. J. Phys.* **42**, 927 (1974).
22. W. Thomson, *Phil. Mag.* [4] **4**, 256 (1852).
23. J. C. Maxwell, *A Treatise on Electricity and Magnetism* (London, 3rd ed., 1891, reprinted by Dover Pubs., New York, 1954), **2**, 440–41.
24. A. E. Woodruff, *Isis* **58**, 188 (1966); *Physics Teacher* **6**, 358 (1968). S. G. Brush and C. W. F. Everitt, *Hist. Stud. Phys. Sci.* **1**, 105 (1969), reprinted as §5.5 in this book.
25. J. C. Maxwell, *Phil. Trans.* **170**, 231 (1880). For more recent work see the papers by Truesdell cited in §13.4, footnote 18; also Harold Grad, *Handbuch der Physik* **12**, 205 (1958).
26. A. Bartoli, *Sopra i movimenti prodotti dalla luce e dal calore e sopra il radiometro di Crookes* (Firenze, 1876).
27. H. T. Eddy, *Sci. Proc. Ohio Mech. Inst.* **1**, 105 (1882); *J. Franklin Inst.* **115**, 182 (1883); abstract by L. Boltzmann in *Ann. Phys. Beibl.* **7**, 251 (1883). See the interview

with P. Debye, September 1965, deposited at the American Institute of Physics, Center for History of Physics, New York.
28. A. Bàrtoli, *Nuovo Cim.* [3] **15**, 193 (1884); *Exner's Rep.* **21**, 198 (1885). This article gives bibliographic details on the book of 1876.
29. L. Boltzmann, *Ann. Phys.* [3] **22**, 31 (1884).
30. L. Boltzmann, *Ann. Phys.* [3] **22**, 291 (1884).
31. The idea that radiation in space could have a temperature was unacceptable to some physicists such as Lord Kelvin; see W. Wien, *Ann. Phys.* [4] **25**, 1 (1908).
32. H. Kangro, *Vorgeschichte des Planckschen Strahlungsgesetzes* (Wiesbaden: Steiner, 1970); M. J. Klein, *Arch. Hist. Exact Sci.* **1**, 459 (1962). A. L. Day and C. E. Van Orstrand, *Ap. J.* **19**, 1 (1904).
33. A. Schleiermacher, *Ueber die Abhängigkeit der Wärmestrahlung von der Temperatur und das Stefan'sche Gesetz* (Karlsruhe, 1885); *Ann. Phys.* [3] **26**, 287 (1885). For later applications of the method to heat conduction measurements see Schleiermacher, *Ann. Phys.* [3] **34**, 623 (1888), **36**, 346 (1889); J. R. Partington, *An Advanced Treatise on Physical Chemistry* (London: Longmans, Green and Co., 1949), **1**, 896–98.
34. P. G. Tait, *Encyclopaedia Britannica* (Edinburgh, 9th ed., 1886), **20**, 212; *Scientific Papers* (Cambridge University Press, 1890–1900), **2**, 457.
35. J. Violle, *Compt. Rend. Acad. Sci. Paris* **105**, 163 (1887).
36. J. T. Bottomley, *Nature* **33**, 85, 101 (1886); *Phil. Trans.* **178**, 429 (1887).
37. A. Schleiermacher, *Ann. Phys.* [3] **34**, 623 (1888).
38. H. F. Weber, *Sitzungsber. Akad. Wiss. Berlin*, p. 933 (1888).
39. L. Graetz, *Ann. Phys.* [3] **36**, 857 (1889).
40. See §13.2, footnote 12.
41. J. Edler, *Untersuchungen über die Abhängigkeit der Wärmestrahlung und der Absorption derselben durch Glimmerplatten von der Temperatur* (Greifswald, 1889); *Ann. Phys.* [3] **40**, 531 (1890).
42. W. Ferrel, *Am. J. Sci.* [3] **38**, 3 (1889). One of Ferrel's earlier papers had been criticized for using the Dulong–Petit law: see *Am. Meterorol. J.* **1**, 375 (1885).
43. W. Ferrel, *Am. J. Sci.* [3] **39**, 137 (1890).
44. J. T. Bottomley, *Phil. Trans.* **184A**, 591 (1893). W. E. Wilson and P. L. Gray, *Phil. Trans.* **185A**, 361 (1894). F. Paschen, *Ann. Phys.* [3] **49**, 50 (1893). H. Le Chatelier, *Compt. Rend. Acad. Sci. Paris* **114**, 737 (1892). O. Chwolson, *Repert. f. Meteorologie* (St. Petersburg) **15**, 1 (1892). W. L. Stevens, *Am. J. Sci.* [3] **44**, 321 (1892).
45. H. Kangro, *op. cit.* (footnote 32), **41**, 149f. K. Amano, cited in T. Hirosige, *Jap. Stud. Hist. Sci.* **9**, 5 (1970).
46. L. Graetz, Winkelmann's *Handbuch der Physik*, art. "Wärmestrahlung" (1896, 1906). There is an extensive discussion of the experimental aspects in a paper by W. Wien and O. Lummer, *Ann. Phys.* [3] **56**, 451 (1895). A similar suggestion about simulating black bodies was made by Lanchester in 1895, according to W. E. Wilson, *Astrophys. J.* **10**, 80 (1899).
47. O. Lummer and E. Pringsheim, *Ann. Physik* [3] **63**, 395 (1897), erratum in *ibid.* [4] **3**, 159 (1900). S. Tereschin, *Zhur. Russk. Fiz.-Khim. Obsh., Chast. Fiz.* **29**, 169, 277 (1897), 30, 15 (1898), abstr. in *Fortschr. Physik*, Abt. 2, p. 354 (1897). C. E. Mendenhall and F. A. Saunders, *Johns Hopkins Univ. Circulars* **17** (135), 55 (1898); *Naturwiss. Rundsch.* **13**, 457 (1898). O. Lummer and F. Kurlbaum, *Verh. Berlin Phys. Ges.* **17**, 106 (1898). W. E. Wilson, *Astrophys. J.* **10**, 80 (1899). O. Lummer, *Rapports presentés au Congrès Int. Phys., Paris, 1900* **2**, 41. J. Violle, *Compt. Rend. Acad. Sci. Paris* **130**, 1658 (1900).

For reviews of experimental tests see: L. Graetz, *Handbuch der Physik* (Breslau, 1896), **2** (2), 246–55, **3**, 371–78 (1906 ed.); C. K. Wagner and G. K. Burgess, *Bull. U.S. Bur. Stds.* **1**, 189 (1904).

48. Anders Ångström, *Smiths. Misc. Coll.* **65**, no. 3 (1915).

13.8 The three modes of heat transfer

By the end of the 19th century physicists had established what might be called a "classical synthesis" of theoretical ideas concerning heat transfer. With only a few modifications this synthesis survives in modern textbooks; this seems to be one area of physics where the revolutionary ideas of quantum mechanics and relativity have had little impact.[1]

A closer look at the theoretical and experimental basis of the classical synthesis reveals a number of defects and limitations, though these are not yet serious enough to shake the complacency of most scientists. In this section I will not attempt to review the 20th-century investigations of heat transfer, but limit myself to an evaluation of the situation as of 1900, with only a few suggestions of problems that may still be unresolved.

In the classical synthesis it is postulated that there are three distinct modes of heat transfer. Contrary to the impression one often gets, these modes are not defined operationally, but in terms of particular theories. (1) *Conduction* is the process described in the kinetic theory, whereby fast molecules drift randomly from hotter to cooler regions of the gas, where they collide with slower molecules and transfer to them some kinetic energy; at the same time slow molecules drift randomly from cooler to hotter regions, collide with faster molecules and acquire kinetic energy from them. The net result is heat conduction in accordance with Fourier's equation, *i.e.* a heat flux proportional to the temperature gradient with a coefficient that is independent of pressure but may vary with temperature in a way that depends on the intermolecular force law. (This process would occur in the absence of any gravitational field or externally imposed bulk motion of the fluid.) (2) *Radiation* is the process of heat transfer by emission and absorption of electromagnetic waves, at a rate which depends on the temperature of the emitter. The total emission rate, integrated over all frequencies, is proportional to the fourth power of the absolute temperature; this result remains valid even if one replaced the 19th-century ether-wave model by the 20th-century photon model; but the latter model gives a more satisfactory way of computing the distribution over frequencies (Planck's law). (3) *Convection* is described as transfer of heat by bulk motion of the fluid; in practice it may be defined as the amount of heat transfer remaining after theoretical contributions of conduction and radiation, defined above, have been subtracted.

Further refinements are explicitly included in the classical synthesis. (1a) At sufficiently low pressures, heat transfer by conduction will no longer be constant but will diminish to zero; if there is no gas left there cannot be any conduction. Presumably the decrease in conduction will occur when the mean free path is no longer very small compared to the dimensions of the apparatus, but it is difficult to do precise calculations of this effect from kinetic theory. (1b) At sufficiently high pressures, when the size of the molecules is no longer very small compared to intermolecular spaces, another mechanism of conduction will be important: "instantaneous" transfer of heat across a finite distance (*i.e.* the molecular diameter) at collisions. The result would be a series of density corrections analogous to the virial expansion for the equation of state; this was predicted on a semi-empirical basis by Gustav Jäger in 1899, and worked out in detail on the basis of the collision-transfer mechanism by Enskog in 1922.[2] (2a) The Stefan–Boltzmann T^4 law applies only to "black-body" radiators; real hot bodies will emit heat radiation at various frequencies in a way determined by their molecular structure. (2b) Radiation heat transfer cannot simply be added to conduction and convection, since some of the radiation will be absorbed by the gas, depending on its own molecular structure; this effect may depend on the radiation frequency.

In contrast to the molecular-theoretic nature of these refinements for the conduction and radiation modes, the standard treatment of convection has a much more phenomenal flavor. First, one distinguishes between "free" or "natural" convection, in which the flow of gas is due only to its own expansion by heating, and "forced" convection, in which the air current is made to flow past the heated surface with some specified speed. (3a) For free convection, the rate of cooling is proportional to $p^c t^b$, where p = pressure and t = temperature difference. The exponent b may have values ranging from 1.0 (corresponding to Newton's law of cooling) to 1.25 (as suggested by Lorenz[3]), while c may have values between 0.2 and 0.5.[4] (3b) For "forced" convection the cooling rate must in addition depend on the velocity.[5] In both cases the actual heat transfer by convection will depend on the geometry of the experimental situation, so that one usually has to solve a boundary-value problem in fluid dynamics.[6]

The value of such a synthesis is not that it solves the entire problem, but rather that it provides a framework within which one may pursue various special research programs. It assures the investigator that if he wants to make a detailed study of one mode of heat transfer, he can reduce the other modes to small calculable corrections by choosing suitable pressures and temperatures. For example, the Stefan–Boltzmann radiation law was used by engineers as early as 1909 as a radiation correction in studies of convection, while at the same time it was frequently asserted that conduction effects could be completely ignored.[7] Physicists, on the other hand, were primarily interested in determining conduction coefficients, so they designed their experiments in such a way that convection effects could be assumed negligible; usually this meant avoiding high pressures.[8] In meteorological applications, convection was considered quantitatively much more important than conduction,[9] and refinement (2b), radiation absorption by certain components of the earth's atmosphere, was of special interest to S. P. Langley and others.[10]

The historian is interested not only in the establishment and application of syntheses or Kuhnian paradigms, but also in the individual scientist who does not organize his ideas within these accepted frameworks; and of course the interest is heightened when the scientist happens to be one's own (somewhat distant) relative. Hence the following brief elaboration of refinement (1a) of the classical synthesis.

In 1881, William Crookes published the results of experiments on heat conduction in rarefied air (part of a research program which had begun with his discovery of the radiometer effect and led eventually to important discoveries concerning electric discharge phenomena). He stated that "the law of cooling in vacua so high that we may neglect convection, has not to my knowledge been determined." With the new types of vacuum apparatus and pumps developed in the second half of the 19th century, it was possible to make accurate measurements at much lower pressures than before. Crookes determined the time required to cool a glass bulb surrounded by air at pressures down to two millionths of an atmosphere ($= 2\,M$ in his notation). The cooling time was found to be nearly constant down to pressures of about $100\,M$, when it began to increase slowly; there was a very sharp increase in cooling time at a pressure of $15\,M$, corresponding (according to the graph presented by Crookes) to a sharp increase in the mean free path. Crookes inferred that heat transfer at low pressures is mainly due to conduction, and that since the time of cooling is rising sharply at $2\,M$ pressure, "in such vacua as exist in planetary space the

§13.8]   THE THREE MODES OF HEAT TRANSFER   527

loss of heat—which in that case would only take place by radiation—would be exceedingly slow."[11] (He did not not remark on the fact that radiation cooling would be much more rapid at higher temperatures.)

Bottomley confirmed Crookes' result that the heat conduction decreases sharply at very low pressures,[12] but there was little interest in exploring this effect further until 1898, when the American electrical engineer C. F. Brush and the Austrian physicist M. von Smoluchowski published experimental papers on heat conduction in rarefied gases.[13] Brush, known earlier for his development of arc-lamps for illumination,[14] repeated the Dulong–Petit experiment with the recently-developed vacuum equipment. He disputed Crookes' conclusion that the loss of heat in high vacua would be extremely small; he found instead that the rate of heat transmission "drops regularly at a rate faster than the diminution of pressure, during 95 per cent of the whole range of pressure from atmospheric to zero. Beyond this point the rate of heat transmission remains substantially constant...down to a pressure of about 0.0003...here the curve suddenly begins to drop again, and falls steadily...until it meets the aether line at the zero of pressure." The same experiment was tried with air, carbon monoxide, ethylene, and hydrogen; Brush said "it is gratifying that the vacuum, or aether line, locates itself exactly the same in all." Brush also reported that no change in the phenomena of heat transmission was noticed when the exhaustion of the container reached the point at which the mean free path of the gas molecules is equal to the distance between the thermometer bulb and the cold walls of the container.

Brush's knowledge of earlier work on heat conduction in gases was rather deficient; he wrote: "In the absence of convection-currents, that part of the heat transmitted by the gas was probably caused by a process analogous to conduction in solids.... But why the conductivity of a gas remains nearly constant through a very wide range of pressure is not very clear. Sir Wm Crookes' explanation of this phenomenon seems to be very unsatisfactory." As this statement indicates, the laws of heat conduction in gases would eventually have been established by experiment in the absence of a molecular theory, but the early publication of theoretical predictions certainly accelerated the progress of the subject by two or three decades.

Brush also suspected that the presence of a gas, even in small amounts, tends to retard or interfere with the heat-transmitting power of the ether, but there is no mention of Tyndall's experiments on diathermancy.

Smoluchowski was working in Warburg's laboratory in Berlin, and was thus acquainted with the effect suggested by Kundt and Warburg in 1875, that there may be a temperature discontinuity at the interface between a solid and a rarefied gas.[15] (As Smoluchowski pointed out, the possibility of such a discontinuity had been considered even earlier in Poisson's theory of heat conduction.[16]) Smoluchowski suggested that the phenomena attributed by Brush to radiation should instead be attributed to such a temperature discontinuity. He argued that the difference in temperature between two bodies in contact should depend on their internal temperature gradients, and developed a theory relating the "coefficient of temperature discontinuity" to the mean free path of molecules in the gas.[17] In another treatment he calculated this coefficient in terms of the fraction of molecules striking a surface that are absorbed and evaporated with a certain velocity, this fraction having been introduced as a parameter by Maxwell in his theory of rarefied gases.[18] Although there was some discrepancy in the numerical values predicted by his two formulae, Smoluchowski stressed the fact that both versions of the theory (corresponding to the "elastic sphere" and inverse fifth power repulsive force models, respectively) predict a temperature discontinuity. Both also predict that the magnitude of the discontinuity should be proportional to the mean free path or inversely as the pressure, in agreement with results obtained from his experiments and those of Brush.

Smoluchowski, writing at the end of the 19th century, was well aware of the positivist attacks on atomism and kinetic theory prevalent at that time, and took some pains to point out that the success of the kinetic theory in explaining heat conduction at low pressures was strong evidence in favor of the validity of the kinetic theory; other theories would not lead one to expect that surface layers of a gas should behave differently from the portions in the interior, or that the physical constants of a gas should depend on the thickness of a layer.[19]

In the meantime Brush, lacking an understanding of the kinetic theory, had been led astray by some of his experimental results and claimed that they proved the existence of a new substance, "Etherion," having enormous thermal conductivity but very small density.[20] According to Smoluchowski, the phenomena attributed by Brush to Etherion were in reality due to water vapor emitted from heated glass in a high vacuum.[21]

The problem of heat conduction in rarefied gases was taken up again around 1910 by Frederick Soddy and Arthur John Berry in Glasgow, and Martin Knudsen in Copenhagen. Their work led to qualitative confirmation of Smoluchowski's conclusions about the temperature-discontinuity, but there was some disagreement on the treatment of energy exchange between molecules and the solid surface.[22] Indeed, this is still a subject of active research.[23]

* * *

There are still a few unresolved problems (at least for me) with the classical synthesis. Perhaps they could be quickly settled if a modern scientist with experience in theoretical and experimental work on heat transfer cared to take the time to analyze the older literature. I have found some indications, in the recent papers mentioned below, that these problems still have some interest.

I will mention briefly the better-known ways in which the classical kinetic theory of gases fails to give an adequate description of heat conduction. First, there are the complexities of internal-energy transfer for polyatomic molecules, usually blamed for the fact that there is quantitative disagreement between theoretical and experimental values of the heat conduction coefficient at any given temperature and pressure. Such difficulties were recognized by Maxwell and Boltzmann, and they would not have been surprised to learn that a more sophisticated theory of molecular structure would be needed to resolve them.

Second, there will clearly be changes in the conductivity at very high temperatures (such that molecules are dissociated and ionized) and these will cause further discrepancies from the values predicted by a kinetic theory of simple particles. Again, once the concepts of dissociation and ionization are accepted, such discrepancies present no difficulties in principle, even though it may be extremely messy to calculate with variable electronic concentrations and long-range Coulomb forces.

The estimation of heat transfer at very low densities has acquired practical significance with the advent of high-speed aerodynamics and reentry vehicles. I have examined only the earliest phases of this subject, in which it is already apparent that qualitative differences in the nature of heat transfer might be expected when equilibration of temperature by collisions among molecules is replaced by gas-surface effects. Some of the results obtained by Smoluchowski and Knudsen, such as the temperature-discontinuity at the interface and linear dependence of conductivity on gas pressure, seem to me to be more compatible with the ideas prevalent *before* the time of Clausius and Maxwell. Within the framework of the caloric theory of heat (assuming a static gas-structure) it seems perfectly reasonable that as the density of a gas decreases, its ability to conduct heat will also decrease, and its temperature near a solid surface need not be the same as the temperature of the surface itself. It is only in the dynamical picture, where molecular properties are quickly equilibrated over distances of a short "mean free path" between collisions, that such properties are no longer seen as normal. This situation gives rise to the speculation that if we could live in a world where atmospheric pressure was 1/100 of its present terrestrial value, our understanding of the nature of gases and of heat transfer might be considerably different. In fact, the concept of heat conduction in gases, as described by Maxwell and Clausius, might be irrelevant since all heat transfer would be effected by radiation or by "molecular convection."

The notion of "molecular convection" (as kinetic heat conduction is sometimes described in textbooks) also suggests that we might re-examine the dead issue: Is convection really distinct from conduction? If conduction is attributed to the random but preferentially directed flow of individual molecules, while convection is described as the flow of large chunks of fluid, might there not be an intermediate region between microscopic and macroscopic–perhaps similar to Brownian movement, which can be thought of as a macroscopic manifestation of microscopic fluctuations? In this connection it is interesting to note that there is an *experimental* situation in which a transition from conduction to convection can be realized by continuous variation of a macroscopic parameter: the onset of Rayleigh–Jeans instability in a shallow horizontal layer of fluid heated from below.[24] Another aspect of this question is the fact that standard engineering formulas for the rate of heat transfer by free convection generally contain a factor depending on the thermal conductivity of the fluid, implying that if there were no conduction there could be no convection either.[25] This question was discussed explicitly by Langmuir, who argued that heat is first carried from a hot body to a thin layer of surrounding gas almost exclusively by conduction, and then propagated further by convection.[26]

At high pressures, Enskog's (1922) theory introduces a new theoretical mechanism for heat conduction: transfer across finite distances at collisions. Although his theory was developed primarily for a dense fluid of elastic spheres (whose diameter is assumed to be *not* negligible compared to the average distances between spheres),[27] nothing much better is available for more "realistic" models, and the theoretical predictions have been qualitatively confirmed by experiments on dense gases.[28] This mechanism of heat transfer is similar to

§13.8] THE THREE MODES OF HEAT TRANSFER 531

that which one might expect to operate in liquids and solids—a molecule can transfer heat to its neighbor even though both are essentially at rest (perhaps vibrating around equilibrium positions)—and thus is in harmony with the static pre-1850 theories of gases, like the results mentioned above for heat transfer at *low* densities.

In the vicinity of the critical point, the experimental difficulty of determining the heat conduction coefficient and of excluding convection becomes especially severe.[29] The temperature gradient in the gas changes in a way not contemplated in the standard methods for reducing the data; one has to ask again "between what points are we measuring the temperature differences" (as in principle one should always ask[30]) and one has to work with very small temperature differences in order to avoid such ambiguities. Moreover, measuring the variation of conductivity with pressure is *not* equivalent to measuring its variation with density since $(\partial p/\partial \rho)_T$ goes to zero at the critical point.

From the discussion of radiation absorption in §13.6, it might have been thought that all modern determinations of heat conduction would include a correction for radiation absorption by the gas. If this were done, one could then go on to ask: What happens to the absorbed energy? Would it not have some further effect on the temperature gradient in the gas and thus on the heat conduction? Would not some of it be reemitted and thus fed back into the radiation flow? My impression is that such questions are generally ignored; to support this impression I can offer only the statement of Leidenfrost (who has discussed these problems in some detail): "no rational adjustment of measurements of thermal conductivity of radiating fluids to allow for radiation absorption and emission characteristics has ever been made."[31] That was in 1964; four years later he concluded that radiant heat-transfer effects in an absorbing and emitting gas must introduce errors ranging from 1% (at 373°K and 10% emissivity of the solid surface) to 35% (at 700°K, 100% emissivity) in the usual method of determining heat conduction.[32]

In the light of the numerous disturbing effects which have been recognized to be important at high or low pressures and at high or low temperatures (including quantum effects in the latter case), it might be worthwhile to look again at the experimental evidence that was alleged to confirm Maxwell's prediction that the heat conduction coefficient is independent of pressure, and to establish a simple temperature-dependence of this coefficient.[33] It is well known in the history of science that the reporting and interpretation of experimental "facts" is often strongly influenced by theoretical preconceptions, and in this respect the investigation of thermal properties of gases is not exceptional.

Acknowledgments. In writing this chapter I have profited considerably from suggestions and research of C. W. F. Everitt, E. W. Garber, H. Kangro, P. E. Liley, E. A. Mason, R. G. Olson, J. S. Rowlinson, J. V. Sengers, and C. Truesdell. My work has been supported by grant GS-2475 from the National Science Foundation, and contract NAS 5-21293 with the National Aeronautics and Space Administration.

Notes for §13.8

1. Aside from affecting numerical values of coefficients, the only significant case where quantum effects produce a qualitatively different kind of heat transfer is the "second sound" phenomenon in superfluid helium. Cf. Fritz London, *Superfluids, Volume II, Macroscopic Theory of Superfluid Helium* (New York: Wiley, 1954), §13.
2. §12.8; D. Enskog, *Kungl. Sv. Vet. Akad. Handl.* **63**, no. 4 (1922); for an English translation and discussion of this work in modern kinetic theory, and of the breakdown of the assumed density series for transport coefficients, see Brush, *Kinetic Theory* **3** (New York: Pergamon Press, 1972).
3. See footnote 12, §13.7.
4. C. H. Lees, *Phil. Mag.* [5] **28**, 429 (1889). S. Tereschin, *Zhur. Russ. Fiz.-Khim. Obsh.* **29**, 169, 277 (1897), **30**, 15 (1898), abstracted in *Fortschr.*, Abt. 2, p. 354 (1897). R. Wagner, *Experimentelle Untersuchungen auf dem Gebiete der inneren und äußeren Wärmeleitung* (Zürich, 1902); *Ann. Phys. Beibl.* **27**, 534 (1903). I. Langmuir, *Phys. Rev.* **34**, 401 (1912); *Proc. Am. Inst. Elect. Engrs.* **31**, 1229 (1912); *Trans. Am. Electrochem. Soc.* **23**, 299 (1913). A. E. Kennelly, C. A. Wright and J. S. van Bylevelt, *Trans. Am. Inst. Elect. Engrs.* **28**, 363 (1909). H. Kraussold, *Forsch. Geb. Ingenieurwesens* **5**, 186 (1934), N. Vargaftik, *Tech. Phys. U.S.S.R.* **4**, 341 (1937). M. Jakob, *Heat Transfer* (New York: Wiley, 1949), 443–50. W. H. McAdams, *Heat Transmission* (New York: McGraw-Hill, 1954), pp. 165–171.
5. J. Boussinesq, *Compt. Rend. Acad. Sci. Paris* **132**, 1382, **133**, 257 (1901) and several other papers; *Théorie Analytique de la Chaleur* (Paris, 1901-3). A. C. Mitchell, *Trans. R.S. Edinburgh* **40**, 39 (1899); A. Russell, *Phil. Mag.* [6] **20**, 591 (1910).
6. For reviews of this subject see H. L. Dryden, F. D. Murnaghan and H. Bateman, *Hydrodynamics* (National Research Council Bulletin 84, reprinted by Dover Pubs., New York, 1956), pp. 400–7; E. U. Schlünder, *Chemie Ingenieur Technik* **42**, 905 (1970).
7. Cf. Kennelly *et al.*, *op. cit.* (footnote 4).
8. As recently as 1953, in a paper on "The thermal conductivity of nitrogen at pressures up to 2500 atmospheres," one finds the statement: "At these pressures the convection effects in the hot wire apparatus are not yet serious enough to affect the accuracy" in reference to data at pressures up to 150 atmospheres obtained by previous workers; a parallel plate method is used for the higher pressures. A. Michels and A. Botzen, *Physica* **19**, 585 (1953). For further discussion see J. V. Sengers, *Thermal conductivity measurements at elevated gas densities including the critical region* (Amsterdam, 1962).

§13.8] THE THREE MODES OF HEAT TRANSFER

9. Cf. H. v. Helmholtz, *Sitzungsber. Akad. Wiss. Berlin* 657 (1888); English trans. in B. Saltzmann, *Selected Papers on the Theory of Thermal Convection* (New York: Dover Pubs., 1962).
10. S. P. Langley, *Researches on Solar Heat and its Absorption by the Earth's Atmosphere*, Professional Papers of the Signal Service No. XV (Washington: Government Printing Office, 1884); *Am. J. Sci.* [3] **24**, 393 (1882), **25**, 169 (1883), **28**, 163 (1884); the last two papers are also in *Phil. Mag.* [5] **15**, 153 (1883), **18**, 289 (1884). Abney and Festing, *Proc. R.S. London* **35**, 80 (1883). H. Rubens and E. Aschkinass, *Ann. Phys.* [3] **64**, 584 (1898). Knut Ångström, *Ann. Phys.* [3] **39**, 267 (1890), [4] **3**, 720 (1900). D. L. Obendorf, *Samuel P. Langley: Solar scientist, 1867–1891* (Dissertation, Berkeley, 1969), E. S. Barr, *Infrared Physics* **3**, 195 (1963).
11. W. Crookes, *Proc. R.S. London* **31**, 239 (1881).
12. J. T. Bottomley, *Proc. R.S. London* **37**, 177 (1884).
13. C. F. Brush, *Phil. Mag.* [5] **45**, 31 (1898); M. v. Smoluchowski, *Ann. Phys.* [3] **64**, 101 (1898), reprinted in *Pisma Marjana Smoluchowskiego* (Krakow: Drukarnia Uniwersytetu Jagiellonskiego, 1924), **1**, 83.
14. Harry J. Eisenman, *Charles F. Brush: Pioneer Innovator in Electrical Technology* (Ph.D. Thesis, Case Institute of Technology, 1967). M. Gorman, *Ohio Historical Quarterly* **70**, 128 (1961).
15. See footnote 6, §13.5.
16. S. D. Poisson, *Théorie mathématique de la chaleur* (Paris, 1933), ch. V.
17. Further details of the calculation were given in another paper, *Sitzungsber. Akad. Wiss. Wien* **107** (Ia), 304 (1898), reprinted in *Pisma* **1**, 113. The coefficient defined by Smoluchowski is now called the "temperature jump distance."
18. J. C. Maxwell, *Phil. Trans.* **170**, 231 (1880). For William Thomson's role in the development of this theory see §5.5.
19. M. v. Smoluchowski, *Phil. Mag.* [5] **46**, 192 (1898); *Öster.-Chem.-Ztg.* **2**, 385 (1899); *Pisma* **1**, 156.
20. C. F. Brush, *Science* **8**, 483 (1898).
21. See also Smoluchowski's note in *Nature* **59**, 223 (1898).
22. F. Soddy and A. J. Berry, *Proc. R.S. London* **83A**, 254 (1910), **84A**, 576 (1911). M. Knudsen, *Ann. Phys.* [4] **34**, 593 (1911); *The Kinetic Theory of Gases* (London: Methuen, 1934, 2nd ed., 1946, 3rd ed., 1950), pp. 46–51. L. Dunoyer, in *Les Idées Modernes sur la Constitution de la Matière*, Coll. Mém. Phys. Soc. Fr. Phys. (Paris, 1913), [2] **3**, 215–71. L. B. Loeb, *Kinetic Theory of Gases* (New York: McGraw-Hill, 1927), pp. 265–79. M. v. Smoluchowski, *Bull. Int. Acad. Cracovie* **A**, 295 (1910); *Ann. Phys.* [4] **33**, 1559 (1910), **35**, 983 (1911); *Phil. Mag.* [6] **21**, 11 (1911); *Pisma* **2**, 128, 134. J. H. Jeans, *An Introduction to the Kinetic Theory of Gases* (New York: Cambridge University Press, 1948), pp. 190–94. E. H. Kennard, *Kinetic Theory of Gases* (New York: McGraw-Hill, 1938), pp. 311–27. P. Lasareff, *Ann. Phys.* [4] **37**, 233 (1912).
23. For surveys of recent work see F. M. Devienne, *Frottement et échanges thermiques dans les gaz raréfiés* (Paris: Gauthier-Villars, 1958); J. P. Hartnett, in *Rarefied Gas Dynamics Symposium* (Berkeley, 1960), p. 1. G. S. Springer, *Advances in Heat Transfer* **7**, 163 (1971).
24. H. A. Thompson and H. H. Sogin, *J. Fluid Mech.* **24**, 451 (1966).
25. See the textbooks by Jakob and McAdams (*op. cit.*, footnote 4); Boussinesq's formula for free convection (*op. cit.*, footnote 5) also has this property.

26. I. Langmuir (*op. cit.*, footnote 4); papers reprinted in *The Collected Works of Irving Langmuir* (New York: Pergamon Press, 1960), **2**, especially p. 69.
27. See footnote 2.
28. See E. McLaughlin, *Chem. Rev.* **64**, 389 (1964).
29. J. V. Sengers, *op. cit.* (footnote 8).
30. M. Jakob, *Heat Transfer*, p. 12.
31. W. Leidenfrost, *Int. J. Heat Mass Trans.* **7**, 447 (1964). I am indebted to Prof. J. V. Sengers for informing me about this and other recent work.
32. W. Leidenfrost, "Critical analysis of the experimental determination of the thermal conductivity of steam," presented at the 7th International Conference on Properties of Steam (Tokyo, 1968); *Thermal Conductivity* (New York: Plenum Press, 1969), p. 213; *Wärme-und Stoffübertragung* **3**, 114 (1970). See also M. Kohler, *Z. angew. Physik* **18**, 356 (1965); H. Poltz, *Int. J. Heat Mass Trans.* **8**, 515 (1965).
33. It is remarkable that even the results of such an eminent investigator as H. Kamerlingh Onnes could be rejected when they failed to support the classical synthesis: according to J. B. Ubbink [*Physica* **14**, 165 (1948)], the measurements of Kamerlingh Onnes, Dorsman, and Weber on hydrogen below 22°K "are not reliable because they found a pressure dependence, not found by other investigators and theoretically not acceptable." Cf. C. C. Minter, "Effect of pressure on the thermal conductivity of a gas," U.S. Naval Research Laboratory report 5907 (Washington, D.C., 1963). The only "modern" study designed specifically to test the pressure-dependence of thermal conductivity is that of H. Gregory and C. T. Archer, *Phil. Mag.* [7], **1**, 593 (1926); after that "Maxwell's law" was not challenged until Minter's work. Practical consequences of this and other difficulties discussed above have been analyzed by J. E. S. Venart, "Uncertainties in the thermal conductivities of liquids and dense gases:—Their effect on heat transfer" (preprint, Department of Mechanical Engineering, The University of Calgary, Calgary, Alberta, Canada, June 1971).

Appendix 13.A: Leslie's Analysis of Heat Transfer

As we have seen (§13.1), it was recognized at the beginning of the 19th century that there might be different molecular or etherial mechanisms for heat transfer, but it was not clear how these mechanisms could be distinguished experimentally. One method was suggested by the chemist Thomas Thomson: he proposed to use the term "radiation" when heat passes through a body without losing any of its velocity; but if heat is slowed down, presumably by being absorbed and reemitted by particles of matter, then one should speak of "conduction." But Thomson did not succeed in carrying out this prescription quantitatively, nor did he try to separate conduction and convection.[1] In contrast with the general level of work being done by his contemporaries, John Leslie's research seems amazingly thorough and sophisticated.[2]

Leslie begins by assuming that Newton's law in cooling is valid, and all of his numerical data seem to have been reduced using this assumption, although he later argues that convection should follow a different law. To clarify the argument (which a modern reader finds expressed rather obscurely in Leslie's text) I will use algebraic symbols. According to Newton's law, if we measure the temperature T from the ambient temperature taken as zero, and if the initial temperature of the hot body is T_0 at $t = 0$, its rate of cooling is

$$\frac{dT}{dt} = -vT$$

and hence its temperature at any later time t will be

$$T = T_0 e^{-vt}$$

It is assumed of course that v is constant. Leslie calls $1/v$ the "range of cooling," *i.e.* the time it would have taken the hot body to cool down to zero temperature if it had continued to cool at a constant rate $-vT_0$. He estimates v by measuring, for example, the time taken to cool down to $T_0/2$ (what we would now call the "half-life"). At this point we have

$$e^{-vt} = \tfrac{1}{2} \quad \text{or} \quad -vt = \log_e \tfrac{1}{2} = 0.693$$

so that v, which Leslie usually calls the "rate" or "velocity" of cooling, can be found by dividing 0.693 by the time required for T to decline to $T_0/2$.[3]

Leslie assumes that the rate of cooling can be represented as the sum of two parts: the cooling due to "pulsations" (radiation) and that due to "aerial contact" (which later turns out to include both convection and conduction)

$$v = v^P + v^A$$

He claims to establish the following general laws by his experiments:

(i) the rate of cooling by air currents is proportional to the velocity of the currents but is independent of the nature of the hot surface;[4]

(ii) the rate of cooling by pulsation is 8 times as great from blackened or paper-covered surfaces as from polished metallic surfaces.[5]

He then argues that the effect of the hot surface on the air current must be equivalent to a mechanical force giving it a certain velocity by acting through some distance. Both the force and the distance should, he claims, be proportional to the change in temperature. Since, according to the principles of mechanics, the square of the acquired velocity is "compounded of the space and the actuating force" it is therefore "as the square of the degree of heat which is absorbed. The velocity of propulsion is hence proportional to the excess of temperature."[6] On combining this with law (i) he concludes that the rate of cooling by air currents must be proportional to the temperature-excess, *i.e.* v is proportional to T (contrary to Newton's law).

On applying these principles to his experimental data, Leslie concluded that there must be two distinct kinds of aerial cooling, one which he calls "regression" or "recession" [convection] and which is subject to law (i); the other he calls "abduction" [conduction] and attributes to the slow diffusion of heat through air without mass motion of the air itself.

> From a comparison of numerous trials made with canisters of various shapes and dimensions, and filled with boiling water, I find, at the equidistant temperatures of 10, 40, and 70 degrees, reckoning from the standard of the external air, that, with a metallic surface, the rates of cooling are very nearly as 2, 3, and 4; but, when the surface is papered or covered with a coat of lamp-black, the rates of cooling are respectively as the numbers 4, 5, and 6. Thus, in either case, the gentle perpendicular flow of heated air; corresponding to an excess of 30° of temperature, has an action as 1; and the double of this, with a similar excess of 60°: it, therefore, exerts effects which are exactly proportioned to its expansion or augmented elasticity.[7]

If we denote by $v_M(T)$ the rate of cooling of a metallic surface with a temperature-excess of T degrees, and by $v_B(T)$ the rate when this surface is covered by paper or lampblack, then we can write

$$v_B(T) = v_B(10) + (T-10)/30; \qquad v_B(10) = 4$$
$$v_M(T) = v_M(10) + (T-10)/30; \qquad v_M(10) = 2$$

The difference between v_B and v_M must be attributed to different radiative power of the paper or lampblack covering the metallic surface, and since $v_B(T) = v_M(T) + 2$ for all T, Leslie points out that the

> energy that a surface of paper communicates by exciting copious pulsations, is constantly denoted by 2; yet its influence is comparatively small in the higher temperatures.[7]

APP. 13A]   LESLIE'S ANALYSIS OF HEAT TRANSFER   537

Since, by law (i), the difference between v_B and v_M is independent of v_B^A and v_M^A, we have

$$v_B^P = v_M^P + 2$$

But according to law (ii),

$$v_B^P = 8 v_M^P$$

Hence, solving for v_M^P, we find

$$v_M^P = \tfrac{2}{7}$$

This value must be independent of T since Newton's law is assumed to apply to cooling by pulsation. Now Leslie compares it with the total rate of cooling at $T = 10°$, $v_M(10) = 2$, and infers that the cooling due to the aerial cooling apart from pulsation is

$$v_M^A(10) = 2 - \tfrac{2}{7} = \tfrac{12}{7}$$

On the other hand, if the aerial cooling rate v^A is indeed proportional to T and can be canceled by setting $T = 0$, then from the equation

$$v_M(T) = v_M(10) + (T - 10)/30$$

one would infer that

$$v_M(0) = v_M^P(0) = 2 - \tfrac{1}{3} = \tfrac{5}{3}$$

and that

$$v_M^A(10) = \tfrac{1}{3}$$

There is clearly a discrepancy here (the difference between $\tfrac{12}{7}$ and $\tfrac{1}{3}$), and thus Leslie concludes that

> there still remains $1\tfrac{8}{21}$, for the power apparently inherent with which every substance tends to an equilibrium of temperature.[8]

After analyzing another experiment which he claims gives a similar result,[9] Leslie describes the nature of this third kind of heat transfer:

> The portion of heat thus consumed is most certainly not annihilated; neither is it transported to a distance, by any species of elastic motion excited in the encircling fluid. It is, therefore, absorbed by the contiguous shell of matter, and afterwards slowly diffused through the extended mass. Air is still the sole medium by which heat endeavours to maintain the balance among remote or

detached bodies; but here its operation is of a passive nature, and it receives and conveys the calorific impressions through its substance in the same manner as a bar of iron or any solid material.[10]

Thus Leslie has not only named three different kinds of heat transfer (pulsation, regression, and abduction, corresponding to the modern terms radiation, convection, and conduction) but he has also given them operational definitions by which their contributions may be estimated quantitatively. In fact, he actually mentions *four* kinds of heat transfer, since he distinguishes between regressive cooling with an externally imposed wind velocity, and regressive cooling in still air where the hot surface itself sets the air in motion.[11] His work can be criticized on both experimental and theoretical grounds—for example, once Newton's law has been abandoned for a particular mode of heat transfer, the "rate of cooling" assigned to that mode can no longer be manipulated as Leslie does in the calculations mentioned above. Moreover, it cannot be assumed that the three modes are *additive*, however obvious that might have seemed in the first half of the 19th century. But these are defects that should have stimulated other scientists to improve Leslie's work rather than ignore it.

Notes for Appendix 13.A

1. T. Thomson, *J. Nat. Phil.* **4**, 529 (1801).
2. I have used the discussion of R. Olson, *Ann. Sci.* **26**, 273 (1970), in conjunction with my own reading of Leslie's *Experimental Inquiry* (London, 1804). For further discussion of Leslie's life and work see: R. Olson, *Am. J. Phys.* **37**, 190 (1969); *Ann. Sci.* **25**, 203 (1969); J. G. Burke, *Isis* **61**, 340 (1970).
3. Leslie, *Inquiry* p. 268.
4. *Ibid.*, p. 282. He expressed his results by the formula $(T/t) = 1 + (V/4.5)$, where T is the "ordinary range" of cooling for the ball in stationary air, and t is the range of cooling when the body moves at a velocity V relative to the air. (V should not be confused with the velocity of cooling v.) Leslie considered this result so exact that he proposed to use it as the principle of a new instrument to measure wind velocity.
5. *Ibid.*, pp. 78–81.
6. *Ibid.*, p. 313.
7. *Ibid.*, p. 315.
8. *Ibid.*, p. 316.
9. He measured the rate of cooling of a mercury thermometer whose bulb was coated with silver leaf, and found $v_M = 1/150$; when the same experiment was repeated with the gilding rubbed off, the result was 1/69, which presumably is to be called v_B. The difference is $v_B - v_M = 1/128$, and by law (i) this should also be the difference $v_B^p - v_M^p$. From law (ii), $v_B^r = v_M^r$, hence $v_M^p = 1/(7 \times 128) \cong 1/900$. This is to be subtracted from the value of v_M, and Leslie says: "But, $1/150 - 1/900 = 1/180$; and

hence, besides the heat abstracted from the ball by the pulsatory and regressive motions of the surrounding air, there is some other mode by which it is dispersed at the rate of the 180th part each second." I think this reasoning is erroneous, and simply shows that $v_M^\wedge = 1/180$, not that "regressive" [convective] motion fails to account for the cooling. The latter conclusion could only be reached by the more elaborate analysis which I have described in the text.

10. Ibid., pp. 318–19.
11. Ibid., p. 319. The modern terms are "forced" and "free" (or natural) convection; see p. 525.

Appendix 13.B: Derivation of the T^4 law

Bartoli assumed that in any space irradiated by heat, there is a small but finite amount of energy in the form of heat radiation, which will be absorbed by a body placed in the space, or by the walls surrounding the space if the space is made smaller (*e.g.* by moving in a piston). Conversely, if the walls are maintained at a particular temperature by a heat reservoir, this reservoir will supply energy to "fill up" the space if it is enlarged by pulling out the piston.

Consider four concentric spherical surfaces A, B, C, and D, with A on the outside and D on the inside; the entire space between A and D is an absolute vacuum. A and D are absolutely black bodies, while B and C are completely reflecting on both sides, and do not conduct heat. Initially a hole is open in surface B so that A irradiates the space between B and C with heat radiation at the same temperature as the space between A and B. Now close the hole in B and open a hole in C. If we assume that D is initially at a temperature *higher* than that of A, we would expect (in accordance with the Second Law of Thermodynamics) that heat energy should be flowing outward from D toward A, but in fact we can make the reverse happen if we simply contract the surface B until the space between it and C is very small. The heat energy which originated in A in the first step of the process will then be "squeezed" through the hole in C into the space between C and D. This additional heat energy must raise the temperature of D. If we now close the hole in surface C and let B expand to its original size, we see that we have transferred heat energy from a colder body (A) to a hotter body (D). Thus we would have violated the Second Law of Thermodynamics *unless* we assumed that radiation exerts a finite pressure, so that the process of contracting surface B and pushing the radiation through the hole in C requires the performance of mechanical work. (The Second Law does allow us to transfer heat from cold to hot provided a compensating amount of mechanical work is performed.)

Bartoli's argument, as summarized and accepted by Boltzmann, leads to the conclusion that radiation must exert a mechanical pressure on a surface. Boltzmann now attempts to find an explicit formula for this pressure, by considering the analog of a "Carnot cycle" with radiation as the working substance of a heat engine.

Consider a hollow cylinder of unit cross-sectional area, in which one can move a piston up and down. The base of the cylinder is maintained at a temperature T_0 while the piston is moved up to a height a. Let E be the energy-density of the heat radiation and P be the pressure, assuming that both are functions *only* of temperature. Then the total energy δU which must be supplied in this process is then the sum of (a) the energy of the radiation which fills the newly-created space, *i.e.* the volume a multiplied by the energy-density E, and (b) the mechanical work which must be done to push back the piston (which would have to be pressed on the other side by an external pressure P to balance the radiation pressure). The mechanical work is the product $P\,dV$, where the change in volume (from zero volume) is again just a. Thus the energy change in the first part of the cycle is

$$\delta U_1 = aE(T_0) + aP(T_0) \qquad (\#1,\text{ isothermal expansion})$$

and the corresponding entropy change is

$$\delta S_1 = \frac{E_1}{T_0}$$

Now assume that we can make an *adiabatic* expansion (with appropriate thermal insulation of the piston and cylinder walls so that we can ignore their energy-change when the temperature varies). Consider a stage in the expansion when the volume of the cylinder is $a + x$ and the temperature is T. For any infinitesimal change in volume dx, radiation energy must be used up in doing work against the external pressure

$$d[(a+x)E(T)] = -P(T)dx \qquad \text{(i)}$$

Since the process is adiabatic, there is no entropy change, $\delta S_2 = 0$.

We then compress the radiation at a constant temperature T back to zero volume, and then heat it from T to T_0. Since it is assumed that all these changes are reversible, we will have returned to the original thermodynamic state of the system, with the same energy and entropy as at the start.

For the isothermal compression, the energy change is computed in the same way as for the isothermal expansion, replacing a by $a + x$

and T_0 by T

$$\delta U_3 = -(a+x)[E(T)+P(T)] \quad (\#3, \text{ isothermal compression})$$

and the entropy change is

$$\delta S_3 = \frac{\delta U_3}{T}$$

In the final heating step there is no entropy change because there can be no radiation energy in zero volume. Thus

$$\delta U_4 = \delta S_4 = 0 \quad (\#4, \text{ heating from } T_0 \text{ to } T)$$

For a complete cycle the net entropy change is zero since all steps are assumed to be reversible. Remembering that $\delta S_2 = \delta S_4 = 0$, we have therefore

$$\delta S_3 + \delta S_1 = 0, \quad \text{or} \quad \frac{(a+x)[E(T)+P(T)]}{T} = \frac{a[E(T_0)+P(T_0)]}{T}$$
$$= \text{const.}$$

We can now consider this last equation as a general condition on the functions $E(T)$ and $P(T)$, in which x and T can have any arbitrary values while a and T_0 are held fixed. (In other words, we could have carried out the adiabatic expansion to any arbitrary extent without affecting the argument.) After some further juggling of the equations,[1] we arrive at the simple relation

$$T\,dP = (E+P)\,dT$$

This is as far as we can go with thermodynamics unless we know the relation between P and E for our working substance. At this point Boltzmann introduces Maxwell's result from the electromagnetic theory of light, that the pressure exerted against a surface is numerically equal to the energy density of the incident waves. However, in thermal radiation there is no preferred direction–the waves will be moving in all directions, with equal probability–and therefore, by analogy with the kinetic theory of gases, Boltzmann assumes that the pressure in a particular direction is only $\frac{1}{3}$ of the energy density,[2]

$$P = \tfrac{1}{3} E$$

Then it follows immediately from the general thermodynamic formula that

$$\frac{dE}{E} = 4\frac{dT}{T}$$

or

$$E = \sigma T^4$$

where σ is a constant, subsequently known as the "Stefan–Boltzmann constant." Hence the energy-density in a vacuum maintained at a fixed temperature – the so-called "black-body" or "Hohlraum" radiation – is proportional to the fourth power of the absolute temperature.[3]

Notes for Appendix 13.B

1. Consider any arbitrary variation of x and T, and compute the corresponding variation of $-\delta S_3$

$$d\left\{\frac{(a+x)[E(T)+P(T)]}{T}\right\}$$
$$= \frac{1}{T} d\{(a+x)[E(T)+P(T)]\} - \frac{(a+x)(E(T)+P(T))}{T^2} dT = 0$$

hence

$$Td\{(a+x)[E(T)+P(T)]\} = (a+x)[E(T)+P(T)]\, dT \qquad \text{(ii)}$$

We can simplify the left-hand side of eq. (ii) by going back to eq. (i) in the text and adding $d[(a+x)P(T)]$ to both sides of it; the right-hand side of eq. (i) would then become

$$-P(T)dx + (a+x)dP(T) + P(T)dx = (a+x)dP(T)$$

and the left-hand side would become the same as the left-hand side of (ii) except for the factor T. Hence, on substituting into (ii) we get

$$T(a+x)dP(T) = (a+x)[E(T)+P(T)]dT$$

or

$$TdP(T) = [E(T)+P(T)]dT$$

2. Boltzmann did not remark that the ratio of energy to pressure is actually twice as large for the "radiation gas" as it is for a gas of material particles in the kinetic theory. Presumably no one would expect the ratio to be the same anyway as long as the wave theory of radiation is the basis for the pressure-energy relation. If, following the ideas of Einstein (1905), one regarded the radiation as a gas of photon particles, for which a relativistic energy–momentum relation $E = pc$ holds in place of the usual $E = p^2/2m$ for non-relativistic particles, the origin of the factor 2 would become apparent.
3. The law can also be derived using an analogy between radiation and a vapor in equilibrium with its condensed phase, by applying the Clapeyron equation. See J. T. Vanderslice, H. W. Schamp, and E. A. Mason, *Thermodynamics* (Englewood Cliffs, N.J.: Prentice-Hall, 1966), p. 149.

CHAPTER 14

Randomness and Irreversibility *

14.1 Introduction: the world-machine and cosmic history

"The most important question, perhaps, of contemporary scientific philosophy is that of the compatibility or incompatibility of thermodynamics and mechanism." That statement was made at the first international physics congress at Paris in 1900 by Bernard Brunhes, Director of the Puy-de-Dôme Observatory, in his discussion of Gabriel Lippmann's paper on the conflict between Carnot's principle and the kinetic theory of gases.[1] At issue was a problem that had been the subject of controversy during the preceding decade: How could one reconcile the basic laws of Newtonian mechanics with the Second Law of Thermodynamics – in particular, how could the principle of irreversibility, apparently grounded quite firmly in experience, be explained by any mechanical model which had to be based on reversible equations of motion? Henri Poincaré had again drawn attention to this difficulty in his opening address to the Paris congress.[2]

In spite of the supposed complacency of physicists at the end of the 19th century,[3] it was clear to many sharp thinkers that this conflict between thermodynamics and mechanics cast serious doubt on the validity and internal consistency of the previously-accepted founda-

---

* Reprinted with minor revisions from *Archive for History of Exact Sciences* **12**, 1–88 (1974), by permission of Springer-Verlag.

tions of physical science. The feeling was reinforced by the publicity given to other difficulties confronting established theories, for example Lord Kelvin's two "clouds over the dynamical theory of heat and light" (the Michelson–Morley experiment and the apparent failure of the equipartition theorem for polyatomic gases).[4] For some, the way out was to reject the mechanistic philosophy and seek to base scientific theories on the principles of "energetics" or electromagnetism.[5] Others, like Ernst Mach, scorned their colleagues' search for theoretical foundations and urged a more empirical approach.[6]

As we know, Kelvin's clouds were soon to be dispersed by the new theories of Planck and Einstein. But the problem of irreversibility was not so easily solved, and is still with us today though in a somewhat different form.[7] Nevertheless the attempt to solve it had already, before 1900, led to the introduction of a statistical viewpoint in molecular physics; or rather, had pushed the earlier statistical viewpoint in the direction of postulating randomness and indeterminacy at the atomic level. Thus the 19th-century attempts to explain irreversibility did much to prepare the way for the stochastic[8] world view that seems now to be an essential part of modern physics. In this chapter I will discuss the concepts of randomness and irreversibility and their interactions in the development of the kinetic theory of gases before 1900.

Since "randomness" is sometimes thought to be a characteristic 20th-century concept, it might be objected that one would be committing the sin of "present-mindedness" or writing "Whig history" by trying to extract its development from 19th-century physics. In defense of the approach followed here, I would point out that while deterministic ideas did dominate 19th-century physics, the older conception of the world as a "fortuitous concourse of atoms" had not been forgotten (certainly not by those who were aware of the origin of kinetic atomism in Greek antiquity) and was occasionally revived in a rather explicit way by 19th century philosophers.[9] On the other hand, a history of randomness would be of little value if it ignored the context of other scientific ideas and theories which were associated with it. In this case the emphasis will be on irreversibility and the problem of justifying thermal equilibrium in the kinetic theory of gases; if we attempted to extend our discussion into the 20th century it would be necessary to review several other areas of physics. The lack of a still wider consideration of the 19th-century context may perhaps be excused here since I have written at some length on this topic elsewhere.[10]

By now there is an abundance of secondary literature on randomness and irreversibility, so this chapter will be in part a summary or critique of what has already been written on the subject;[11] but I will examine more closely certain important but neglected aspects.

* * *

The introduction of statistical methods in 19th-century kinetic theory is often seen against the background of an orthodox viewpoint supposedly prevailing in the 18th century. This viewpoint could be characterized as the Newtonian mechanical philosophy or "clockwork universe" picture, in which all motions are in principle determined by specifying them at some initial time, and all changes are cyclic; thus randomness and irreversibility are both completely absent from the main body of accepted physical laws. Unfortunately for the conventional accounts, things are not quite so simple: first because Newton himself had quite firmly rejected this view, second because geophysical speculations had already introduced the notion of irreversible heat flow by the end of the 18th century, and third because statistical considerations were by no means excluded from theories of natural phenomena in 1800. Thus the assertions of determinism and cyclic stability found in the writings of Laplace and his colleagues at the beginning of the 19th century must not be read as expressions of a monolithic world view that had been accepted in all areas of science, but rather as admittedly hypothetical descriptions of an ideal world, of strictly limited value in dealing with the real world.

The clockwork universe of the 17th-century mechanical philosophers such as Descartes and Boyle[12] was deeply repugnant to Newton on theological grounds, and moreover seemed to him inconsistent with certain obvious facts about the physical world. In the *Opticks* he pointed out that irreversible processes such as viscosity of fluids and imperfect elasticity of solids tend to make the world-machine run down: "motion is much more apt to be lost than got, and is always upon the decay."[13] In order to prevent the total quantity of motion in the world from decreasing to nothing, there must be "active principles" that operate to renew motion. Otherwise everything would freeze and life would cease; moreover, mutual gravitational perturbations of planets in the solar system would accumulate over long periods of time "till this system wants a reformation" which God perhaps accomplishes by feeding in comets with appropriately chosen masses and orbits.[14]

Newton's suggestion that the laws of physics by themselves are insufficient to ensure the proper functioning of the world over long periods of time without divine intervention was attacked by Leibniz, and was one of the major issues in the famous Leibniz–Clarke debate of 1715–16.[15] As Leibniz put it, Newton's view meant that "God almighty needs to wind up His watch from time to time; otherwise it would cease to move," implying that God was such a poor craftsman that He could not make a machine that would run forever without repairs.[16] Leibniz, on the contrary, believed that "the same force and vigour remains always in the world, and only passes from one part of matter to another, agreeably to the laws of nature and the beautiful pre-established order." It was Leibniz's opinion that in processes such as the generation of heat by mechanical friction the total "force" (*i.e.* some quantity equivalent to the *vis viva*, mv^2) would still be conserved, although it might be converted into the invisible motion of atoms.[17] Presumably such a process would not have to be irreversible, though Leibniz did not explicitly address that point.

For anyone who accepted Newton's concept of atoms as being hard bodies that could never change their size or shape, it would seem that atomic collisions must be inelastic and hence irreversible; if two atoms meet head-on they would simply stop and not rebound. This theoretical difficulty seemed to make Newtonian atomism incompatible with any kind of conservation law for motion (either momentum or energy). The debate on this point has been comprehensively analyzed by Wilson Scott.[18] Its significance for our story is that even though irreversibility in physical processes had been recognized by Newton, it was not yet possible to talk about irreversibility in the modern sense (*e.g.* as involving entropy increase) because a more basic kind of irreversibility – decrease of motion – had not yet been excluded by a conservation law. This dilemma corresponds to the circumstance that logically one cannot have a Second Law of Thermodynamics until after one has established a First Law, yet historically the Second Law (or something that looks like it) came earlier.

For Newton, as for many later scientists, the present state of the physical universe could not be explained as a result of "blind chance" because there were too many evidences of intelligent design.[19] Nor could it be attributed to the deterministic action of natural laws and initial conditions established by God, in the sense of the clockwork universe, for that would make it too easy to eliminate divine providence entirely.[20] Nevertheless by proposing a system of physical laws that could be successfully applied to the motions of planets and satellites, and to the shape of the earth, Newton had provided the

§14.1]   THE WORLD-MACHINE AND COSMIC HISTORY   547

essential basis for the idea that physical laws could in principle explain in a deterministic fashion the motions of all matter in the universe.

* * *

Newton's assertion that the solar system might be unstable because of the accumulated effect of gravitational perturbations was taken up as a challenge to the ingenuity of the greatest mathematicians of the 18th and early 19th centuries. The "three-body problem" was attacked by Euler, Laplace, Lagrange, and Poisson, in the hope of getting at least a good approximation to the long-term effects of perturbations. Their conclusion was that Newton had misconstrued the effects of planetary interactions, and that all deviations from the present orbits would oscillate cyclically between fixed limits, so that the solar system would be stable for an indefinitely long time.[21] Hence, just as Newton had feared, celestial mechanics could dispense with divine providence; in Laplace's celebrated phrase, "I have no need for that hypothesis."[22]

This conclusion did not imply that there are no irreversible processes in astronomy. Despite the success of Newton's law of gravity, treated as if it were pure action-at-a-distance with no need for propagation through a medium, the continental theorists could not entirely dispense with the hypothesis that interplanetary space is filled with an ethereal fluid, and it seemed improbable that the planets would not somehow be retarded as they moved through this fluid. Laplace concluded that the effect of such frictional action would be to make elliptical orbits more nearly circular,[23] without changing the mean distance of the planet from the sun. It seemed likely that the tides of terrestrial oceans would have some effect on the rate of the earth's rotation, perhaps forcing it ultimately to present always the same face to the moon, as the moon does to the earth.[24] Thus, not for the last time, the ether was assigned the duty of irreversibily dragging ponderable matter toward a state of final equilibrium (cf. Culverwell's suggestion, §14.6 below), in collaboration with other natural processes that seemed to have a similar effect.

Even if the present arrangement of the solar system should continue more or less the same for the indefinite future, as a stable equilibrium state, that did not mean that it had never changed in the past. Rather than postulate that God had created the planets in their present orbits (perhaps placed so that they would receive the right amount of the sun's heat in proportion to their density[25]) it seemed to at least a few philosophers—Kant, Laplace, and their followers—more reasonable to assume a gradual evolution of the solar system from a whirling chaos of primal matter, with the planets being formed as hot liquid balls.[26] As traditional theology loosened its grip on scientific speculation, such naturalistic schemes of cosmic evolution became more popular.

But, granted that things do not always remain the same, how did it happen that during the 19th century "evolution"–an apparently neutral term but one loaded with optimistic connotations–had to compete with "dissipation" and "degeneration"? How could Humphry Davy say, as early as 1829, that "human science... has discovered the principle of the decay of things"?[27]

Notes for §14.1

1. *Travaux Cong. Int. Phys.*, *Paris*, *1900* (Paris: Gauthier-Villars, 1901), **4**, 29. For the text of Lippmann's paper see *Rapports Cong. Int. Phys., Paris, 1900* (Paris: Gauthier-Villars, 1900), **1**, 546. (The fourth volume of the "Rapports" was published with the title "Travaux.")
2. Poincaré, *Rapports Cong. Int. Phys., Paris, 1900* **1**, 1.
3. L. Badash, *Isis* **63**, 48 (1972) has given a number of examples of English and American scientists who suggested in the 1880's and 1890's that all the basic principles of physics were known and only the details remained to be worked out. Dissatisfaction with foundations, and discoveries leading to new foundations, were found more frequently among German and French scientists. See for example Max Planck, *The New Science* (New York: Meridian Books, 1959), pp. 4–5 [trans. from "Theoretische Physik," 1930]; S. Petruccioli in *Alcuni Aspetti dello Sviluppo delle teorie fisiche 1743–1911* (Pisa: Domus Galilaeana, 1972), pp. 169, 238; and Boltzmann's remarks cited below near the end of §14.7.
4. *Proc. Roy. Inst.* **16**, 363 (1900); *Phil. Mag.* [6] **2**, 1 (1901); *Baltimore Lectures on Molecular Dynamics and the Wave Theory of Light* (London: Clay, 1904) p. 486.
5. A survey of writings on energetics has recently been given by E. N. Hiebert in *Perspectives in the History of Science and Technology*, ed. D. H. D. Roller (Norman: University of Oklahoma Press, 1971) p. 67. Further references may be found in my notes to the English translation of Boltzmann's *Lecture on Gas Theory* (Berkeley: University of California Press, 1964), pp. 24, 215. On the electromagnetic view see R. McCormmach, *Isis* **61**, 459 (1970); also A. M. Bork, *Science* **152**, 597 (1966). Bork concludes his account with the comment: "We cannot but be impressed with the great activity and restlessness which characterize the period before 1905. On all sides the physicist found his Newtonian universe floundering, not only in its details but even in its underlying mechanistic assumptions. The stage was set for the revolution to come."
6. Cf. S. G. Brush, *Graduate Journal* **7**, 477 (1967), especially the works mentioned on pp. 533–34 and 564–65. The best general account of Mach's position is in the recent

§14.1]   THE WORLD-MACHINE AND COSMIC HISTORY   549

   book by J. T. Blackmore, *Ernst Mach* (Berkeley: University of California Press, 1972).
7. W. Büchel, *Philosophia Naturalis* **6**, 167 (1960). G. J. Whitrow, *The Natural Philosophy of Time* (London: Nelson, 1961), pp. 10–12, 268–310. P. Morrison, in *Preludes in Theoretical Physics in honor of V. F. Weisskopf* (New York: Interscience, 1966), p. 347. M. Gardner, *Sci. Am.* **216** (1), 98 (1967). T. Gold, ed., *The Nature of Time* (Ithaca: Cornell University Press, 1967). E. M. Henley, *Ann. Rev. Nuclear Sci.* **19**, 367 (1969). M. Dako, *Studium Generale* **22**, 965 (1969). R. E. Peierls, in *Methods and Problems of Theoretical Physics*, ed. J. E. Bowcock (New York: American Elsevier, 1970), p. 3. P. C. W. Davies, *Physics Bull.* **22**, 211 (1971). J. Biel and J. Rae, eds., *Irreversibility in the Many-Body Problem* (New York: Plenum Press, 1972). P. T. Landsberg, in *The Study of Time*, eds. J. T. Fraser, F. C. Haber, and G. H. Müller (New York: Springer, 1972), p. 59. B. Gal-Or, *Science* **176**, 11, **178**, 1119 (1972). R. G. Sachs, *Science* **176**, 587, **178**, 1119 (1972). S.-T. Hwang, *Found. Phys,* **2**, 315 (1972). J. Mehra and E. C. G. Sudarshan, *Nuovo Cimento* [11] **11B**, 215 (1972).
8. I do not want to use the word "statistical" here since a statistical theory may or may not assume that individual atomic behavior is deterministic. "Stochastic" means "random" but "stochastic theory" does not have quite the same connotations as "random theory"–it is the event or process that is said to be random, not the theory about it. "Stochastic" also implies that deterministic or lawful elements as well as random elements may be present. (I am indebted to Prof. Octave Levensput for advice on this usage.)
9. C. S. Peirce, *Monist* **1**, 162 (1891), **2**, 321 (1892). Antoine-Augustin Cournot, *Essai sur les fondements de nos connaissances* (Paris, 1851); English trans. with introd. by M. H. Moore, *An Essay on the Foundations of our Knowledge* (New York: Liberal Arts Press, 1956), pp. xxvii, 41, *etc.* There was also Darwinian evolution with its postulate of random variation; cf. §14.5, note 19.
10. S. G. Brush, *Graduate Journal*, **7**, 477 (1967).
11. H. Bernhardt, *NTM, Z. Ges Naturwiss. Tech. Med.* **4** (10), 35 (1967), **6** (2), 27 (1969). B. Brunhes, *La Dégradation de l'Énergie* (Paris: Flammarion, 1922). C. Brunold, *L' Entropie* (Paris: Masson, 1930). E. Daub, *Isis* **60**, 318 (1969); *Hist. Stud. Phys. Sci.* **2**, 321, 165 (1970); *Stud. Hist. Phil. Sci.* **1**, 213 (1970). R. Dugas, *La Théorie Physique au sens de Boltzmann* (Neuchatel: Griffon, 1959). Adolf Grünbaum, *Archiv f. Philos.* **7** (1957). D. ter Haar, *Elements of Statistical Mechanics* (New York: Rinehart, 1954), Appendix I. E. N. Hiebert, *The Conception of Thermodynamics in the Scientific Thought of Mach and Planck* (Freiburg i. Br.: Ernst-Mach-Institut, 1968); in *Perspectives in the History of Science and Technology*, ed. D. H. D. Roller (Norman: University of Oklahoma Press, 1971), p. 67. L. Janossy, in *Max-Planck-Festschrift 1958* (Berlin: VEB Deutscher Verlag der Wissenschaften, 1959), p. 389. M. J. Klein, *Natural Philosopher* **1**, 83 (1963); *Am. Scient.* **58**, 84 (1970); *Paul Ehrenfest* (New York: American Elsevier, 1970), **1**, ch. 6; in *The Boltzmann Equation*, eds. E. G. D. Cohen and W. Thirring (New York: Springer-Verlag, 1973), p. 53. W. Köhler, *Erkenntnis* **2**, 336 (1932). V. G. Lenzen, *Univ. Calif. Publ. Philos.* **10**, 119 (1928). H. Reichenbach, *The Direction of Time* (Berkeley: University of California Press, 1956). A. Rey, *La Théorie de la Physique chez les physiciens contemporains* (Paris: Alcan, 2nd ed., 1923); *Le Retour Éternel et la Philosophie de la Physique* (Paris: Flammarion, 1927). L. Rosenfeld, *Acta Phys. Polon.* **14**, 3 (1955); in *Max-Planck-Festschrift 1958* (Berlin: VEB Deutscher Verlag der Wissenschaften, 1958), p. 203;

in *Irreversibility in the Many-Body Problem*, eds. J. Biel and J. Rae (New York: Plenum Press, 1972), p. 1. R. Schlegel, *Time and the Physical World* (New York: Dover Pubs., 1968 reprint of the 1961 ed.).

12. M. Boas [Hall], *Osiris* **10**, 412 (1952); *Robert Boyle on Natural Philosophy* (Bloomington: Indiana University Press, 1965). E. A. Burtt, *The Metaphysical Foundations of Modern Physical Science* (Garden City, N.Y.: Doubleday, 1954, reprint of the 2nd ed.). E. J. Dijksterhuis, *The Mechanization of the World Picture* (New York: Oxford University Press, 1961, trans. of the Dutch ed., 1950).
13. *Opticks* (4th London ed., 1730; New York: Dover Pubs., 1952), p. 398.
14. See D. Kubrin, *J. Hist. Ideas* **28**, 325 (1967). Newton's suggestion that perturbations might eventually cause the earth to fall into the sun was echoed in Swift's *Voyage to Laputa*; see M. Nicolson, *Science and Imagination* (Ithaca: Cornell University Press, 1956), pp. 123–27, reprinted from an article by M. Nicolson and N. M. Mohler, *Annals of Science* **2**, 299 (1937). (I thank Professor C. Truesdell for this reference.)
15. See H. G. Alexander, *The Leibniz–Clarke Correspondence* (Manchester, Eng.: Manchester University Press, 1956). For evidence that Clarke was really expressing Newton's opinions see A. R. Hall and M. B. Hall, *Isis* **52**, 583 (1961); A. Koyré and I. B. Cohen, *Arch. Int. Hist. Sci.* **15**, 63 (1962).
16. Alexander, *op. cit.*, pp. 11–12.
17. Alexander, *op cit.*, pp. 87–88.
18. W. L. Scott, *The Conflict between Atomism and Conservation Theory 1644–1860* (New York: Elsevier/London: Macdonald, 1970).
19. *Opticks*, p. 402; for a general survey of Newton's opinions on this point see O. B. Sheynin, *Arch. Hist. Exact Sci.* **7**, 217 (1971). Another article by Sheynin presents quotations from several authors illustrating the concept of randomness [*ibid.* **12**, 97 (1974)].
20. Alexander, *op. cit.*, pp. 13–14. According to Newton's friend, the theologian Richard Bentley, there is really not much difference between randomness and determinism if one does not accept divine guidance in natural phenomena; randomness merely implies lack of knowledge of mechanical causes (see the passage quoted by Sheynin, *op. cit.*, p. 232).
21. P. S. de Laplace, *Traité de Mécanique Celeste* (Paris, 1825, reprinted by Chelsea Pub. Co., Bronx, N.Y., 1969, together with the Bowditch translation of the first four volumes), V, Livre XV, Chapitre I. See also Laplace's *Exposition du Système du Monde* (1796, 5th ed., 1824), English trans., *The System of the World* (Dublin, 1830), pp. 328–32; A. Pannekoek, *A History of Astronomy*, (New York: Interscience Pubs./London: Allen & Unwin, 1961, translated from Dutch edition of 1951), ch. 30.
22. A. de Morgan, *A Budget of Paradoxes* (Chicago; Open Court, 1915, reprint of 2nd ed.), II, 1–2. The essence of this legend is supported by Laplace's published criticisms of Newton's theological assumptions, *e.g.* in *System of the World* (Dublin, 1830), pp. 331–33.
23. Laplace, *Mécanique Celeste*, IV, Livre X, Chap. VII.
24. Immanuel Kant, *Wöchentliche Frag- und Anziehungs-Nachrichten* (1754), reprinted in *Allgemeine Naturgeschichte und Theorie des Himmels* (1755); partial English trans. by W. Hastie (1900), reprinted with new introd. by M. K. Munitz, *Universal natural History and Theory of the Heavens* (Ann Arbor, University of Michigan Press, 1969).
25. See I. B. Cohen, in *Philosophy, Science, and Method*, ed. S. Morgenbesser (New York: St. Martin's Press, 1969), p. 523.

26. Kant, *op. cit.*; Laplace, *Exposition du Systeme du Monde* (1796).
27. H. Davy, *Consolations in Travel; or, the Last Days of a Philosopher* (Boston, 5th ed. 1870), p. 273. (The preface is dated 1829, the year of Davy's death.)

14.2 The cooling of the earth

It is difficult to conceive of a time when people did not know that heat flows from hot bodies to cold bodies. Our problem is to understand how this apparently trivial example of irreversibility was translated into an illustration of a general law of nature, the Principle of Dissipation of Energy, and as such was seen to be in conflict with Newtonian mechanics. The usual explanation is that this came about as a result of Sadi Carnot's analysis of steam engines, leading to the Second Law of Thermodynamics as a condition on the interconversion of heat and mechanical work; thus irreversibility was associated with the operation of *real* steam engines in which any flow of heat through a finite temperature-difference meant the loss of a possible transformation of some of that heat into useful mechanical work. Perhaps that is how irreversible heat flow acquired its unpleasant moral connotation – the term "dissipation" being a synonym for "intemperate, dissolute, or vicious mode of living...squandering, waste..."[1] – in the culture of industrialized Victorian Britain. But the idea that the natural behavior of heat entails a distinction between past and future time directions, and even the extrapolation to an ultimate "heat death" of the world, goes back to the 18th century where it has nothing to do with steam engines. Indeed we must look to the literature of geophysical or planetary science, where so many fundamental discoveries and theories of 19th-century physics originated.[2]

Many speculators in earlier centuries had suggested that the earth was originally formed in a hot molten state and has subsequently cooled down. In the late 18th and early 19th centuries, this theory was most commonly attributed to Leibniz, who advanced it in his *Protogaea*.[3] Leibniz assumed that the outer surface of a molten proto-earth would solidify in a somewhat irregular manner, with bubbles bursting through at various places, while the inner part would still remain liquid. The evidence for this hypothesis consisted of familiar phenomena such as vitrified rocks, volcanos, hot springs, and the gradual increase in temperature noted as one goes down into the earth.[4]

Another argument for the central heat theory was provided by Dortous de Mairan, who analyzed the seasonal variations of surface temperature in various parts of the world and concluded that the heat received from the sun is not sufficient to account for these temperatures. In particular, he estimated that the difference between summer and winter temperatures is so small compared to the average absolute temperature of the earth's surface that the latter must be maintained by a continual flow of heat from the inside of the earth.[5]

Buffon incorporated the notion of gradual refrigeration of the earth into his grand theory of the Epochs of Nature. He conducted a series of laboratory experiments on the rate of cooling of heated spheres of iron and other substances of various diameters, and on the time required for molten iron to solidify; by extrapolating his results he estimated that it would take 1342 years for a globe of iron the size of the earth to solidify. Taking account of several other corrections he concluded that the time elapsed from the earth's molten state to the present must be about 74000 years.[6] The recent discovery of ivory trunks of elephants in northern regions where these animals cannot now survive was taken as further evidence that these regions must formerly have been much warmer – one would expect that the earth's polar regions would cool down and become fit for organic life a little sooner than the equatorial regions. Hence the hypothesis of a cooling earth seemed to gain support from paleontology.[7]

Although Buffon is also known for his stochastic method for determining π (the "Buffon needle problem"), I am not aware that he attempted to make any connection between randomness and irreversibility.

From the theory of central heat and gradual refrigeration of the earth it was but a short step to the conjecture that all bodies in the universe are cooling off and will eventually become too cold to support life. This step seems to have been first taken by the French astronomer Jean-Sylvain Bailly (1736–93) in his writings on the history of astronomy and the ensuing correspondence with Voltaire. According to Bailly, all the planets must have an internal heat and are now at some particular stage of cooling: Jupiter, for example, is still too hot for life to arise for several thousand more years; the moon, on the other hand, is already too cold. The final state is described as one of "equilibrium" where all motion has ceased. Thus the modern concept of the "heat death" of the universe, usually attributed to the 19th-century thermodynamic speculations of Thomson, Clausius, and Helmholtz, was actually published as early as 1777.[8]

The theory of cooling of the earth was attacked by a few writers in the 18th century,[9] the strongest opposition coming from the influential geologist James Hutton. While Hutton accepted the concept of internal heat he did not like Buffon's physical approach to the subject [10] and denied that there had been any substantial cooling of the earth's interior since its original formation. It is the debate on this particular aspect of Hutton's theory that is of interest in connection with the development of the concept of irreversibility, and in fact it is Hutton's critics (now forgotten) who used as a weapon the postulate that heat cannot stay concentrated in one place but must flow to cooler regions.

In reply to a criticism of Richard Kirwan, who complained that there seemed to be no plausible means for continually generating heat inside the earth to replace that which diffused toward the surface, Hutton disclaimed any obligation to explain the source of subterraneous fire but merely inferred its presence from the general appearances of mineral bodies which "must necessarily have been in a state of fusion."[11] Yet Hutton claimed not only that "subterraneous fire had existed previous to, and ever since, the formation of this earth," but also "that it exists in all its vigour at this day."[12] Indeed, as Gordon Davies has recently pointed out, Hutton had a cyclic view of the earth's history in which denudation processes leading to the destruction of continents must be followed by consolidation of sediments and uplift of new continents (powered by the internal heat engine) in order to reconcile destruction with benign deity.[13] Thus a terrestrial version of the Newtonian world-machine was confronted directly with the problem of irreversible heat flow, in the first round of a continuing debate on this theme.

Hutton's disciple John Playfair also felt the need to justify the assumption that the internal fire continues to burn with undiminished intensity throughout the geological ages, but failed to come up with more than a vague suggestion: nature may have "the means of producing heat, even in a very great degree, without the assistance of fuel or of vital air. Friction is a source of heat, unlimited, for what we know, in its extent, and so perhaps are other operations, both chemical and mechanical...."[14] This passage reminds us again of the historical confusion between irreversible processes that conserve energy and those that supposedly do not; primitive ideas of energy conservation could still be invoked to counter apparent examples of irreversibility.

In another passage that was to become notorious in the 19th century, Playfair emphasized that the Huttonian theory implies a cyclic view of the earth's history. As long as the central fires keep burning, new mineral strata can be formed and thrust upwards by volcanic action to replace those that are worn away by erosion; and there is nothing in the laws of nature to prevent this sequence from repeating itself indefinitely.[15] For Playfair there is a clear connection between this cyclic terrestrial history (denying the possibility that the earth may be irreversibly cooling off) and the cyclic nature of the solar system. In a supplementary note he cites the mathematical investigations of Lagrange and Laplace which show that the effects of perturbations are confined within fixed limits, so that it can last forever in its present state as long as planetary motions are governed only by presently-known laws.[16] (Obviously the validity of the geological theory depends on the assumption that there are never any drastic changes in the mean earth–sun distance or sharp fluctuations in the amount of heat which the earth receives from the sun.)

For Playfair, both the Huttonian system of the earth and the Newtonian system of the world behave like intelligently designed machines. The possibility that the earth-machine might be more nearly comparable to the steam engines of Hutton's friend James Watt, as some later commentators have suggested, was probably only dimly realized at the time, and in any case the irreversible character of real heat engines had not yet been clearly pointed out.[17]

Contemplation of the significance of the earth's internal heat did lead some skeptics to formulate more or less explicitly the principle of irreversible heat flow. Thus John Hunter, in 1788, wrote that

> it is well known that heat in all bodies has a tendency to diffuse itself equally through every part of them, till they become of the same temperature.[18]

A few years later John Murray asserted that

> The essential and characteristic property of the power producing heat, is its tendency to exist everywhere in a state of equilibrium, and it cannot hence be preserved without loss or without diffusion, in an accumulated state.... If a heat, therefore, existed in the central regions of the earth, it must be diffused over the whole mass; nor can any arrangement effectually counteract this diffusion. It may take place slowly; but it must always continue progressive, and must be utterly subversive of that system of infinitely renewed operations which is represented as the grand excellence of the Huttonian theory.[19]

In reply to this, Playfair argued that if heat is "communicated to a solid mass, like the earth, from some source or reservoir in its interior," the equilibrium state will not be one of uniform temperature, if the heat can escape from the surface of the body into infinite space. Instead, there will be an equilibrium state in which the temperature is high near the center of the body but decreases going outward.[20]

Murray challenged Playfair's assumption that heat is "supplied" at the center, which he considered merely an arbitrary hypothesis dragged in to avoid his previous objection. On the other hand, Murray argued that the earth's heat would not be lost into space through its atmosphere but that the atmosphere would tend to retain the heat supplied by the sun. The temperature would gradually rise until the earth and its atmosphere are not enough to produce a balance by reradiation of heat; the final state would therefore be a stable one of constant temperature, in contrast to the opinions of certain writers (he cites Bailly's *Histoire de l'Astronomie Moderne*) who had claimed that planets are extinct suns and that all heavenly bodies are gradually cooling down, being destined to reach eventually a "state of Ice and Death."[21] Thus by 1814 the Heat Death was not only known in Britain but subject to attack.

The problem of terrestrial heat flow was obviously ripe for quantitative treatment by this time, and in fact Playfair proposed such a treatment in the last paper mentioned above. But he restricted himself to the very special case of a steady temperature distribution maintained by a heat source at the center of a sphere, which he computed on an *ad hoc* basis without attempting to formulate a general equation for heat flow.

The attempt to formulate that equation was apparently first made by J. B. Biot in 1804, but Biot failed to obtain a consistent differential equation.[22] He did at least recognize that a general law of heat conduction might be based on the fundamental assumption that the rate of heat flow between two bodies is proportional to their temperature difference, and he succeeded in deducing the exponential variation of equilibrium temperature with length on a long bar heated at one end.

The modern theory of heat conduction was established by Joseph Fourier in a series of publications beginning in 1808.[23] Irreversibility was explicit from the beginning:

> When heat is unequally distributed among the different points [parts] of a solid body, it tends to come to equilibrium and pass successively from hotter to colder parts. At the same time the heat dissipates itself at the surface and loses itself in the surroundings or the vacuum. This tendency toward a uniform distribution, and this spontaneous cooling which takes place at the surface of the

body, are the two causes which change at every instant the temperature of the different points.

In the later debates on irreversibility, it was often suggested that there is a basic contradiction between Newtonian mechanics and any theory, such as Fourier's, which is not symmetrical with respect to past and future time directions. It was pointed out that in Newton's second law, $F = ma$, the substitution of $-t$ for t leaves the right-hand side invariant, whereas this is certainly not the case with Fourier's heat conduction equation which contains a first derivative with respect to time. But the alleged contradiction rests on the assumption that Newtonian mechanics deals only with *forces* that are time-reversible, *i.e.* with what are now called conservative systems. This seemed a natural assumption for those scientists who assumed that all macroscopic laws must ultimately be reducible to theories of atomic interactions, and that these interactions could not involve dissipation or velocity-dependent forces. Yet for many other scientists in the 18th and 19th centuries, there was no reason to impose such restrictions, and thus no contradiction. Far from perceiving a conflict between his theory and Newtonian physics, Fourier recognized that he was in a sense improving and generalizing Newton's "law of cooling," though the route from Newton's assumption about the heat lost from an object to the surrounding space, to Fourier's assumption about the flow of heat among infinitesimal portions of matter within a body, was by no means an easy one, as Biot had already discovered.[22,24] Nevertheless the qualitative idea of irreversible heat flow was common to Fourier and Newton.

While Newton's law of cooling, as it applied to finite temperature differences, was frequently challenged during the 18th century and finally rejected during the time Fourier was developing his theory, Fourier's equation became the basis for a successful and widely-used mathematical theory, and this theory in turn permeated the theoretical physics of the 19th and 20th centuries.[25] But at the same time, and especially after Helmholtz formulated the principle of energy conservation in terms of forces between point particles, and Clausius and Maxwell revived the kinetic theory of gases, the assumption that Newtonian physics is fundamentally a time-reversible theory was becoming prevalent. The conflict between these two streams of theoretical physics became evident in the last quarter of the 19th century.

In understanding the development of Fourier's theory it is essen-

§14.2] THE COOLING OF THE EARTH 557

tial to keep in mind his interest in its geophysical applications. This interest should not be a surprise to anyone who reads carefully the "Preliminary Discourse" but it is somewhat under-represented in the main body of the text of the *Théorie Analytique*. Unfortunately the standard English translation of this work[26] omits the supplement on the problem of terrestrial temperatures and heat flow inside a sphere which was published separately in a later volume of the *Mémoires* of the Paris Academy for 1821–22. It is therefore worth noting that Fourier himself stated that the geophysical problem had been a prime source of motivation.[27]

Fourier's conclusions about the cooling of the earth have to be interpreted within the context of early 19th-century geophysical speculation. First, he showed that the periodic temperature variations at the surface due to solar heating would be washed out at a depth of less than 100 meters, and if there were no internal source of heat the temperature would be constant down to the center of the earth. Since existing data showed that there *is* an increase of temperature with depth below 100 meters, there must be an internal reservoir of heat, left over from the original formation of the earth. But the effect of this internal heat on the surface temperature is at present negligible, and it cannot have had any significant effect on the climatic variations during the past several thousand years, contrary to what had been assumed by Buffon and other 18th century writers. Fourier's results were viewed by some contemporary scientists as a *refutation* of the main features of the refrigeration theory; in any case they allowed the possibility that external causes might have produced much *lower* surface temperatures at some time in the past. (In this sense Fourier had to come before Agassiz!)

Fourier derived a theoretical formula for the time required for a sphere to cool down from an initial temperature b to its present temperature (taken as zero), in terms of the present temperature gradient at the surface, $\Delta = \partial T/\partial r$, and a ratio CD/K depending on the heat capacity and conductivity of the sphere

$$t = \frac{b^2}{\pi \Delta^2} \frac{CD}{K}$$

Curiously, though he suggested numerical values for all the quantities on the right-hand side of this equation, he did not actually work out an estimate for t. Perhaps he considered the value obtained by such a calculation—which could be as great as 200 million years—so absurdly large that it was not even worth writing down.[28] The outcome in any case was that while the earth does have an internal heat that has been diminishing slowly over a very long period of time, the actual amount of heat passing through the surface is so small as to have no significance on the time-scale of interest to geologists in the *early* 19th century, though with the greatly enlarged time-scales contemplated later in the century the same quantitative results took on an entirely different significance.

* * *

With the work of Charles Lyell we enter a new phase of geological speculation, in which–freed from many of the earlier theological restraints–scientists began to talk about periods of the order of millions or even hundreds of millions of years for the age of the earth.[29] Lyell's "Uniformitarian" geology was based on the assumption that all present features of the earth's surface should be explained by invoking only those physical causes now seen to be in operation. For many geologists such an assumption was essential if geology was to become a science in the same sense as physics and chemistry–for otherwise there would be no limit to the introduction of *ad hoc* catastrophic hypotheses to explain each special feature. But it was precisely in this attempt to become more scientific that the geologists collided head-on with the physicists, with both sides dissipating a considerable amount of energy.

In his *Principles of Geology* (1830), Lyell could not avoid discussing the still-popular theory of the cooling of the earth, but refused to give up the Hutton–Playfair doctrine of a *constant* internal heat. In the absence of any evidence that the internal heat is variable in quantity, he thinks it is "more consistent with philosophical caution, to assume that there is no instability in this part of the solar system."[30] As Martin Rudwick has recently noted, it was essential to Lyell's basic strategy in geological theory to deny any overall directional tendency or irreversibility in the history of the earth.[31] But other geologists could be Uniformitarians while at the same time claiming a gradual *progression* in the earth's history resulting from physical causes that might have been either constant or quantitatively more important in past epochs. (The term "progressive" was used in almost the same sense as "irreversible.")

Lyell's theory encountered the familiar criticism that internal heat simply could not remain constant. Thus George Greenough (founder and first president of the Geological Society of London) declared in 1834:

If there be heat in the centre of the globe, it must have the properties of heat and none other. I ask not how the Heat originally was lodged in that situation, for the origin of all things is obscure; but I ask why, in the countless succession of ages which the Huttonian requires, the Heat has not passed away by conduction, and if it has passed away, by what other heat has it been replaced.[32]

In the 1840's the situation changed somewhat, partly because of the researches of William Hopkins. Hopkins (1793–1866) is now best known as the tutor of William Thomson, James Clerk Maxwell, and other young mathematical physicists at Cambridge University–he is one of the few scientists to have acquired a reputation by teaching rather than research!–but was also one of the founders of British geophysics.[33] Hopkins claimed that the precession and nutation of the earth–moon–sun system would be affected by the physical state of the earth's interior, and that the existing astronomical data could be accounted for only by assuming that the solid crust of the earth has a thickness at least one-fifth of its radius.[34] This was a rather impressive and unexpected argument, which convinced many scientists of the period that the liquid interior of the earth does not come nearly as close to the surface as had previously been thought. As a result, geological theories relying on the direct action of a molten interior in volcanic and other phenomena were less appealing. Lyell's argument that climatic changes have resulted from geographical changes rather than cooling of the earth as a whole gained favor, with the help of Fourier's proof, mentioned above, that cooling could have little effect on the surface temperature even in a million years.[35] Considerable interest was shown in Louis Agassiz' theory of glacial epochs (*Étude sur les Glaciers*, 1840) which assumed that the surface temperature must have been rising rather than falling at some periods in the past.

Hopkins was very much involved in these geological discussions of the 1840's and went along with the tendency to depreciate the importance of the earth's internal heat in accounting for most geological phenomena. But in an address to the Geological Society of London in 1852, after surveying recent research and speculation, he did assert very strongly the ultimate significance of terrestrial refrigeration in the "progressive development" of inorganic matter. On a long enough time scale, the cooling of the earth *is* important, and imposes an overall irreversibility on all processes. The unavoidable fact that heat flows from high temperatures to low means that no geological theory can legitimately be based on the assumption of a permanent or even cyclically changing high temperature inside the earth. Cyclic changes could be a result only of external action – of periodically changing irradiation from the sun or stars. But, if we may assume that heat has the same properties elsewhere in the universe as it does on earth, we may exclude such external causes since the sun and stars would also have to lose their heat eventually by radiation. Thus, contradicting the Huttonian doctrine, Hopkins announced that he was "unable in any manner to recognize the seal and impress of eternity stamped on the physical universe, regarded as subject to those laws alone by which we conceive it at present to be governed."[36]

For Hopkins, the physical properties of heat imply progressive geological change in the long run, but do not exclude uniformity in the short run, simply because the cooling of the earth takes place so slowly. Hence cosmic irreversibility is an axiom, not an hypothesis subject to test; no evidence of approximate uniformity during geological epochs can refute this idea of "progressive change towards an ultimate limit."[37]

* * *

William Thomson's proficiency in the mathematical theory of heat conduction was a major factor in his rapid rise to eminence in British science in the middle of the 19th century. He first learned Fourier's theory in 1840,[38] and soon became the leading British expert on it; in fact, just before his 17th birthday in 1841, he published a paper pointing out a mistake committed by Professor Philip Kelland, of Edinburgh, who had criticized one of Fourier's statements in his book on heat.[39] At about the same time (April 1841) Thomson entered Peterhouse at Cambridge University. His father, James Thomson, had probably chosen Peterhouse so that William could have the advantage of being tutored by Hopkins, who was already famous for the number of successful "wranglers" he had coached for the mathematical examinations. During his years at Cambridge, Thomson continued his original work in mathematical physics, going off in various directions from his original interest in the theory of heat conduction but always returning to that subject. Curiously enough there seems to be no record of any discussion between Hopkins and Thomson on the subject of heat conduction inside the earth; yet it seems hardly possible that this topic of mutual concern should not have figured in many of their conversations.

§14.2] THE COOLING OF THE EARTH 561

At the end of his fourth paper on Fourier's theory, in 1842, Thomson pointed out that when negative values of the time are substituted into the solution of the heat equation for a specified temperature distribution at $t = 0$, there is in general no meaningful solution. In other words, an arbitrary initial distribution cannot in general be produced by evolution from some previous possible distribution.[40] Many years later, as Lord Kelvin, he referred to this result as a mathematical deduction that there must have been a creation.[41]

Thomson was elected to the Chair of Natural Philosophy at the University of Glasgow in 1846, on the strength of testimonials from Hopkins and many other distinguished scientists.[42] It was generally recognized that he had already embarked on a brilliant career in theoretical physics, although there was some doubt about whether he could effectively teach the practical side of science to ordinary students; it was partly to improve his experimental competence that Thomson spent some time working on the properties of steam in Victor Regnault's laboratory in Paris after his graduation from Cambridge. It was his understanding of the properties of steam that was to enable him to place the irreversibility of heat flow in a wider context (see next section).

Thomson's inaugural dissertation at Glasgow dealt with the theory of distribution of heat inside the earth, and in particular the problem of the earliest time to which the solution of Fourier's equation could be extended, going backwards from a specified temperature distribution.[43] He suggested "that a perfectly complete geothermic survey would give us data for determining an initial epoch in the problem of terrestrial conduction," and this proposal was later put into effect by a committee of the British Association for the Advancement of Science. As he noted in 1881, it was this dissertation "which, more fully developed afterwards, gave a very decisive limitation to the possible age of the earth as a habitation for living creatures; and proved the untenability of the enormous claims for TIME which, uncurbed by physical science, geologists and biologists had begun to make and to regard as unchallengable."[44] Thus Thomson's work on heat conduction was the prelude to his attack on Uniformitarian Geology and (indirectly) Darwinian Evolution, "one of the best known of the fierce scientific battles that enlivened Victorian times"[45] which I have reviewed in another essay.[46]

The geological context of Thomson's irreversibility principle, first published in 1852 in a short note entitled "On a Universal Tendency in nature to the Dissipation of Mechanical Energy," should now be clear enough from his own words:

1. There is at present in the material world a universal tendency to the dissipation of mechanical energy.
2. Any *restoration* of mechanical energy, without more than an equivalent of dissipation, is impossible in inanimate material processes, and is probably never effected by means of organized matter, either endowed with vegetable life, or subjected to the will of an animated creature.
3. Within a finite period of time past the earth must have been, and within a finite period of time to come the earth must again be, unfit for the habitation of man as at present constituted, unless operations have been, or are to be performed, which are impossible under the laws to which the known operations going on at present in the material world are subject.[47]

We must now examine the reasons why Thomson's statement could become incorporated into the select company of "laws of physics" while Hopkins' statement, along with earlier assertions about the irreversibility of natural processes,[48] never reached that status.

Notes for §14.2

1. *Oxford English Dictionary* (London: Oxford University Press, 1933, reprinted 1961). For the association of energy-dissipation with "degeneration" see L. Pfaundler, *Die Physik des täglich Lebens* (Stuttgart, 1904), p. 268; W. S. Franklin, *Phys. Rev.* **30**, 766 (1910). While it is generally assumed that dissipation of energy has pessimistic connotations, this is not necessarily the case. Ludvig Colding phrased it as the principle that "nature strives to realize an ever more perfect liberation of the forces of nature," and associated the scattering of matter and energy throughout infinite space with increasing freedom of the human soul. See *Oversigt over Det Kgl. Danske Videnskabernes Selskabs Forhandlinger,* No. 4–6, p. 136 (1856); English trans. in Per F. Dahl, *Ludvig Colding and the Conservation of Energy Principle* (New York: Johnson Reprint Corp., 1972). Herbert Spencer, a decade later, saw dissipation as only one aspect of a general principle of Evolution, which "can end only in the establishment of the greatest perfection and the most complete happiness" (quoted in Brush, *Graduate J.* **7**, 512 (1967).
2. See S. G. Brush, *Proc. XIII Int. Cong. Hist. Sci.* (Moscow, 1971), §VI, 343 (pub. 1974). Neglect of the geophysical literature accounts for statements such as that of Barnett, that the concept of irreversibility was absent from "modern physical philosophy as it was developed in the seventeenth and eighteenth century" and was foreign to physical thought before 1850. M. K. Barnett, *Osiris* **13**, 327 (1958).
3. An extract was published in the *Acta Eruditorum* in 1693, and larger portions were available in various editions during the 18th century. The original complete work was

§14.2]   THE COOLING OF THE EARTH   563

first published in 1949 by W. E. Peuckert as Volume I of the new edition of Leibniz' *Werke*. See C. C. Beringer, *Geschichte der Geologie* (Stuttgart: Enke, 1954), p. 25ff; B. Sticker, *Sudhoffs Arch.* **51**, 244 (1967). A brief English extract may be found in K. F. Mather and S. L. Mason, *A Source Book in Geology* (New York: McGraw-Hill, 1939, reprinted by Harvard University Press), pp. 45–46.
4. R. Boyle, "Of the temperature of the subterraneal regions as to hot and cold," in *Tracts written by the Honourable Robert Boyle* (Oxford, 1671), reprinted in *The Works of the Honourable Robert Boyle*, ed. T. Birch (London, new ed., 1772), **3**, 326. E. C. Bullard, in *Terrestrial Heat Flow*, ed. W. H. K. Lee (Washington: American Geophysical Union, 1965), p. 1.
5. J. J. Dortous de Mairan (variously alphabetized as Dortous, de, or Mairan), *Mem. Math. Phys. Acad. Roy. Sci.*, p. 104 (1719); p. 143 ($\overline{1765}$). See F. $\overline{C}$. Haber, *The Age of the World* (Baltimore: Johns Hopkins Press, 1959) for a general discussion of Mairan's work and its relation to geophysical speculation in the 18th century.
6. G. L. Leclerc, Comte de Buffon, *Introduction à l'Histoire de Mineraux* (Paris, 1774); *Oeuvres Complètes de Buffon*, ed. M. Flourens, nouv. ed., **9**, 82, 89, 307–8, 348–453 (Paris: Garnier Freres, n.d.); Haber, *op. cit.*, pp. 116–22; J. Roger, *Dict. Sci. Biog.* **II**, 576 (New York: Charles Scribner's Sons, 1970); S. Toulmin and J. Goodfield, *The Discovery of Time* (New York: Harper & Row, 1965), pp. 143–49.
7. Buffon, *Oeuvres* **9**, 455–660.
8. "Jupiter, où regne encore une chaleur brûlante, où les élémens travaillent pour atteindre l'équilibre; à la lune, déjà glacée, et où tout est équilibre, parce que tout est sans mouvement"–J. S. Bailly, *Lettres sur l'Origine des Sciences* (London and Paris, 1777), p. 342; see also his *Histoire de l'Astronomie Moderne* (Paris, nouv. ed. 1785), **II**, 726, 729. For further discussion of the Bailly–Voltaire correspondence see Haber, *Age of the World*, pp. 132–135; E. B. Smith, *Trans. Am. Philos. Soc.* [n.s.] **44**, 427 (1954).
9. J. B. L. Romé de l'Isle, *L'Action du feu central bannie de la surface du globe, et le soleil retablie dans ses droits, contre les assertions de MM. le Comte de Buffon, Bailly, de Mairan, et c.* (Stockholm, 1779), cited by Haber, *Age of the World*, p. 135. who says that "Romé de l'Isle expressed the sentiments of a large number of the naturalists" in his critique. Condorcet and d'Alembert were also skeptical, according to Smith, *op. cit.*
10. Hutton advises that we should not "suppose the wise system of this world to have arisen from the cooling of a lump of melted matter which had belonged to another body. When we consider the power and wisdom that must have been exerted, in the contriving, creating, and maintaining this living world which sustains such a variety of plants and animals, the revolution of a mass of dead matter according to the laws of projectiles, although in perfect wisdom, is but like a unit among an infinite series of ascending numbers"–J. Hutton, *Theory of the Earth* (Edinburgh, 1795), p. 272.
11. Hutton, *Theory of the Earth*, p. 236. The opinions of Hutton and Kirwan have been reviewed by Haber, *op. cit.*, pp. 164–71.
12. *Ibid.*, p. 244. The only evidence he gives for this contention is that "the fires, which we see almost daily issuing with such force from volcanos, are a continuation of that active cause which has so evidently been exerted in all times, and in all places, so far as have been examined of this earth" (*ibid.*, pp. 246–47).
13. *Ann. Sci.* **22**, 129 (1966); *The Earth in Decay* (London: Macdonald, 1969).
14. *Illustrations of the Huttonian Theory of the Earth* (1802, reprinted by University of Illinois Press, 1956, and by Dover Pubs., New York, 1964), pp. 185–86.

15. "How often these vicissitudes of decay and renovation have been repeated, is not for us to determine; they constitute a series, of which, as the author of this theory has remarked, we neither see the beginning nor the end; a circumstance that accords well with what is known concerning other parts of the economy of the world. In the continuation of the different species of animals and vegetables that inhabit the earth, we discern neither a beginning nor an end; and, in the planetary motions, where geometry has carried the eye so far both into the future and the past, we discover no mark, either of the commencement or the termination of the present order. It is unreasonable, indeed, to suppose, that such marks should anywhere exist. The Author of nature has not given laws to the universe, which, like the institution of men, carry in themselves the elements of their own destruction. He has not permitted, in his works, any symptom of infancy or of old age, or any sign by which we may estimate either their future or their past duration. He may put an end, as he no doubt gave a beginning, to the present system, at some determinate period; but we may safely conclude, that this great *catastrophe* will not be brought about by any of the laws now existing, and that it is not indicated by any thing which we perceive." Playfair, *Illustrations*, pp. 119–20.
16. *Ibid.*, pp. 437–38.
17. E. Bailey, *Charles Lyell* (London: Nelson, 1962), p. 18.
18. J. Hunter, *Phil. Trans.* **78**, 53 (1788).
19. J. Murray, *System of Chemistry*, p. 49 (see also p. 51), as quoted by Playfair.
20. *Trans. R. S. Edinburgh*, **6**, 353 (1812 [read 1809]).
21. J. Murray, *Trans. R. S. Edinburgh*, **7**, 411 (1815 [read 1814]).
22. J. B. Biot, *J. Mines* **17**, 203 (1804); *Bibl. Brit.* **27**, 310 (1804). According to J. R. Ravetz (as quoted by M. P. Crosland in the *Dict. Sci. Biog.* article on Biot), Biot, "was unable to present the differential equation corresponding to this physical model because of his inability to find plausible physical reasons for dividing a second difference of temperatures by the square of the infinitesimal element of length. Hence he could not convert his second difference into a second derivative." Cf. J. Fourier, *The Analytical Theory of Heat*, trans. A. Freeman (1878, reprinted by Dover Pubs., New York, 1955), pp. 59, 459–60. For a comprehensive discussion of the relation between Biot's and Fourier's formulations see I. Grattan–Guinness, *Joseph Fourier 1768–1830* (Cambridge, Mass.: MIT Press, 1972), pp. 83–87 and elsewhere.
23. J. Fourier, *Bull. Soc. Philomath.* **1**, 112 (1808) [summary by S. D. Poisson]; *Mem. Acad. Sci. Paris* **4**, 185 (1824 [submitted 1811]), published in somewhat different form as *Théorie Analytique de la Chaleur* (Paris, 1822); *Mem. Acad. Sci. Paris* **5**, 153 (1826) [submitted 1811]). These and other papers are reprinted in *Oeuvres de Fourier*, ed. G. Darboux (Paris: Gauthier-Villars, 1888, 1890). The original manuscript of the 1807 paper was first published in full with critical apparatus by Grattan–Guinness, *op. cit.*, p. 33, from which I have translated the quotation in the text; cf. Freeman trans., p. 14.
24. I. Newton, *Phil. Trans.* **22**, 824 (1701); English trans. in the abridged ed. of *Phil. Trans.* (London, 1809), **4**, 572; both reprinted in I. B. Cohen (ed.), *Isaac Newton's Papers and Letters on Natural Philosophy* (Cambridge, Mass.: Harvard University Press, 1958), pp. 259–68; for further discussion and references see §2 of the previous chapter. Fourier's acknowledgement of his debt to Newton may be found in the manuscript of the 1807 paper (footnote on p. 92 of the Grattan–Guinness edition, *op. cit.*) although it was omitted in later published versions. A somewhat vaguer reference to Newton did appear in ch. IX, art. 429 of *Théorie Analytique de la Chaleur*: "Newton

§14.2]   THE COOLING OF THE EARTH   565

a considéré le premier la loi du refroidissement des corps dans l'air . . ." as compared to "Newton a connu le premier le principe précédent et il en a fait usage pour déterminer la loi du refroidissement d'un corps exposé à un courant d'air," in the 1807 paper.
25. The importance of Fourier's theory as an inspiration for Franz Neumann and others involved in the development of the physics discipline in 19th-century Germany has recently been stressed by R. McCormmach, *Hist. Stud. Phys. Sci.* **3**, ix (1971).
26. Freeman's translation cited in note 22, above.
27. Fourier, *Ann. Chim. Phys.* **27**, 136 (1824); *Mém. Acad. Roy. Sci. Paris* **7**, 570 (1827); see *Oeuvres* **2**, 114, where the first person plural is changed to first person singular.
28. "Quant au nombre T [number of centuries since beginning of cooling] il est évident qu'on ne peut l'assigner; mais on est du moins certain qu'il surpasse la durée des temps historiques, telle qu'on peut la connaître aujourd'hui par les annales authentiques les plus anciennes: ce nombre n'est donc moindre qui soixante ou quatre-vingts siècles. On en conclut, avec certitude, que l'abaissement de la temperature pendant un siécle est plus petit que 1/57600 d'un degré centesimal. Depuis l'École grecque d'Alexandrie jusqu'à nous, la déperdition de la chaleur centrale n'a pas occasionné un abaissement thermométrique d'un 288e de degré. Les températures de la superficie du globe ont diminué autrefois, et elles ont subi des changements très grandes et assez rapides; mais cette cause a, pour ainsi dire, cessé d'agir à la surface: la longue durée du phénomène en a rendu le progrès insensible, et le seul fait de cette durée suffit pour prouver la stabilité des temperatures." *Bull. Soc. Philomath.* **58** (1820); quotation from *Oeuvres* **2**, 286.
29. Haber, *Age of the World*; C. C. Gillispie, *Genesis and Geology* (Cambridge, Mass.: Harvard University Press, 1951).
30. C. Lyell, *Principles of Geology* (2nd ed., 1832), **1**, 162.
31. M. J. S. Rudwick, *Isis* **61**, 5 (1970); *Perspectives in the History of Science and Technology*, ed. D. H. D. Roller (Norman: University of Oklahoma Press, 1971), p. 209. M. Bartholomew, *Brit. J. Hist. Sci.* **6**, 261 (1973). Lyell himself cited Humphry Davy [cf. §1, ref. 27] as a major exponent of the theory of "progressive development," in *Principles of Geology* **1**, 145–46 (1830 ed.). Lyell's acceptance of the alternative cyclic theory is illustrated by the remark in his letter to Gideon Mantell, 12 February 1830: "All these changes are to happen in the future again, and iguanodons and their congeners must as assuredly live again in the latitude of Cuckfield as they have done so. "[Quoted by S. Herbert, *The Logic of Darwin's Discovery* (Ph.D. Dissertation, Brandeis University, 1968), p. 11a, from *Life, Letters and Journals of Sir Charles Lyell, Bart.* (London, 1881).]
32. G. Greenough, *Proc. Geol. Soc. London* **2**, 42 (1838) (quotation from p. 64).
33. See W. F. Cannon, *Isis* **51**, 38 (1960).
34. W. Hopkins, *Phil. Trans.* **132**, 43 (184). See also *Trans. Camb. Phil. Soc.* **6**, 1 (1838) [read 1835]; *Phil. Trans.* **129**, 381 (1839), **130**, 193 (1840). Thomson initially accepted this argument but was persuaded by Simon Newcomb that it is inconclusive. Later, with G. H. Darwin, he developed other arguments based on tidal phenomena, leading to the conclusion that the earth behaves as if it were a solid, as rigid as steel.
35. C. Lyell, *Principles of Geology* (London, 1830–33), **1**, ch. VII.
36. *Quart. J. Geol. Soc. London* **8**, pt. I, xxi (1852), quotation from p. lxxiv. This was Hopkins' "Anniversary Address" as President of the Society, and it may be noted that Lyell had reiterated his non-progressionist views in a similar address the previous year; see *Quart. J. Geol. Soc. London* **7**, xxv (1851).

37. Hopkins, *op. cit.*, p. lxxv.
38. From Professor John Pringle Nichol at Glasgow University, where Thomson studied before going up to Cambridge; see S. P. Thompson, *The Life of William Thomson, Baron Kelvin of Largs* (London: Macmillan, 1910), I, 14.
39. P. Kelland, *Theory of Heat* (Cambridge, 1837), p. 64; P. Q. R. [William Thomson], *Cambridge Math. J.* 2, 258 (1841), reprinted in his *Mathematical and Physical Papers* (Cambridge, 1882–1911), I, 1. This collection will be cited as Thomson's *Papers*.
40. *Cambridge Math. J.* 3, 170 (1842); *Papers* 1, 10.
41. S. P. Thompson, *op. cit.*, pp. 42, 111, 186.
42. Hopkins' testimonial is reprinted along with others, by Thompson, *op. cit.*, pp. 170–71.
43. *Ibid.*, p. 188; see *Rept. Brit. Ass. Adv. Sci.* 25, 18 (1855), 29, 54 (1859); Thomson's *Papers* 2, 175, 3, 291.
44. *Papers* 1, 39.
45. A. Holmes, *The Age of the Earth* (London: Nelson, 2nd ed., 1937), p. 31.
46. *Graduate J.* 7, 477 (1967). See also: J. W. Gregory, *Trans. Geol. Soc. Glasgow* 13 (2), 170 (1908); H. I. Sharlin, *Annals of Science* 29, 271 (1972); J. D. Burchfield, *Lord Kelvin and the Age of the Earth* (New York: Science History Pubs., 1975).
47. *Proc. R. S. Edinburgh* 3, 139 (1852).
48. In addition to the irreversibility statements mentioned in the text I have found the following:

(a) "The sort of retardation which fluids experience in gliding over the surface of a solid obstacle is, therefore, distinct from resistance on the one hand, and from friction on the other, though more allied to the former. But clearly to trace its origin and mode of operation, will require a careful analysis of those several means wherewith Nature speedily extinguishes every motion upon earth, and seems to diffuse a principle of silence and repose; which made the ancients ascribe to matter a sluggish inactivity, or rather an innate reluctance and inaptitude to change its place. [A footnote here refers to a note discussing the views of Kepler and Galileo on inertia.] We shall perhaps find, that this prejudice, like many others, has some semblence of truth." J. Leslie, *An experimental inquiry into the nature and propagation of heat* (London, 1804), pp. 298–99.

(b) "Without reference to any theory, I venture to propose the following as the simple experimental law: All bodies of *unequal* temperature tend to become of equal temperature." B. Powell, *Rept. Brit. Ass. Adv. Sci.* 2, 259 (1832) [this is in the context of a discussion of radiant heat].

(c) "Heat has a constant tendency to diffuse itself over all bodies, till they are brought to the same temperature." W. Enfield, *Institutes of Natural Philosophy* (London, 2nd ed., 1799), p. 399.

14.3 The Second Law of Thermodynamics and the concept of entropy

We now come to the area of science which is, according to almost all commentators, the sole source of the modern idea of irreversibility; the development of steam-engine theory from Carnot through Clapeyron to Clausius, Rankine, and Thomson.[1] More recently, historians have explored the roots of Sadi Carnot's theory in the discussions of the efficiency of various kinds of machines by his father Lazare Carnot and other writers on engineering.[2] In these discussions there is always implicit or explicit the notion that in a poorly designed machine something is "lost" or "wasted" but until Sadi Carnot's 1824 memoir – or, strictly speaking, until Carnot wrote his later notes renouncing the caloric theory – it was not understood how this loss can be consistent with a conservation law.[3] This situation contrasts sharply with the geophysical speculations on heat flow, in which it was generally assumed that the total heat in the universe or in a closed system is conserved, so that if heat flows out through the surface of the earth the amount remaining inside must *therefore* decrease correspondingly.

Carnot posed the question:

> Is the motive power of heat invariable in quantity, or does it vary with the agent employed to realize it as the intermediary substance, selected as the subject of action of the heat?[4]

He had already stated that

> The production of motive power is then due in steam-engines not to an actual consumption of caloric, but *to its transportation from a warm body to a cold body*, that is, to its re-establishment of equilibrium....[5]

Whenever there is a difference of temperature one has the *opportunity* of producing motive power in a steam engine; the question is whether this opportunity or potentiality depends only on the temperatures of the warm and cold bodies, or also on the working substance used in the engine. Carnot's celebrated answer is that it does depend only on the temperatures. Of more interest to us, however, are his sketchy remarks on *why* the theoretical opportunity is never realized in practice:

> Since every re-establishment of equilibrium in the caloric may be the cause of the production of motive power, every re-establishment of equilibrium which shall be accomplished without production of this power should be considered as an actual loss. Now, very little reflection would show that all changes of temperature which is not due to a change of volume of the bodies can be only a useless re-establishment of equilibrium in the caloric.[6]
>
> ... Every change of temperature which is not due to a change of volume or to chemical action ... is necessarily due to the direct passage of the caloric from a more or less heated body to a colder

body. This passage occurs mainly by the contact of bodies of different temperature; hence such contact should be avoided as much as possible. It cannot probably be avoided entirely...."[7]

This is the necessary condition for a *reversible* heat flow, but the term is not used yet. On the next page Carnot refers again to the "loss of motive power" caused by contact between bodies at different temperatures, and says "This kind of loss is found in all steam-engines." Thus ordinary heat conduction, which previously seemed innocent enough by itself, has now been identified as the cause of inefficiency in steam engines.

In Clapeyron's reformulation of Carnot's theory, "loss of motive power" became "loss of force" or "loss of *vis viva*" and was again attributed to the direct passage of heat which naturally occurs whenever two bodies at different temperatures are in contact; in order to obtain the maximum efficiency one must try to avoid such contact.[8]

According to Mendoza, "as late as the 1830's the term "Carnot's theorem" denoted a statement that in any machine the accelerations and shocks of the moving parts all represented losses of ... useful work done," and even in 20th-century textbooks one occasionally finds this usage – the Carnot in question being Lazare rather than Sadi.[9] But there are also scattered statements in the engineering literature before 1850 concerning the waste of motive power in steam engines, though there is usually no reference to Sadi Carnot in this connection.[10]

In 1848 William Thomson published his first paper on Carnot's theory of the motive power of heat. At that time he had not been able to find a copy of Carnot's original memoir, and was acquainted with the theory only through Clapeyron's paper which had been translated into English in the first volume of Taylor's *Scientific Memoirs*.[11] Moreover, he had not yet accepted the principle of convertibility of heat and mechanical work, and assumed with Carnot and Clapeyron that the quantity of heat is conserved when motive power is produced. There is no mention of irreversibility here except indirectly when Thomson alludes to engines 'in which the economy is perfect" with the implication that for imperfect engines a smaller amount of mechanical effect would be obtained by the transmission of a given quantity of heat.[12]

A few months after that paper was published, Thomson finally obtained a copy of Carnot's book from Lewis Gordon, and on January 2, 1849, he read another "Account of Carnot's Theory of the Motive Power of Heat" to the Royal Society of Edinburgh. In this paper Thomson was grappling with the crucial problem of whether to retain

Carnot's published assumption that heat is conserved in the steam-engine cycle, or to accept Joule's proposal that it is actually converted into mechanical work; and this intellectual struggle seems to have blotted out the other problems such as irreversibility. He did state that "engines may be constructed in which the whole, or any portion of the thermal agency is wasted" when heat flows from one body to another by conduction, but a footnote attached to this sentence indicates that he was still unclear as to what is meant by "wasted":

> When "thermal agency" is thus spent in conducting heat through a solid, what becomes of the mechanical effect which it might produce? Nothing can be lost in the operations of nature–no energy can be destroyed. What effect then is produced in place of the mechanical effect which is lost? A perfect theory of heat imperatively demands an answer to this question; yet no answer can be given in the present state of science....[13]

Here we see vividly how the Second Law of Thermodynamics, already born, cannot be christened until it has dragged its brother the First Law out of the womb.

The scene now shifts momentarily to Germany (Thomson having missed his chance to be first in formulating the laws of thermodynamics) where Rudolf Clausius has taken up the problem of the motive power of heat, stimulated by the papers of Clapeyron, Thomson, and Holtzmann. Like Thomson, Clausius considered Carnot's work to be the most important even though so far he knew of it only through the writings of Clapeyron and Thomson.

To Clausius in 1850 it was already clear that heat is not only interconvertible with mechanical work but in fact actually "consists in a motion of the least parts of bodies"[14]–and that this latter conclusion leads us to adopt the equivalence of heat and work, rather than the other way around. But he has not yet grasped the irreversibility implications of Carnot's theory, for in reviewing Thomson's position he says

> Heat can be transferred by simple conduction, and in all such cases, if the mere transfer of heat were the true equivalent of work, there would be a loss of working power in Nature, which is hardly conceivable.[15]

(In the context it appears that this is a report of Thomson's opinion but Clausius at least does not dispute it.) Later in this paper, in order to demonstrate that one substance cannot be used to produce more work with a given amount of heat than another, Clausius invokes the argument that if this were not true, one could transfer heat from a cold to a hot body, which "is not in accord with the other relations of heat, since it always shows a tendency to equalize temperature differences and therefore to pass from *hotter* to *colder* bodies."[16] While there is some justification for the usual view that this paper contains a complete formulation of the laws of thermodynamics, Clausius has not yet put together the pieces of the complete Second Law as we now know it.

Back to Scotland, where on February 4, 1850, W. J. M. Rankine has read to the Royal Society of Edinburgh his paper "On the Mechanical Action of Heat, especially in Gases and Vapours." Rankine's paper contains much of the content of the Clausius–Thomson thermodynamics but gives the appearance of being restricted to deductions from a special molecular-vortex model. It contains two brief statements of irreversibility which, because they follow a recognition of the equivalence of heat and mechanical work, are more significant than earlier statements about the inefficiency of steam engines:

> ... the true mechanical equivalent of heat is considerably less than any of the values deduced from Mr. Joule's experiments; for in all of them there are causes of loss of power the effect of which it is impossible to calculate. In all machinery, a portion of the power which disappears is carried off by waves of condensation and expansion, along the supports of the machine, and through the surrounding air; this portion cannot be estimated, and is, of course, not operative in producing heat within the machine....[17]
>
> Dr. Lyon Playfair, in a memoir on the Evaporating Power of Fuel,[18] has taken notice of the great disproportion between the heat expended in the steam-engine and the work performed. It has now been shown that this waste of heat is, to a great extent, a necessary consequence of the nature of the machine....[19]

There is a similar statement in another paper in 1851, but no attempt to generalize from the limitations of steam engines to a law of nature.[20] Indeed, at this point Rankine has not admitted that Carnot's theorem is independent of the principle of equivalence of heat and work.[21]

By March 1851 Thomson had been converted to convertibility of heat and work and hastened to catch up with Clausius and Rankine by giving his own formulation of thermodynamics. The custom of calling Carnot's principle the "Second Law" of thermodynamics probably originated in this paper; but the way Thomson initially phrased it was not the same as what is now generally called the "Kelvin statement of the Second Law:"[22]

> PROP II. (Carnot and Clausius).–If an engine be such that when it is worked backwards, the physical and mechanical agencies in every part of its motion are all reversed, it produces as much mechanical effect as can be produced by any thermodynamic engine, with the same temperatures of source and refrigerator, from a given quantity of heat.[23]

This proportion was based on an "axiom" which is itself more nearly the usual version of the "Kelvin statement":

> It is impossible, by means of inanimate material agency, to derive mechanical effect from any portion of matter by cooling it below the temperature of the coldest of the surrounding objects.[24]

The second proposition follows from this axiom because of the postulated *reversibility* of the parts of the engine; this is one of the first uses of the *term* "reversibility" in thermodynamics, although the *concept* obviously goes back at least as far as Sadi Carnot.[25]

Thomson then gave "the axiom on which Clausius' demonstration is founded," *i.e.* the "Clausius statement of the Second Law," in a somewhat more explicit form than Clausius himself had yet published:

> It is impossible for a self-acting machine, unaided by any external agency, to convey heat from one body to another at a higher temperature.[26]

Thomson asserted that "it is easily shown" that either axiom is a consequence of the other.[27] In any case it is clear that both are negative statements and do not assert any tendency toward irreversibility. An explicit statement about irreversibility comes in only later on when Thomson discusses a *perfect* Carnot engine operating over an infinitesimal temperature range; in that case he states that the mechanical effect is the largest possible "although it is in reality only an infinitely small fraction of the whole mechanical equivalent of the heat supplied; the remainder being irrecoverably lost to man, and therefore "wasted," although not *annihilated*."[28] Hence the paradox that even the completely reversible engine must include an irreversible process (except when the surroundings are at absolute zero temperature). But that process is not "wasting" energy since the maximum amount of mechanical effect has been extracted.

The lack of any general statement about irreversibility in this paper is puzzling. In fact, there had been one in an earlier draft dated March 1, 1851:

> Everything in the material world is progressive. The material world could not come back to any previous state without a violation of the laws which have been manifested to man; that is, without a creative act or an act possessing similar power.... I believe the tendency in the material world is for motion to become diffused ....[29]

The use of the term "progressive" where we might expect "irreversible" or even "degenerative" or "regressive" is another clear indication of the geological background of Thomson's thinking; for it was Hopkins and the other anti-Lyellians who proposed a "progressive" tendency, resulting from the gradual cooling of the earth, in opposition to the cyclic view.

Thomson's paper of April 19, 1852, on the tendency toward dissipation of energy, begins with the following statement clarifying the relation of this tendency to thermodynamics:

> The object of the present communication is to call attention to the remarkable consequences which follow from Carnot's proposition, that there is an absolute waste of mechanical energy available to man when heat is allowed to pass from one body to another at a lower temperature, by any means not fulfilling his criterion of a "perfect thermo-dynamic engine," established, on a new foundation, in the dynamical theory of heat. As it is most certain that Creative Power alone can either call into existence or annihilate mechanical energy, the "waste" referred to cannot be annihilation, but must be some transformation of energy.[30]

Unfortunately the arguments by which Thomson proceeds from his axiom (the "Kelvin statement" quoted above) to this consequence are presented in an extremely obscure and incomplete manner. For example, there remains some confusion as to whether energy is indeed "dissipated" in a perfect Carnot engine, or whether the dissipation is only a result of the friction of steam rushing through pipes. The only part of the exposition that seems at least qualitatively valid in the absence of detailed calculation is the assertion that some mechanical work could be obtained by using a perfect thermodynamic engine to equalize a nonuniform temperature distribution without allowing heat conduction; hence if the equalization is accomplished by heat conduction alone there must be a waste of mechanical effect.[31]

Thomson listed four distinct processes, all involving heat, in this paper: (1) reversible creation of heat; (2) creation of heat by an "unreversible process (such as friction)" (3) diffusion of heat by conduction; and (4) absorption of radiant heat or light except by vegetation or chemical action. The last three involve dissipation of energy. There is no mention of what was later to be seen as one of the most fundamental of all irreversible processes: *mixing* of two kinds of molecules at constant temperature. In this respect the "principle of dissipation of energy" is less general than the "principle of irreversibility."

So far Thomson has not mentioned any *molecular* basis for the dissipation of energy except in the vague allusion to the tendency for motion to become diffused, in his draft of 1 March 1851. A somewhat more concrete presentation of his views appeared in a short note "On Mechanical Antecedents of Motion, Heat, and Light" read to the British Association meeting in 1854. Here Thomson stated that gravitational potential energy[32] is continually being expended to produce motion and heat. If we trace these actions forwards in time,

> we find that the end of this world as a habitation for man, or for any living creature or plant at present existing in it, is *mechanically inevitable*....[33]

There is a presumption (but not a very clear statement) that thermodynamics is reducible to mechanics.

Clausius did not comment immediately on Thomson's statement of the dissipation principle; in fact, he did not refer to it in print until 1864.[34] But Rankine, in a paper read to a meeting of the British Association at Belfast on 2 September 1852, challenged its universal validity. While admitting that the tendency for all other forms of energy to be converted into heat at uniform temperature "so that there will be an end of all physical phenomena... appears to be soundly based on experimental data, and to represent truly the present condition of the universe, so far as we know it," Rankine pointed out that *radiant* heat is the "ultimate form to which all physical energy tends." According to Rankine, radiant heat is conducted by an "interstellar medium" which cannot convert radiant heat into the "fixed or conductible form" and therefore cannot have a "temperature." But if we assume that this interstellar medium" has bounds beyond which there is empty space, then on reaching those bounds the radiant heat of the world will be totally reflected, and will ultimately be reconcentrated into foci." If a star (or lump of inert matter) happens to arrive at one of these foci, radiant heat can be converted into chemical power and thus "the world, as now created, may possibly be provided within itself with the means of reconcentrating its physical energies, and renewing its activity and life." Thus it is possible to imagine that in the distant future the mechanical energy of the universe could be reconcentrated and the world come to life again.[35]

* * *

The next major step in formulating the concept of irreversibility is found in a paper of Clausius, published in December 1854. Clausius introduced the concept of "equivalence of transformations" which he based on the principle:

Heat can never pass from a colder to a warmer body without some other change, connected therewith, occurring at the same time. Everything we know concerning the interchange of heat between two bodies of different temperatures confirms this; for heat everywhere manifests a tendency to equalize existing differences of temperature, and therefore to pass in a contrary direction, *i.e.* from warmer to colder bodies. Without further explanation, therefore, the truth of the principle will be granted.[36]

In a footnote he amplified the phrase "without some other change," explaining that it would of course be possible for heat to be transferred from a colder to a warmer body if this transfer were intimately associated with the passage of at least as much heat in the opposite direction. One example would be heat transfer by radiation; a body at any temperature is continually radiating heat, some of which may be observed by a *warmer* body, but it is necessary that the amount absorbed by the *cold* body from the warm body be even greater. Clausius also mentions that in addition to the simple transfer of heat, "another permanent change may occur which has the peculiarity of not being reversible" unless it is replaced by another similar permanent change or by a flow of heat from a warmer to a colder body. But Clausius did not indicate at this point the nature of these nonthermal irreversible processes.

In describing the various cyclic processes that may be undergone by a gas, returning it eventually to its initial state, Clausius distinguished between reversible ones that could be run backwards, and non-reversible ones.[37] He then defined the "equivalence-value of a transformation of work into the quantity of heat Q, of the temperature t," as $Q \cdot f(t)$, where $f(t)$ is a function of temperature. He selected the sign convention to be such that conversion of work into heat and passage of heat from a higher to a lower temperature will be *positive* transformations.

If heat is transferred from temperature t_1 to t_2, the equivalence-value of the transformation must depend on both t_1 and t_2, so Clausius writes it $Q \cdot F(t_1, t_2)$. If the two temperatures are interchanged the value must have opposite sign, by definition, hence $F(t_2, t_1) = -F(t_1, t_2)$. For a reversible cyclic process the total equivalence-values of all transformations must be zero; this implies that[38]

$$F(t, t') = f(t)' - f(t)$$

Then Clausius introduced the symbol T as "an unknown function of the temperature" defined as the reciprocal of f

$$f(t) = 1/T$$

Further, instead of placing the temperature t in parentheses he used subscripts to denote the values of T at particular values of t: T_1, T_2, etc. (This notation is psychologically preparing the reader to accept T as actually being equal to the absolute temperature, though Clausius does not want to commit himself to this yet.) Finally, he introduced the symbol N for the total value of all transformations in a cycle

$$N = \frac{Q_1}{T_1} + \frac{Q_2}{T_2} + \cdots = \sum \frac{Q}{T}$$

or, if the transfer takes place at continuously varying temperatures,

$$N = \int \frac{dQ}{T}$$

He could then state the theorem: in a reversible cyclic process $N = 0$.

The proof of the theorem relied on the irreversibility principle: if N were negative, this would mean in effect that heat was passing from a colder body to a warmer body without compensation, contrary to the principle stated above. If N were positive and the cycle were reversible, then one could run it backwards and obtain a negative value of N, which is forbidden for the same reason.

This was now Clausius' way of stating the Second Law of Thermodynamics: for all reversible cyclic processes $\int dQ/T = 0$.[39] In the case of nonreversible cyclic processes, which he treated much more briefly at the end of the paper, the theorem was modified to read:

"The algebraical sum of all transformations occurring in a cyclical process can only be positive," *i.e.* $N > 0$. Such a transformation he called an "uncompensated" one. He stated that there are numerous kinds of such transformations, although they do not differ essentially: the transmission of heat by mere conduction; production of heat by friction, or by the passage of an electric current against a resistance; and "all cases where a force, in doing mechanical work, has not to overcome an equal resistance, and therefore produces a perceptible external motion, with more or less velocity, the *vis viva* of which afterwards passes into heat."[40] The notable feature of this list is that every case involves the production of heat; this is a severe limitation on the concept of irreversibility as articulated by almost all writers up to and including Clausius. (The purely mechanical examples discussed earlier by Newton and others usually involved violations of energy conservation unless one assumed that the "lost" *vis viva* was converted into molecular *vis viva* or heat.)

Clausius has practically reached the modern formulation of the entropy concept at this point, except that he cannot yet prove that his "unknown function" T is really the absolute temperature. For this, he points out, it is necessary to assume that "*a permanent gas, when it expands at a constant temperature, absorbs only so much heat as is consumed by the exterior work thereby performed.*" That would be true for an *ideal* gas obeying the laws of Mariotte and Gay-Lussac; in that case one could write simply $T = a + t$, where t is the centigrade temperature and $a = 273°$. Clausius believed that this assumption had been verified by Regnault's experiments and therefore could be adopted "without hesitation."[41]

It would seem that one should date the discovery or invention of the entropy concept from this 1854 paper, since the change in terminology from "equivalence-value of a transformation" to "entropy" can have no effect on the physical meaning of the concept itself. One might object that the physical definition of entropy has not yet been clearly established since T cannot rigorously be identified with absolute temperature except for ideal gases (Clausius has not yet adopted Thomson's definition of absolute temperature based on Carnot's theorem). Nevertheless it is certainly incorrect to ignore the 1854 paper entirely and so state, as is sometimes done, that Clausius first introduced the entropy concept in 1865.[42]

Clausius was able to provide a theoretical foundation for his assumption about T in his first paper on kinetic theory in 1857. There he explained that no interior work has to be performed to change the volume of a perfect gas since molecular attractions are assured to be insignificant at large distances.[43]

Although Clausius did not want to put himself in the awkward position of basing his (macroscopic) mechanical theory of heat on microscopic assumptions, he continued in his later writings to state that T is simply the absolute temperature without providing a proof.[44] He did introduce in 1862 the concept of "disgregation," defined as a quantity dependent on molecular arrangements, but this quantity was never clearly related to actual positions and velocities of molecules, and it did not seem to imply any degree of randomness in those positions and velocities.[45] In stating that in nature there is "a general tendency, to transformations of a definite direction" he meant only that the equivalence-value of uncompensated transformations is positive, with no further explanation of the physical significance of this tendency.[46]

By 1863 Clausius was finding that his "equivalence value of transformations," though still lacking a clear meaning, was a useful concept in describing various problems and in confounding objections to his theory. In particular, his principle that heat "incessantly strives to pass from warmer to cold bodies" and therefore cannot of itself pass from a colder to a warmer body had seemed obvious to him, but was doubted by G. A. Hirn and earlier by Rankine in the paper mentioned above.[47] This criticism was valuable since it stimulated Clausius to clarify and refine his ideas and to show that the schemes proposed by Hirn and Rankine could not in fact lead to violations of the Second Law.[48] Presumably it was this experience that encouraged him to replace the original clumsy phrase by a handy new one, and so in 1865 we see at last the famous term "entropy" introduced for the first time by the equation $dS = dQ/T$.[49] Clausius has at last recognized the significance of Thomson's dissipation principle,[50] and sees that his entropy concept provides a convenient way to state the directional character of cosmic processes. So the 1865 paper concludes with the celebrated statement of the "two fundamental theorems of the mechanical theory of heat":

1. The energy of the universe is constant.
2. The entropy of the universe tends to a maximum.

In a lecture in Frankfurt in 1867, Clausius gave a more elementary discussion of disgregation and the equivalence-value of transformations. He noted that the Second Law contradicts the idea that (as "one hears it often said") the world is cyclic and may go on forever in the same way. On the contrary, the entropy of the universe tends toward a maximum, which has the consequence that:

> The more the universe approaches this limiting condition in which the entropy is a maximum, the more do the occasions of further changes diminish; and supposing this condition to be at last completely obtained, no further change could evermore take place, and the universe would be in a state of unchanging death.[51]

This was a definitive statement of the "heat death" concept which was so widely discussed in the late 19th century and afterwards.[52]

* * *

It has occasionally been noted[53] that Rankine also introduced in 1854 a "thermodynamic function" equivalent to Clausius' "equivalence-value of a transformation." Unfortunately Rankine's theory appeared to be so deeply entangled with his hypothesis of molecular vortices that he never received much credit for his contributions to thermodynamics, and that is certainly the case with his proto-entropy concept. His thermodynamic function was defined by the equation $\delta F = \delta H/Q$, where δH is the heat consumed in passing from one "curve of no transmission" (*i.e.* adiabat) to another lying indefinitely close to it, and Q is the "actual heat" contained in the substance.[54] Thus $F = $ constant could be used as the equation for a particular adiabatic change of state. Rankine also gave an explicit formula for F for a perfect gas.[55] But the relation between Q and absolute temperature was obscured because at this time Rankine was trying to maintain a distinction between the "absolute zero of gaseous tension" and the "point of absolute cold." He suggested that the difference between these two points might be determined from the Joule–Thomson experiment.[56] Rankine did not use his thermodynamic function to give a formulation of the Second Law in the same way that Clausius did at this time. But in 1865, in a paper on "The Second Law of Thermodynamics," Rankine did point out that his thermodynamic function was identical to the entropy of Clausius.[57] He did not perceive irreversibility as an essential aspect of the Second Law and made no statement about a unidirectional change in his thermodynamic function. Moreover, he asserted that the Second Law could be *derived* from his hypothesis of molecular vortices, provided only that one assumes the molecular motion is *regular*.

It seems to me that both Clausius and Rankine missed their opportunities to give a satisfactory theory of irreversibility based on the entropy concept. In his later papers Clausius concerned himself with the mechanical interpretation of the Second Law but never tried to give a mechanical explanation of irreversibility.[58] In the third edition of his treatise on the mechanical theory of heat he even eliminated the statement that the entropy of the world tends toward a maximum.[59]

Notes for §14.3

1. A typical statement is: "the narrow range of technical interests relative to the economy of heat engines, which constitutes the sole historical source of the Second Law, presents a sharp contrast to the wide variety of roots leading to the Energy Principle–a contrast in full conformity with the novel character of the ideas of restricted convertibility and irreversibility. At the same time, there follows, happily for the historian of thermodynamics, the result that, whereas, because of the great variety of interests involved, the chain of discoveries and enunciations constituting the history of the First Law is exceedingly difficult to expose, the development of the Second Law, on the contrary, takes a direct and relatively simple course which begins definitely and undisputedly with Sadi Carnot and culminates, twenty-five years later, in the systematic treatments of Clausius and Thomson." M. K. Barnett, *Osiris* **13**, 327 (1958) (quotation from p. 335). Other examples of the standard treatment of the history of the Second Law–generally well-written and accurate as far as they go–are V. V. Raman, *J. Chem. Ed.* **47**, 331 (1970); F. O. Koenig, in *Men and Moments in the History of Science*, ed. H. M. Evans (Seattle: University of Washington Press, 1959), p. 57; M. Mott-Smith, *The Concept of Energy Simply Explained* (New York: Dover Pubs., 1964, reprint of *The Story of Energy*, 1934); D. S. L. Cardwell, *From Watt tc Clausius* (Ithaca: Cornell University Press, 1971); O. B. Mathias, *An examination of the evolution of the first two laws of thermodynamics, being an attempt to discover the significance of conceptual changes accompanying their development* (Dissertation, University of Kansas City, Missouri, 1962).
2. T. S. Kuhn, *Arch. int. Hist. Sci.* **13**, 251 (1960); *Isis* **52**, 567 (1961); M. Kerker, *Isis* **51**, 257 (1960); W. Scott, *The Conflict between Atomism and Conservation Theory 1644–1860* (London: Macdonald/New York: Elsevier, 1970); Cardwell, *op. cit.*, ch. 6.
3. W. Scott, *op. cit.*
4. *Reflections on the Motive Power of Fire by Sadi Carnot...*, ed. E. Mendoza (New York: Dover Pubs., 1960), p. 9. In a footnote on this page Carnot states that he uses the terms "quantity of caloric" and "quantity of heat" indifferently, in the sense that should be familiar to the reader from elementary textbooks. This note seems to have been ignored by some later writers who claimed that Carnot was giving a new meaning to caloric, perhaps equivalent to that of entropy.
5. *Ibid.*, p. 7. In the posthumous manuscript notes, translated in an appendix to the *Reflections* in Mendoza's edition. Carnot stated that heat is in fact consumed when motive power is produced (pp. 62–63, 68–69). He had already asked in the *Reflections*, "is it possible to conceive the phenomena of heat and electricity as due to anything else than some kind of motion of the body, and as such should they not be subjected to the general laws of mechanics?" (footnote on page 12). Later, he noted that to deny the conservation of heat in a cycle of operations involving production of motive power "would be to overthrow the whole theory of heat"–yet

"the main principles on which the theory of heat rests require the most careful examination. Many experimental facts appear almost inexplicable in the present state of this theory" (footnote on page 19).
6. *Ibid.*, pp. 12–13.
7. *Ibid.*, p. 13.
8. *Ibid.*, p. 75 (translated from Clapeyron's memoir in *J. École Polyt.* **14**, 153 (1834).
9. *Ibid.*, p. x. (Cf. L. A. Pars, *A treatise on analytical dynamics* (London: Heinemann, 1965), p. 23: "Carnot's theorem. Loss of energy due to the imposition of an inert constraint. When an inert constraint is imposed there is a loss of energy which is equal in value to the energy of the relative motion."
10. M. Seguin, *De l'influence des chemins de fer* (Paris, 1839), pp. xvi–xviii, 378–422 [or Bruxelles, 1839, pp. ix–x, 243–271]; H. de la Beche and L. Playfair, *Mem. Geol. Surv. Great Britain* **2** (II), 539 (1848).
11. W. Thomson, *Proc. Cambridge Phil. Soc.* **1**, 66 (1848); *Phil. Mag.* [3] **33**, 313 (1848); *Papers* **1**, 100. For his own version of the early history of the Second Law see Kelvin, *Popular Lectures and Addresses* (London, 1894), **2**, 451.
12. *Papers* **1**, 103.
13. *Trans. R. S. Edinburgh* **16**, 541 (1849); *Papers* **1**, 113 (quotation from pp. 118–19).
14. R. Clausius, *Ann. Phys.* [2] **79**, 368, 500 (1850); English trans. by W. F. Magie, reprinted in *Reflections on the Motive Power*, etc., ed. Mendoza (quotation from p. 110 of this edition).
15. *Ibid.*, p. 111. Mathias [*op. cit.*, note 1] suggests that it is fortunate that Clausius didn't realize at this time the contradiction between this statement and the irreversibility implied by the Second Law because it might have hindered his development of the latter (p. 96).
16. *Ibid.*, p. 134. Mathias points out that in Magie's translation "other relations of heat" loses the anthropomorphic character of the original German (*Verhalten*, conduct). C. Truesdell has criticized several aspects of Clausius' formulation in this paper, in his recent book *The Tragicomedy of Classical Thermodynamics* (Vienna and New York: Springer-Verlag, 1973).
17. W. J. M. Rankine, *Trans. R. S. Edinburgh* **20**, 147 (1850); *Miscellaneous Scientific Papers* (London, 1881), p. 234 (quotation from p. 245).
18. This may refer to the report by de la Beche and Playfair cited in note 10 above.
19. Rankine, *Papers*, p. 278.
20. *Papers*, p. 304. Maxwell, in 1878, remarked that "In his earlier papers" Rankine "appears as if battling with chaos, as he swims, or sinks, or wades, or creeps, or flies, 'And through the palpable obscure finds out/His uncouth way'" and followed this with similar jibes about particular thermodynamic statements which Rankine had made [Maxwell's *Scientific Papers* **2**, 663]. On the other hand, Kelvin himself criticized Rankine's molecular-vortex theory for not being concrete enough; this provided the occasion for his famous assertion, "I never satisfy myself until I can make a mechanical model of a thing. If I can make mechanical model I can understand it." *Notes of Lectures on Molecular Dynamics and the Wave Theory of Light* (Baltimore: Johns Hopkins University, 1884), p. 270.
21. *Papers*, p. 301.
22. F. O. Koenig, *op. cit.* (note 1), p. 76.
23. W. Thomson, *Trans. R. S. Edinburgh* **20**, 261 (1851); *Papers* **1**, 174 (quotation from p. 178).
24. *Ibid.*, p. 179. Thomson adds a footnote: "If this axiom be denied for all temperatures,

it would have to be admitted that a self-acting machine might be set to work and produce mechanical effect by cooling the sea or earth, with no limit but the total loss of heat from the earth and sea, or, in reality, from the whole material world." The use of the phrase "inanimate material agency" alludes to Kelvin's belief that the Second Law of Thermodynamics may not apply to living beings.

25. *Reflections*, ed. Mendoza, pp. 11, 15.
26. *Papers* 1, 181.
27. On the equivalence of the two forms see C. N. Hamtil, *Am. J. Phys.* 22, 93 (1954); N. L. Balazs, *Am. J. Phys.* 22, 495 (1954).
28. *Papers* 1, 189.
29. Quoted by Harold Sharlin in ch. 7 of his forthcoming biography of Thomson; draft of paper on "Dynamical Theory of Heat" dated 1 March 1851, at Cambridge University.
30. W. Thomson, *Proc. R. S. Edinburgh* 3, 139 (1857) [read 1852]; *Phil. Mag.* [4] 5, 102 (1853); *Papers* 1, 554.
31. Further details of the calculations for this case were given in a paper published the following year: *Phil. Mag.* [4] 5, 102 (1853); *Papers* 1, 554.
32. The term *potential energy* was introduced by Rankine around this time and immediately adopted by Thomson. W. J. M. Rankine, *Proc. Glasgow Phil. Soc.* 3, 276 (1853); *Papers*, p. 203; footnote on p. 554 of Thomson's *Papers* 1.
33. *Rept. Brit. Ass. Adv. Sci.* 24, (II), 59 (1854); *Edinburgh New Phil. J.* 1, 90 (1855); *Papers* 2, 34 (quotation from p. 37). In 1862, Thomson stated that since the universe is infinite, the Second Law does *not* imply a state of universal death. *Popular Lectures and Addresses* (London, 1891), 1, 356.
34. R. Clausius, *Ann. Phys.* [2] 121, 1 (1864).
35. W. J. M. Rankine, *Phil. Mag.* [4] 4, 358 (1852); *Papers*, p. 200. Some other suggestions along this line are cited by M. Čapek, *The Philosophical Impact of Contemporary Physics* (Princeton: Van Nostrand, 1961), p. 128.
36. R. Clausius, *Ann. Phys.* [2] 93, 481 (1854); *Phil. Mag.* [4] 12, 81 (1856); *Mechanical Theory of Heat*, trans. Hirst (London, 1867), p. 111 (quotation from pp. 117–18).
37. *Mechanical Theory*, 121.
38. *Ibid.*, p. 123–25.
39. *Ibid.*, p. 129. As anyone who has taken a course in thermodynamics is well aware, the mathematics used in proving Clausius' theorem is of a very special kind, having only the most tenuous relation to that known to mathematicians. "Six times have I tried to follow the argument of Clausius in the last quarter century, and six times has it gravelled me" (Truesdell, *Tragicomedy*, p. 30).
40. *Ibid.*, pp. 134–35.
41. *Ibid.*, p. 135.
42. P. Fong, *Foundations of Thermodynamics* (New York: Oxford University Press, 1963), p. 17; E. O. Hercus, *Elements of Thermodynamics and Statistical Mechanics* (Melbourne: University Press, 1950), p. 19; E. Hoppe, *Geschichte der Physik* (Braunschweig: Vieweg, 1926), p. 225; The opposite mistake is made by M. Tribus, who says Clausius coined the word entropy in 1850 [*Encyclopedia of Physics*, ed. R. Besancon (New York: Reinhold, 1966), p. 239]. The confusion that can arise from reliance on secondary sources is well illustrated in the article by M. Dutta, *Physics Today*, p. 75 (Jan. 1968).
43. R. Clausius, *Ann. Phys.* [2] 100, 353 (1857), English trans. reprinted in S. G. Brush, *Kinetic Theory* 1 (New York: Pergamon Press, 1965); see end of §9.

44. Clausius, *Ann. Phys.* [2] **116**, 73 (1862); see *e.g.* p. 217 of the English trans., *Mechanical Theory of Heat* (1867).
45. *Ibid.*, p. 220. See the discussions of "disgregation" by E. E. Daub, *Isis* **58**, 293 (1967); M. J. Klein, *Hist. Stud. Phys. Sci.* **1**, 127 (1969); C. Weiner, "Clausius and the 'Internal' explanation of entropy" (unpublished).
46. *Ibid.*, p. 247.
47. G. A. Hirn, *Exposition analytique et expérimentale de la Théorie Mécanique de la Chaleur* (Paris and Colmar, 1862). P. de Saint-Robert, *Cosmos, Revue Enc.* **22**, 200 (1863). G. A. Hirn, *Cosmos, Revue Enc.* **22**, 283, 413, 734 (1863). R. Clausius, *Cosmos, Revue Enc.* **22**, 560 (1863). Rankine, paper cited in note 35; also Rankine's article on "Heat" in *A Cyclopedia of the Physical Sciences* (London, 1857, 2nd ed., 1860), esp. the statement of the Second Law on p. 413 (2nd ed.).
48. R. Clausius, *Ann. Phys.* [2] **120**, 426 (1863) (*Mechanical Theory of Heat*, p. 267); *Ann. Phys.* [2] **121**, 1 (1864) (*Mechanical Theory of Heat*, p. 290).
49. R. Clausius, *Ann. Phys.* [2] **125**, 353 (1865) (*Mechanical Theory of Heat*, p. 327; see p. 357 for definition of entropy, with change of sign from earlier definitions). Modern authors generally assume that the word is supposed to mean "energy transformation" but on the basis of its Greek roots, it could also signify "modesty," "humiliation" or "tricks." See H. G. Liddell and R. Scott, *A Greek–English Lexicon* (Oxford: Clarendon Press, new ed., 1940), pp. 577–78.
50. See note 33; further brief mention in *Mechanical Theory of Heat*, p. 364.
51. R. Clausius, *Phil. Mag.* [4] **35**, 405 (1868).
52. See S. G. Brush, *Graduate Journal*, **7**, 477 (1967). In addition to the literature cited there, the following works may be of interest to anyone who wants to study the various ramifications and influences of the "heat death" concept. Adolf Fick, *Die Naturkräfte in ihrer Wechselbeziehung* (Würzburg, 1869). H. F. Walling, *Proc. Am. Assoc. Adv. Sci.* **22**, 46 (1873); *Pop. Sci. Mon.* **4**, 430 (1874). A. Ritter, *Anwendungen der mechanischen Wärmetheorie auf kosmologische Probleme* (Hannover, 1879) 64 and the lecture of E. du Bois-Reymond which he cites. S. T. Preston, *Phil. Mag.* [5] **8**, 152 (1879), **10**, 338 (1880); *Nature* **19**, 460, 555 (1879), **20**, 28 (1879). W. Muir, *Nature* **20**, 6 (1879). T. H. Huxley, "The struggle for existence in human society" (1888), reprinted in *Selections from the Essays of T. H. Huxley* (New York: Appleton-Century Crofts, 1948). A. W. Bickerton, *Trans. New Zealand Inst.* **27**, 538 (1895). G. Hirth, *Entropie der Keimsysteme und erbliche Entlastung* (München, 1900). S. Arrhenius, *Lehrbuch der kosmischen Physik* (Leipzig: Hirzel, 1903); *The Life of the Universe*, trans. H. Borns (London and New York: Harper, 1909), **2**, 230–41. K. S. Trincher, *Biology and Information*, trans. from Russian (New York: Consultants Bureau, 1965), appendix by Kuznetsov discussing writings of Timiryazev, Umov, and Auerbach, 1901–5. R. C. Tobey, *The American Ideology of National Science, 1919–30* (Pittsburgh: University of Pittsburgh Press, 1971), on R. A. Millikan's views. O. Lodge, Nature **128**, 722 (1931). P. W. Bridgman, in his *Reflections of a Physicist* (New York: Philosophical Library, 1950), p. 150 (reprinted from *Bull. Am. Math. Soc.* 1932). E. A. Milne, *Relativity, Gravitation and World Structure* (Oxford: Clarendon Press, 1935), pp. 285–86. H. N. Russell, in *Time and its Mysteries*, Series III (New York: NYU Press, 1949), p. 1 (lecture given in 1940). J. Barzun, *Science the Glorious Entertainment* (New York: Harper & Row, 1964), p. 117. J. H. Buckley, *The Triumph of Time* (Cambridge, Mass.: Harvard University Press, 1966), ch. 5. N. Georgescu-Roegen, *The Entropy Law and the Economic Process* (Cambridge: Harvard University Press, 1971).

§14.4] THE INTRODUCTION OF STATISTICAL IDEAS

53. In his review of Poincaré's text on thermodynamics, P. G. Tait wrote: "We look in vain for any mention of Rankine or his Thermodynamic Function; though we have enough, and to spare, of it under its later *alias* of Entropy"–*Nature* **45**, 245 (1892). Rankine is cited by J. T. Merz, *A History of European Thought in the Nineteenth Century* (Edinburgh: Blackwood, 1904–14), **1**, 316, **2**, 169; J. Swinburne, *Electrician* **50**, 442 (1903); V. V. Raman, *J. Chem. Educ.* **50**, 274 (1973). "Rank" was used as a unit for entropy by J. Perry, *Electrician* **50**, 398 (1902).
54. W. J. M. Rankine, *Phil. Trans.*, p. 115 (1854). *Papers*, p. 339 (see p. 351 for the definition of *F*).
55. $F = \text{hyp. log } Q - \frac{Nh}{nq+h} + N \text{ hyp. log } V$, where h represents the deviation from the ideal gas law, $PV = NQ + h$. See *Papers*, pp. 363–64.
56. *Papers*, p. 376. See also p. 390 where he assumes that the difference (κ) between the two zero points is 2.1°C, but says in a footnote that "it is probable that κ may be found to be inappreciably small."
57. Rankine, *Phil. Mag.* [4] **30**, 241 (1865); *Papers*, p. 427. This appears to be one of the first uses in print of the phrase "Second Law of Thermodynamics"–Clausius was still calling it the second theorem of the mechanical theory of heat.
58. E. Daub, *Dict. Sci. Biog.* **3**, 309, 310 (1971).
59. R. Clausius, *Die Mechanische Wärmetheorie*, 3. Aufl. (Braunschweig, 1887). For further details on later thermodynamic discussions of entropy and irreversibility see E. E. Daub, *Hist. Stud. Phys. Sci.* **2**, 321 (1970). Extensive bibliographies on thermodynamics may be found in J. R. Partington, *An Advanced Treatise on Physical Chemistry* (London: Longmans, Green and Co., 1949), **I**, 115–233.

14.4 The introduction of statistical ideas in kinetic theory

The failure of Clausius to develop a statistical theory of irreversibility is all the more remarkable since, in addition to inventing entropy, he was the first to find an effective use for statistical methods in the kinetic theory of gases. But before we discuss that topic we must review the context of scientific thinking about molecular motion as it had developed up to the mid-19th century.

The suggestion of Lucretius that atoms swerve randomly in their paths, thereby permitting the possibility of free will, was probably familiar to all educated men in the 17th and 18th centuries.[1] Randomness played some role in debates about the nature of the world in the time of Newton, as well as in the development of probability theory.[2] For those who were deeply concerned about the place of God in the world, both randomness and determinism were distasteful. With the triumph of Newtonian mechanics, it was recognized that molecular motions are "in principle" determined, so that a super-intelligence that could know all the positions and velocities of all molecules in the universe at one instant could know both the past and the future. This assertion is now generally referred to as "Laplacean determinism" because Laplace popularized it in an especially vivid way in his essay on probability theory.[3] As has recently been noted by Roger Hahn, similar statements can be found in earlier writings of Laplace, indicating the probable influence of Condorcet.[4] Other scientists such as Boscovich asserted the determinism of mechanical motions,[5] so that one should probably regard this as an accepted position at the beginning of the 19th century; yet it by no means excluded the application of probability theory to all kinds of physical phenomena, and in fact it was just at the beginning of the 19th century that one notes a flowering of many branches of statistics.[6]

Cassirer has claimed that little attention was paid to the broader implications of Laplace's statement on determinism until the "ignorabimus" speech of Emil du Bois–Reymond in 1872.[7] While one does find occasional discussions of Laplacean determinism in a philosophical context between 1814 and 1872,[8] Cassirer's view meshes with my interpretation that there was little serious debate on the issue of determinism until after the effectiveness of statistical methods had been demonstrated in kinetic theory. Even after the debate on the reversibility objection to the H-theorem had strongly suggested a need for assuming that molecular motions are "disordered" (see below), it was difficult for scientists to abandon the view of Laplace that one assumes phenomena to be random merely because of lack of knowledge rather than because of any inherent indeterminism.

An example of this view in midcentury Britain is furnished by a letter from R. L. Ellis to J. D. Forbes, in connection with the debate on the application of statistical theory to observations of double stars. Ellis says that "random" means nothing except with reference to the knowledge of the observer and his system of classifying the phenomena; "for everything which exists there is a definite reason why it is what it is" so the notion of *fundamental* randomness is meaningless.[9]

Laplacean determinism has also been taken to imply the elimination of the concept of "time" (except as a mere mathematical parameter) in Newtonian physics. Insofar as the equations of mechanics are time-reversible, there is no qualitative difference between past and future, only the difference between a plus and a minus sign.[10] This argument may be a source of confusion since it suggests a (false) converse; logically it would be quite possible to design an irreversible theory which is also deterministic (Fourier's theory of heat conduction is an example).

* * *

The published writings of scientists who identified heat with molecular motion before 1856 rarely state that this motion is in any way irregular or random. Herapath in 1821 postulated that gases consist "of atoms, or particles, moving about, and among one another, with perfect freedom" but also stated that different temperatures of the same body depend on the "velocity of vibration" of its particles.[11] As he noted, this was the usual definition proposed by "the advocates for the theory of heat by intestine motion" and indeed the "vibration" of atoms was often conceived as a regular back-and-forth motion.[12] The replacement of the caloric theory of heat by the wave theory[13] reinforced this idea by associating heat with vibrations of the ether, and those writers who talked about thermal molecular motion in the early 1850's often explicitly identified heat with vibrations of atoms.[14] Joule, adopting Herapath's kinetic theory in 1847–48, emphasized that molecular motion is *rapid*, though he also remarked that the molecules are "constantly flying about in every direction". Only Waterston, in his 1845 paper that remained generally unknown until 1892, stressed the idea that the particles are "moving in all directions" and "encounter one another in every possible manner" during an infinitesimal time period. The earliest statements identifying heat with molecular motion by Clausius, Thomson, and Tyndall in the years 1852–53 are remarkably noncommittal about what kind of motion it is.[15]

According to Krönig, who revived the kinetic theory of gases in 1856, the molecules of a gas move at constant speed until they strike another molecule or the side of the container. Since the smoothest wall is very rough on the molecular level the resulting path of a molecule must be quite irregular, but according to the laws of probability this complete irregularity leads to complete regularity of behavior.[16] But in fact Krönig makes no explicit use of probability concepts at all in this paper.

Clausius, sharing the same assumptions in his more substantial treatment in 1857, did not put any emphasis on the irregularity of molecular motion. He did invoke the "laws of probability" in arguing that "there are as many molecules whose angles of reflexion fall within a certain interval, *e.g.* between 60° and 61°, as there are molecules whose angles of incidence have the same limits, and that, on the whole, the velocities of the molecules are not changed by the side."[17] But here the stress is on the uniform distribution of directions of motion among all possible values, an idea that is not essentially connected with randomness.

As a result of the criticism of the Dutch meteorologist C. H. D. Buys-Ballot,[18] Clausius introduced his "mean-free-path" [*mittlere*

Weglänge] concept. Rather than assume that a molecule can move several meters in a straight line before hitting a macroscopic object, Clausius preferred to attribute a finite size, or rather a finite sphere of action, to the molecules, so that intermolecular collisions would be frequent enough to cause each molecule to change its direction of motion before it can go more than a very short distance. In order to compute the relation between the average distance traveled by a molecule between successive collisions and the effective molecular diameter, Clausius suggested that we should "imagine a great number of molecules moving irregularly about amongst one another"[19] and then fix our attention on one particular molecule to see how often it collides with another one. The probability that a molecule will strike another one in passing through a layer of thickness x is asserted to be simply the ratio of the cross-sectional area corresponding to the average number of molecules to be found in such a layer, to the total area of the layer.

Up to this point the reasoning is compatible with an "ignorance" concept of probability: if we know nothing about how the molecules are arranged in space it is reasonable to make such an assumption about the probability of a collision in the first infinitesimal layer, even if there is actually a regular lattice structure. But now, without any further discussion, Clausius assumes that if the molecule passes through the first such layer without suffering a collision, it must risk the same chance of a collision in the second layer; or, better, the probability that it does *not* suffer a collision in either the first or the second layer is the square of the probability that it does not suffer a collision in a single layer. Clausius did recognize that in order for this calculation to be valid one must at least exclude the possibility that the molecules are regularly arranged in space; but he didn't worry about any more subtle types of correlation.[20]

As was noted in the preceding chapter, Clausius used statistical methods in a rather limited and clumsy fashion in his theory of heat conduction in gases and was reluctant to take full advantage of Maxwell's theory of the velocity distribution. The same is characteristic of his other writings on kinetic theory; he does not seem to want to recognize that any physically significant consequences might follow from the assumption of randomness,[21] and he is eager to replace molecular quantities by their average values in every calculation, sometimes prematurely.

We recall the remark of Gibbs, in his obituary of Clausius:

In reading Clausius we seem to be reading mechanics; in reading Maxwell, and in much of Boltzmann's most valuable work, we seem rather to be reading in the theory of probabilities.[22]

* * *

With the entrance of James Clerk Maxwell, the kinetic theory finally draws on the mainstream of the development of probability theory; in fact, Maxwell at first goes overboard in assuming what amounts to complete randomness of molecular motion, and later has to retreat to a more deterministic approach in order to comply with the accepted physical viewpoint.

As Charles Gillispie has noted, Maxwell probably was influenced by John Herschel's review of Quetelet's books in the *Edinburgh Review* (1850); here Herschel provided a derivation of the normal law of errors quite similar to that which Maxwell himself later used in presenting his velocity distribution.[23] Maxwell's correspondence with Lewis Campbell in 1850, and with R. B. Litchfield in 1858, indicates that he was probably familiar with Herschel's article; he may also have discussed it with W. F. Donkin and J. D. Forbes, who participated in a debate on this subject in the *Philosophical Magazine* in 1850–51.[24] In any case Maxwell initially felt no need to justify his use of the law of errors for molecular velocities, and it was his failure to explain what this law had to do with the motions and collisions of molecules (assumed to obey Newtonian laws) that prevented other kinetic theorists from appreciating the validity of his law when it was first published in 1860.

By analogy with Herschel's assumption that deviations (for example of a ball dropped from a height, aimed at a mark) in perpendicular directions are independent, Maxwell assumed in his 1860 paper that the probability of a molecule having a certain value of the x-component of velocity is not affected by knowledge of its y-component of velocity. He did not recognize that this assumption cannot be true in a finite system with fixed total energy (if one component of velocity is so large that it corresponds to nearly the entire kinetic energy of the system, then the other components cannot have similarly unrestricted values[25]). Maxwell did give a generalized treatment that takes account of this situation in a much later paper, written near the end of his life.[26] But in his memoir "On the Dynamical Theory of Gases" published in 1867, he merely noted that "this assumption may appear precarious," and tried instead to derive the distribution law in a way explicitly involving molecular collisions.

Maxwell's 1867 treatment avoids the use of terms suggesting randomness, asserting merely that in a gas there are a certain number of molecules having a specified value of the velocity vector. The number of encounters of molecules having two particular values of the velocity vector is then assumed to be proportional to $n_1 n_2$, the product of the numbers having those values separately; but Maxwell does not explain why such an assumption of independence of the velocities of two molecules is any more acceptable than his previous assumption of the independence of different components of the velocity of the *same* molecule. There is one important difference which becomes clear only in Boltzmann's later work: Maxwell's second assumption makes it possible to describe an irreversible time evolution of the velocity distribution function. Thus a connection between randomness and irreversibility emerged mathematically from Maxwell's attempt to prove that his velocity distribution law represents a stable equilibrium in a gas of colliding molecules. The development of this connection will be our major concern in the rest of this chapter, but there is only the barest hint of its significance in Maxwell's conclusion that his distribution "is therefore a possible form of the final distribution of velocities" (because once attained it is not altered by further collisions); "it is also the only form" (because otherwise the direct and inverse collisions would not balance.[27]

The same memoir contains a discussion of another topic, seemingly unrelated to the problem of irreversibility, which was involved in the later Loschmidt–Boltzmann discussion of the reversibility paradox. Under the heading "Final Equilibrium of Temperature" Maxwell stated that, after some difficulty, he had managed to prove that a column of gas under gravitational forces must have the same temperature throughout. This result seemed to some scientists at the time contrary to common sense as well as to experience, since it was by then "well known" that the air gets colder as you go up in the earth's atmosphere.[28] If Fourier's law of heat conduction were applicable here, one would expect this temperature gradient to be associated with a flow of heat from the earth out into space. But in many of the discussions of this problem, the temperature gradient was attributed *not* to thermal conditions alone, but rather (or mainly) to the action of the earth's gravity on air molecules at different heights. Herapath and Waterston had proposed explanations of the temperature gradient, based on kinetic theory, invoking such action, but their reasoning can now be seen as fallacious.[29] Maxwell, after discussing the problem with

William Thomson, decided that the Second Law of Thermodynamics requires a uniform temperature distribution. As justification for this conclusion, however, Maxwell advanced only a much weaker principle: "if the temperature of any substance, when in thermic equilibrium, is a function of the height, that of any other substance must be the same function of the height" (otherwise it would be possible to rig up an engine that could take heat from the colder substance and give it to the hotter substance at the same height). Then, having shown that the temperature is independent of height for gases, Maxwell argues that it must also be independent of height for all other substances. Since he thought that the proof for gases depends on the precise form of the velocity distribution law, Maxwell wrote: "we may regard this law of temperature, if true, as in some measure a confirmation of the law of distribution of velocities."[30] This is a rather curious statement since all observational evidence at that time indicated that the uniform-temperature law is *not* true. Of course the only experimental data available was for the atmosphere, where it was not evident how to disentangle the effects of differential heat input at the top and bottom of the imaginary column from the effect of gravity. Until the discovery of the isothermal layer (tropopause) by Teisserenc de Bort around 1900, it was generally thought that there is a uniformly linear decrease of temperature with height, and Maxwell's theory (as reinforced by Boltzmann) provided the main support for the contention that this decrease is to be attributed entirely to the fact that more heat is supplied at the bottom of the atmosphere than at the top.

In 1867 began the correspondence with P. G. Tait leading to the concept of "Maxwell's Demon."[31] Maxwell used this device to show that the Second Law of Thermodynamics cannot be an absolute law of nature, since one can conceive of violating it by sorting out individual molecules into fast and slow categories. Thus the Second Law, according to Maxwell, "has only a statistical certainty"[32]–it is valid only as long as we consider very large numbers of molecules which we cannot deal with individually.

It must not be assumed that "statistical" here implies randomness at the molecular level, for it is crucial to the operation of the Maxwell Demon that he be able to observe and predict the detailed course of motion of a single molecule. This point is made clear by Maxwell in his *Theory of Heat*:

> ... in adopting this statistical method of considering the average number of groups of molecules selected according to their velocities, we have abandoned the strict kinetic method of tracing the

exact circumstances of each individual molecule in all its encounters. It is therefore possible that we may arrive at results which, though they fairly represent the facts as long as we are supposed to deal with a gas in mass, would cease to be applicable if our facilities and instruments were so sharpened that we could detect and lay hold of each molecule and trace it through all its source.[33]

This statement comes near the beginning of the chapter in which the Demon makes his first public appearance, and is obviously intended to lay the groundwork for the discussion that follows. For Maxwell it is our *knowledge* of the world that is statistical, not the world itself; and in fact he flatly states that we should not suppose that the masses, for example, of hydrogen molecules have a statistical distribution of which we only observe the average; on the contrary he thinks that "the equality which we assert to exist between the molecules of hydrogen applies to each individual molecule, and not merely to the average of groups of millions of molecules."[34]

While Maxwell's Demon is generally cited in connection with the possibility of violating the Second Law of Thermodynamics, it seems equally important to note that by making the *mixing* of different molecules the fundamental irreversible process, Maxwell has really strengthened the concept of irreversibility, especially for those who seek molecular explanations for all phenomena. In this context it was only a short step (though not a trivial one) to Boltzmann's identification of entropy with disorder, and the idea that irreversibility is simply a tendency for things to get more chaotic.

Maxwell's first explicit suggestion of a connection between irreversibility and randomness is found not in his discussion of the Demon but in a letter to the editor of the *Saturday Review*, 13 April 1868. In an earlier letter (7 April 1868) Maxwell had commented on an article in the April 4 issue on "Science and Positivism" discussing Caro's treatment of "the doctrine of the gradual conversion of all kinds of energy into the form of heat, and the ultimate uniform distribution of temperature over all matter." In reply to a request for further information, Maxwell wrote that the tendency for a gas to acquire a statistical distribution of velocities is an irreversible operation similar to a process in which black and white balls are "jumbled together" in a box: "the operation of mixing is irreversible."[35] In contrast to the previous examples of "statistical" interpretations of the Second Law, the mixing is attributed not to the natural deterministic motions of the balls left to themselves, but to an external agent. From the viewpoint of

§14.4] THE INTRODUCTION OF STATISTICAL IDEAS   591

an observer who sees the balls but not the external agent, their motion is random.

But are molecules in a gas really moving randomly? In his lecture on "Molecules" to the British Association meeting at Bradford in 1873, Maxwell noted Lucretius' hypothesis that the atoms "deviate from their courses at quite uncertain times and places, thus attributing to them a kind of irrational free will, which on his materialistic theory is the only explanation of that power of voluntary action of which we ourselves are conscious."[36] But Maxwell rejected the materialistic view, which would make all motions cyclic if such randomness were not present, while maintaining that the motions of individual molecules are deterministic: "As long as we have to deal with only two molecules, and have all the data given us, we can calculate the result of their encounter." In our conception of molecules, "we leave the world of chance and change, and enter a region where everything is certain and immutable." It is only because we lack the necessary data that we must use the statistical method in dealing with a gas containing a large number of molecules.

Yet by the time he delivered the Bradford address Maxwell was already beginning to move away from this position in expounding his ideas to nonscientists. In February 1873 he had read an essay to a faculty discussion club at Cambridge University on the question, "Does the progress of Physical Science tend to give any advantage to the opinion of Necessity (or Determinism) over that of the Contingency of Events and the Freedom of the Will?"[37] Maxwell suggested that "recent developments of Molecular Science seem likely to have a powerful effect on the world of thought" by calling attention to the distinction between the Dynamical and the Statistical kinds of knowledge. While the emphasis is still on the epistemological side of this distinction, there is a significant shift in Maxwell's conception of what really does happen at the molecular level when he writes: "Our free will at the best is like that of Lucretius's atoms—which at quite uncertain times and places deviate in an uncertain manner from their course." Here he stands against the "Determinist" who asserts that some cause other than the Ego determines the result of every action. In any case the doctrine that "from like antecedents follow like consequents" is of little use in a world where antecedents can never be established with sufficient precision, and we know that frequently a small error in the data leads to a large error in the result. Thus a pragmatist must renounce determinism.

In his correspondence with Herbert Spencer later the same year,

Maxwell stated that he had used the word "agitation" for the deviation of the actual velocity of an individual molecule from the mean velocity of the group in order to avoid the connotation of "rhythm." Spencer was surprised that Maxwell had rejected his notion that molecular motion is rhythmic, and was not much inclined to incorporate statistical notions into his own philosophy.[38]

Further evidence of the drift of Maxwell's thinking may be found in his 1875 lecture to the Chemical Society of London, in which he remarked:

> The peculiarity of the motion called heat is that it is perfectly irregular; that is to say, that the direction and magnitude of the velocity of a molecule at a given time cannot be expressed as depending on the present position of the molecule and the time.[39]

That this irregularity is essential to irreversibility was explicitly recognized in Maxwell's article "Atom" for the *Britannica*:

> The constancy and uniformity of the properties of the gaseous medium is the direct result of the inconceivable irregularity of the motion of agitation of its molecules. Any cause which could introduce regularity into the motion of agitation, and marshal the molecules into order and method in their evolutions, might check or even reverse that tendency to diffusion of matter, motion, and energy which is one of the most invariable phenomena of nature, and to which Thomson has given the name of the dissipation of energy.[40]

This seems to me a stochastic as opposed to a statistical explanation of irreversibility, though Maxwell himself does not point out the distinction.

In another article, "Diffusion," written for the *Britannica*, Maxwell again emphasized that molecular motion is "irregular."[41] He also discussed the question of whether diffusion leads to an irreversible increase of entropy. This was a crucial point in the development of a general theory of irreversibility, going beyond the special case of heat flow. Maxwell observed that the answer depends on whether the gases which interdiffuse are the same, or whether they are different and can be separated by a reversible process. In the first case there is no entropy increase, but in the second there is. But how can we be sure that two gases are really the same? This is the famous "Gibbs paradox" and Maxwell's discussion is probably influenced by that of Gibbs though he does not mention him.[42] Maxwell's conclusion goes beyond

Gibbs, for he points out that it is quite possible that we might mix two gases which we *thought* were identical, and later discover that they could be separated by a reversible process; in this case we would have to correct the entropy-increase assigned to the original mixing from zero to a positive value. But this means that entropy is not an observable property of the system itself but depends on our knowledge about the system:

> Dissipated energy is energy which we cannot lay hold of and direct at pleasure, such as the energy of the confused agitation of molecules which we call heat. Now, confusion, like the correlative term order, is not a property of material things in themselves, but only in relation to the mind which perceives them.

So once again Maxwell draws back from the position that molecular motions are random in themselves, giving in the process a remarkable anticipation of the modern "information theory" interpretation of entropy.

Maxwell's critique of the attempts of Boltzmann, Clausius, Szily and others to reduce the Second Law to a purely mechanical principle is well known and is cited here only for the sake of completeness, and to reiterate that his own interpretation of the Second Law in most of these remarks is statistical rather than stochastic.[43] While Boltzmann and Helmholtz later revived the attempts to find mechanical analogies for thermodynamics in their papers on monocyclic systems in the 1880's, both recognized that the irreversibility aspect of the Second Law had to be based on statistical rather than purely mechanical foundations.[44]

Notes for §14.4

1. T. L. Carus, *De Rerum Natura* (London, 1886), 1, Bk. 2, 11. 216–24, 251–62, 292–93; English trans. by A. D. Winspear (New York: S. A. Russell, The Harbor Press, 1956), p. 56. On the occasion of the Belfast meeting of the British Association in 1874, Maxwell wrote a poem on "Molecular Evolution" which begins:

   At quite uncertain times and places,
      The atoms left their heavenly path,
   And by fortuitous embraces,
      Engendered all that being hath.
   And though they seem to cling together.
      And form "associations" here,
   Yet soon or late, they burst their tether,
      And through the depths of space career...

L. Campbell and W. Garnett, *The Life of James Clerk Maxwell* (London, 1882, reprinted with a selection of letters from the 2nd ed. 1884 by Johnson Reprint Corp., New York, 1969), p. 637. Cf. F. M. Turner, *Vict. Stud.* **16**, 329 (1973).
2. O. B. Sheynin, *Arch. Hist. Exact Sci.* **7**, 217 (1971).
3. P. S. de Laplace, *Essai Philosophique sur les Probabilités* (Paris, 1814, reprinted by Gauthier-Villars, Paris, 1921), p. 3: "Nous devons donc envisager l'état présent de l'universe comme l'effet de son état antérieur, et comme la cause de celui qui va suivre. Une intelligence qui pour un instant donné connaîtrait toutes les forces dont la nature est animée et la situation respective des êtres qui la composent, si d'ailleurs elle était assez vaste pour soumettre ces données à l'analyse, embrasserait dans la même formule les mouvements des plus grands corps de l'universe et ceux du plus léger atome: rien ne serait incertain pour elle, et l'avenir comme le passé serait présent a ses yeux." A similar statement, with a more astronomical flavor, may be found in *The System of the World* (Dublin, 1830), p. 24.
4. R. Hahn, *Actes XIe Cong. Int. Hist. Sci.*, Cracow, 1965 (pub. 1968), **2**, 167. Hahn quotes the following passage from Laplace's memoir of 1773, which may be set beside the quotation in the preceding note: "L'état présent du système de la Nature est évidemment une suite de ce qu'il étoit au moment précédent, et si nous concevons une Intelligence qui, pour un instant donné, embrasse tous les rapports des êtres de cet Univers, elle pourra déterminer pour un temps quelconque pris dans le passé ou dans l'avenir, la position respective, les mouvements, et généralement les affections de tous ces êtres." Hahn notes that in Condorcet's *Lettre à d'Alembert* (1768) one finds a similar passage: "si la loi de la continuité n'etoit point violée dans l'univers, on pourroit regarder ce qu'il est à chaque instant, comme le résultat de ce qui devoit arriver à la matiere arrangée, une fois dans un certain ordre, et abandonnée ensuite à elle-memê.... Une intelligence qui connoîtroit alors l'état de tous les phénomenes dans un instant donné, les loix auxquelles la matiere est assujettie, et leur effet au bout d'un tempts quelconque, auroit une conncissance parfaite du Système du Monde."

Hahn has given a more extensive discussion of the development of Laplace's ideas on determinism and probability in a paper in *Proc. XIII Int. Cong. Hist. Sci.* (Moscow, 1971), §I, 170 (pub. 1974). I am indebted to Professor Hahn for sending me a preprint of this paper.

On the relation of Laplace's work in probability to various astronomical and political problems, see C. C. Gillispie, *Proc. Am. Phil. Soc.* **116**, 7 (1972).
5. R. J. Boscovich, *Theoria Philosophiae Naturalis* (Vienna, 1758; Venice, 1763). English trans. by J. M. Child (from the 1st Venetian ed., 1763), *A Theory of Natural Philosophy* (Chicago: Open Court, 1922; reprinted by MIT Press, Cambridge, Mass., 1966), pp. 141–42: "Any point of matter, setting aside free motions that arise from the action of arbitrary will, must describe some continuous curved line, the determination of which can be reduced to the following general problem. Given a number of points of matter and given, for each of them, the point of space that it occupies at any given instant of time; also given the direction and velocity of the initial motion if they were projected, or the tangential velocity if they are already in motion; and given the law of forces expressed by some continuous curve, such as that of Fig. 1, which contains this theory of mine; it is required to find the path of each of the points.... Now, although a problem of such a kind surpasses all the powers of the human intellect, yet any geometer can easily see thus far, that the problem is determinate ... a mind which had the powers requisite to deal with such a problem in a proper manner and was brilliant enough to perceive the solutions of it (and such a mind might even be finite, provided the number of points were finite, and the notion

of the curve representing the law of forces were given by a finite representation), such a mind, I, say, could, from a continuous arc described in an interval of time, no matter how small, by all points of matter, derive the law of forces itself.... Now, if the law of forces were known, and the position, velocity and direction of all the points at any given instant, it would be possible for a mind of this type to foresee all the necessary subsequent motions and states, and to predict all the phenomena that necessarily followed from them." This passage was pointed out by K. Stiegler in *Proc. XIII Int. Cong. Hist. Sci.* (Moscow, 1971), §6, 307 (pub. 1974). On the difference between Laplacean and Boscovichean determinism see O. B. Sheynin, *Arch. Hist. Exact. Sci.* **9**, 306 (1973).

6. J. T. Merz, *A History of European Thought in the Nineteenth Century*, **2** (Edinburgh and London: Blackwood, 2nd ed., 1912), ch. XII. C. C. Gillispie, *op. cit.* (note 4) and earlier paper in *Scientific Change*, ed. A. C. Crombie (New York: Basic Books, 1963), p. 431. H. M. Walker, *Studies in the History of Statistical Method* (Baltimore: Williams & Wilkins, 1929), p. 19. H. L. Westergaard, *Contributions to the History of Statistics* (New York: Agathon Press, 1968, reprint of 1932 ed.), chs. XII and XIII. V. John, *Geschichte der Statistik* (Wiesbaden: Sändig, 1968, reprint of 1884 ed.), 1. Teil, 314ff.

7. E. Cassirer, *Götteborgs Högskolas Årsskrift* **42** (3) (1936); English trans. by O. T. Benfey, *Determinism and Indeterminism in Modern Physics* (New Haven: Yale University Press, 1956), p. 4; E. du Bois-Reymond, *Tageblatt* 1872 Vers. Deutsch, *Naturf. u. Aerzte*, p. 85, English trans. in *Pop. Sci. Mont.* **5**, 17 (1874).

8. A. Cournot, *Essai sur les fondements de nos connaissances* (Paris, 1851), **1**, 62. C. Babbage, *The Ninth Bridgewater Treatise* (London, 1837, 2nd ed., 1838), p. 111.

9. R. L. Ellis to J. D. Forbes, 10 October 1850, in *Life and Letters of James David Forbes* by J. C. Shairp *et al.* (London, 1873), p. 481.

10. M. Čapek, *Philosophical Impact of Contemporary Physics* (Princeton: Van Nostrand, 1959), ch. VIII. E. Meyerson, *Identity and Reality* (New York: Dover Pubs., 1962, reprint of the English trans. by K. Loewenberg, 1930, of the 3rd French ed., 1926), ch. VI.

11. J. Herapath, *Ann. Phil.* [2] **1**, 273 (1821), esp. p. 281. E. Mendoza, *Brit. J. Hist. Soc.* **8**, 155 (1975).

12. R. Hooke (1678), quoted in S. G. Brush, *Kinetic Theory* (Oxford and New York: Pergamon Press, 1965), **1**, 6; M. V. Lomonosov, *Nov. Comm. Acad. Sci. Imp. Petrop.* **1**, 230 (1750), English trans. in *Mikhail Vasil'evich Lomonosov on the Corpuscular Theory*, by H. M. Leicester (Cambridge, Mass.: Harvard University Press, 1970), p. 203. But Lomonosov also refers to "disordered motion" of atoms (*ibid.*, p. 215).

13. See ch. 9.

14. L. Wilhelmy, *Versuch einer mathematisch-physikalischen Wärme-Theorie* (Heidelberg, 1851), p. 16; N. Dellingshausen, *Versuch einer speculativen Physik* (Leipzig, 1851), 57–58; Z. Allen, *Philosophy of the Mechanics of Nature* (New York, 1852), **41**, 344, 349, 355. L. Colding, *Kgl. Danske Vid. Selsk. Skr.* [5] **3**, 1 (1852), English trans. in *Ludvig Colding and the Conservation of Energy Principle* by P. F. Dahl (New York: Johnson Reprint Corp., 1972) p. 80. See also C. F. Mohr, *Ann. Chem. Pharm.* **24**, 141 (1837); *Z. Phys.* [2] **5**, 419 (1837), English trans. in *Phil. Mag.* [5] **2**, 110 (1876); Babinet, *Compt. Rend. Acad. Sci. Paris* **7**, 781 (1838).

15. J. P. Joule, *Mem. Manchester Lit. Phil. Soc.* [2] **9**, 107 (1851), read 1848); see also his 1847 lecture reprinted in Brush, *Kinetic Theory* **1**, 78. J. J. Waterston, *Phil. Trans.* **183**A, 5 (1893); the quoted phrases also appear in the abstract of his paper, published in *Proc. R. S. London* **5**, 604 (1846).

"... heat consists in a motion of the ultimate particles of bodies, and is the measure of the vis viva of this motion"–R. Clausius, *Ann. Phys.* [2] **86**, 337 (1852), English trans. in *Scientific Memoirs*, eds. J. Tyndall and W. Francis (London, 1853), p. 1, quotation from p. 342 and p. 5, resp.

"The work which any external forces do upon [a substance], the work done by its own molecular forces, and the amount by which the half *vis viva* of the thermal motions of all its part is diminished, must together be equal to the mechanical effect produced from it..."–W. Thomson, *Trans. R. S. Edinburgh* **20**, 261 (1851), *Phil. Mag.* [4] **4**, 8 (1852) (quotation from p. 12 of the latter).

"Assuming the hypothesis which is now gaining ground, that heat, instead of being an agent apart from ordinary matter, consists in a motion of the material particles...."–J. Tyndall, lecture 11 February 1853 at the Royal Institution, London, reprinted in *The Royal Institution Library of Science, Physical Sciences* **1**, 78 (New York: American Elsevier, 1970).

16. A. K. Krönig, *Ann. Phys.* [2] **99**, 315 (1856); G. Ronge, *Gesnerus* **18**, 45 (1961); E. E. Daub, *Isis* **62**, 612 (1971).
17. R. Clausius, *Ann. Phys.* [2] **100**, 353 (1857); English trans. in *Phil. Mag.* [4] **14**, 108 (1857), reprinted in Brush, *Kinetic Theory* **1**, 111. See chapter 4.
18. C. H. D. Buys-Ballot, *Ann. Phys.* [2] **103**, 240 (1858). There is some indication that Clausius had been thinking about mean-free-path ideas in an earlier paper on the scattering of light by water drops in the atmosphere, *Ann. Phys.* [2] **76**, 161 (1849).
19. R. Clausius, *Ann. Phys.* [2] **105**, 239 (1858), English trans. in *Phil. Mag.* [4] **17**, 81 (1859), reprinted in Brush, *Kinetic Theory* **1**, 135 (quotation from p. 139).
20. He also stated that if the target molecules were not stationary but themselves moving with various velocities, then the mean free path would be different; but here he thought it was permissible to assume that each molecule is moving at the average velocity in order to do the calculation, and obtained the result that the mean free path would be $\frac{3}{4}$ as great as in the hypothetical case when all molecules but one are at rest. It was left for Maxwell to generalize this calculation to the case where the molecules have a statistical distribution of velocities, and Clausius resisted for some time Maxwell's replacement of the factor $\frac{3}{4}$ by $1/\sqrt{2}$. Clausius, *op. cit.*, p. 140; see §12.2.
21. The Ehrenfests emphasized that Clausius did introduce in his treatment of mean free paths the assumption about the number of collisions, the "Stosszahlansatz," which was later to play an important role in Boltzmann's theory. P. and T. Ehrenfest, *Enc. math. Wiss.* **IV** (2, II, 6) (Leipzig: Teubner, 1912), English trans. by M. J. Moravcsik, *The Conceptual Foundations of the Statistical Approach in Mechanics* (Ithaca, N.Y.: Cornell University Press, 1959), p. 5.
22. J. W. Gibbs, *Proc. Am. Acad.* (n.s.) **16**, 458 (1889); *The Scientific Papers of J. Willard Gibbs* (London and New York: Longmans, Green & Co., 1906. **2**, 261.
23. J. Herschel, *Edinburgh Rev.* **92**, 1 (1850), reprinted in his *Essays* (London, 1857), p. 365; C. C. Gillispie, in *Scientific Change*, ed. A. C. Crombie (New York: Basic Books, 1963), p. 431; E. Garber, *Centaurus* **17**, 11 (1972). For a discussion from the viewpoint of statistical theory, see O. B. Sheynin, *Biometrika* **58**, 234 (1971).
24. J. D. Forbes, *Phil. Mag.* [3] **37**, 401 (1850); W. F. Donkin, *Phil. Mag.* [4] **1**, 353, 458, **2**, 55 (1851); R. L. Ellis, *Phil. Mag.* [3] **37**, 321, 462 (1850). For correspondence relating to this debate, including an "ignorance" definition of randomness by Ellis, see J. C. Shairp *et al.*, *Life and Letters of James David Forbes* (London, 1873), ch. XIV (by P. G. Tait).
25. Cf. L. Boltzmann, *Phil. Mag.* [5] **23**, 305 (1887); *Wissenschaftliche Abhandlungen*

(Leipzig: Barth, 1909), III, 255–56; J. L. F. Bertrand, *Calcul des Probabilités* (Paris: Gauthier-Villars, 1889), pp. 29–32; *Compt. Rend. Acad. Sci. Paris* **122**, 963, 1083, 1174, 1314 (1896); L. Boltzmann, *Compt. Rend. Acad. Sci. Paris* **122**, 1173, 1314 (1896); *Wissenschaftliche Abhandlungen* III, 564, 566.
26. J. C. Maxwell, *Trans. Camb. Phil. Soc.* **12**, 547 (1879), reprinted in *The Scientific Papers of James Clerk Maxwell* (Cambridge University Press, 1890, reprinted by Dover Pubs., New York, 1952, 1965), **2**, 713.
27. J. C. Maxwell, *Phil. Trans.* **157**, 49 (1867); *Papers* **2**, 26; reprinted in S. G. Brush, *Kinetic Theory* **2**, 23 (see p. 48).
28. See, *e.g.* J. Herschel's article on "Meteorology" in the *Encyclopedia Britannica* (Edinburgh, 1861), or earlier reviews of the problem such as that by J. Ivory, *Phil. Mag.* **66**, 81, 241, (1825); *Phil. Trans.* **113**, 409 (1823).
29. J. Herapath, *Times*, 10 Jan. 1826, quoted in S. G. Brush, *Notes and Rec. Roy. Soc. London* **18**, 173 (1963); Herapath, *Railway Mag.* **1**, 109, 260 (1836); *Mathematical Physics* (London, 1847, reprinted by Johnson Reprint Corp., New York, 1972), **2**, 142–63. J. J. Waterston, *Phil. Trans.* **183**, 1 (1892, submitted 1845), reprinted in *The Collected Scientific Papers of John James Waterston*, ed. J. S. Haldane (Edinburgh: Oliver & Boyd, 1928) (see p. 250ff).
30. Maxwell, 1867 paper reprinted in Brush, *Kinetic Theory* **2** (see p. 85); further discussion in *Nature* **8**, 527, 753 (1873). For a direct proof without *assuming* a Maxwell distribution see C. Truesdell, *Mathematical Aspects of the Kinetic Theory of Gases*, Notas de Matemática Fisica, vol. 3, Instituto de Matemática, Univ. Federal do Rio de Janeiro, 1973 (see ch. IX).
31. Tait to Maxwell, 6/12/67 and subsequent correspondence, at Cambridge University. For those readers not familiar with the voluminous secondary literature on Maxwell's Demon, the article by M. J. Klein, *Am. Sci.* **58**, 84 (1970) is especially recommended. Some interesting aspects of the subject are revealed in a paper by E. E. Daub, *Stud. Hist. Phil. Sci.* **1**, 213 (1970).
32. Maxwell's "Catechism on Demons," published in C. G. Knott, *Life and Scientific Work of Peter Guthrie Tait* (Cambridge, Eng.: Cambridge University Press, 1911), pp. 214–15. According to Knott, this is an "undated letter, which must have been written about this time," *i.e.* shortly after Maxwell's letter to Tait of 11 December 1867 in which the Demon idea is first introduced (*ibid.*, pp. 213–14).
33. J. C. Maxwell, *Theory of Heat* (London, 7th ed., 1883), pp. 308–9.
34. *Ibid.*, p. 329.
35. Letters held in the Pattison Collection, Bodleian Library, Oxofrd University; I thank Dr. Thomas Simpson for providing copies.
36. J. C. Maxwell, *Nature* **8**, 437 (1873); *Phil. Mag.* [4] **46**, 453 (1873); *Pop. Sci. Monthly* **4**, 276 (1874); *Papers* **2**, 361 (quotations from p. 373). Cf. Maxwell's poem cited above, note 1, this section.
37. Campbell and Garnett, *op. cit.* (note 1), p. 434. Similar views were expressed in Maxwell's 1871 lecture on experimental physics, in his *Papers* **2**, 241.
38. D. Duncan, *Life and Letters of Herbert Spencer* (New York: Appleton, 1809), **2**, 161–63; letters from Maxwell to Spencer, 17 December 1873, and Spencer to Maxwell, 30 December 1873, at Cambridge University.
39. Maxwell, *Nature* II, 357 (1875); see *Papers* **2**, 436 for quotation.
40. *Encyclopedia Britannica* (Edinburgh, 8th ed.) **3**, 36 (1875). See *Papers* **2**, 462 for quotation.
41. *Encyclopedia Britannica*, 9th ed., **7**, 214 (1878); see *Papers* **2**, 628 for quotation.

42. J. W. Gibbs, *Elementary Principles in Statistical Mechanics* (New York: Scribner, 1902), pp. 206–7; reprinted in *The Collected Works of J. Willard Gibbs* (New Haven: Yale University Press, 1948; New York: Dover Pubs., 1960), **2**. See also Gibbs, *Trans. Conn. Acad.* **3**, 108 (1875), *Works*, **1**, 55, esp. p. 167. (This paper is discussed in §14.5, below.)
43. Knott, *op. cit.* (note 31), pp. 115–16; *Nature* **17**, 257, 278 (1878); *Papers* **2**, 660 (see esp. 669–71).
44. Helmholtz published at least three statements (1882, 1885, and 1886) to the effect that heat is random molecular motion, that entropy is a measure of disorder, and that irreversibility is not an inherent property of nature but is due to our inability to reverse atomic motions. See his *Wissenschaftliche Abhandlungen* (Leipzig: Barth, 1895), **2**, 972, **3**, 209, 593.

14.5 Boltzmann's statistical theory of entropy

Ludwig Boltzmann is usually credited with establishing the connection between randomness and irreversibility, though much of Boltzmann's early work was anticipated or stimulated by the publications of Maxwell.[1] In one of his first papers Boltzmann attempted to reduce the Second Law of Thermodynamics to the mechanical principle of least action, but did not pay special attention to the aspect of irreversibility.[2] However, this attempt did lead him in 1871 to introduce a statistical distribution function for molecular positions,[3] based on his own earlier generalization of Maxwell's distribution to cases where forces are present.[4] Thus, as Edward Daub has pointed out, Boltzmann successfully used probability concepts in the reduction of the Second Law to mechanics, but this reduction was not fruitful since "it failed to evoke ideas which were not contained in the laws which it explained"[5] and in particular it did not deal with the problem of irreversibility.

A major breakthrough came in 1872 with Boltzmann's paper "Weitere Studien über das Wärmegleichgewicht unter Gasmolekülen," which despite its bland title is one of the most important and influential works in the entire history of kinetic theory.[6] The introductory paragraph comes very close to postulating that molecular motions are random, arguing that the "most irregular" [*regellosesten*] events, "when they occur in the same proportions, give the same average value," hence we can observe "completely definite laws of behavior of warm bodies."[7] But when Boltzmann proceeds to his mathematical derivations involving the distribution function *f*, he always refers to this as giving the *number* of molecules having some specified velocity or other characteristic quantity.[8] The transition from a stochastic back to a statistical approach occurs in the following sentences:

§14.5] BOLTZMANN'S STATISTICAL THEORY OF ENTROPY

If one does not merely wish to guess a few occasional values of the quantities that occur in gas theory, but rather desires to work with an exact theory, then he must first of all determine the probabilities of the various states which a given molecule will have during a very long time or which different molecules will have at the same time. In other words, one must find the number of molecules out of the total number whose states lie between any given limits.[9]

Boltzmann assumes that in the initial state, each direction of the molecular velocity is equally probable, and that the distribution function does not depend on the space coordinates. (Thus heat flow due to an externally imposed temperature gradient, the most important example of an irreversible process in the earlier discussions reviewed in this paper, is excluded.) He then asserts that the number of collisions in time τ between pairs of molecules in which the two molecules have kinetic energies between x and $x + dx$, and between x' and $x' + dx'$, before the collision, and the first molecule has kinetic energy between ξ and $\xi + d\xi$ after the collision, is given by the expression[10]

$$dn = \tau f(x, t)\, dx \cdot f(x', t)\, dx'\, d\xi \cdot \psi(x, x', \xi)$$

where $\psi(x, x', \xi)$ depends on the nature of the collision and the force law. There are only three independent variables since conservation of total kinetic energy is assumed, $x + x' = \xi + \xi'$.

Thus, without any discussion Boltzmann assumes (as did Maxwell in his 1867 paper) that there is no correlation between the two molecules before the collision so that the joint distribution can be written as a product of the single-molecule distributions.

Each such collision will reduce by one the number of molecules having kinetic energy x; thus the rate of change of $f(x, t)$ will depend on the integral of the above expression for dn over all permitted values of the other energies x' and ξ. A corresponding expression is then written down for the increase in $f(x, t)$ due to collisions in which one molecule acquires energy x after the collision, so that the net change in f is given by an equation of the form

$$f(x, t + \tau)\, dx = f(x, t)\, dx - \int dn + \int d\nu$$

After some transformations, the expression for $\int d\nu$ is reduced to

$$\int d\nu = \tau\, dx \int_0^\infty \int_0^{x+x'} f(\xi, t) f(x + x' - \xi, t) \psi(\xi, x + x' - \xi, x)\, dx'\, d\xi$$

The function ψ must then be proved to satisfy certain properties corresponding to permutations of the variables x, x' and ξ for inverse collisions; Boltzmann has to assume at this point that the force between two point-particles is a function of their distance and acts in the direction of the line of centers, and that action and reaction are equal. The result is that f satisfies the integro-differential equation

$$\frac{\partial f(x,t)}{\partial t} = \int_0^\infty \int_0^{x+x'} \left[\frac{f(\xi,t)}{\sqrt{\xi}} \frac{f(x+x'-\xi,t)}{\sqrt{x+x'-\xi}} - \frac{f(x,t)}{\sqrt{x}} \frac{f(x',t)}{\sqrt{x'}} \right] \sqrt{xx'} \, \psi(x,x',\xi) \, dx' \, d\xi$$

This is in fact the famous "Boltzmann equation" which is widely used as the basis for solving problems of kinetic theory, plasma physics, and solid state physics–though it is derived originally here in a somewhat unfamiliar form because energies rather than velocities have been taken as the variables, and of course the terms corresponding to spatial nonuniformity and external forces are omitted.

From the Boltzmann equation it follows immediately that Maxwell's distribution,

$$f(x,t) = C\sqrt{x} \, e^{-hx},$$

represents an equilibrium state in the sense that one gets $\partial f(x,t)/\partial t = 0$ by direct substitution. That is about as far as Maxwell himself was able to go in justifying his distribution function: he could argue plausibly that once this state had been attained, subsequent collisions would not change it. But Boltzmann could now set himself a more ambitious task: suppose $f(x,t)$ is not initially Maxwellian (but still, for the moment, subject to the conditions mentioned above); prove that it will inevitably tend toward the Maxwell function.

For this purpose Boltzmann had the brilliant inspiration (probably the result of educated guesses based on his previous work with entropy formulae, combined with some trial-and-error work) to define a functional

$$E = \int_0^\infty f(x,t) \left\{ \log\left[\frac{f(x,t)}{\sqrt{x}}\right] - 1 \right\} dx$$

This is the Boltzmann "H-function," written in terms of energy rather than velocity; the actual use of the letter H was still two decades in the future.

By a procedure familiar to students of kinetic theory but of little interest to others, Boltzmann then computed the time-derivative of E

using the expression derived earlier for $\partial f(x,t)/\partial t$, and found that

$$\frac{dE}{dt} \leq 0$$

where the equality sign holds only when f is the Maxwell distribution.[11] This is Boltzmann's H-theorem.

The quantity E, when evaluated for the Maxwell–Boltzmann form of the distribution function f, is the same (within a constant factor) as the expression Boltzmann had previously found for the "well-known integral $\int dQ/T$."[12] The proof of the H-theorem has therefore "prepared the way for an analytical proof of the second law in a completely different way from those previously investigated" and in particular will allow a proof that this integral is negative for irreversible processes – previous work had only attempted to show that the integral is zero for reversible cyclic processes.

So far the H-theorem applies only to the special case of a dilute monatomic gas of point atoms interacting with central forces, in which only binary collisions need be considered, and for cases where external forces and spatial nonuniformities are absent. Boltzmann now has his work cut out for him: to remove these restrictions one by one, so as to establish the molecular basis of the Second Law for the most general case possible. He takes the first step in this direction in the last section of the same paper, by considering a system of polyatomic molecules (still assuming central forces between the molecules), but is able to complete the proof of the H-theorem only for the case of diatomic molecules which interact like elastic spheres. (He never did give a completely satisfactory treatment of polyatomic molecules.)

There was an interval of three years between Boltzmann's completion of his 1872 paper and the next one on this subject, read to the Vienna Academy in October 1875. During this interval Boltzmann was occupied with experimental work on electrical problems (stimulated in part by Maxwell's electromagnetic theory) and developed a theory of elastic aftereffects. But he now returned to his unfinished business, and tackled the problems of generalizing the proof of thermal equilibrium to systems in which external forces are present. This involved first a detailed proof of the integro-differential equation for f, which had been written down without proof in the 1872 paper and used to compute transport coefficients, more or less by analogy with Maxwell's 1867 theory. Then followed a proof of the H-theorem by routine manipulations similar to those of the 1872 paper. The conclusion was that in spite of the action of external forces, each direction of the molecular velocity is equally probable, and in each spatial element the velocity distribution is the same as it would be for a gas of the same density and temperature on which no external forces act. The effect of external forces consists only in causing the density of the gas to vary from one place to another in the manner already known from the laws of hydrostatics.[13]

Boltzmann's conclusion clearly implied the theorem, stated earlier by Maxwell, that the temperature is the same throughout a vertical column of gas. It was this theorem that soon attracted the criticisms of Boltzmann's colleague Josef Loschmidt, and led Boltzmann, in his defense of it, towards a clearer physical interpretation of the relation between molecular motion and irreversibility.

* * *

The term "reversibility paradox" was invented by Paul and Tatiana Ehrenfest in 1907 for an argument which they attributed to Loschmidt.[14] But before Loschmidt published his very brief remark on the reversal of molecular motions, subsequently elaborated on by Boltzmann, the paradox had been discussed extensively by Maxwell with his friends Tait and Thomson. Maxwell's first letter to Tait on how his Demon could violate the Second Law, dated 11 December 1867, has a pencilled addition which reads: "Very good. Another way is to reverse the motion of every particle of the Universe and to preside over the unstable motion thus produced." According to Tait's biographer C. G. Knott, the addendum is by William Thomson, but to me it looks more like Tait's handwriting.[15]

In his letter to the editor of the *Saturday Review*, 7 April 1868, Maxwell said that the materialist believes that if every motion in the world were accurately reversed, everything would run backwards, water would collect out of the sea and run up the rivers, all living things would regress from the grave to the cradle, and so forth—but that our experience of irreversible processes leads us to expect that no such thing would happen.[16] Similar thoughts were expressed in Maxwell's letter to Strutt (later Lord Rayleigh) in 1870.[17]

The culmination of this discussion (of which only fragmentary records survive) was Thomson's paper "On the kinetic theory of the dissipation of energy," published in 1874.[18] Thomson drew a distinction between "abstract dynamics" which is perfectly reversible, and "Physical dynamics" which is not. Like Maxwell, he associated the hypothesis that life processes are governed by abstract dynamics with materialism which he of course rejected. While speculation about the reversal of life processes is "utterly unprofitable," Thomson thought that consideration of the consequences of reversal of the motion of inanimate matter could clarify the theory of energy dissipation. For this purpose he invoked first an army of Maxwell Demons, with instructions to turn back selected molecules as they reach an interface between hot and cold regions of a gas. He showed that without changing the pressure, the Demons can either maintain a temperature difference in the presence of diffusion, or create a temperature difference where none existed before.

But the most important part of Thomson's discussion does not involve the Maxwell Demon at all. He simply supposes that, starting from an initial unequal distribution of temperature, we allow diffusion to occur until after a finite time interval the temperature is very nearly equal throughout the gas, and then instantaneously reverse the motion of each molecule.

> Each molecule will retrace its former path, and at the end of a second interval of time, equal to the former, every molecule will be in the same position, and moving with the same velocity, as at the beginning; so that the given unequal distribution of temperature will again be found, with only the difference that each particle is moving in the direction reverse to that of its initial motion.

While it might appear that this process is contrary to the principle of dissipation of energy, Thomson points out, first, that if the reversed motion continues, there will be an "instantaneous subsequent commencement of equalization," so that the unequal distribution of temperature will be short-lived. Second, if we looked at a gas in thermal equilibrium, there would be no way to pick out the particular arrangement that could evolve into a nonequilibrium state if the velocities were reversed. It is true that if any gas be left for a sufficiently long time in a perfectly rigid vessel with no external influences, it will inevitably happen that, for example, more than 90% of the energy will be in one half of the vessel. But the probability of this happening at any particular time is enormously smaller than the probability of a more or less equal distribution. The odds against an unequal distribution become even greater if the gas interacts with an external heat reservoir.

To clinch the argument (and to give precise meaning to Maxwell's statement that the validity of the Second Law is a "statistical certainty") Thomson calculated the probability than in a jar containing 2×10^{12} molecules of oxygen and 8×10^{12} molecules of nitrogen, all of the oxygen molecules are found in a specified part of the jar whose volume is $\frac{1}{5}$ of the whole: "The number expressing the answer in the Arabic notation has about 2 173 220 000 000 of places of whole numbers."

While Thomson's explanation of irreversibility is statistical, it is not stochastic; there is no question of any fundamental randomness at the atomic level. In this connection it may be recalled that Thomson's main objection to Darwin's theory of evolution was that it was based on randomness rather than purposeful divine guidance.[19]

* * *

Before turning to the more famous discussion of the reversibility paradox by Loschmidt and Boltzmann, we must mention a frequently-quoted remark of J. Willard Gibbs, first published in 1875. At the beginning of his memoir "On the equilibrium of heterogeneous substances," Gibbs placed Clausius' 1865 formulation of the two laws of thermodynamics ("Die Energie der Welt ist constant, Die Entropie der Welt einem Maximum zu"), and based his own formulation of thermodynamics on energy and entropy as fundamental quantities.[20] But when he discussed the entropy increase associated with the mixing of two gases, Gibbs noted that this increase depends on the existence of a difference between the gases; for if they were identical in all respects, there would be no change in total entropy before and after mixing (the so-called "Gibbs paradox"). But, he speculated, it is conceivable that two gases might be "absolutely identical in all the properties (sensible and molecular) which come into play while they exist as gases either pure or mixed with each other, but which should differ in respect to the attractions between their atoms and the atoms of some other substances, and therefore in their tendency to combine with such substances." In this case their mixing *would* involve an entropy increase but there would be no way to distinguish this situation experimentally from the mixing of two identical gases. (As Maxwell was to point out a little later in the passage already mentioned in the previous section, this means that entropy is not strictly an observable quantity but depends on knowledge or theories possessed by the observer.) So, Gibbs concluded,

> when such gases have been mixed, there is no more impossibility of the separation of the two kinds of molecules in virtue of their ordinary motions in the gaseous mass without any especial external influence, than there is of the separation of a homogeneous gas into the same two parts into which it has once been divided, after

these have once been mixed. In other words, the impossibility of an uncompensated decrease of entropy seems to be reduced to improbability.[21]

As can be seen from the preceding context (usually ignored when the last sentence is quoted[22]) Gibbs' suggestion that the Second Law has only statistical validity is not based on a specific atomic-kinetic model of matter, nor was it intended to apply to most situations in which energy is dissipated. His paper was entirely phenomenological in nature, and it is rather misleading to drag it into discussions of the statistical interpretation of irreversibility.

* * *

According to textbook accounts, following the Ehrenfests,[23] the reversibility paradox was first proposed by Josef Loschmidt in discussions with Boltzmann in Vienna, and was published in a series of papers in 1876–77.[24] There is also an embellishment of the story circulating among modern physicists, to the effect that when Loschmidt told Boltzmann that his system would simply run backwards if all the molecular velocities were reversed, Boltzmann replied, "Well, *you* just try to reverse them!"[25] Actually Loschmidt's published discussion of the paradox consists of only a single sentence in the context of a long discussion of the problem mentioned above in connection with the equilibrium under gravitational forces. Loschmidt did not accept Maxwell's conclusion[26] that a column of gas would have constant temperature throughout, but claimed instead that thermal equilibrium was possible without equality of temperature. In this way he hoped to demonstrate that the heat death of the universe is not inevitable. He claimed that the second law could be correctly formulated as a mechanical principle without reference to the sequence of events in time; he thought he could thus "destroy the terroristic nimbus of the second law, which has made it appear to be an annihilating principle for all living beings of the universe; and at the same time open up the comforting prospect that mankind is not dependent on mineral coal or the sun for transforming heat into work, but rather may have available forever an inexhaustible supply of transformable heat."[27]

After proposing a model which supposedly violated Maxwell's constant-temperature theorem, Loschmidt noted that in any system "the entire course of events will be retraced if at some instant the velocities of all its parts are reversed."[28] His application of this reversibility principle to the validity of the Second Law was somewhat obscurely stated, but Boltzmann (perhaps as a result of private discussions) quickly got the point and published a reply,[29] in which he gave a thorough discussion of the reversibility paradox, as well as a 50-page paper elaborating his theory of molecular motion in gases subject to external forces.[30]

Boltzmann conceded that it is impossible to prove that the entropy of a system always increases without taking account of the initial conditions. Moreover, such a statement cannot be true for *all* initial conditions since it is certainly possible to find a special initial state (obtained by reversing all the molecular velocities of a system which has evolved from a nonuniform one) for which succeeding states will have lower entropy. The crucial point, however, is that "since there are infinitely many more uniform than non-uniform distributions, the number of states which lead to uniform distributions after a certain time t_1 is much greater than the number that lead to non-uniform ones, and the latter are the ones that must be chosen, according to Loschmidt, in order to obtain a non-uniform distribution at t_1."[31]

There follows the very important remark:

> One could even calculate, from the relative numbers of the different state distributions, their probabilities, which might lead to an interesting method for the calculation of thermal equilibrium.

Following up his own suggestion, Boltzmann developed soon afterward his statistical method for calculating equilibrium properties, based on the relation between entropy and probability articulated in this discussion of the reversibility paradox.[32] So it appears that Loschmidt has followed the tradition of Franciscus Linus and C. H. D. Buys-Ballot by stimulating a major advance in gas theory through his criticism.[33]

It is curious that Boltzmann, who was apparently unaware of Thomson's discussion published three years earlier,[34] chose exactly the same example to illustrate the statistical nature of irreversibility: he notes that a spontaneous decrease in entropy is "extraordinarily improbable and can be considered impossible for practical purposes; just as it may be considered impossible that if one starts with oxygen and nitrogen mixed in a container, after a month one will find chemically pure oxygen on the lower half and nitrogen in the upper half, although according to probability theory this is merely very improbable but not impossible."

Finally, Boltzmann mentioned

§14.5] BOLTZMANN'S STATISTICAL THEORY OF ENTROPY

> a peculiar consequence of Loschmidt's theorem, namely that when we follow the state of the world into the infinitely distant past, we are actually just as correct in taking it to be very probable that we would reach a state in which all temperature differences have disappeared, as we would be in following the state of the world into the distant future.... If perhaps this reduction of the second law to the realm of probability makes its application to the entire universe appear dubious, yet the laws of probability theory are confirmed by all experiments carried out in the laboratory.

If the world is to end in a Heat Death, it must have begun in a Heat Birth. Having got to this point Boltzmann, it would seem, is now prepared to give an interpretation of the "recurrence paradox" but in fact he did not do so until challenged by Zermelo nearly 20 years later (see below).

Boltzmann was still unaware of another paradox: he has reached his conclusions by reasoning from what he calls "probability theory" while assuming that exact deterministic laws still apply to molecular motions and collisions.

* * *

In 1877 Boltzmann, inspired according to his own account by the reasoning involved in his reply to Loschmidt's reversibility objection, proposed a new method for determining the state of thermal equilibrium of a system. This method, which is applicable to any system, not only gases, consists in enumerating all possible "complexions"–for example, all the ways in which a given total amount of energy can be distributed among a specified number of molecules–and assuming that the probability of a macroscopic state is proportional to the number of corresponding molecular complexions. (Each complexion is assigned equal probability.) The entropy of the system is directly related to this probability, and in the later forms of the theory is simply proportional to the logarithm of the probability, $S = k \log W$ in modern notation. The state of thermal equilibrium is then asserted to be the one that has the greatest probability.

Using this relation between entropy and probability, Boltzmann proposed the following interpretation of the physical significance of the Second Law:

> In most cases the initial state will be very improbable; the system passes from this through ever more probable states, reaching

finally the most probable state, that is the state of thermal equilibrium.[35]

Thus irreversibility is simply a tendency to go from less probable to more probable states.

Ernest Nagel has suggested that "perhaps the greatest triumph of probability theory within the framework of nineteenth-century physics was Boltzmann's interpretation of the irreversibility of thermal processes."[36] Others might feel that this triumph was achieved only at the cost of muddying the concept of "probability." There has been considerable confusion about how one should interpret the quantity denoted by W, which Planck and others have called the "thermodynamic probability" of a state of the system.[37] It cannot be determined by a routine combinatorial procedure as Boltzmann's remark seems to imply, for two reasons. First, in classical physics the particles of the system are permitted a continuous range of positions and velocities, so the actual "number" of complexions is infinite for any macroscopically defined state. If one tries to convert W into a proper fraction by dividing it by the *total* number of complexions, the result will be $W = 0$ unless the "total" is limited in some special way. Second, since the entropy S is a function of temperature, either the number of complexions corresponding to a state, or the total number, or both, must depend on temperature. Yet temperature is a derived average property of the system from the viewpoint of kinetic theory, not part of its original specification, so it is not clear how this temperature-dependence can be consistently introduced into the model. If one takes too literally the frequently-made assertion that the equilibrium thermodynamic state of the system corresponds to the overwhelming majority of all microstates accessible at a given temperature (or fixed total energy), one would end up with the result $W = 1$ for *all* temperatures. As J. R. Partington remarked, "thermodynamic states" are what Francis Bacon would have called "Idols of the Market Place."[38]

As this is an account of 19th-century theories, not a monograph on the foundations of statistical mechanics from the modern viewpoint, I shall not attempt to resolve these difficulties, but can only call attention to them. Within the framework of classical physics, it seems to have been generally agreed that one must retreat from the formula for "absolute entropy," $S = k \log W$, and talk only about the relative entropy of two states: $S - S' = k \log W/W'$. According to Gibbs and Fowler, it is possible to justify such a formula by regarding the system under consideration as a random sample from a very large number of

Fig. 14.5-1. The Boltzmann memorial in Vienna (reproduced by courtesy of W. Flamm).

systems with certain hypothetical properties. Quantum physics, however, does give a procedure for computing absolute entropy, and does provide some justification for Boltzmann's postulated relation between entropy and probability.[39]

From this perspective it is of great interest that Boltzmann himself evaded the problem of counting a continuum of microstates by assuming first that each molecule can have only a finite number of energy-values,

$$0, \epsilon, 2\epsilon \quad \text{up to} \quad p\epsilon$$

and then afterwards letting $\epsilon \to 0$ and $p \to \infty$ in such a way that $p\epsilon$ approaches a finite number, the specified total energy of the system. Boltzmann wrote that "this fiction corresponds to no realizable mechanical problem, but rather a problem which is mathematically much easier to treat, and which reduces at once to the problem we have to solve" when the indicated limits are taken.[40] But it is evident from this why Boltzmann's method could so easily be taken over into quantum theory by Max Planck in 1900.

While Boltzmann's relation between entropy and probability was invoked to account for irreversibility (a tendency to go from "less probable" to "more probable" states) and thus suggested a physical meaning for the qualitative directionality of time, it also had the effect of *eliminating* time as a variable in the description of the system. The process of seeking the most probable state becomes a mathematical exercise which may have no relation at all to the physical time-development of the system. It is also a process in which the deterministic dynamics of molecular collisions is replaced by random choice, for in carrying out the calculation of state probabilities, Boltzmann assumed "that the kinetic energy of each individual molecule is determined, as it were, by a lottery, which is selected completely impartially from a collection of lotteries which contains all the kinetic energies that can occur in equal numbers."[41] This is the key to the power of the new method: it is not restricted to special molecular models for which collision mechanisms can be worked out in detail; it can be used for any system, including (with slight modifications) those governed by quantum mechanics, for which the spectrum of possible energies is known.

In one sense Boltzmann has simply come back to Maxwell's 1860 viewpoint, in which a velocity distribution was derived directly from probabilistic arguments without regard to the particular molecular processes that might bring it about. That viewpoint was considered inadequate at the time by Maxwell and others, and it had to be justified by calculations based on special molecular models, culminating in Boltzmann's H-theorem of 1872. By 1877 Boltzmann was able to build on a solid foundation of molecular theory, so his method is not quite as simple-minded as the above summary makes it appear. For example, he is well aware that one cannot just postulate equal probabilities for all kinetic energies of a molecule (even though the postulate may be conditioned on fixed total energy for all molecules in the system) for that leads to the wrong answer in three-dimensional problems. Instead one has to insert a weighting factor, equivalent to the assumption that the distribution is uniform with respect to the momentum variable rather than the energy variable. The proof of this fact goes back to the theory of molecular collisions, reminding us that the emancipation from Newtonian dynamics is not yet complete.[42]

The extent of Boltzmann's acceptance of a probabilistic view of molecular behavior at this time is circumscribed by his comments on what is known known as the "ergodic hypothesis." As I have already discussed the history of this subject in some detail,[43] I need only summarize here the main point: the assumption that one can simply average over all possible states of a system in order to calculate its thermodynamic properties could be justified by proving that the system will eventually pass through all those states before returning to its initial condition. If that were the case, then it would be clearly understood that the use of probabilistic methods is only a matter of convenience, and does not contradict the belief that the behavior of the system is ultimately deterministic on the molecular level. The use of an "ensemble" of systems (as we now say, following the terminology of J. Willard Gibbs) is an equivalent but more abstract (and often more convenient) way of applying statistical calculations to deterministic systems. Thus in his review of Maxwell's paper "On Boltzmann's theorem on the average distribution of energy in a system of material points,"[44] Boltzmann wrote:

> There is a difference in method between Maxwell and Boltzmann, inasmuch as Boltzmann measures the probability of a condition by the time during which the system possesses this condition on the average, whereas Maxwell considers innumerable similarly constituted systems with all possible initial conditions. The ratio of the number of systems which are in that condition to the total number of systems determines the probability in question.[45]

(In 1894 Boltzmann repeated these two definitions of probability, but by this time he had made considerable use of the definition attributed to Maxwell.[46])

In 1886 Boltzmann discussed his interpretation of the Second Law in less technical terms, in a lecture at the Vienna Academy. After stating that he would make no attempt to rescue the universe from the heat death, he said that the molecular interpretation of the Second Law depends on the law of large numbers, in the same way that the number of "so-called voluntary [*freiwilligen*] acts, marriages at a certain age, crimes, and suicides" remains constant in a sufficiently large population.[47] The implication of this statement is a little ambiguous, but it certainly suggests that molecular motions are individually at least unpredictable if not inherently random. But in any case Boltzmann's statistical interpretation of the Second Law has not yet reached its final stage, since toward the end of this lecture he remarks:

> Since a given system of bodies can never by itself pass into an absolutely equally probable state, but rather always into a more probable one, so it is not possible to construct a system of bodies which, after passing through different stages, periodically returns to its original state: a perpetuum mobile.[48]

Notes for §14.5

1. See chs. 6, 10, 12 and 13.
2. Boltzmann, *Sitzungsberichte, K. Akademie der Wissenschaften, Wien Mathematisch-Naturwissenschaftliche Klasse* [this journal will be cited as *Wien. Ber.* in the sequel] **53**, 195 (1866); reprinted in *Wissenschaftliche Abhandlungen von Ludwig Boltzmann*, hrsg. F. Hasenöhrl (Leipzig: Barth, 1909; reprint, New York: Chelsea Pub. Co., 1968), I, 9 [this collection will be cited as *Wiss. Abh.*].
3. L. Boltzmann, *Wien. Ber.* **63**, 712 (1871); *Wiss. Abh.* I, 288.
4. L. Boltzmann, *Wien. Ber.* **58**, 517 (1868), **63**, 397 (1871); *Wiss. Abh.* I, 49, 237.
5. E. E. Daub, *Isis* **60**, 318 (1969).
6. L. Boltzmann, *Wien. Ber.* **66**, 275 (1872); *Wiss. Abh.* I, 316; English trans. in S. G. Brush, *Kinetic Theory* (Oxford and New York: Pergamon Press, 1966), 2, 88.
7. Quoted from my translation, *op. cit.*, except that *regellos* has been translated as "irregular" instead of "random" so as not to prejudice the issue; Boltzmann presumably could have chosen the word *zufällig* if he had wanted to approximate the meaning that is conveyed in English by "random."
8. For detailed discussion of the derivation of the *H*-theorem see Boltzmann's *Vorlesungen über Gastheorie* I. Teil (Leipzig: Barth, 1896); English trans. by S. G. Brush, *Lectures on Gas Theory* (Berkeley: University of California Press, 1964), or R. C. Tolman, *The Principles of Statistical Mechanics* (London and New York: Oxofrd University Press, 1938), Part One. Most of the abbreviated derivations given in modern textbooks are quite unsatisfactory, as I discovered some years ago in teaching a course covering this subject.

9. My translation, in *Kinetic Theory* **2**, 90.
10. The factor τ was omitted by a misprint in eq. (2), p. 96 of my translation.
11. *Ibid.*, p. 116. The last step in the derivation deserves to be recorded here because of its similarity to an equation used later by Planck: dE/dt is equal to an integral over the expression

$$\log\left(\frac{ss'}{\sigma\sigma'}\right)(\sigma\sigma' - ss')$$

If it is not the case that $ss' = \sigma\sigma'$ (corresponding to the Maxwell distribution) then either $ss' > \sigma\sigma'$ or $ss' < \sigma\sigma'$. "In the first case, $\log(ss'/\sigma\sigma')$ is positive but $\sigma\sigma' - ss'$ is negative, and in the second case the converse is true; in both cases the product $\log(ss'/\sigma\sigma')(\sigma\sigma' - ss')$ is negative.... Therefore E must necessarily decrease."
12. L. Boltzmann, *Wien. Ber.* **63**, 712 (1871); *Wiss. Abh.* **I**, 288. It has been argued that under certain circumstances the *H*-theorem does not imply the entropy principle (*e.g.* in shearing flow): see C. Truesdell, *J. Rational Mech. Anal.* **5**, 55 (1956), §50.
13. L. Boltzmann, *Wien. Ber.* **72**, 427 (1875); *Wiss. Abh.* **II**, 1.

    In 1887, H. A. Lorentz pointed out that there is a gap in Boltzmann's proof of the *H*-theorem for polyatomic molecules, due to the fact that in collisions between non-spherical molecules, inverse collisions may not exist. Boltzmann admitted the defect and showed that the damage could be repaired by constructing a cycle of collisions that would still produce the same effect; hence this objection was not a serious challenge to the validity of the theorem. See H. A. Lorentz, *Wien. Ber.* **95**, 115 (1887), reprinted in his *Collected Papers* (The Hague: M. Nijhoff, 1934–39), **6**, 74; L. Boltzmann, *Wien. Ber.* **95**, 153 (1887); *Wiss. Abh.* **III**, 272. For further discussion and diagrams of the collisions in question see Tolman, *op. cit.* (note 8), pp. 119–20.
14. P. and T. Ehrenfest, *Phys. Z.* **8**, 311 (1907), reprinted in Paul Ehrenfest's *Collected Scientific Papers*, ed. M. J. Klein (New York: Interscience/Amsterdam: North-Holland, 1959), p. 146: *Enc. math. Wiss.* **IV** (2 II, 6) (1912) reprinted in Ehrenfest's *Papers*, p. 213; English trans. by M. J. Moravcsik, *The Conceptual Foundations of the Statistical Approach in Mechanics* (Ithaca, N.Y.: Cornell University Press, 1959). See also M. J. Klein, *Paul Ehrenfest* (Amsterdam: North-Holland/New York: American Elsevier, 1970), **1**, ch. 6; H. Bernhardt, *NTM, Z. f. Gesch. Naturwiss., Tech. Med.* **4** (1) 35 (1967).
15. C. G. Knott, *Life and Scientific Work of Peter Guthrie Tait* (Cambridge University Press, 1911), p. 213–14. I thank Dr. C. W. F. Everitt for providing a photocopy of the original letter preserved at Cambridge University.
16. Bodleian Library, Oxford, Pattison MSS (copy supplied by Dr. Thomas Simpson).
17. R. J. Strutt, *Life of John William Strutt, Third Baron Rayleigh* (London: Arnold, 1924; reprint with additions by J. N. Howard, Madison, Wisc.: University of Wisconsin Press, 1968), p. 47.
18. William Thomson, *Proc. R. S. Edinburgh* **8**, 325 (1874), reprinted in S. G. Brush, *Kinetic Theory* **2**, 176.
19. W. Thomson, *Rept. Brit. Ass. Adv. Sci.* **41**, lxxxiv (1871), reprinted in his *Popular Lectures and Addresses* (London: Macmillan, 1894), **2**, 132; see also G. Basalla et al., eds., *Victorian Science* (Garden City, N.Y.: Anchor Books, 1970), p. 128. Thomson seems to have derived his objection from that of J. Herschel, *Physical Geography of the Globe* (Edinburgh, 1861), p. 12. Cf. K. E. von Baer, *Augsburger Allgemeine Zeitung*, **130**, 1968 (1873), English trans. in *Darwin and his Critics*, ed. D. L. Hull (Cambridge, Mass.: Harvard University Press, 1973), p. 116. On dislike for random-

ness as a source of neo-Lamarckian hypotheses see E. F. Gerson, *Synthesis* **1** (2), 13 (1973).
20. J. W. Gibbs, *Trans. Conn. Acad.* **3**, 108 (1875), reprinted in *The Collected Works of J. Williard Gibbs* (New Haven: Yale University Press, 1948), **1**, 55; German trans. by W. Ostwald, *Thermodynamische Studien* (Leipzig: Engelmann, 1892); French trans. by G. Matisse, *L'Equilibre des substances hétérogènes* (Paris: Gauthier-Villars, 1919).
21. Gibbs, *Collected Works* **1**, 167.
22. L. Boltzmann, citation below, §14.6, note 20; P. S. Epstein, in *A Commentary on the Scientific Writings of J. Willard Gibbs*, ed. A. Haas (New Haven: Yale University Press, 1936), **2**, 59 (see p. 106, 112).
23. *Op. cit.* (note 14), esp. *Conceptual Foundations*, pp. 14–15; Tolman, *op. cit.* (note 8), p. 152; K. F. Herzfeld, *Kinetische Theorie der Wärme* (Braunschweig: Vieweg, 1925) (Müller–Pouillets Lehrbuch der Physik, Elfte Aufl., Dritter Band, Zweite Hälfte), pp. 353–54; D. ter Haar, *Elements of Statistical Mechanics* (New York: Rinehart, 1954), p. 340f; M. J. Klein, *op. cit.* (note 14), p. 102; M. Kac, *Probability and Related Topics in Physical Sciences* (New York: Interscience, 1959), p. 61; H. Bernhardt, *op. cit.* (note 14).
24. J. Loschmidt, *Wien. Ber.* **73**, 128, 366 (1876), **75**, 287, **76**, 209 (1877). R. Dugas, *La Théorie Physique au sens de Boltzmann* (Neuchatel: Griffon, 1959), pp. 158–84. For Loschmidt's earlier ideas on this subject see E. E. Daub, *Stud. Hist. Phil. Sci.* **1**, 213 (1970).
25. See *e.g.* J. E. Mayer, in *Isotopic and Cosmic Chemistry*, eds. H. Craig *et al.*, (Amsterdam: North-Holland, 1964), 10; Kac, *op. cit.* (note 23).
26. See note 29, §14.4. In the second paper of his series, Loschmidt mentioned the continuing controversy about Maxwell's conclusion in England, See R. C. Nichols, *Nature* **11**, 486 (1875); J. J. Murphy, *Nature* **12**, 26 (1875); R. C. Nichols, *Nature* **12**, 67 (1875); S. H. Burbury, *Nature* **12**, 107 (1875).
27. Loschmidt, *op. cit.* (note 24), first paper, p. 135; see also the third paper, p. 293.
28. "Denn wenn wir im obigen Falle, nachdem eine zur Herstellung des stationären Zustandes vollkommen ausreichende Zeit τ verstrichen ist, plötzlich die Geschwindigkeiten aller Atome in entgegengesetzter Richtung annehmen, so würden wir damit am Beginne eines Zustandes stehen, dem ebenfalls der Charakter des Stationären zuzukommen scheinen würde." *Ibid.*, p. 139.
29. L. Boltzmann, *Wien. Ber.* **75**, 67 (1877); *Wiss. Abh.* **II**, 116; English trans. in Brush, *Kinetic Theory* **2**, 188.
30. L. Boltzmann, *Wien Ber.* **74**, 503 (1876); *Wiss. Abh.* **II**, 55. Boltzmann discussed the gravitational-equilibrium problem again in the second section of his paper in *Wien. Ber.* **78**, 7 (1878); *Wiss. Abh.* **II**, 250. Much later he dismissed the "ausgebreitete Literatur" on this problem in a few lines, with a footnote citing nine authors; see L. Boltzmann and J. Nabl, *Enc. math. Wiss.* **V**, (i) 516 (1905).
31. *Ibid.*, p. 192.
32. L. Boltzmann, *Wien. Ber.* **76**, 373 (1877).
33. Linus criticized Boyle's theory of gas pressure, forcing Boyle to defend it and present quantitative evidence which he had not done earlier. Buys-Ballot criticized Clausius' kinetic theory (based on the assumption that the molecules have negligible size) by pointing out that the theory predicted diffusion at a rate much more rapid than is observed; this led Clausius to introduce his "mean free path" concept (attributing a finite but small diameter to his molecules). See §§1.2 and 4.4.
34. He did however cite it much later in his review article with Nabl, *op. cit.*
35. L. Boltzmann, *Wien. Ber.* **76**, 373 (1877); *Wiss. abh.* **II**, 164.

According to Edward Daub, the success of Boltzmann's reduction of the Second Law to probability considerations rested on his application of the results to the thermodynamics of diffusion. However, it appears to me that in the paper in question, published in 1878, Boltzmann refers to his earlier statistical calculation of entropy only to justify his assumption that the entropy of a mixture is the sum of the entropies of its components, and that the rest of the argument does not involve probability in any essential way. See E. E. Daub, *Isis* **60**, 318 (1969).

36. E. Nagel, in *International Encyclopedia of Unified Science*, ed. O. Neurath (Chicago: University of Chicago Press, 1939, 1955), 1 (6), 355 [= p. 13 in the separate edition of this number, *Principles of the Theory of Probability*].
37. M. Planck, *The Theory of Heat Radiation*, trans. by M. Masius from the 2nd ed. (1913) of *Waermestrahlung* (New York: Dover Pubs., 1959), p. 120. A comprehensive discussion of the problem may be found in R. H. Fowler, *Statistical Mechanics* (Cambridge, Eng.: Cambridge University Press, 2nd ed., 1936), pp. 189–207.
38. J. R. Partington, *An Advanced Treatise on Physical Chemistry*, (London: Longmans, Green and Co., 1949), **1**, 293. Despite his professions of skepticism Partington ends up by accepting Boltzmann's relation between entropy and probability.
39. Fowler, *op. cit.*, pp. 203, 230. J. W. Gibbs, *Elementary Principles of Statistical Mechanics* (1902), ch XV; see *The Collected Works of J. Willard Gibbs* (New Haven: Yale University Press, 1948), **2**, 203. R. C. Tolman, *The Principles of Statistical Mechanics* (London: Oxford University Press, 1938), p. 562. A. I. Khinchin, *Mathematical Foundations of Statistical Mechanics*, trans. from Russian by G. Gamow (New York: Dover Pubs., 1949), pp. 139–45. P. G. Wright, *Contemp. Phys.* **11**, 581 (1970).
40. *Wien. Ber.* **76**, 376.
41. *Wien. Ber.* **76**, 382.
42. *Ibid.*, p. 404. For the proof of the required theorem Boltzmann referred to H. W. Watson's *Treatise on the Kinetic Theory of Gases* (Oxford: Clarendon Press, 1876), p. 12. A further indication that the use of the new method is a little tricky is provided by Boltzmann's reference to O. E. Meyer's attempt to apply it in his textbook on kinetic theory [*Die Kinetische Theorie der Gase* (Breslau, 1877), 262]. According to Boltzmann, Meyer made several mathematical errors that somehow cancel out in such a way that he gets the desired Maxwell distribution law as a result. For further discussion see Boltzmann, *Wien. Ber.* **78**, 7 (1878); Meyer, *Ann. Phys.* [3], **10**, 296 (1884); Boltzmann, *Ann. Phys.* [3] **11**, 529 (1880).
43. See ch. 10; also S. G. Brush, *Transp. Theory and Stat. Phys.* **1**, 287 (1971). H. Bernhardt, *NTM, Z. Gesch. Naturwiss., Tech., Med.* **8**, (1) 13 (1971).
44. J. C. Maxwell, *Trans. Camb. Phil. Soc.* **12**, 547 (1879); *Papers* **II**, 713.
45. L. Boltzmann, *Ann. Phys. Beibl.* **5**, 403 (1881); *Wiss. Abh.*, **II**, 582; English trans. in *Phil. Mag.* [5] **14**, 299 (1882).
46. L. Boltzmann, *Rept. Brit. Ass. Adv. Sci.* **64**, 102 (1894); *Wiss. Abh.* **III**, 521.
47. L. Boltzmann, *Populäre Schriften* (Leipzig: Barth, 1905), p. 25 (quotation translated from p. 34).
48. *Ibid.*, p. 48. Cf. Boltzmann's letter to Ernst Mach, 1893: "... Ich glaube, dass die Unmöglichkeit des perpetuum mobile ein reiner Erfahrungssatz ist, der in noch nicht geprüften Fällen jeden Augenblick durch die Erfahrung widerlegt werden kann. Dass ich dies bezüglich des s.g.1. Hauptsatzes für enorm unwahrscheinlich, bezüglich des 2. Hauptsatzes für nicht einmal zu unwahrescheinlich halte, ist eine rein subjektive unbeweisbare Meinung." K. D. Heller, *Ernst Mach* (Wien: Springer, 1964), p. 27.

14.6 Molecular disorder

The problem of irreversibility was revived in England in the 1890's as part of a more general discussion of the conditions for validity of the equipartition theorem. Initially P. G. Tait and others had tried simply to improve Maxwell's proof of his velocity distribution law.[1] There were a few attempts to give quantitative descriptions of the approach to equilibrium for special models.[2] G. J. Stoney, in 1887, rediscovered the reversibility paradox and concluded that the Second Law of Thermodynamics is not a "true dynamical law," but also suggested that time itself does not exist apart from events in the universe,[3] thus anticipating in a rudimentary way Boltzmann's suggestion made a decade later. Also at this time there was published the interesting suggestion of L. Gouy that Brownian movement may be considered as a visible violation of the Second Law, though this idea did not attract much attention until Poincaré mentioned it in 1900.[4]

Most of these scientists were either unaware of Boltzmann's earlier discussion of the same problems, or reluctant to plow through his lengthy memoirs. Instead, they preferred to discuss general principles on the basis of simple arguments and short calculations. Finally Boltzmann himself entered the debate, visiting a meeting of the British Association in 1894 and later replying to some of the letters in *Nature*. The outcome of Boltzmann's participation in this discussion was the concept of "molecular disorder," first pinpointed by S. H. Burbury (1831–1911), a barrister who had turned to mathematics after becoming deaf; Burbury was responding to a deceptively simple criticism of kinetic theory published by E. P. Culverwell (1855–1931) of Trinity College, Dublin. Another stimulus for Boltzmann was a brief exchange with Max Planck concerning the assumptions involved in Kirchhoff's derivation of the state of thermal equilibrium in a gas. For some reason these penetrating discussions, which indicated a need for assuming randomness in mechanistic theories, have been overshadowed by the somewhat more exotic debate on the "recurrence paradox," although this involved some of the same issues. It is also of interest to follow the development of Max Planck's ideas on randomness and irreversibility in radiation theory, in the years just before he arrived at the quantum theory. (These topics will be discussed in §§14.7 and 14.8).

* * *

In 1890 E. P. Culverwell published a "Note on Boltzmann's Kinetic Theory of Gases, and on Sir W. Thomson's Address to Section

A, British Association, 1884."[5] Using the example of a system of particles interacting with forces proportional to their distance, he claimed that (since in this case the motion is strictly periodic) it is impossible to prove *in general* that a set of particles will tend to the "Boltzmann configuration, in which the energy is equally distributed among all the degrees of freedom." Appealing to the reversibility principle, he asserted that "for every configuration which tends to an equal distribution of energy, there is another which tends to an unequal distribution." In order to explain the fact that temperature equilibrium does nevertheless occur, he suggested that there must be some kind of interaction of molecules with the ether. After all, "one of the most important purposes for which the existence of the aether is required," he wrote, is heat transfer leading to thermal equilibrium. Conversely, Culverwell asserted that if a system of particles in a vessel *not* containing ether did in fact attain equilibrium in all cases, "it would be to my mind a proof that the ultimate particles of matter did not individually obey those laws which they are known to obey when collected in the enormous numbers which compose the bodies for which the laws of motion have been experimentally proved." Explaining his views further at a British Association meeting the same year, he argued that molecular motions might be inherently irreversible, yet obey the Newtonian laws of motion when taken *en masse*.[6] It is also conceivable, he thought, that there might be periodic deviations from Newton's laws, "the period being so short that no observations could detect it."

Responding to the widespread interest in such problems, the British Association appointed a committee, consisting of J. Larmor and G. H. Bryan, to investigate "the present state of our knowledge of Thermodynamics, specially with regard to the Second Law," and Bryan gave the first report on this subject at the 1891 meeting.[7] This dealt primarily with the attempts of Clausius, Szily, Helmholtz and Boltzmann to reduce the Second Law to purely mechanical principles,[8] though even here Bryan suggested that it might be necessary to introduce some kind of statistical element in order to justify certain assumptions in the proofs. He insisted that "A system which is irreversible will certainly not be monocyclic according to the definition of Helmholtz," so, to the extent that this viewpoint is valid, "it seems necessary to accept the principle of degradation of energy as a statistical property and not as a dynamical principle."[9] Bryan discussed one of Boltzmann's mechanical models of a monocyclic system[10] but criticized Boltzmann's attempt to extend the dynamical analogy to irreversible processes, on the grounds that Boltzmann's argument holds only if friction is present, which is "not allowable in forming a purely dynamic analogue of the properties of heat."[11]

As for Culverwell's critique, Bryan noted, first, that systems with forces directly proportional to the distance (what we now call the "harmonic oscillator") have special properties as regards periodicity, and conclusions based on such systems cannot be generalized to other systems. (This seems to miss the point, that a *general* theorem cannot be valid if it fails in a particular case.) Second, although it is true that a conservative dynamical system is always reversible, the reversed motion is often "dynamically unstable in the highest degree" as is readily discovered when one tries to ride a bicycle backwards. The slightest disturbance of the reversed motion leads to the Maxwellian "special state" again, rather than to the original ordered state. (A similar argument has frequently been used in attempts to dispose of the reversibility paradox.[12] Yet it is not really a question of whether an exact reversal of molecular motions is *physically* possible, but rather whether in deriving the H-theorem one may group together direct and inverse collisions, making certain assumptions about their probability of occurrence. For this reason I will make a distinction, below, between the "reversibility paradox" and the "reversibility objection to the H-theorem.")

Bryan's first report did not present any definite conclusions about the physical origin of irreversibility. He appealed to the example of mixing two different substances "in a minute state of subdivision" – after mixing them it is obviously impossible to separate them by simply stirring them up. For the same reason it is understandable how it *could* be proved on statistical grounds that heat cannot pass from a cold body to a hot one, but such an argument is admittedly not a proof. While Bryan thought it possible that the presence of the ether will "facilitate the dissipation of energy" he defended the attempt to explain irreversibility *without* invoking the ether, on the grounds that such an attempt fulfills "what should be the highest object of scientific inquiry – namely, of helping us to 'judge the unknown from the known.'"[13]

The second part of Bryan's report was presented at the Oxford meeting of the British Association in 1894, with an appendix by Boltzmann who attended this meeting. There is a review of the extensive literature on attempts to prove Maxwell's distribution law for various systems, including the "test-cases" against the equipartition theorem proposed by William Thomson (now Lord Kelvin) in the 1890's. A short section on "Boltzmann's Minimum Theorem" is based on recent work of Burbury (who introduced the letter H in 1890[14] for the functional which Boltzmann called E in 1872) and H. W. Watson.[15] Burbury was following the Maxwellian 1860 tradition of using probability theory, explicitly treating the molecular coordinates as random variables, in contrast to the later Maxwell–Boltzmann developments based on consideration of molecular collisions. But Burbury also suggested that the process of redistribution of energy between the molecules may be effected by waves transmitted through the ether.

In conclusion, Bryan wrote:

> The proof of the [Maxwell–Boltzmann distribution] law and the assumptions involved in it are fairly satisfactory for gases whose molecules collide with each other to a certain extent at random, but in a medium in which the molecules never escape from each other's influence the subject still presents very great difficulties.[16]

This is quite a fair summary of the state of the subject (even today) but it appears that Bryan did not realize the full implications of his use of the word "random."

The discussion which followed the presentation of Bryan's report was apparently concerned mainly with the problem of specific heats of gases. In response to various questions about the equipartition theorem, Boltzmann made a statement to the effect that he had primarily regarded the kinetic theory from the mathematical viewpoint, and while he had chosen his postulates in the light of the experimental data to be explained—for example, the specific heats of diatomic molecules, which could be accounted for by a model with five mechanical degrees of freedom—he did not feel obliged to prove that his theory was in agreement with *all* properties of gases.[17] In reply to G. F. Fitzgerald's objection that the equipartition theorem, if true, ought to apply to *everything* in the universe, including the ether, Bryan indicated that Boltzmann thought this was still an open question, but Bryan was willing to give his own opinion:

> In the absence of mutual action between the various solar systems, this would *not* be the only permanent distribution, nor would there be any tendency to assume such a distribution. If, however, the different solar systems were to collide with or encounter one another *at random* in such a way that transference of energy was liable to take place between any of the coordinates of any one system and any of the coordinates of any other system, the Boltzmann–Maxwell distribution *would* probably be unique

and there would be a tendency to assume such a distribution as the ultimate result of a great number of encounters taking place.[18]

But Bryan does not indicate that there is any conflict between the assumption of randomness and the assumption that one is dealing with a deterministic mechanical system.

The discussion in the columns of *Nature* soon turned to the problems of proving "Boltzmann's minimum theorem." Culverwell led off by criticizing a step in Watson's proof, and then remarked: "I do not know Boltzmann's proof, but while I suppose it is all right, I find it very hard to understand how any proof can exist." Because of the reversibility of molecular motions, he thought it would be impossible to prove that dH/dt would be negative for all initial configurations, although possibly "by striking some kind of average" among configurations which approach the Maxwellian state and those that recede from it (obtained by reversing velocities) the average value of dH/dt might be shown to be negative. He concluded by asking: "Will some one say exactly what the H-theorem proves?"[19]

Culverwell's modest inquiry seems to me to mark the transition from the "reversibility paradox" of the 1870's which–according to Thomson's and Boltzmann's discussion–was resolved by a statistical but non-stochastic conception of molecular motions–and the more subtle "reversibility objection to the H-theorem." In the latter, but not in the former, attention is directed to identifying the stage in the proof of this particular theorem where irreversibility sneaks in. Culverwell is not quite able to do this himself but he smells something.

Burbury replied by stating that the proof of the H-theorem depends on the statement that

> If the collision coordinates be taken at random, then the following condition holds, *viz.*:–For any given direction of R [relative velocity vector of two colliding spheres] before collisions, all directions after collision are equally probable. Call that condition A.

> But in the case of the reverse motion condition A is not fulfilled; hence the proof is not applicable. (In other words, Maxwell's original assumption that the number of collisions is proportional to $n_1 n_2$ does not hold for reverse collisions.)

> Somebody may perhaps say that by this explanation I save the mathematics only by sacrificing the importance of the theorem, because I must (it will be said) admit that there are, after all, as

many cases in which H increases as in which it diminishes. I think the answer to this would be that any actual material system receives disturbances from without, the effect of which, coming at haphazard, is to produce that very distribution of coordinates which is required to make H diminish. So there is a general tendency for H to diminish although it may conceivably increase in particular cases. Just as in matters political, change for the better is possible, but the tendency is for all change to be from bad to worse.[20]

It is this particular paper of Burbury that Boltzmann himself cites as the origin of his "molecular disorder" assumption, before mentioning the earlier discussion with Loschmidt; but it does of course throw some new light on the reversibility paradox too.[21] The need for a postulate of randomness did not clearly emerge from the *general* discussion of the statistical nature of the Second Law by Maxwell, Thomson, Tait, Loschmidt, and Boltzmann, but only in response to a detailed technical criticism of the H-theorem by Culverwell, who must now be added to our list of worthy dissenters.[22,23]

Culverwell did not quite get the point of Burbury's new postulate, though he recognized that something new was being added to the theory which he characterized as "some amount of assumption as to an average state having been already attained."[24] Burbury then restated his postulate that the coordinates are "taken at haphazard" with respect to each other even for molecules that have just collided (in analyzing the reversed collision). This assumption, he thought, was sufficient though perhaps not necessary for the proof and is "the most useful assumption, because the distribution of coordinates assumed to exist is that which would tend to be produced by any disturbances acting on the system from without."[25]

Bryan, replying two weeks after Burbury to Culverwell's question, missed the point in another way. He said that Culverwell's assumption about the reversed motion amounted to endowing his molecules "with the power of forethought and the prediction regarding their future state necessary to enable them to regulate their movements" whereas if the molecules "are allowed to take their own natural course, and nothing special is known about them," the only reasonable assumption to make is that which is conventionally made in proving the H-theorem, namely that the two molecules are uncorrelated before the inverse collision.[26] But in fact Culverwell's assumption had been nothing more than the validity of deterministic laws of motion, so that the implication of

Bryan's quoted phrase is either that the "natural course" of a molecule is random, or that it appears to be random because we do not know all its coordinates.

Larmor's resolution of the problem was similar to the original discussion of Thomson and Boltzmann:–the number of states which, on reversal, would return to an ordered state is small compared to the total number; and, moreover, if the reversed motion is continued the order will quickly disappear. But he was also willing to concede that "if the whole universe were thus reversed, the aberration would be permanent." However, he immediately ruled out this possibility by asserting that "the whole universe is a permanently dissipative system, and there is no question of a steady state being attained by it in measurable time."[27]

Culverwell, having digested the various responses to his challenge published by Larmor, Burbury, Bryan, and Watson,[20,23,26,27] was now prepared to knock all the heads together. Bryan's argument, as he pointed out, is no proof that H will decrease since it "depends on the previous assumption that the particles do 'naturally' tend to move in the desired way." Watson, by stating that H decreases even when the system is *receding* from its equilibrium state, has given up the physical meaning of the H-theorem. Burbury still depends on an additional assumption which he has not yet stated very clearly (according to Culverwell) for the general case. Culverwell essentially agreed with Larmor's interpretation, but improved it significantly with his own statement that if the proof of the H-theorem "does not somewhere or other introduce some assumption about averages, probability, or irreversibility, it cannot be valid."[28]

Finally, on 28 February 1895, the British wranglers received an authoritative message from Vienna. Boltzmann, in a letter published in *Nature*, presented a discursive exposition of his answer to two questions: "(1) Is the Theory of Gases a true physical theory as valuable as any other physical theory? (2) What can we demand from any physical theory?" The second part of this letter[29] is a direct reply to Culverwell, in which Boltzmann reiterated his 1877 position that the Second Law is based on probability theory and can never be proved mathematically from dynamics alone. But while the casual reader might get the impression that Boltzmann is only repeating in simpler terms the view which he had already published before Culverwell raised his objections, there are some subtle changes. In particular, Boltzmann introduced a description of the "H-curve" (graph of H vs. time) in which he now asserted, contrary to what he had said in 1886,[30] that even if one starts from an initial state which "is not specially arranged for a certain purpose, but haphazard governs freely," the value of H must occasionally rise above its minimum value. While "the probability that H decreases is always greater than that it increases" it is also certain that H must *sometimes* increase. This does not disprove the H-theorem, he now says, but only illustrates its probabilistic nature. Unfortunately Boltzmann's use of the term "probability" is still ambiguous, and he does not state whether he agrees with Burbury's interpretation.

Boltzmann concluded the letter with "an idea of my old assistant, Dr. Schuetz":

> We assume that the whole universe is, and rests for ever, in thermal equilibrium. The probability that one (only one) part of the universe is in a certain state, is the smaller the further this state is from thermal equilibrium; but this probability is greater, the greater is the universe itself. If we assume the universe great enough, we can make the probability of one relatively small part being in any given state (however far from the state of thermal equilibrium), as great as we please. We can also make the probability great that, though the whole universe is in thermal equilibrium, our world is in its present state....
>
> If this assumption were correct, our world would return more and more to thermal equilibrium; but because the whole universe is so great, it might be probable that at some future time some other world might deviate as far from thermal equilibrium as our world does at present. Then the afore-mentioned H-curve would form a representation of what takes place in the universe. The summits of the curve would represent the world where visible motion and life exist.

In accordance with the "Matthew effect"[31] this idea was subsequently ascribed to Boltzmann rather than Schuetz.[32]

Culverwell, commenting on Boltzmann's letter, insisted that they were actually in agreement on the main point: a purely dynamical proof of the H-theorem is impossible, but it can be proved by making certain probabilistic assumptions. Though Culverwell is still somewhat vague on just what those assumptions are, he does give a helpful explanation of the paradox that "while there are as many configurations for which dH/dt is positive as there are for which it is negative" nevertheless H is much more likely to descend to its minimum value than to rise still higher. The point is that one should not "set off a configuration for which H increases against one for which it decreases, *although the values of H for each are different.*" He suggests the analogy of a tree turned upside down with an infinite number of branches passing through each point of the trunk. If you start from any point above the bottom, there are more paths leading down than up, since every upward branch finally turns downward; whereas if you start at the bottom you are more likely to choose a branch running upwards to the trunk. In this way it can be seen how, starting from any value of H above the minimum, dH/dt is more likely to be negative than positive.[33]

Burbury developed his idea of random external disturbances a little more in two letters published in *Nature* in 1895. The thrust of these letters was to emphasize that his "condition A" (see above) is unlikely to be true for the reversed motion unless the gas is already in an equilibrium state, and that there is no particular reason that the condition will always be satisfied, even apart from reversals, in a system left to itself for an indefinite time free of external influences.[34] He did not specify the reason for his doubts, but in view of his later work in kinetic theory (or, anticipating Boltzmann's response) we might guess that he had in mind the possibility that two molecules cannot be completely uncorrelated before a collision if they have collided at some time in the past; and the magnitude of this effect would be greater in a dense gas. Thus the H-theorem, and perhaps other results of the kinetic theory, are valid only for rarified gases.[35]

Boltzmann conceded the force of this reasoning (which he was developing himself to some extent independently but obviously stimulated by the debate with the British) and admitted that the proof of the H-theorem is valid only if the mean free path of a molecule is very long compared to the average distance of two neighboring molecules. In this case the assumption of external disturbances is not necessary.[36] Bryan, enlarging on this suggestion, remarked that in the opposite case of liquids and solids where the molecules are crowded closely together, we know that the system can exist simultaneously in either of two states, hence the distribution cannot be unique, and it is just as well that we do not try to prove too much with the theorem. The same would be true for molecules immersed in a continuous ether. Yet we do know that solids and liquids obey the Second Law even if we do not know whether they obey the Maxwell–Boltzmann distribution law; and this may simply be because Burbury's "condition A" is always satisfied when we bring together a hot and a cold body which we can assume to have no initial statistical correlation.[37] Burbury agreed with Bryan that "contact with the refrigerator or with the reservoir, such as is supposed

§14.6] MOLECULAR DISORDER 625

to take place in thermodynamics, is for this purpose a disturbance" which can bring about condition A.[38]

In the first volume of his *Vorlesungen über Gastheorie* Boltzmann adopted Burbury's postulate, calling it the assumption that the state of the gas is "molecular disordered" [molekular-ungeordnet].[39] In addition to giving credit to Burbury, he stated that Kirchhoff also made the assumption, although Boltzmann had criticized the way the assumption was used in the posthumous edition of Kirchhoff's lectures published by Max Planck (see below). The following passage shows that Boltzmann recognized the importance of "condition A" as an *assumption* in gas theory, yet was not ready to go so far as to make it a *postulate* about molecular motions:

> That it is necessary to the rigor of the proof to specify this assumption in advance was first noticed in the discussion of my so-called H-theorem or minimum theorem. However, it would be a great error to believe that this assumption is necessary only for the proof of this theorem. Because of the impossibility of calculating the positions of all the molecules at each time, as the astronomer calculates the positions of all the planets, it would be impossible without this assumption to prove the theorems of gas theory. The assumption is also made in the calculation of the viscosity, heat conductivity, etc. Also, the proof that the Maxwell velocity distribution is a possible one – *i.e.* that once established it persists for an infinite time – is not possible without this assumption. For one cannot prove that the distribution always remains molecular-disordered. In fact, when Maxwell's state has arisen from some other state, the exact recurrence of that other state will take place after a sufficiently long time.[40]

As can be seen from the last sentence of this quotation, Boltzmann is well prepared to meet the next attack on his H-theorem.

* * *

Notes for §14.6

1. P. G. Tait, *Proc. R. S. Edinburgh*, **13**, 21 (1884); *Trans. R. S. Edinburgh* **33**, 65 (1886), reprinted in *Phil. Mag.* [5] **21**, 343 (1886) and in his *Scientific Papers* (Cambridge University Press, 1890–1900), II, 124.
2. P. G. Tait, *Trans. R. S. Edinburgh* **33**, 65 (1887), Part V. Rayleigh, *Phil. Mag.* [5] **32**, 424 (1891), reprinted in his *Scientific Papers* (New York: Dover Pubs., 1964), III, 473, S. H. Burbury, *Phil. Trans. Roy. Soc. London* **183A**, 407 (1892).

3. G. J. Stoney, *Proc. R. S. Dublin* **5**, 448 (1887); *Phil. Mag.* [5] **23**, 544 (1887).
4. L. Gouy, *J. de Physique* [2] **7**, 561 (1889); see below, §15.3. H. Poincaré, *Rapports Cong. Int. Phys. Paris, 1900* (Paris: Gauthier-Villars, 1900) **1**, 1; *Congress of Arts and Science, Universal Exposition, St. Louis, 1904* (Boston: Houghton, Mifflin & Co., 1905), **I**, 604. H. V. Helmholtz, *Vorlesungen über Theorie der Wärme*, hrsg. F. Richarz (Leipzig: Barth, 1903), p. 260.
5. E. P. Culverwell, *Phil. Mag.* [5] **30**, 95 (1890).
6. E. P. Culverwell, *Rept. Brit. Ass. Adv. Sci.* **60**, 744 (1890).
7. G. H. Bryan, *Rept. Brit. Ass. Adv. Sci.* **61**, 85 (1891).
8. For further discussion and references on mechanical analogies for the Second Law of Thermodynamics see G. H. Bryan, *Enc. math. Wiss.* **5**, (1, 3), 73 (1903), §IV; L. de Broglie, *La Thermodynamique de la particule isolée* (Paris: Gauthier-Villars, 1964); I. Fenyes, *Z. Phys.* **132**, 140 (1952). M. J. Klein, *Centaurus* **17**, 58 (1972).
9. Bryan, *op. cit.*, pp. 107-8.
10. L. Boltzmann, *Vorlesungen über Maxwell's Theorie der Elektrizität und des Lichtes* (Leipzig: Barth, 1891), **I**, 3-23.
11. Bryan, *op. cit.*, p. 109.
12. See Borel's discussion quoted by R. Dugas, *La Théorie Physique au sens de Boltzmann* (Neuchatel: Griffon, 1959), pp. 186-87.
13. Bryan, *op. cit.*, p. 120. Cf. J. R. Mayer's opinion (1848), quoted by O. B. Mathias, *An Examination of the Evolution of the First Two Laws of Thermodynamics* (Dissertation, University of Kansas City, Kansas City, Missouri, 1962), from *Phil. Mag.* [4] **25**, 241 (1863).
14. S. H. Burbury, *Phil. Mag.* [5] **30**, 301 (1890). According to S. Chapman [*Nature* **139**, 931 (1937)] this was the first use of the letter H instead of E. It has sometimes been suggested that H is intended as a capital Greek eta, but no definite evidence for this has ever come to my attention [see S. G. Brush, *Am. J. Phys.* **35**, 892 (1967)].

    In another paper in *Nature* **49**, 150 (1893), Burbury calls this "Boltzmann's minimum function" and denotes it as B, but Bryan, while retaining the phrase, went back to the letter H [G. H. Bryan, *Rept. Brit. Ass. Adv. Sci.* **64**, 64 (1894); see p. 86]. Burbury also calls it B in another paper, *Phil. Mag.* [5] **37**, 143 (1892).

    Some further information about Burbury is given by Ross Hesketh, [London] *Times Higher Education Supplement*, p. 11 (20 April 1973). (I thank Professor E. A. Mason for this reference.)
15. S. H. Burbury **183**, 407 (1892); H. W. Watson, *A Treatise on the Kinetic Theory of Gases* (Oxford: Clarendon Press, 2nd ed., 1893).
16. Bryan, *op. cit.*, p. 98; this conclusion is quoted in the summary in *Nature*, **50**, 406 (1894).
17. G. H. Bryan, *Nature* **51**, 31, 152 (1894); E. P. Culverwell, *Nature* **51**, 78 (1894).
18. G. H. Bryan, *Nature* **51**, 152 (1894); G. F. FitzGerald, *Nature* **51**, 221 (1895).
19. E. P. Culverwell, *Nature* **50**, 617 (1894).
20. S. H. Burbury, *Nature* **51**, 78 (1894).
21. L. Boltzmann, *Vorlesungen über Gastheorie* I. Teil (Leipzig: Barth, 1896); see English trans. by S. G. Brush, *Lectures on Gas Theory* (Berkeley: University of California Press, 1964), pp. 40, 58.
22. Boltzmann's closest approach to an independent discovery of the need for this hypothesis, which he cites in the same footnote in *Gastheorie*, is in the concluding pages of one of his replies to Loschmidt, *Wien. Ber.* **78**, 7 (1878); *Wiss. Abh.* **II**, 250. The discussion there is more concerned with the relation between the excessively

high kinetic energies of a single molecule and that of its neighbors as a source of statistical correlation, rather than with the correlation between two molecules in an inverse collision.
23. To illustrate the technical nature of the objection it may be noted that Watson, whose proof provoked Culverwell's original complaint, could not see why the value of H would necessarily be the same at the end of the reversed motion when the system has returned to its original state [but with reversed velocities], and argued instead that *because* the H-theorem is true, it could *not* have the same value. See H. W. Watson, *Nature* **51**, 105 (1894).
24. E. P. Culverwell, *Nature* **51**, 105 (1894).
25. S. H. Burbury, *Nature* **51**, 175 (1894).
26. G. H. Bryan, *Nature* **51**, 176 (1894), **52**, 29 (1895).
27. J. Larmor, *Nature* **51**, 152 (1894).
28. E. P. Culverwell, *Nature* **51**, 246 (1895).
29. L. Boltzmann, *Nature* **51**, 413 (1895); *Wiss. Abh.* **III**, 535. (The first few paragraphs of the letter are quoted in my introduction to Boltzmann's *Lectures on Gas Theory*, English trans.)
30. See note 44, §14.5.
31. R. K. Merton, *Science* **159**, 56 (1968).
32. There is no reference to Schuetz in Boltzmann's later discussions of the idea; but he is mentioned by J. Nabl, *Nat. Rund.* **21**, 337 (1906).
33. E. P. Culverwell, *Nature* **51**, 581 (1895). Boltzmann accepted the tree analogy in a somewhat different sense: see *Nature* **51**, 581 (1895); *Wiss. Abh.* **III**, 545.
34. S. H. Burbury, *Nature* **51**, 320, **52**, 104 (1895).
35. S. H. Burbury, *Nature* **52**, 316 (1895), *Proc. London Math. Soc.* **26**, 431 (1895); *Phil. Trans.* A**187**, 1 (1896); *Rept. Brit. Ass. Adv. Sci.* **66**, 716 (1896); *Proc. London Math. Soc.* **28**, 331 (1897), **29**, 225 (1898); *A Treatise on the Kinetic Theory of Gases* (Cambridge University Press, 1899), esp. p. 33: *Phil. Mag.* [5] **50**, 584 (1900).
36. L. Boltzmann, *Nature* **52**, 221 (1895); *Wiss. Abh.* **II**, 546.
37. G. H. Bryan, *Nature* **52**, 244 (1895).
38. Burbury, *Nature* **52**, 316 (1895).
39. Boltzmann, *Vorlesungen über Gastheorie* I, 20–21 [the preface is dated September 1895]; see pp. 40–41 in the English trans. On the distinction between "molecular chaos" and the "*Stosszahlansatz*" assumption, see P. and T. Ehrenfest, *The Conceptual Foundations of the Statistical Approach in Mechanics* (Ithaca: Cornell University Press, 1959), p. 40f; T. P. Eggarter, *Am. J. Phys.* **41**, 874 (1973). T. S. Kuhn (private communication) has pointed out to me that Boltzmann's assumption is somewhat different from Burbury's insofar as he wishes to *exclude* certain ordered states which would be expected to occur in a stochastic system. While this exclusion violates the meaning of a stochastic process, one could argue that it is consistent with the ordinary usage of the term "random"–"not ordered."
40. Quoted from my translation, *Lectures on Gas Theory*, pp. 41–42; see also pp. 58–59 for a summary of the outcome of Boltzmann's discussion with the British physicists.

14.7 The recurrence paradox

The notion that history repeats itself – that there is no progress or decay in the long run, but only a cycle of development that always returns to its starting point–has been inherited from ancient philosophy and primitive religion. It has been noted by some scholars that belief in recurrence, as opposed to unending progress, is intimately connected with man's view of his place in the universe, as well as with his concept of history. Starting, in some cases, from a pessimistic view of the present and immediate future, it denies the reality or validity of human actions and historical events by themselves; actions or events are real only insofar as they can be understood as the working out of timeless archetypal patterns of behavior in the mythology of the society. This attitude is said to be illustrated in classical Greek and Roman art and literature, where there is no consciousness of past or future, but only of eternal principles and values. By contrast the modern Western view, as a result of the influence of Christianity, is deeply conscious of history as progress toward a goal.[1] Nevertheless, the cyclical view has by no means died out, and its traces may have something to do with the persistent tendency to draw historical analogies and comparisons as well as the frequent revival of "oscillating universe" theories.[2]

The suggestion that eternal recurrence might be proved as a theorem of physics, rather than as a religious or philosophical doctrine, seems to have occurred at about the same time to the German philosopher Friedrich Nietzsche and the French mathematician Henri Poincaré. Nietzsche encountered the idea of recurrence on his studies of classical philology, and again in a book by Heine. It was not until 1881 that he began to take it seriously, however, and then he devoted several years to studying physics in order to find a scientific-sounding formulation of it.[3] Poincaré, on the other hand, was led to the subject by his attempts to complete Poisson's proof of the stability of the solar system, though he was also concerned with the difficulty of explaining irreversibility by mechanical models such as Helmholtz's monocyclic systems.[4] Poincaré's theorem belongs to the history of theoretical physics, Nietzsche's speculations to the history of philosophical culture, and they are not usually discussed in the same context. Yet I find it necessary to consider them together since it was just at the end of the 19th century that developments in science were strongly coupled to the philosophical-cultural background. Both Nietzsche and Poincaré were trying, though in very different ways, to attack the "materialist" or "mechanist" view of the universe.

Nietzsche's doctrine of the Eternal Return, as described in his book *Der Wille zur Macht* and elsewhere, has generally been treated by literary and philosophical commentators as a purely symbolic or metaphorical expression of his apocalyptic world view. As Rose

§14.7]   THE RECURRENCE PARADOX   629

Pfeffer wrote recently, "Nietzsche's theories and hypotheses have found no recognition and acceptance and have, on the whole, not been taken seriously by either scientists or philosophers."[5] Professor Pfeffer herself claims that the notion of eternal recurrence is of central importance in Nietzsche's philosophy, but she wishes to interpret the basic recurring units not as classical atoms but as quanta of energy: recurrence means only recurrence of "simultaneously occurring values of energy," not "configurations of simultaneously existing material, static, immutable elements." Thus Nietzsche is credited with having advanced a "dynamic world view" which "is in contrast to the mechanistic, materialistic principles of his time." I think a somewhat different conclusion will emerge if we interpret Nietzsche's effort as a qualitative anticipation of Poincaré's theorem.

Nietzsche's "proof" of the necessity of eternal recurrence (written during the period 1884-88 but not published until after his death in 1900) is as follows: "If the universe has a goal, that goal would have been reached by now" since the universe, he thinks, has always existed; the concept of a world "created" at some finite time in the past is considered a meaningless relic of the superstitious ages. He absolutely rejects the idea of a "final state" of the universe, and further remarks that "if, for instance, materialism cannot consistently escape the conclusion of a finite state, which William Thomson has traced out for it, then materialism is thereby refuted." He continues:

> If the universe may be conceived as a definite quantity of energy, as a definite number of centres of energy–and every other concept remains indefinite and therefore useless–it follows therefrom that the universe must go through a calculable number of combinations in the great game of chance which constitutes its existence. In infinity, at some moment or other, every possible combination must once have been realized; not only this, but it must have been realized an infinite number of times. And inasmuch as between every one of these combinations and its next recurrence every other possible combination would necessarily have been undergone, and since every one of these combinations would determine the whole series in the same order, a circular movement of absolutely identical series is thus demonstrated: the universe is thus shown to be a circular movement which has already repeated itself an infinite number of times, and which plays its game for all eternity.[6]

Nietzsche thought that his doctrine was not materialistic because materialism entailed the irreversible dissipation of energy and the ultimate heat death of the universe. But the discussion of Poincaré's theorem by Poincaré, Zermelo, and Boltzmann showed that, on the contrary, it is precisely the mechanistic view of the universe that has recurrence as its inevitable consequence. Since Zermelo and some other scientists believed that the Second Law must have absolute rather than merely statistical validity, they thought that the mechanistic theory was refuted by the "recurrence paradox."[7] The implication of Nietzsche's conclusion, as proved mathematically by Poincaré, is actually just the opposite of what he thought it should be: if there *is* eternal recurrence, so that the Second Law of Thermodynamics cannot always be valid, then the materialist view (as represented by Boltzmann's interpretation) would be substantiated.

Poincaré's recurrence theorem was first published in his memoir "Sur le problème des trois corps et les équations de dynamique" which was awarded the prize of King Oscar II of Sweden on 21 January 1889.[8] Poincaré considered a mechanical system governed by a set of differential equations

$$\frac{dx_1}{dt} = X_1, \quad \frac{dx_2}{dt} = X_2, \ldots, \quad \frac{dx_n}{dt} = X_n$$

where the $X_1 \ldots X_n$ are given functions of the variables $x_1 \ldots x_n$. For the case $n = 3$, (x_1, x_2, x_3) are the coordinates of a point P in space, and this point describes a certain curve (the "trajectory") as we vary the time t. In general, if the functions X_1, X_2, and X_3 are "uniform," we know that "one and only one trajectory will pass through every point in space" because of the determinism of Newtonian mechanics. It is for this purely mechanistic system that Poincaré seeks to prove "stability" in the sense defined by Poisson[9]:

> the point P should return after a sufficiently long time, if not to its initial position, at least to a position as close as one wishes to this initial position.

Poincaré recognized that such a proof could not hold for *all* solutions, and in fact he noted that there will be an infinity of "asymptotic" solutions which are not stable in this sense; nevertheless he hoped to establish not only that there are also an infinity of solutions that *are* stable, but that the unstable ones are so much less numerous that they can be regarded as "exceptional." In the language of modern mathematics, the group of measure-preserving transformations of the phase space resulting from the dynamical equations of motion has the property that almost all points in a set of positive measure are carried back into that set infinitely many times. Thus while a completely rigorous proof of Poincaré's theorem had to wait for the development of the theory of measure of point sets, by Lebesgue and others at the beginning of the 20th century,[10] Poincaré's own proof of the theorem was essentially correct; the finishing touches were added by Carathéodory in 1919.[11]

In his proof Poincaré had to *assume* that the point P remains at a finite distance, *i.e.* that it does not leave a bounded region R having volume V. This restriction would appear to make it of less interest to the theory of stability of the solar system, but quite relevant to the problem of a gas in a finite container, provided the effect of the walls can be described in a non-singular way. But Poincaré did not give any indication in 1889 that he was going to be concerned with the latter application of his theorem.

In a brief paper in 1893 addressed to philosophers, Poincaré discussed the consequences of his theorem for the mechanistic conception of the universe.[12] Mechanism, he says, implies that all phenomena must be reversible, yet experience shows that many irreversible phenomena exist in nature.[13] To escape the contradiction, physicists have postulated "hidden movements": for example, if we did not know that the earth rotates we would regard the motion of the Foucault pendulum as "irreversible" but having discovered that the earth does rotate, we can *imagine* that it might just as well be rotating in the opposite direction. Hence we do not consider this a contradiction of the principle of reversibility.[14]Similarly one might suppose that there are motions in the molecular world which account for macroscopic irreversibility, and which are "in principle" reversible.

Poincaré alluded briefly to Maxwell's demon, and the argument that "the apparent irreversibility of natural phenomena is . . . due to the fact that the molecules are too small and too numerous for our gross senses to deal with them." Yet, while the kinetic theory of gases based on this premise is, according to Poincaré, "up to now the most serious attempt to reconcile mechanism and experience," it still has not overcome the difficulties: his recurrence theorem, which would seem to apply to the entire world if the kinetic theory is valid, contradicts the "heat death" theory. If one attributed absolute validity to the Second Law, then the universe, instead of returning to its initial state, would tend toward a final state of uniform temperature.

One could reconcile the two theories by assuming that the heat death is not permanent but only lasts a very long time, so that the universe, after slumbering for millions of millions of centuries, will eventually reawaken. Then, as Poincaré puts it, "to see heat pass from a cold body to a warm one, it will not be necessary to have the acute vision, the intelligence, and the dexterity of Maxwell's demon; it will suffice to have a little patience."

* * *

In 1896 the mathematician Ernst Zermelo (at that time a student of Max Planck) published a paper in the *Annalen der Physik* in which he claimed that Poincaré's theorem makes it impossible for the mechanical view of nature to explain irreversible processes.[15] While the recurrence paradox applies in the first instance to the kinetic theory of a system of mass-points interacting with conservative forces, Zermelo argued that any other model within the framework of Newtonian mechanics would be subject to the same objections. Thus one must either give up the validity of the Second Law of Thermodynamics or the mechanical theory of nature.

Boltzmann, who had apparently been unaware of Poincaré's earlier publications, replied that while the theorem is correct, it cannot be used as an objection to the molecular interpretation of the Second Law, since (as he had repeatedly emphasized) the validity of the Second Law is only statistical, not absolute.[16] As a result of the earlier discussion of the reversibility objection to his *H*-theorem, Boltzmann had already stated not only that entropy may sometimes decrease, but also that a system may eventually return to its initial state.[17] Thus the recurrence theorem seemed to be completely in harmony with his statistical viewpoint.

Boltzmann could decisively refute the contention that the mechanical viewpoint contradicts *experience* because the term "experience" had been improperly extended by Zermelo (and Poincaré) to include *theoretical predictions* about what will happen to the universe in the remote future. The heat death is not a fact of experience but only an extrapolation from the observation that heat "always" flows from hot to cold. From the kinetic theory Boltzmann could estimate the time needed for an approximate recurrence of the positions and velocities of all the molecules in 1 cc of gas at ordinary density; it is a number so large that it would take trillions of digits even to write it down. Thus the recurrence paradox has nothing to do with the behavior of gases in the laboratory –

> when Zermelo concludes, from the theoretical fact that the initial states in a gas must recur, – without having calculated how long a

§14.7]   THE RECURRENCE PARADOX   633

time this will take–that the hypotheses of gas theory must be rejected or else fundamentally changed, he is just like a dice player who has calculated that the probability of a sequence of 1000 one's is not zero, and then concludes that his dice must be loaded since he has not yet observed such a sequence!

It is interesting to note that Boltzmann is quite willing to jettison the "theory of central forces"–"the hypothesis that all natural phenomena can be explained by means of central forces between mass points"–while keeping the kinetic theory, which depends only on the assumption that the Lagrange equations of motion apply to the molecular collisions with sufficient accuracy for the explanation of thermal phenomena.[18] The difficulties about the equipartition theorem may be in Boltzmann's mind when he says here that "gas theory does not assume that either the properties of the aether or the internal constitution of molecules can be explained by centres of force," or perhaps he is simply indicating his allegiance to the mechanistic Cartesian as opposed to the dynamic Kant-Boscovich tradition in natural philosophy.

In support of his statistical interpretation of the Second Law, Boltzmann cited "famous scientists, such as Helmholtz"[19] and quoted the remark of Gibbs, "The impossibility of an uncompensated decrease of entropy seems to be reduced to an improbability."[20] But his own theory of the H-curve is still only qualitative, and somewhat unsatisfactory from a mathematical viewpoint. Boltzmann asserted that the curve runs along very close to its minimum value most of the time, with occasional peaks corresponding to significant deviations from the equilibrium state. The probability of a peak decreases rapidly as the height of the peak decreases, and if the initial state lies on a very high peak, the state of the system will drop down toward the equilibrium state (minimum value of H) "with enormously large probability, and during an enormously long time it will deviate from it by only vanishingly small amounts." On the other hand if one waits an even longer time the initial state will eventually recur. Yet Boltzmann insisted that for any state with a value of H above the minimum, H is more likely to decrease than increase. No evidence was presented for any of these statements other than the original (qualitative) H-theorem.

In a second paper, Zermelo protested that the properties attributed to the H-curve by Boltzmann are not only unproved, but incompatible with the laws of mechanics; and that probability theory cannot resolve this contradiction.[21] First, the overall periodicity of the system implies that every decrease in H must be balanced by an increase at some other time. Second, the probability of occurrence of a certain value of H should be measured by the volume in phase space of all states having this value; but from the equations of motion it can be shown that this volume is independent of time (this is called "Liouville's theorem" by physicists[22]). Hence there cannot be any tendency for H to increase or decrease. (The same argument was developed in more detail by Gibbs in 1902.[23]) While these objections might apply to an ensemble of systems over a long period of time, Zermelo realized that Boltzmann had based his case for the H-theorem on more specific assumptions about the short-time behavior of the H curve for individual systems, and so he must also attack these assumptions.

If we assume that H has occasional peaks, and we choose the initial state to have a value of $H(=H_0)$ greater than its minimum value, then it would seem that H_0 can just as well lie on a rising as a falling part of the curve and therefore can either increase or decrease. If we assume that the increasing branch occupies a smaller time interval, so that the probability of landing there is smaller, it would still appear that the increase observed when one *does* land there is steeper and thus must be given a correspondingly greater weight. Zermelo interpreted Boltzmann's argument as an attempt to avoid this objection by postulating that H_0 is always at a *maximum* of the H-curve, so that one only observes it to decrease. But Zermelo says he "cannot conceive of such a curve" which consists only of maxima, nor can anyone else. As he says, it would only make sense if the maxima are not mathematical points but flat portions of the curve; but this again contradicts our experience of rapid dissipations of temperature inequalities or other ordered states.

Zermelo also revived the reversibility argument, which he contended makes it impossible ever to derive irreversibility. Any alleged deduction must depend on errors or fallacious assumptions, in particular the "unprovable (because untrue) assumption that the molecular state of a gas is always, in Boltzmann's expression, 'disordered.'" According to Zermelo, only the initial state may be assumed to be disordered; the probability of a later state must depend on the initial state. It is not surprising that when Zermelo refers to the "mechanical view of nature" he still has in mind a deterministic mechanical system in which randomness plays no role in the molecular motions themselves, but only in the observer's description of these motions. What is more remarkable is that Boltzmann, after having introduced the "molecular disorder" postulate, does not challenge this view; as we noted earlier, molecular disorder is for Boltzmann an assumption that may or may not be true—not a postulate.

Boltzmann's reply to Zermelo's second paper is in one sense a reiteration of his earlier arguments, but it is at the same time a retreat from his contention that irreversibility follows *in general* from the kinetic theory.[24] We are concerned, he says, with what will happen in the present state of the world, which happens to be a state of low entropy; therefore we can say that H is a maximum in the initial state without having to claim (as Zermelo suggested) that *all* points of the H-curve are maxima. If, however, we selected a completely arbitrary state of the universe, there are four possibilities. First, and most likely, the state is one of thermal equilibrium, so there will be no significant change of H at all from its minimum value. Second, H is above its minimum, and will "almost immediately" decrease if we follow it either forwards or backwards in time. Third, H is above its minimum, on an increasing branch, so the system passes to more improbable states as one goes forward in time. Fourth, H is above its minimum, on a decreasing branch, so the system passes to more improbable states as one goes backwards in time. The third and fourth cases have equal probability but both are "much rarer" than the second, which is in turn much rarer than the first.

From the description of the second case we can perhaps see why Boltzmann persists in saying that most parts of the H-curve above its minimum are maxima, even though he admits that this cannot be literally true. The key word is "almost" – if the peak is very narrow in time, then even if the initial state is slightly to the left of the actual maximum, one will quickly get over the top and further down the other side in a short time interval.[25] Thus it is a maximum with respect to finite differences but not with respect to infinitesimal differences. Some of the confusion might have been avoided if Boltzmann had stated this more explicitly.

But the second case cannot be taken as the "initial state" in laboratory experiments, for if we look at the value of H in 1896 and follow it backwards in time we expect it to increase not decrease. Thus Boltzmann is really forced into accepting the fourth case as the typical one, which means that the third case is equally likely to be found somewhere, sometime, elsewhere in the universe. He therefore proposes that one should really *define the direction of time* as the direction in which one goes from less to more probable states. This would make the direction of time dependent on the individual observer, and would be different for different parts of the universe at different epochs:

> This viewpoint seems to me to be the only way in which one can understand the validity of the Second Law and the Heat Death of

each individual world without invoking a unidirectional change of the entire universe from a definite initial state to a final state.[26]

It is ironic that Boltzmann has now adopted the viewpoint[27] of another of his critics, Ernst Mach, who in 1894 wrote:

> If we could really determine the entropy of the world it would represent a true, absolute measure of time. In this way is best seen the utter tautology of a statement that the entropy of the world increases with time. Time, and the fact that certain changes take place only in a definite sense, are one and the same thing.[28]

Boltzmann's conception of alternating time directions in the universe, and the idea that the direction of time is determined by human experience, has been revived in recent years by philosophers and cosmologists.[29] But from the viewpoint of this book, it is more significant to note that the proposal was motivated by Boltzmann's desire to push the deterministic (though statistical) mechanical world view to its furthest extreme, perhaps not entirely seriously. He chose *not* to make the alternative assumption that "molecular disorder" is continually maintained by random or external causes acting at the molecular level, as had been suggested by Burbury.[30] While it is true that the statistical interpretation, like Poincaré's deterministic calculation, predicts recurrences, it is not hard to conceive of a postulate of continual or repeated randomization that would enforce irreversibility much more strongly; in fact this is just what Wolfgang Pauli's proof of the quantum-mechanical H-theorem involves.[31] With such a postulate, recurrence is not impossible but neither is it certain. Here again we must stress the distinction between a statistical and a stochastic explanation of the Second Law.[32]

At the same time Boltzmann was willing to speculate in another direction, more in line with the "hidden variables" interpretation which would attribute randomness in molecular motion to determinism at a still lower level:

> Since today it is popular to look forward to the time when our view of nature will have been completely changed, I will mention the possibility that the fundamental equations for the motion of individual molecules will turn out to be only approximate formulas which give average values, resulting according to the probability calculus from the interactions of many independent moving entities forming the surrounding medium—as for example in meteorology the laws are valid only for average values obtained

by long series of observations using the probability calculus. These entities must of course be so numerous and must act so rapidly that the correct average values are attained in millionths of a second.[33]

* * *

I do not want to leave the impression that the hypothesis of randomness provides a logically satisfactory explanation for irreversibility in modern physics. In the context of the kinetic theory of gases and attempts to prove an H-theorem (either classical or quantal), it is not randomness itself but the way it is introduced into the equations that leads to irreversibility; one still has the choice of regarding irreversibility as an inherent property of the world or as a feature of our method of describing the world. The historical importance of the debates at the end of the 19th century was not that they led to a final solution of the problem, but that they popularized among scientists a new set of ideas, some of which were to assist the transition from classical to quantum physics that took place in the following decades.

Notes for §14.7

1. E. Meyerson, *Identity and Reality* (New York: Dover Pubs., 1962, trans. of the French ed. of 1926), ch. VIII. A. Rey, *Le Retour Éternal et la Philosophie de la Physique* (Paris: Flammarion, 1927). P. Sorokin, *Social and Cultural Dynamics* (New York: American Book Co., 1937), II, ch. 10. L. White, *J. Hist. Ideas* 3, 147 (1942). J. Baillie, *The Belief in Progress* (New York: Oxford University Press, 1950), §10. M. Elaide, *The Myth of the Eternal Return* (New York: Pantheon Books, 1954). M. Čapek, *J. Phil.* 57, 289 (1960) (discussion of 1896 MS. of C. S. Peirce). S. L. Jaki, *Science and Creation: From Eternal Cycles to an Oscillating Universe* (New York: Science History Pubs., 1974). For evidence against the usual statement that the Greeks accepted ancient and Oriental cyclic views while Jews and Christians rejected them, see A. D. Momigliano, in *History and the Concept of Time*, ed. G. H. Nadel (Middletown, Conn.: Wesleyan University Press, 1966) [*History and Theory, Beiheft* 6], p. 1. Recent ideas are surveyed in the articles by G. I. Whitrow and others, in *The Study of Time*, eds. J. T. Fraser, F. C. Haber and C. H. Müller (New York: Springer, 1972).
2. E. A. Poe, *Eureka* (1848), reprinted in *Selected Prose, Poetry, and Eureka*, ed. W. H. Auden (New York: Holt, 1950), p. 483. J. Delevsky, in *Studies and Essays in the History of Science and Learning* (New York: Schuman, 1944), p. 375. E. J. Öpik, *The Oscillating Universe* (New York: New American Library, 1960). H. Schmidt, *J. Math. Phys.* 7, 494 (1966). For the current cosmological literature see the compilations prepared by the Astronomisches Rechen-Institut Heidelberg, *Astronomy and Astrophysics Abstracts* (Berlin: Springer, 1969–), 1–, subject category 162.
3. W. A. Kaufman, *Nietzsche: Philosopher, Psychologist, Antichrist* (Princeton: Princeton University Press, 1950), ch. 11. C. Andler, *Nietzsche, sa Vie et sa Pensée*

(Paris: Gallimard, 1958), **4**, Livre 2, Chap. I; Livre 3, Chap. I. O. Becker, *Blättern für Deutsche Philosophie* **9**, 368 (1936), reprinted in his *Dasein und Dawesen: Gesammelte Philosophische Aufsätze* (Pffullingen: Verlag Neske, 1963), p. 41. J. Stambaugh, *Nietzsche's Thought of Eternal Return* (Baltimore: Johns Hopkins Press, 1972. An important influence on Nietzsche may have been the passage on recurrence in J. G. Vogt, *Die Kraft* (Leipzig, 1878), pp. 89–90.
4. H. Poincaré, *Compt. Rend. Acad. Sci. Paris* **108**, 550 (1889). According to G. H. Bryan [*Rept. Brit. Ass. Adv. Sci.* **61**, 106 (1891)], Poincaré's critique is irrelevant because he assumed in effect that the temperature is equal to zero.
5. R. Pfeffer, *Rev. Metaphysics* **19**, 276 (1965).
6. F. Nietzsche, *Der Wille zur Macht*, in his *Gesammelte Werke* (München: Musarion Verlag, 1926), **19**, Book 4, Part 3; English trans. by O. Manthey-Zorn in *Nietzsche, an Anthology of his Works* (New York: Washington Square Press, 1964), p. 90.
7. See also F. Wald, *Die Energie und ihre Entwerthung* (Leipzig, 1889), p. 104. E. Mach, *Die Prinzipien der Wärmelehre* (Leipzig, 1896), p. 362. G. Helm, *Die Lehre von der Energie historisch-kritisch entwickelt* (Leipzig, 1887); *Grundzüge der mathematischen Chemie* (Leipzig, 1898). P. Duhem, *Traité d'Énergetique* (Paris: Gauthier-Villars, 1911). H. Poincaré, *Thermodynamique* (Paris, 1892); *Nature* **45**, 414, 485 (1892).
8. *Acta Math.* **13**, 1 (1890), reprinted in *Oeuvres de Henri Poincaré* (Paris: Gauthier-Villars, 1952). **VII**, 262; English trans. of the section on the recurrence theorem in S. G. Brush, *Kinetic Theory* **2**, 194.
9. S. D. Poisson, *Nouv. Bull. Sci. Soc. Philomath. Paris* **1**, 191 (1808); *Mém. Acad. Roy. Sci. Inst. France* **7**, 199 (1827).
10. A comprehensive account of the history of this subject is given by T. Hawkins, *Lebesgue's Theory of Integration* (Madison: University of Wisconsin Press, 1970).
11. C. Carathéodory, *Sitzungsber. Preuss. Akad. Wiss. Berlin*, 579 (1919); English trans. by S. G. Brush, *On Poincaré's Recurrence Theorem*, UCRL Trans.-871 (L), University of California, Lawrence Radiation Laboratory, Livermore, California. Professor Truesdell has persuaded me that my earlier statement in *Kinetic Theory* **2**, 17, to the effect that Poincaré's proof was not much better than Nietzsche's, is unfair to Poincaré. He has shown me a proof which is essentially the same as Poincaré's but uses the concepts of measure theory (unpublished lecture notes).
12. H. Poincaré, *Rev. Métaphys. Morale* **1**, 534 (1893), English trans. in Brush, *Kinetic Theory* **2**, 203. This paper seems to be unknown to the physicists who have discussed the recurrence paradox. Zermelo, in the paper cited below (note 15) explicitly stated that Poincaré "does not seem to have noticed [his theorem's] applicability to systems of molecules or atoms and thus to the mechanical theory of heat." While one does not expect scientists to read journals on metaphysics and morals, it is surprising that Dugas omits this paper from his comprehensive account and bibliography in *La Théorie Physique au sens de Boltzmann* (Neuchatel: Griffon, 1959).
13. See the paper cited in note 4; this criticism was repeated in his textbook *Thermodynamique* (Paris, 1892), pp. xviii, 414–23. See also Poincaré's exchange with Tait, who criticized the book for ignoring the statistical basis of the Second Law [*Nature* **45**, 245, 414, 485 (1892)]. For Poincaré's later views on the relation between thermodynamics and kinetic theory see *La Valeur de la Science* (Paris: Flammarion, 1904), 180–85; *J. de Physique* [4] **5**, 369 (1906); *Compt. Rend. Acad. Sci. Paris* **143**, 989 (1906). Poincaré's view, that the apparent incompatibility of the principle of irreversibility and the reversibility of atomistic-mechanistic theories tended to undermine the accepted foundations of physics, was shared by some other scientists

at the time; see the remarks of Brunhes cited in note 1, §1, and also his book *La Dégradation de l'Énergie* (Paris, Flammarion, 1922), quatrième partie.
14. Poincaré attributes this example to Helmholtz without giving a specific reference.
15. E. Zermelo, *Ann. Phys.* [3] **57**, 485 (1896), English trans. in Brush, *Kinetic Theory* **2**, 208.
16. L. Boltzmann, *Ann. Phys.* [3] **57**, 773 (1896); *Wiss. Abh.* **III**, 567; English trans. in Brush, *Kinetic Theory* **2**, 218. "Professor G. E. Uhlenbeck once told me that he had heard from Ehrenfest that Boltzmann always referred to Zermelo as 'Dieser Halunke' (rogue or villain)."–J. Mehra, *Physica* **79A**, 468 (1975).
17. See the quotation at the end of §14.6, above.
18. See the translation in Brush, *Kinetic Theory* **2**, 225; Cf. Boltzmann's remarks in *Nature* **51**, 413 (1895) [*Wiss. Abh.* **III**, 535] where he claims that "this simple conception of Boscovich is refuted almost in every branch of science."
19. *Sitzungsber. Akad. Wiss. Berlin* **17**, 172 (1884).
20. See above, §14.5, for the context of Gibbs' remark. Boltzmann cites both the original publication, *Trans. Conn. Acad.* **3**, 229 (1875), and Ostwald's German ed., p. 198. The same quotation appears at the beginning of the second part of Boltzmann's *Vorlesungen über Gastheorie* (Leipzig: Barth, 1898) [English trans., *Lectures on Gas Theory*, p. 215]. According to Erwin Hiebert, both sides in the energetics controversy of 1895–96 tried to claim Gibbs' support [*The Conception of Thermodynamics in the Scientific Thought of Mach and Planck* (Freiburg i. Br.: Ernst-Mach-Institut, 1968), p. 53].
21. E. Zermelo, *Ann. Phys.* [3] **59**, 793 (1896), English trans. in Brush, *Kinetic Theory* **2**, 229.
22. Cf. Brush, *Kinetic Theory* (Oxford and New York: Pergamon Press, 1972), **3**, 59; Zermelo, though a mathematician, seems to have adopted the label from Kirchhoff, *Vorlesungen über die Theorie der Wärme* (Leipzig: Teubner, 1894), p. 144.
23. J. W. Gibbs, *Elementary Principles in Statistical Mechanics* (New York: Scribner, 1902), ch. XII; *Collected Works* **II**, 139–64. See also R. C. Tolman, *The Principles of Statistical Mechanics* (London: Oxford University Press, 1938), pp. 165–79.
24. L. Boltzmann, *Ann. Phys.* [3] **60**, 392 (1897); *Wiss. Abh.* **III**, 579; English trans. in Brush, *Kinetic Theory* **2**, 238.
25. Cf. Boltzmann, *Math. Ann.* **50**, 325 (1898); *Wiss. Abh.* **III**, 629; P. and T. Ehrenfest, *Conceptual Foundations*, p. 34.
26. Quoted from the translation in Brush, *Kinetic Theory* **2**, 242; the statement is repeated with some further discussion in *Gas Theory* **II**, §91.
27. As noted above, the idea was to some extent anticipated by Stoney in 1887 (see note 3, §14.6).
28. E. Mach, *Monist* **5**, 22 (1894), reprinted in his *Popular Scientific Lectures* (LaSalle, Ill.: Open Court, 5th ed., 1943), quotation from p. 178. For further discussion of Mach's views see ch. 8, E. N. Hiebert, *op. cit.* (note 20).
29. W. S. Franklin, *Phys. Rev.* **30**, 766 (1910). A. S. Eddington, *Nature* **127**, 447 (1931). M. Bronstein and L. Landau, *Phys. Z. Sowjetunion* **4**, 114 (1933); English trans. in *Collected Papers of L. D. Landau*, ed. D. ter Haar (Oxford and New York: Pergamon Press, 1965), p. 69. C. F. von Weizsäcker, *Ann. Phys.* [5] **36**, 277 (1939). E. Schrödinger, *Proc. R. Irish Acad.* **53A**, 189 (1950). K. G. Denbigh. *Brit. J. Phil. Sci.* **4**, 183 (1953). H. Reichenbach, *The Direction of Time* (Berkeley: University of California Press, 1956). W. Büchel, *Philosophia Naturalis* **6**, 108 (1960). H. Schmidt, *J. Math. Phys.* **7**, 494 (1966). W. J. Cocke, *Phys. Rev.* [2] **160**, 1165 (1967). H. Zanstra, *Vistas in Astronomy* **10**, 23 (1968). G. K. Berger, *Time and Thermodynamics*

(Dissertation, Columbia University, 1971). Y. P. Terletskii, *Zh. Eksp. Teor. Fiz.* **22**, 506 (1952). A Dauvillier, *Les Hypotheses Cosmogoniques* (Paris: Masson, 1963). K. Popper, in *The Philosophy of Karl Popper*, ed. P. A. Schilpp. (LaSalle, Ill.: Open Court, 1974), pp. 124–30.

30. For further discussion and reformulation of the "molecular disorder" hypothesis see J. H. Jeans, *Phil. Trans.* **196A**, 397 (1901); *Phil. Mag.* [6] **5**, 597 (1903); P. and T. Ehrenfest, *Conceptual Foundations*, pp. 40–42.

31. W. Pauli, in *Probleme der modernen Physik, Arnold Sommerfeld zum 60. Geburtstage gewidmet von seinen Schülern* (Leipzig: Hirzel, 1928). R. C. Tolman. *op. cit.* (note 23), p. 455. D. ter Haar, *Elements of Statistical Mechanics* (New York: Rinehart, 1954), p. 368; *Rev. Mod. Phys.* **27**, 289 (1955).

   In their analysis, P. and T. Ehrenfest stress the need for making the assumption about the number of collisions ("Stosszahlansatz") after every short time interval Δt; see *Conceptual Foundations*, p. 16.

32. Cf. W. Köhler, *Erkenntnis* **2**, 336 (1931).

33. *Lectures on Gas Theory*, §91. Cf. D. Bohm, *Causality and Chance in Modern Physics* (New York: Harper, 1961, reprint of 1957 ed.), pp. 110–13; *Observation and Interpretation in the Philosophy of Physics*, ed. S. Körner (New York: Dover Pubs., 1962, reprint of the 1957 ed.), p. 33.

14.8 Toward quantum theory: Planck's irreversible radiation processes

One of the best-known quotations about the nature of science is Max Planck's remark,

> An important scientific innovation makes its way by gradually winning over and converting its opponents.... What does happen is that its opponents gradually die out, and that the growing generation is familiarized with ideas from the beginning.[1]

I suppose most people who read (or repeat) this quotation think Planck is referring to his quantum theory, but in fact he was talking about his struggle to convince scientists in the 1880's and 1890's that the Second Law of Thermodynamics involves a principle of irreversibility, and that the flow of energy from hot to cold is *not* analogous to the flow of water from a high level to a low one, as Ostwald and the energetists claimed. He goes on to lament that his own efforts were fruitless, but the battle was eventually won because of advances from another direction: the statistical interpretation of entropy based on kinetic theory.

As is well known, Planck's quantum theory was developed with the help of Boltzmann's statistical theory of entropy; but it is only within the last few years that we have been reminded by historians of science such as Martin Klein and Hans Kangro that Planck's work in radiation up to 1900 was done from a completely different viewpoint, and that considerations such as the Rayleigh "ultraviolet catastrophe" (posthumously baptized by Paul Ehrenfest) played no significant role in his thinking during this period. Instead of approaching the problem of the frequency-distribution of black-body radiation by the methods of statistical mechanics (which leads to the difficulty of understanding why the ether does not take its proper share of energy as predicted by the equipartition theorem), Planck was attempting to develop a fundamental macroscopic theory based on thermodynamics and electromagnetic theory.[2] He hoped to establish the principle of irreversibility as part of this theory. We are concerned with Planck's development of radiation theory only insofar as it brought him into conflict with Boltzmann, and thereby led him to believe in a need for postulating randomness in order to explain irreversibility; but we shall not follow any of the subsequent development of quantum theory.

In his *Inauguraldissertation* (1879), Planck introduced a distinction between two kinds of processes: (1) those in which Nature has the same preference [*Vorliebe*] for the final state as for the initial state – such processes he calls *neutral*; (2) those in which Nature prefers the final state to the initial one – such are *natural* processes.[3] The neutral processes include "reversible" processes but also others, such as the motion of a freely falling body. The neutral processes are not associated with entropy change; but entropy increases for natural processes. The former are only "ideal" since all actual processes occurring in nature are attended by heat conduction or friction or percussion, which are *natural* in the above sense; thus it appears that nature proceeds toward a certain goal, namely to maximize the total entropy, as Clausius has indicated in his statement of the Second Law.[4]

Planck's hostility to atomistic theories was evident as early as 1882 in a paper on evaporation, melting, and sublimation.[5] He emphasized that his results are independent of any molecular hypothesis, and argued that one should go as far as possible with thermodynamics before introducing assumptions about the interior constitution of bodies. At the end of the paper he wrote:

> The second law of thermodynamics, logically developed, is incompatible with the assumption of finite atoms.[6] Hence it is to be expected that in the course of the further development of the theory, there will be a battle between these two hypotheses, which will cost one of them its life. It would be premature to predict the result of this battle with certainty; yet there seem to be at present

many kinds of indications that in spite of the great successes of atomic theory up to now, it will finally have to be given up and one will have to decide in favor of the assumption of a continuous matter.

The following year, in a paper on the thermodynamic equilibrium of gas mixtures, Planck showed that Dalton's law of partial pressures, which Maxwell and Stefan had deduced from kinetic theory, can be derived without the help of that theory.[7] In his conviction that the Second Law could be used as a research tool, and that the kinetic-atomic theory of matter was an erroneous or at best superfluous hypothesis, Planck might have seemed a likely recruit to the Ostwald–Duhem school of "Energetics," but in fact he eventually refused to follow that path.[8]

In a substantial three-part series, "Ueber das Princip der Verhaltung der Entropie" (1887), Planck announced at the beginning that he would abstain from special assumptions about the nature of molecular motions. He was interested only in developing methods for calculating the entropy function, which would determine what reactions actually occur, and in illustrating the fruitful applications of the Second Law to physical chemistry.[9]

It is surely not without significance that Planck served as co-editor of the third volume of the second edition of Clausius' treatise, *Die Mechanische Wärmetheorie*, dealing with the kinetic theory of gases, following Clausius's death in 1888.[10] This would have been a forceful reminder that the physicist whom Planck respected for his work in founding thermodynamics had also grappled with the problem of finding a molecular interpretation of the thermal properties of bodies.

When Planck became interested in the theory of solutions, he seemed quite willing to discuss molecular hypotheses and the analogy between solute molecules and gas molecules.[11] Yet the following year, at the meeting of German scientists at Halle (1891), he still insisted that kinetic theory was of little use, since this analogy had been discovered independently of kinetic theory and could not be explained or further developed with the aid of that theory.[12] Planck and Ostwald joined in defending the macroscopic thermodynamic viewpoint in a discussion with Boltzmann at this meeting.[13]

Our earlier discussion of Maxwell's ideas suggested that a crucial step in the development of an atomistic interpretation of irreversibility was the recognition that mixing of different kinds of molecules, rather than flow of heat, is the fundamental irreversible process. With this in mind we are interested to see that in 1892 Planck criticized "English physicists" for describing the Second Law too narrowly in terms of "dissipation of energy," pointing out that processes such as the interdiffusion of two ideal gases involve no dissipation of energy but rather a "dissipation of matter." This is for Planck another instance of the need for formulating the Second Law in terms of entropy rather than deriving it merely from energy considerations as Ostwald and his Energetics group wished to do.[14] As Erwin Hiebert notes, this paper is soon followed by a more sympathetic view of atomism, and prepares the way for an open attack on Energetics.[15]

By this time Planck had become involved in another editing task touching on the atomistic interpretation of thermodynamics. In 1889, following the death of Gustav Robert Kirchhoff, Planck was called to Berlin to occupy Kirchhoff's chair. He was the logical person to edit Kirchhoff's lectures on heat for publication as a volume in the *Vorlesungen über Mathematische Physik*. As it happened Kirchhoff, though not personally very enthusiastic about the kinetic theory of gases, had felt obliged to include a thorough treatment of this subject in his lectures, and as a result Planck was forced to become somewhat familiar with it.

Shortly after the publication of Kirchhoff's lectures on heat, edited by Planck, Boltzmann criticized the derivation of the collision integral given in the book, implying that it contained an error that might be the fault of the editor.[16] In particular, Kirchhoff seemed to be assuming that the probability of collision of two molecules could be calculated as if their coordinates before the collision were statistically independent, despite the fact that the variables have been defined in such a way that the molecules must have previously collided. This is almost exactly the same as Culverwell's reversibility objection against the proof of the H-theorem, and seems to require an additional postulate of "molecular disorder" for its justification. Planck recognized this problem in his reply, and argued that the same objection applied to any proof, not only Kirchhoff's. While as editor he had not felt it proper to criticize the validity of Kirchhoff's derivation, limiting himself to reproducing it accurately from the manuscript, he now suggested that the only way to avoid the difficulty was to assume that the Maxwell distribution already is established, since it is only for this distribution that the probability that two molecules separate with certain velocities before the collision. Thus the proof of the Maxwell distribution for thermal equilibrium might be based on the fact that it is the only one that satisfies the reversibility criterion.[17]

While Boltzmann did not accept Planck's conclusion,[18] the exchange of views reinforced the impact of the debate with the English physicists in the same period (1894–95) and must therefore be considered as one of the contributing factors leading to Boltzmann's "molecular disorder" assumption.

* * *

Another consequence of Planck's move to Berlin was his contact with the experimental work being done there on black-body radiation by Otto Lummer and E. Pringsheim at the Physikalisch-Technischen Reichsanstalt.[19] Planck's interest in demonstrating the manifold applications of thermodynamics, combined with this stimulus, led him to write a series of papers "Über irreversible Strahlungsvorgänge" in 1897–1900, culminating in the discovery of the quantum theory of radiation. These papers are relevant to our subject primarily because they led Planck to realize (as a result of Boltzmann's criticism) that the principle of irreversibility could not be derived, as he had originally thought, from electromagnetic theory alone, but required an additional postulate of randomness.[20]

In the preface to his *Vorlesungen über Thermodynamik*, dated April 1897, Planck hinted that the Second Law might ultimately have to be based on electromagnetism, rather than on the kinetic theory.[21] He began the series on radiation with the statement that irreversible processes cannot be explained satisfactorily by the kinetic theory, assuming point-molecules interacting with conservative forces, because of the recurrence objection of Zermelo.[22] But a resonator which can absorb and emit electromagnetic radiation can introduce irreversibility even though Maxwell's equations themselves indicate that radiation in empty space or reflected from smooth walls behaves reversibly.[23]

Such a suggestion could hardly pass unchallenged by Boltzmann, who in addition to being the chief defender of kinetic theory was one of the leading authorities on Maxwell's electromagnetic theory and its connection with thermodynamics. He immediately pointed out that any process of interaction between resonators and electromagnetic waves must be described by reversible equations, and that the apparent irreversibility in the process invoked by Planck was due only to the choice of special initial conditions. Just as in the case of a sphere fixed in space, bombarded by smaller spheres: if the latter are initially moving in parallel paths, they will be scattered in all directions; but if one then reverses their motions, the entire process will run backwards.

Aside from the possibility of Joule heating there is no essential difference between purely mechanical and electrical processes. Boltzmann claimed to have satisfactorily answered the objections of Loschmidt, Culverwell, Poincaré, and Zermelo, thereby reinstating his molecular explanation of irreversibility, and argued that any explanation of irreversibility based on electromagnetic processes must rely on exactly the same kind of assumptions.[24]

In his second paper in this series, Planck stated that the reversibility theorem does not apply to the particular process he had in mind, since he assumed "that the intensity of the primary exciting wave at the location of the resonator (always assumed to be infinitely small) has at all times finite and continuous values." But the secondary spherical waves emitted by the resonator must necessarily have an unlimited intensity in the neighborhood of the resonator, if the resonator is infinitesimal in size. If the process is reversed, the primary wave (now a spherical wave converging on the resonator) no longer fulfills the condition. Boltzmann's objection therefore applies only to a singular case explicitly excluded in the theory.[25]

Since no process has yet been found in nature in which irreversible changes are produced by the action of conservative forces, according to Planck, it is important to investigate the laws of radiation which seem to offer such a possibility. There is a hint that the electromagnetic world view, currently being advocated by Lorentz and other physicists, may thereby gain an advantage over the mechanical world view.[26]

Boltzmann was of course unable to accept the implication that his own attempts to explain irreversibility had failed; nor could he accept the special assumption about the resonator which Planck used to avoid Boltzmann's earlier objection. Boltzmann noted that if one adopted the physically-unrealistic assumption that the resonator has infinitesimal size, he would still have to postulate that it is surrounded by a region of very strong electric vibrations in order to get any scattering of the incoming wave at all; and then the process would still be reversible.[27]

As for the recurrence paradox against kinetic theory, Boltzmann insisted that this applies only to systems of a finite number of molecules, and that it is reasonable to expect agreement with the Second Law only in the limiting case of an infinite number of molecules, when the recurrence time also becomes infinite. The same relation between the finite and the infinite situation should hold if one replaces the differential equations of electromagnetism by finite-difference equations. Conversely, since the equations of electromagnetic theory have been derived from a purely mechanical model (even if an artificial one), it follows that if electric or even acoustic resonators can give rise to irreversible processes, then one would have a contradiction of Poincaré's statement[28] that irreversible processes cannot in principle be derived from the differential equations of pure mechanics.

Boltzmann suggested that just as in gas theory, one can determine for radiation a "most probable state" or rather a general formula that includes all states in which the waves are not ordered but "run through each other" in all possible ways. This state would be expected to evolve in a space filled with resonators of sufficient multiplicity. It would happen only in relatively few cases that a disordered state changes back into an ordered one. But just as in the case of kinetic theory, one cannot prove that the latter process is impossible. Moreover, if one replaces the differential equations by finite difference equations (thinking of the ether as a large but finite number of vector atoms) then the recurrence theorem would also apply here.

In his next paper, without explicitly mentioning Boltzmann's criticism, Planck proposed to exclude those radiation processes which he called "synchronized with the system" ["auf das System abgestimmt"]–namely, those for which the intensity associated with one or more Fourier components of the wave is comparable to the total intensity.[29] Such waves would have regularly-recurring gaps. The other (and more general) kind of wave would be associated with irreversible processes; and the distinction is clearly between "ordered" and "disordered" waves. Planck claimed to prove that his disordered waves cannot be reversed.[30] He admitted the possibility that the phase constants of the partial waves have values such that an initially disordered radiation process will appear to become ordered at a later time; whether such processes occur in nature or not depends on the conditions satisfied by the initial state.[31]

Boltzmann still was not satisfied, and claimed that Planck's proof that his disordered waves could not be reversed was incorrect.[32] Planck conceded an error in this proof, and proposed yet a further refinement of his theory, in which he introduced the concept of "natural" radiation and postulated that it was the only kind found in nature.[33] His definition of "natural" radiation depended somewhat on the details of his description of the resonator and the waves which it emits and absorbs, but, roughly speaking, may be seen as a generalization of his earlier concept of radiation processes "not synchronized with the system."[34]

Planck could then show that if the radiation is *always* "natural," a quantity analogous to entropy (defined in terms of the logarithm of the energy-density for each frequency) always decreases. The proof depended on the symmetry between ingoing and outgoing waves interacting with the resonator, *i.e.* on the same kind of *microscopic reversibility* assumption that underlies the proof of Boltzmann's *H*-theorem. Planck was thus forced to recognize that in a deterministic system in which the radiation does not remain for an indefinitely long time (*e.g.* for a system of waves interacting with a resonator in a closed space) the property of irreversibility cannot be established without additional assumptions. Hence, he wrote, "this indeterminacy lies in the nature of the subject"–not perhaps in nature itself, but at least in any rational theory we can construct about nature.[35] In such a theory, "If a resonator is at any time stimulated by natural radiation of variable intensity, then the occurrence of the inverse process is absolutely excluded for all later times, provided that the exciting wave retains the properties of natural radiation."[36] But the door is still left open to a future theory in which more detailed information about the Fourier components of the radiation, and the response of the resonator to these components, might be used to derive irreversibility.

By 1899, when he published the fifth paper in this series on irreversible radiation processes, Planck had molded his theory even more closely along the lines of Boltzmann's kinetic theory. He defined the entropy of a resonator with vibration frequency ν and energy U as

$$S = -(U/a\nu) \log (U/eb\nu)$$

and described the basic interaction between the incoming wave and a resonator in terms of two intensities rather than one–distinguishing between two directions of polarization of the incoming and outgoing waves. Thus the formula for entropy contained four terms, just like the Boltzmann *H*-function which has terms for the two colliding molecules before and after the collision; and the proof that $dS/dt \geq 0$ turns on exactly the same property of logarithmic expressions of the form $\alpha \log \alpha + \beta \log \beta - \gamma \log \gamma - \delta \log \delta$.[37] The same kind of mathematics leads to exponential formulae for the radiation intensity in the stationary state, and to Wien's formula for the energy distribution over frequencies.

In a lecture summarizing his theory at the Naturforscherversammlung in Munich later in 1899, Planck explained his hypothesis of natural radiation in somewhat clearer fashion. He appealed to the fact that it is not possible to find an absolutely sharp line in the spectrum, *i.e.* there is no such thing as absolutely monochromatic radiation.

Rather, even the most homogeneous ray is spread over a finite region of frequencies. But this "indeterminacy" [Unbestimmtheit] implies that for example in the visible spectrum an interval whose endpoints have frequencies in the ratio 1:1000001 would correspond to all vibration numbers between 510 billion and 510000510000000, *i.e.* a range of 510 million different frequency numbers. Thus in the Fourier decomposition of the ray we have to deal with 510 million "unknown quantities." Moreover it is in principle impossible to determine these components experimentally, since the terms in the Fourier series depend on the choice of a basic period. On the other hand, one can hardly believe that these details affect the measurable physical properties of the radiation. Hence one must add to Maxwell's theory a new hypothesis, based on the concept of "natural radiation." This hypothesis states that the energy of the radiation is distributed completely *irregularly* [*unregelmässig*, italics in original] over the partial vibrations. It is this assumption that leads to the irreversibility of the Second Law of Thermodynamics. Planck now recognizes that this assumption of "natural radiation" is precisely analogous to Boltzmann's assumption of "molecular disorder" in the kinetic theory, and that the latter must also be added as a special hypothesis rather than derived from the original model of atoms bouncing around in a container with perfectly reflecting walls. (He adds that by modifying the assumption of perfectly reflecting walls, which seems physically unrealistic in any case, one could probably get over this difficulty and derive irreversibility.[38])

In view of the fact that Planck soon afterwards developed his quantum theory of radiation with the help of Boltzmann's statistical theory of entropy, one might be tempted to suggest that the concept of randomness was derived from Boltzmann by Planck and thence passed into the world view of modern physics.[39] That things are not quite so simple is indicated by some remarks in Planck's 1932 lecture, "Causality in Nature":

> The determinists... look for a rule behind every irregularity, and it is their task to formulate a theory of the laws of gases on the assumption that the collision between any two molecules is causally determined. The solution of this problem was the lifework of the great physicist, Ludwig Boltzmann, and it is one of the finest triumphs of theoretical investigation.... The new world image of quantum physics is due to the desire to carry through a rigid determinism in which there is room for quanta. For this

purpose the material point which had hitherto been a fundamental part of the world image had to lose this supremacy. It has been analyzed into a system of material waves, and these material waves are the elements of the new world image.... It is an essential fact, however, that the magnitude which is characteristic for the material waves is the wave function, by means of which the initial conditions and the final conditions are completely determined for all times and places.... We see then that there is fully as rigid a determinism in the world image of quantum physics as in that of classical physics.[40]

* * *

We have now followed the discussion up to the eve of the invention of quantum theory. After 1900 we encounter a rapidly changing situation, in which many new factors appear: the phenomenon of radioactivity, in which an atom seems to explode at a time that is absolutely unpredictable on an individual basis though subject to statistical regularity when many atoms are involved; the Einstein–Smoluchowski theory of Brownian movement, invoking observable statistical fluctuations of just the type that Boltzmann and his opponents had assumed would be extremely unlikely to occur during the lifetime of a human observer; and of course the development of quantum theory itself, leading to what some would call indeterministic laws of subatomic behavior.[41] The place to survey 20th-century ideas about randomness and irreversibility is certainly not this chapter, which attempts to cover (already at too great a length) only the role of such ideas in the classical kinetic theory of gases.[42] So we must conclude by mentioning only a handful of post-1900 contributions which do not seem to have been substantially affected by the above-mentioned developments and thus still pertain to the 19th-century universe of discourse.

Boltzmann,[43] in his last major publication on kinetic theory (a survey written jointly with J. Nabl), started by stating the assumption that the smallest particles of bodies are in continuous irregular motion [steter unregelmässiger Bewegung].[44] But the word "irregular" cannot perhaps be taken too literally in view of his philosophical conviction, expressed a year earlier in a lecture at the St. Louis Congress, that "the regularity of the phenomena is the fundamental condition for all cognition."[45] Nevertheless, in order to prove the H-theorem Boltzmann must assume that the state of the gas is molecular-disordered not only initially but also remains so throughout the course of time; he knows that the latter does not necessarily follow from the former, though he does not put as much stress as did Planck on the fact that an extra assumption is needed.[46] The fact that Boltzmann still retains his concept of alternating time directions in the universe may indeed be an indication that he is willing to accept a world in which molecular disorder does not always prevail, but, on the contrary, recurrences of ordered states are to be expected.[47]

By this time Gibbs' *Statistical Mechanics*[48] had appeared and was beginning to attract attention among those scientists concerned with fundamental or mathematical aspects of gas theory. In his discussion of what later became known as the generalized H-theorem (or "Gibbs H-theorem"), Gibbs ascribed the approach of an ensemble to equilibrium not to any element of randomness in molecular behavior, but to the fact that the flow in phase space (determined by the equations of motion) produces a mixing which progressively deprives a macroscopic observer of information about the system. He suggested the analogy of a colored liquid or dye, initially separate from a body of water, which is then allowed to mix as the water is stirred. Assuming conservation of the amount of colored liquid and incompressibility (as well as mutual insolubility) of both liquids, one sees that the "average density" of colored liquid remains constant, as does the mean-square density (which one might expect to provide a measure of the deviation from completely uniform mixing), *if* one defines density for sufficiently small spatial elements. On the other hand if one fixes the size of the spatial element and continues the mixing indefinitely long, then the mean-square density does decrease to a minimum.[49] In order to sharpen this distinction the Ehrenfests introduced the terms "fine-grained density" and "coarse-grained density" for these two conceptions.[50] One may then say that Boltzmann's H-function defined in terms of fine-grained density (which represents the "real behavior" of the systems on the microscopic level, as seen by a Maxwell demon) is a constant of the motion; but the H-function defined in terms of the "coarse-grained density" (which is more like what a macroscopic observer could actually measure) does decrease as a result of the mixing process.[51] Of course this makes irreversibility an attribute of the interaction between nature and the observer, rather than an intrinsic property of nature itself–as indeed Maxwell had observed in his remarks quoted in §14.4.[52] In the same way we could complain that the word "random" is often used to characterize a limitation on our knowledge of nature rather than a property of nature itself.[53] These would be considered irrelevant objections by those who accept the

"Copenhagen" philosophy, in which the observer is no longer, even in principle, considered separate from what he is observing.

It was S. H. Burbury who again emphasized in 1903 the need for assuming randomness in order to derive irreversibility from the classical kinetic theory, and insisted that this is what Boltzmann's "molecular disorder" hypothesis must mean. Commenting on J. H. Jeans' reformulation of the problem in the language of Gibbsian statistical mechanics, Burbury agreed with Jeans that any such assumption is "mathematically impossible if the motion is continuous, that is, if the state of the system at any instant is a necessary consequence of its past history."[54] It could only be made for a system "which is continually receiving disturbances at haphazard, which in fact takes a fresh start from chaos at every instant. It is fair to say that in nature disturbances are very frequently taking place. The isolated system, with its $6N$ variables left to its own forces, hardly exists in practice."[56]

Jeans put the opposite interpretation on the same conclusion: the assumption favored by Burbury, he said,

> merely amounts to a licence to misapply the calculus of probabilities. It is, if I was right, as illogical to base a kinetic theory on this assumption, coupled with the laws of dynamics, as it would be to base a system of dynamics on the assumption that there is no causation in nature, coupling this assumption with the fundamental laws of dynamics.[56]

Just as illogical – but just as logical, perhaps, as what actually happened a few years later to the foundations of microscopic physics.

Notes for §14.8

1. M. Planck, *A Scientific Autobiography and other Papers* (London: Williams & Norgate, 1950), pp. 33–34; *Philosophy of Physics* (New York: Norton, 1936), p. 97. On one occasion the quotation has been attributed to Keynes – see the query in the *New York Times Book Review*, 31 August 1969, p. 23 and replies in the issue of October 5.
2. M. J. Klein, *Arch. Hist. Exact Sci.* 1, 459 (1962); H. Kangro, *Vorgeschichte des Planckschen Strahlungsgesetzes* (Wiesbaden: Steiner, 1970).
3. M. Planck, *Über den zweiten Hauptsatz der mechanischen Wärmetheorie* (München, 1879), reprinted in Planck's *Physikalische Abhandlungen und Vorträge* (Braunschweig: Vieweg, 1958), I, 1. For a general survey of Planck's views on thermodynamics see E. N. Hiebert, *The conception of thermodynamics in the scientific thought of Mach and Planck* (Freiburg: Ernst-Mach-Institut, 1968). An

abbreviated version of this report has been published in *Perspective in the History of Science and Technology*, ed. D. H. D. Roller (Norman: University of Oklahoma Press, 1971), p. 67.
4. Planck, *Phys. Abh.* **I**, 42. Cf. E. Bauer's assertion [*Ann. Chim.* [8] **29**, 377 (1913)] that the word "irreversible" was *invented* for the case of luminescence.
5. Planck, *Ann. Phys.* [3] **15**, 446 (1882); *Phys. Abh.* **I**, 134.
6. A footnote to the first sentence refers to Maxwell's discussion of his "demon" in *Theory of Heat* (1871).
7. Planck, *Ann. Phys.* [3] **19**, 358 (1883); *Phys. Abh.* **I**, 164.
8. See his recollections recorded in the "Scientific Autobiography" cited above in note 1, and Hiebert's account, *op. cit.*, pp. 41–50, 55–64.
9. Planck, *Ann. Phys.* [3] **30**, 562, **31**, 189, **32**, 462 (1887); *Phys. Abh.* **I**, 196, 217, 232.
10. *Die Mechanische Wärmetheorie* von R. Clausius, zweite umgearbeitete und vervollständigte Auflage des unter dem Titel „Abhandlungen über die mechanische Wärmetheorie" erschienenen Buches. Dritter Band. Entwickelung der besonderen Vorstellungen von der Natur der Wärme als einer Art der Bewegung. Herausgegeben von Dr. Max Planck und Dr. Carl Pulfrich (Braunschweig: Vieweg, 1889–1891). The second title page carries the title *Die Kinetische Theorie der Gase*.
11. Planck, *Ann. Phys.* [3] **39**, 161 (1890); *Phys. Abh.* **I**, 330 (see esp. p. 342).
12. Planck, *Z. phys. Chem.* **8**, 372 (1891); *Phys. Abh.* **I**, 372.
13. W. Ostwald, *Lebenslinien, Eine Selbstbiographie*, Zweiter Teil, Leipzig, *1887–1905* (Berlin 1927), pp. 487–88, quoted by Hiebert, *op. cit.* (note 3), pp. 33–34.
14. Planck, *Ann. Phys.* [3] **46**, 162 (1892); *Phys. Abh.* **I**, 426; Hiebert, *op. cit.*, pp. 41–42. See also Planck's next paper on the Second Law, *Z. phys. chem. Un.* **6**, 217 (1893); *Phys. Abh.* **I**, 437, in which he insists on the importance of irreversible processes in which there is no change of temperature (Hiebert, p. 45).
15. Hiebert, *op. cit.*, p. 46.
16. L. Boltzmann, *Sitzungsber. k. Bayer. Akad. Wiss. München* **24** (3), 207 (1894); *Ann. Phys.* [3] **53**, 955 (1894); Boltzmann's *Wiss. Abh.* **III**, 528.
17. Planck, *Sitzungsber. k. Bayer. Akad. Wiss. München* **24** (4), 391 (1895); *Ann. Phys.* [3] **55**, 220 (1895); Planck's *Phys. Abh.* **I**, 442.
18. Boltzmann, *Ann. Phys.* [3] **55**, 223 (1895); *Sitzungsber. k. Bayer. Akad. Wiss. München* **25**, 25 (1896); *Wiss. Abh.* **III**, 532. See also *Gas Theory* I, §6; II, §92.
19. See the comprehensive discussion in Kangro, *op. cit.* (note 2).
20. Klein, *op. cit.* (note 2); *Natural Philosopher* **1**, 83 (1963); Kangro, *op. cit.*, pp. 133–34. For Planck's own version of the exchange see his *Nobel-Vortrag* (1920), in *Phys. Abh.* **III**, 121; English trans. reprinted in *A Survey of Physical Theory* (New York: Dover Pubs., 1960), p. 102.
21. In this preface, he distinguished three approaches to the development of the theory of heat. The first, the kinetic theory, "penetrates deepest into the nature of the processes considered, and, were it possible to carry it out exactly, would be characterized as the most perfect." But "Obstacles, at present unsurmountable, however, seem to stand in the way of its further progress. These are due not only to the highly complicated mathematical treatment, but principally to essential difficulties, not to be discussed here, in the mechanical interpretation of the fundamental principles of thermodynamics." The second method is that of Helmholtz, based on the general principle that heat is due to motion but refusing to make special hypotheses as to the nature of this motion; Planck feels that this viewpoint, while "safer" than the first, "does not as yet offer a foundation of sufficient breadth." The

§14.8]   TOWARDS QUANTUM THEORY   653

third method, which he calls the most fruitful so far, proceeds directly from empirical facts and deduces physical and chemical laws of extensive application; while it is the best one available (and is to be used exclusively in this book) it "cannot be considered as final . . . but may have in time to yield to a mechanical, or perhaps an electro-magnetic theory." *Treatise on Thermodynamics*, reprint of the 3rd ed. (1926), English trans. by A. Ogg from the 7th German ed. (1922) (New York: Dover Pubs., n.d.), p. viii. See Hiebert, *op. cit.*, pp. 66–67 for an extensive quotation from the original German version.

Lest someone be tempted to assume that Planck refers, in this quotation, to the Boltzmann–Zermelo debate of 1896–97, it should be noted that he used almost the same words to characterize the status of kinetic theory in his introduction to Clausius' posthumous treatise on kinetic theory in 1889 (see p. viii of the book cited in note 10).

22. For Planck's views on the Zermelo–Boltzmann debate see his letter to Leo Graetz, 23 May 1897, quoted in Kangro, *op. cit.*, p. 131.
23. Planck, *Sitzungsber. Preuss. akad. Wiss. Berlin*, p. 57 (1897); *Phys. Abh.* I, 493.
24. Boltzmann, *Sitzungsber. Preuss. Akad. Wiss. Berlin*, 660 (1897); *Wiss. Abh.* III, 614.
25. Planck, *Sitzungsber. Preuss. Akad. Wiss. Berlin* 715 (1897); *Phys. Abh.* I, 505.
26. Planck, *Sitzungsber. Preuss. Akad. Wiss. Berlin*, 641 (1894); *Phys. Abh.* III, 1. Cf. R. McCormmach, *Isis* **61**, 459 (1970); S. Goldberg, *Arch. Hist. Exact Sci.* **7**, 7 (1970), §IIa.
27. Boltzmann, *Sitzungsber. Preuss. Akad. Wiss. Berlin* 1016 (1897); *Wiss. Abh.* III, 618.
28. Boltzmann does not give a specific citation for this opinion of Poincaré; for the probable source see the publications cited in notes 4 and 13, §14.7.
29. Planck, *Sitzungsber. Preus. Akad. Wiss. Berlin* 1122 (1897); *Phys. Abh.* I, 508 (see esp. p. 518).
30. *Ibid.* pp. 524–25.
31. *Ibid.* p. 531.
32. Boltzmann, *Sitzungsber. Preuss. Akad. Wiss. Berlin* 182 (1898); *Wiss. Abh.* III, 622.
33. Planck, *Sitzungsber. Preuss. Akad. Wiss. Berlin* 449 (1898); *Phys. Abh.* I, 532.
34. See Planck's *Phys. Abh.* I, 550–52.
35. "Diese Unbestimmtheit liegt übrigens in der Natur der Sache" (*Ibid.*, p. 557).
36. "Wenn ein Resonator zu irgend einer Zeit durch natürliche Strahlung von veränderlicher Intensität erregt wird, so ist der Eintritt des umgekehrten Vorgangs für all späteren Zeiten absolut ausgeschlossen, so lange die erregende Welle der Eigenschaften der natürlichen Strahlung behält," (*Ibid.*, p. 559).
37. Planck, *Sitzungsber. Preuss. Akad. Wiss. Berlin* 440 (1899); *Phys. Abh.* I, 560 (see esp. pp. 585, 588 and compare Boltzmann's derivation as indicated above, §14.5, note 11).
38. M. Planck, *Ann. Phys.* [4] **1**, 69 (1900); *Phys. Abh.* I, 614. *The Theory of Heat Radiation*, trans. from German ed. of 1913 (New York: Dover Pubs., 1959), pp. 116–17.
39. M. J. Klein, *Arch. Hist. Exact Sci.* **1**, 459 (1962); *Natural Philosopher* **1**, 83 (1963). L. Rosenfeld, in *Max-Planck-Festschrift* 1958, hrsg. B. Kockel *et al.* (Berlin: VEB Deutscher Verlag der Wissenschaften, 1959), p. 203.
40. M. Planck, *The Philosophy of Physics* (New York: Norton, 1936, reprinted 1963), pp. 58, 64–65, trans. from *Der Kausalbegriff in der Physik* (Leipzig: Barth, 1932). See also Planck's lecture in *Naturwissenschaften* **14**, 249 (1926); *Phys. Abh.* III, 159.
41. M. J. Klein, *Hist. Stud. Phys. Sci.* **2**, 1 (1970). M. Jammer, *The Conceptual*

Development of Quantum Mechanics (New York: McGraw-Hill, 1966), pp. 281–293, 323–345.
  On Brownian movement, see ch. 15. The Ehrenfests' article, for example, which contains a fascinating (though sometimes historically misleading) discussion of Boltzmann's work, introduces ideas about the "determinancy of visible states" (pp. 36–37) based on Brownian movement, and thus belongs really to a later period than the one with which we are concerned.
42. Cf. E. Cassirer, book cited in note 7, §14.4; A. M. Bork, *Antioch Review* **27**, 40 (1967); P. Forman, *Hist. Stud. Phys. Sci.* **3**, 1 (1971); J. C. Greene, *Proc. Am. Phil. Soc.* **103**, 716 (1959); S. G. Brush, *J. Hist. Ideas* (in press).
43. According to Lise Meitner, Boltzmann in his Vienna lectures from 1902 to 1906 never mentioned Planck's quantum theory or Einstein's theory of Brownian movement. *Advancement of Science* **20**, (99), 39 (1964); *Bull. Atomic Sci.*, Nov. 1964, p. 2.
44. L. Boltzmann and J. Nabl, *Enc. math. Wiss.* V, (I), 493 (1905).
45. L. Boltzmann, *Congress of Arts and Sciences Universal Exposition, St Louis, 1904* ed. H. J. Rogers (Boston: Houghton, Mifflin & Co., 1905), I, 591 (quotation from p. 598); the original German text was published in Boltzmann's *Populäre Schriften* (Leipzig: Barth, 1905), p. 345.
46. Boltzmann and Nabl, *op. cit.*, p. 513.
47. *Ibid.* pp. 521–22.
48. J. W. Gibbs, *Elementary Principles in Statistical Mechanics* (New York: Scribner, 1902), reprinted in *The Collected Works of J. Willard Gibbs* (New York: Dover Pubs., 1960), II.
49. Gibbs, *Works* II, 144–51. Cf. E. Zermelo, *Jahresber. D. Math. Ver.* **15**, 232 (1906) for discussion of Gibbs' theory of irreversibility.
50. P. and T. Ehrenfest, *Conceptual Foundations*, p. 52.
51. A clear explanation may be found in Tolman, *Principles of Statistical Mechanics*, pp. 165–79.
52. "... der zweite Hauptsatz nur in bezug auf die Unvollkommenheit unserer technischen Mittel definiert ist"–M. V. Smoluchowski, *Festschrift Ludwig Boltzmann* (Leipzig: Barth, 1904), p. 626, reprinted in *Pisma Marjana Smoluchowskiego* (Krakowie: Drukarnie Uniwersytetu Jagiellonskiego, 1924), I, 421 (quotations from p. 426).
53. "We speak of chance in nature, when small variations in the initial data occasion considerable variations in the final elements, because we cannot observe those small variations"–A. Pannekoek, *Proc. Sect. Sci. K. Akad. Wet. Amsterdam* **6**, 42 (1903) (quotation from p. 48). According to J. D. van der Waals, Jr., adoption of the statistical view does not require us to abandon determinism or the mechanical view of nature. *Phys. Z.* **4**, 508 (1903).
54. S. H. Burbury, *Phil. Mag.*[6] **6**, 529 (1903). The same point was made by W. F. Magie: any state of the system is ordered in the sense that it is determined by the initial state, *Science* **23**, 161 (1906).
55. Burbury, *op. cit.*
56. J. H. Jeans, *Phil. Mag.* [6] **6**, 720 (1903). Cf. his discussion in *The Dynamical Theory of Gases* (Cambridge: Cambridge University Press, 1940), **14**, 32, 49–58.

CHAPTER 15

Brownian Movement *

Modern science differs from Newtonian science in its emphasis on processes. Psychologists study the development of personality from infancy through adolescence rather than merely analyzing the faculties of the adult human; biologists deal with evolution and life cycles of organisms rather than concentrating on the classification of species and descriptive anatomy; chemists and physicists are interested more in the reactions of molecules, atoms, and particles than in the static properties of substances; and astronomers have turned away from mapping the heavens and analyzing the cyclic motions of planets to speculating about the evolution of stars and of the universe. The new concerns of science have also stimulated new developments in mathematics, notably the theory of random ("stochastic") processes.

In this chapter we examine the history of observations and theories of Brownian movement. Starting with Robert Brown–who was apparently the first to recognize, in 1828, that the irregular movements of all kinds of small particles in fluids have a physical rather than a biological cause–we review various unsuccessful attempts to explain these movements. This survey is a prelude to a more detailed discussion of the theories of Einstein and Smoluchowski.

It is surprising that Brownian movement played almost no role in physics until 1905, and was generally ignored even by the physicists

---

* Reprinted with minor revisions from *Arch. Hist. Exact Sci.* 5, 1 (1968) by permission of Springer-Verlag.

who developed the kinetic theory of gases, though it is now frequently remarked that Brownian movement is the best illustration of the existence of random molecular motions. Moreover, the existence of Brownian movement itself has been interpreted to imply that the Second Law of Thermodynamics does not have absolute validity on the microscopic level; yet this argument was almost completely overlooked at the time when the validity and consequences of the Second Law were being hotly disputed in the 1890's.

Early attempts to apply kinetic theory to Brownian movement employed the theorem that all particles have equal average kinetic energy in a state of thermal equilibrium (Waterston–Maxwell equipartition theorem). Observers attempted to measure the speed of particles in Brownian movement, but none of the results came anywhere near the theoretical prediction. Only after Einstein had developed his theory was the reason for the discrepancy realized; three-quarters of a century of experimentation produced almost no useful results, simply because no theorist had told the experimentalists what quantity should be measured!

Einstein's theory of Brownian movement combined two postulates that seemed to have nothing to do with each other (perhaps this was characteristic of Einstein). He took a formula from hydrodynamics, for the force on a sphere moving through a viscous fluid; and another formula, from the theory of solutions, for the osmotic pressure of dissolved molecules. Inserting these (apparently incompatible) physical characterizations into his description of the random motion of a particle, he arrived at his famous result for the mean-square displacement of the particle. Smoluchowski, shortly afterwards, published a more comprehensible derivation of a similar result, using a theoretical model taken over from gas theory.

Jean Perrin quickly attempted to establish Einstein's theory by experimental tests of the displacement formula and of another formula for the vertical distribution of particles in a fluid. Perrin also claimed that the confirmation of Einstein's theory proved the real existence of the atom, previously considered a merely hypothetical entity. He was remarkably successful in putting over this argument, and in fact the anti-atomistic "Energetics" movement never recovered from this blow.

From the viewpoint of modern physics, Brownian movement theory is not just another application of the Maxwell–Boltzmann statistical-atomistic approach, for it played an important role in the transition from classical to modern physics by reconfirming the validity of that approach at a time when it was in doubt. But one does not have to rely on a present-minded "Whig" view of the history of science to attribute such a role to the theory. Einstein himself was not looking for an explanation of known phenomena but, as he explicitly stated at the beginning of his first paper, was seeking a physical situation in which the predictions of the statistical-atomic approach would differ from those of the macroscopic thermodynamic approach and thus would provide an opportunity for a decisive test.

15.1 Robert Brown's observations and interpretations thereof

The modern reader of Brown's papers may be puzzled by his use of the word "Molecule." The significance of this term for biologists in the early 19th century was strongly influenced by the theories of the French naturalist Georges Louis Leclerc, Comte de Buffon (1707–86). Buffon taught that all plants and animals developed out of a basic supply of "organic molecules"; these molecules were like interchangeable parts, and their formation into an organism of a particular kind was simply guided by an "interior mold."[1] Buffon's ideas seemed to be supported by microscopic observations of John Needham, Lazzaro Spallanzani, and others in the 18th century. While these naturalists did not necessarily agree with Buffon or with each other about the generation and constitution of organisms, they did report having seen tiny particles from organic substances that appeared to be self-animated when placed in a fluid. Later in the 19th century, the doctrine of organic molecules was to be swallowed up by the cell theory, but in the 1820's it was still being defended by some biologists.[2] Brown's discovery, therefore, was not his observation of the motion of microscopic particles in fluids; that observation had been made many times before; instead, it was his emancipation from the previously current notion that such movements had a specifically organic character. What Brown showed was that almost any kind of matter, organic or inorganic, can be broken into fine particles that exhibit the same kind of dancing motion; thus he removed the subject from the realm of biology into the realm of physics.

Robert Brown (1773–1858) was one of England's greatest botanists. He is best known for his discovery of the nuclei of plant cells, as well as for classifying a large number of unfamiliar plants which he brought back from an Australian expedition in 1801–5. The first volume of his *Prodromus Florae Novae Hollandiae et Insulae Van Diemen* was published in 1810, and in the same year he became librarian to Joseph Banks, President of the Royal Society. Though offered a university chair he preferred to stay with Banks where he had the use of valuable collections. The library and collections were bequeathed to him by Banks on the latter's death in 1820, with the condition that they were to go to the British Museum after Brown's death. Brown arranged for the British Museum to take over the collections in 1827, with himself as keeper of the botanical department on an assured stipend.[3]

In 1828, Brown wrote a pamphlet entitled "A brief account of microscopical observations made in the months of June, July and August, 1827, on the particles contained in the pollen of plants; and on the general existence of active molecules in organic and inorganic bodies." This work is a bibliographic curiosity, for it was never officially "published" even though it was printed in the *Edinburgh Journal of Science* in 1828 and reprinted elsewhere.[4] It was originally intended only for private circulation, and Brown avoided the formal manner of presentation customarily found in articles written for scientific journals. Although Brown's informal style pleases those of us who like to see how one step leads to another in a scientific investigation, his inclusion of preliminary conjectures led to some later misunderstandings about his views on the cause of the movement he observed.

The researches originated, Brown tells us, in an attempt to find the mode of action of pollen in the process of impregnation. The first plant examined was *Clarckia pulchella*, whose pollen contains particles varying from $\frac{1}{4000}$th to $\frac{1}{5000}$th of an inch in length.

> While examining the form of these particles immersed in water, I observed many of them very evidently in motion; their motion consisting not only of a change of place in the fluid, manifested by alterations in their relative positions, but also not infrequently by a change of form of the particle itself.... In a few instances the particle was seen to turn on its longer axis. These motions were such as to satisfy me, after frequently repeated observation, that they arose neither from currents in the fluid, nor from its gradual evaporation, but belonged to the particle itself.

Brown then examined particles (or "Molecules" as he now started to call them) from several other plants, not only living ones but also some that had been preserved in an herbarium for not less than a century. He observed similar movements in all cases. At this point he recorded rather recklessly his first guess about the origin of the motion:

> Reflecting on all the facts with which I had now become acquainted, I was disposed to believe that the minute spherical particles or Molecules of apparently uniform size... were in reality the supposed constituent or elementary Molecules of organic bodies, first so considered by Buffon and Needham, then by Wrisberg with greater precision, and very recently by Dr. Milne Edwards, who has revived the doctrine.... I now therefore expected to find these molecules in all organic bodies: and accordingly on examining the various animal and vegetable tissues, whether living or dead, they were always found to exist....

But after studying several mineralized vegetable remains, Brown began to suspect that moving molecules could also be obtained from inorganic sources. It turned out that practically every conceivable substance, from a piece of window glass to a fragment of the Sphinx, could be made to yield particles that moved in water.

Brown's memoir attracted considerable attention, and several other scientists reported observations of a similar kind. But there was almost universal condemnation of Brown, at least on the Continent, for what was thought to be his opinion that the molecules are self-animated. All kinds of physical explanations for the motion were suggested: unequal temperatures in the strongly illuminated water, evaporation, air currents, heat flow, capillarity, motions caused by the hands of the observer, and so forth.[5]

Michael Faraday gave a Friday evening lecture on Brownian movement at the Royal Institution on 21 February 1829, in which he defended Brown. According to Faraday, Brown's experiments were carefully done and sufficed to show that the movements could not be explained by any of the causes so far suggested. In fact, Brown had simply admitted that he could not account for the motions; but by using the term "molecule" (which Faraday was careful to distinguish from "ultimate atoms") Brown had laid himself open to misunderstanding, "because the subject connects itself so readily with general molecular philosophy that all *think* he must have meant this or that...."[6]

The physicist David Brewster (who was the editor of the *Edinburgh Journal of Science*, in which Brown's original memoir had been printed) thought that physical causes would probably be found sufficient to explain the motion. But even if a complete physical explanation were not yet possible, he thought it quite improper to attribute the motion to animal life. There was nothing surprising in the fact that molecules should have their own characteristic motion:

> Why should not the molecules of the hardest solids have their orbits, their centres of attraction, and the same varied movements which are observed in planetary and nebulous matter? The existence of such movements has already been recognized in mineral and other bodies. A piece of sugar melted by heat, and without any regular arrangement of its particles, will in process of time gradually change its character, and convert itself into regular crystals.... In these changes the molecules must have turned round their axes, and taken up new positions within the solid.... Before another century passes away, the laws of such movements will probably be determined; and when the molecular world shall thus have surrendered her strongholds, we may look for a new extension of the power of man over the products of inorganic nature.[7]

Brewster is the intellectual ancestor of the modern physicist who is supremely confident that all problems in chemistry, and perhaps even biology, can "in principle" be reduced to the solution of the appropriate Schrödinger equation; we need only build a big enough computer to solve that equation. At the same time, Brewster was here representing rather accurately the attitude of 19th-century physicists toward Brownian movement: it was a phenomenon that would certainly find its ultimate explanation in a future theory of molecular motion, but was not worth the trouble of a detailed investigation at the present time. The leading natural philosophers of the day had more important problems to be solved first.

In a second memoir, "Additional remarks on active molecules," Brown replied to some of his critics and reported further experiments.[8] He disclaimed the view that the molecules are animated, admitting that some readers may have misunderstood him because he had "communicated the facts in the same order in which they occurred, accompanied by the views which presented themselves in the different stages of the investigation." He now wished to prove that the motion of particles in a fluid cannot be due, as others had suggested, to "that intestine motion which may be supposed to accompany its evaporation." To accomplish this proof, he mixed water containing particles with almond-oil. After being shaken, the mixture contained drops of water ranging from $\frac{1}{50}$th to $\frac{1}{200}$th of an inch in diameter. Being surrounded by almond-oil, these drops of water do not evaporate for a considerable time. Some of the drops contained only a single particle. "But in all the drops thus formed and protected, the motion of the particles takes place with undiminished activity, while the principal causes assigned for that motion, namely, evaporation, and their mutual attraction and repulsion, are either materially reduced or absolutely null."

Brown was aware that the evaporation of liquids was considered by physicists to be somehow connected with "intestine motions," but he misunderstood the connection; he thought that if evaporation were suppressed by some external cause, then the intestine motion would also have to stop.

The other cause of motion that Brown thought he had excluded by this experiment—mutual attraction or repulsion of the particles—was occasionally proposed later in the 19th century. No one seems to have noticed Brown's refutation of this explanation: the fact that a single particle in a drop of water will exhibit the same motion as it does when other particles are present.

In this second memoir, Brown also referred to the previous observations of Leeuwenhoek, Stephen Grant, Needham, Buffon, and Spallanzani, but he said that all these writers confused "Molecular" motion with animalcular motion. He also cited Gleichen, Wrisberg, Müller, and James Drummond. He noted that in 1819, Bywater of Liverpool had published an account of microscopical observations "in which it is stated that not only organic tissues, but also inorganic substances consist of what he terms animated or irritated particles." But, according to Brown, Bywater was susceptible to "optical illusions." Thus Brown disposed of a possible claimant to the discovery of the generality of Molecular motion.[9]

It would take us too far from the main subject of this chapter to pursue the effect of Brown's writings on biological speculations about "organic molecules" later in the 19th century. We merely note in passing that by the 1870's, at least, it was becoming common for authors of books on the microscope to include warnings about Brownian movement, in case observers should mistake it for the motion of living beings and attempt to build fantastic theories on it.[10,11]

Before leaving Brown, I think it may not be superfluous to quote the recollections of Charles Darwin from the 1830's:

> I saw a good deal of Robert Brown, "facile Princeps Botanicorum," as he was called by Humboldt. He seemed to me to be chiefly remarkable for the minuteness of his observations and their perfect accuracy. His knowledge was extraordinarily great,

and much died with him, owing to his excessive fear of ever making a mistake. He poured out his knowledge to me in the most unreserved manner, yet was strangely jealous on some points. I called on him two or three times before the voyage of the *Beagle* [1831], and on one occasion he asked me to look through a microscope and describe what I saw. This I did, and believe now that it was the marvelous currents of protoplasm in some vegetable cell. I then asked him what I had seen; but he answered me, "That is my little secret."[12]

Notes for §15.1

1. G. L. L. Buffon, *Histoire Naturelle, Générale et Particulière, avec la Description du Cabinet du Roy*, Tome Second, Chapitre IV et VI; Tome Quatrième, "Le Boeuf" (p. 437ff). (Paris: L'Imprimerie Royale, 1749, 1753). See also P. Flourens, *Histoire des Travaux et des Idées de Buffon*, 2nd ed., (Paris: Hachette, 1850), p. 67ff.
2. On the history of Brownian movement before Brown, see F. I. F. Meyer, "Historische-physiologische Untersuchungen über selbstbewegliche Molecüle der Materie," pp. 327–498 in vol. **4** of *Robert Brown's Vermischte Botanische Schriften*, ed. C. G. Nees von Esenbeck. (Leipzig: F. Fleischer & Nürnberg: L. Schrag, 1825–34). See also P. W. van der Pas, *Actes XII Cong. Int. Hist. Sci. 1968* **8**, 143 (1971): D. C. Goodman, *Episteme* **6**, 12 (1972).
3. See the article "Robert Brown" by J. B. Farmer in *Makers of British Botany*, ed. F. W. Oliver (Cambridge University Press, 1913).
4. Robert Brown, *Edinburgh New Phil. J.* **5**, 358 (1828); *Annales des Sciences Naturelles* (Paris) **14**, 341 (1828); *Froriep's Notizen aus dem Gebiete der Natur- und Heilkunde* **22**, 161 (1828); Oken's *Isis* **21**, 1006 (1828); *Phil. Mag.* **4**, 161 (1828); *Ann. Phys.* [2] **14**, 294 (1828); *The Miscellaneous Botanical Works of Robert Brown* (London, 1866), **1**, 465.
5. G. W. Muncke, *Ann. Phys.* [2] **17**, 159 (1829). Francois Raspail, *Mem. Soc. Hist. Nat. Paris* **4**, 347 (1828); *Edinburgh J. Sci.* **10**, 96 (1828). C. A. S. Schultz, *Mikroskopische Untersuchungen ueber des Herrn Robert Brown Entdeckung lebender, selbst in Feuer unzerstoerbarer Teilchen* ... (Carlsruhe & Freiburg, 1828). C. M. Marx, *Schweigger's Journal für Chemie und Physik* **61**, 121 (1831). F. Unger, *Flora* **15**, 713 (1832). R. E. Grant, *Edinburgh J. Sci.* **10**, 346 (1829). R. Bakewell, *Magazine of Natural History* **2**, 1 (1829).
6. Bence Jones, *The Life and Letters of Faraday* (London, 1870), **1**, 403.
7. D. Brewster, *Edinburgh J. Sci.* **10**, 215 (1829).
8. Robert Brown, *Edinburgh J. Sci.* **1**, 314 (1829); *Annales des Sciences Naturelles* (Paris) **19**, 104 (1830); *Edinburgh New Phil. J.* **8**, 41 (1830); *Froriep's Notizen aus dem Gebiete der Nature- und Heilkunde* **25**, 305 (1829); *Phil. Mag.* **6**, 161 (1829); *The Miscellaneous Botanical Works of Robert Brown* **1**, 479.
9. The only publication of Bywater which I have been able to find is an article in *Phil. Mag.* **49**, 283 (1817). In this article Bywater does not mention movements in any substances other than those derived from animals and plants. The British Museum

catalogue lists a book, *Physiological Fragments; or, Sketches of various subjects intimately connected with the study of Physiology* (London, 1819), which is probably the one Brown refers to; this was reissued in 1824 with an addendum, *Supplementary Observations to show that vital and chemical energies are of the same nature and both derived from solar light* (London). Brown says he has not seen the original memoir of 1819 but only a later edition published in 1828.

10. See for example C. P. Robin, *Traité du Microscope, son mode d'emploi* (Paris, 1871), p. 526; W. B. Carpenter, *The Microscope and its revelations* (London, 5th ed., 1875), p. 199.
11. On biological speculations, see F. Rádl, section "Spekulationen über kleinere Lebensteilchen als die Zelle," in *Geschichte der Biologische Theorien* II. Teil (Leipzig: Engelmann, 1909). (The English trans. of Rádl's book, *The History of Biological Theories*, published by Oxford University Press in 1930, lacks many of the references for this section.)

    The history of "abiogenesis," "biogenesis," "organic molecules," and Pasteur's germ theory is discussed by T. H. Huxley in his Presidential Address to the British Association in 1870: *B. A. Rep.* **40**, lxxiii (1870).
12. *Charles Darwin: His Life told in an autobiographical Chapter, and in a selected series of his published letters*, ed. by his son, Francis Darwin (London, 1892; New York: Schuman, 1950), p. 46.

15.2. Miscellaneous observations and qualitative explanations, 1840-78

After the initial flurry of excitement caused by Brown's publications in 1828–29, interest in Brownian movement dropped off to almost nothing for about thirty years.[1] But with the development of thermodynamics and the revival of the kinetic theory of gases in the 1850's, there was a new stimulus for researches into the relation between heat and microscopic motion. This was certainly a factor in the explanations proposed for Brownian movement later in the 19th century. However, until Nägeli's paper in 1879 (see next section), there was no serious attempt to develop a quantitative theory of Brownian movement, based on the mechanical-atomic theory of heat. What is perhaps most significant about the history of this period is the absence of any publications on Brownian movement by the kinetic theorists–Clausius, Maxwell, and Boltzmann.

Notions of the old caloric theory of heat were still evident in the discussion of the British biologists Griffith and Henfrey, in their *Micrographic Dictionary* published in 1856:

> Heat is the only agent which affects [the motion]; this causes the motion to become more rapid. Hence it might be attributed to the various impulses which each particle receives from the radiant

heat emitted by those adjacent. Or, as it takes place when the temperature is uniform, may it not arise from the physical repulsion of the molecules, uninterfered with by gravitation, hence free to move? The effect of heat would then be explicable, because this increases the natural repulsion of the particles of matter, as in the conversion of water into vapour....[2]

In 1858, Jules Regnauld (Professor of Physics at the École de Pharmacie in Paris, and later Professor of Pharmacology) did some experiments which convinced him that Brownian movement was due to the absorption of heat by the particles from rays of light falling on the suspension; the transmission of this heat to the surrounding fluid sets up currents which move the particles.[3]

Christian Wiener (Professor of Descriptive Geometry and Geodesy at Karlsruhe), reporting on his experiments in 1863, gave various arguments to show that Brownian movement cannot be due to external causes but must be attributed to internal motions in the fluid. Wiener's concept of atomic motion, however, derived from the period before Clausius and Maxwell. He believed that matter consists not only of material atoms but also of aether atoms; heat is the kinetic energy of both kinds of atoms. The essential difference between solids and liquids, he said, is that in solids the direction of vibration of the molecules is opposite to that of the aether atoms, whereas in liquids it is the same; heat of melting is the energy needed to reverse the relative directions of these two vibrations. Wiener's explanation of Brownian movement is closely related to these ideas about aether vibrations, and he says that the wavelength of red light is about the same as the diameter of the smallest groups of molecules that move together in the liquid.[4]

Wiener's theory was criticized, in 1894, by Meade Bache, who said:

> The theory of Herr Wiener, that the movements are due to the action of the red-wave of light and heat is refuted by the single fact that, as I have proved by experiment, one may interpose at pleasure between the source of light or heat and the particles, either a violet glass or a red glass, without being able to observe the slightest alteration in the movements.[5]

Nevertheless, Wiener is credited by some later writers[6] with being the first to discover that Brownian movement is due to molecular motions of molecules in the liquid.

Giovanni Cantoni, an Italian physicist, also attributed Brownian movement to thermal motions in the liquid, and considered that this phenomenon provides a "beautiful and direct experimental demonstration of the fundamental principles of the mechanical theory of heat."[7] The honor of having discovered the true cause of Brownian movement has also been claimed for him.[8]

In 1868, J. B. Dancer, in Manchester, said:

The cause of the phenomenon is not yet satisfactorily accounted for. Some have imagined that it is the physical repulsion of the particles when uninfluenced by gravitation. The author ... thinks that the movement may possibly be connected with the absorption and radiation of heat.[9]

W. Stanley Jevons, the British writer on political economy and scientific method, claimed that Brownian movement is an electrical phenomenon and is related to osmosis, "for if a liquid is capable of impelling a particle in a given direction, the particle if fixed is capable of impelling the liquid in an opposite direction by an equal force."[10] Lacking van't Hoff's theory of osmotic pressure, however, Jevons was unable to quantify this suggestion or relate it to diffusion theory in the way that Einstein did in 1905 (see §15.4 below). Ignoring Dancer's protest that "the results of many experiments point to heat as a probable cause,"[11] Jevons elaborated his electrical explanation and even proposed a new name, *pedesis*, for the effect. (Brown, after all, was not the first to see dancing particles under the microscope; he was merely a good publicist.) Jevons suggested that pedesis plays an important role in sewage treatment, geological processes, and so forth, by maintaining particles in suspension; it is also involved in the detergent action of soap. Yet, despite a brief reference to pedesis as a random process that might be treated by probability theory, Jevons made no real progress toward a quantitative theory.[12] Only one other scientist, William Ramsay, adopted his term "pedesis."[13]

During the 1870's, the opinion that Brownian movement is somehow related to heat was frequently expressed. But there was still considerable confusion and vagueness in such explanations, stemming mainly from the lingering belief that motion is connected only with a *change* of temperature. It was understandable that the absorption of radiation coming from outside the fluid might cause unequal heating within it, and thereby set up currents which would produce motion of the particles. An atomic explanation, on the other hand, would have to account for the fact that the particles move even in fluids of uniform temperature, simply because of bombardments by individual molecules. The suggestion was occasionally made that several molecules might move together to cause the visible motion of the particle, and that such cooperative motions could be viewed as a possible fluctuation from the average state of motion.[14] One might think that such a suggestion would be countered by the argument that the Second Law of Thermodynamics denies the possibility of such conversions of thermal energy into mechanical energy if there is no temperature-difference in the system. While this argument may have operated unconsciously or privately to discourage explanations based on fluctuations, there is no explicit reference to it in print until later on.[15] It certainly did not prevent scientists like Joseph Delsaulx, in Brussels, from talking about the "Thermodynamic origin of Brownian motions."[16]

To summarize the situation in 1878: first, the phenomenon of Brownian movement was becoming widely known, as can be seen from the many references to it in scientific journals and even in fiction.[17] Second, while a minority of scientists still attributed its cause to electrical effects, osmosis, or surface tension, most seemed to think that it must be connected with thermal molecular motions. Third, there was still no quantitative theory that could be tested against experiment.

Notes for §15.2

1. J. D. Botto, *Memorie della Reale Accademia delle Scienze di Torino* [2] **2**, 457 (1840). E. H. Weber, *Leipzig. Ber.* p. 57 (1854).
2. J. W. Griffith and A. Henfrey, art. "Molecular Motion" in *Micrographic Dictionary* (London, 1856), p. 428; (3rd ed., 1875), **1**, 498.
3. J. Regnauld, *Fortschr.* **14**, 9 (1860) [this is a summary by Wilhelmy; I have not been able to verify the original citation, given as *Journ. de Pharm. et de Chim.* **34**, 141 (1858)].
4. Chr. Wiener, *Ann. Phys.* [2] **118**, 79 (1863).
5. R. Mead Bache, *Proc. Am. Phil. Soc.* **33**, 163 (1894).
6. The Svedberg, *Ion* **1**, 373 (1909). J. Perrin, *Ann. Chim. Phys.* [8] **18**, 1 (1909).
7. Giovanni Cantoni, *Reale Istituto Lombardo di Scienze e Lettere* (Milano), *Rendiconti* [2] **1**, 56 (1868). See also *ibid.* **22**, 152 (Sem. 1, 1889).
8. Ic. Guareschi, *Isis* **1**, 47 (1913).
9. J. B. Dancer, *Manchester Proc.* **7**, 162 (1868).
10. W. S. Jevons, *Manchester Proc.* **9**, 78 (1870).
11. J. B. Dancer, *ibid.*, p. 82.
12. W. S. Jevons, *Journal of Science and Annals of Astronomy* **8**, 167, 514 (1878).
13. See William Ramsay's paper cited below, §15.3, note 5.
14. E. Budde, *Bonn Ber.* **27**, 108 (1870). J. Thirion, *Revue des Questions Scientifiques* **7**, 5 (1880).

15. Gouy, *J. Phys.* [2] **7**, 561 (1888).
16. J. Delsaulx, *Monthly Microscopic Journal* **18**, 1 (1877).
17. See for example George Eliot's *Middlemarch* (1872), Book I, ch. XVII, p. 181, in which Robert Brown's memoir is offered to the Rev. Mr. Farebrother by the surgeon Lydgate.

15.3 Criticisms of the molecular-impact theory

In 1879, the German botanist Karl Nägeli published a long memoir in which he attempted to disprove the suggestion that Brownian movement is caused by the collisions of the particles with molecules in the surrounding fluid.[1] Nägeli was an authority on microscopic methods of observation, and also had some knowledge of physics which he frequently tried to apply to biology.[2] The immediate stimulus for this particular work was a question that had come up in a discussion at the Munich Academy of Sciences on the spreading of fungus infections by the wind. Nägeli, motivated partly by the need to defend his own theory of fungus action, developed a general theory of the motions of small particles in air, in which the role of collisions could be compared with the role of intermolecular forces. The main part of his argument applied to sun-motes being bombarded by air molecules, but he claimed that it extended also to Brownian movement of particles in liquids.

Nägeli had one great advantage over all the scientists who had previously written on Brownian movement: he was familiar with the estimates of molecular masses and speeds that had been obtained from the kinetic theory of gases.[3] He could therefore argue as follows:

> Since the idea that the molecules of a gas travel past each other with large velocities has entered physics and, because of its irrefutable proof, has found general agreement, one might also suppose that the "dancing motion" of sun-motes is caused by the frequent and variously directed impulses which they receive from gas molecules.... [But] because of their greater weight, they [sun-motes] are like bodies completely at rest among the air molecules flying hither and yon, and there can be no question of a dancing or quivering of the sun-motes resulting from molecular collisions.
>
> This can easily be proved by a calculation of the number and energy of molecular collisions which a particle of definite size under definite conditions experiences in the air. Such a calculation

has a firm base, since, thanks to the mechanical theory of gases, one has a rather accurate idea of the weight and velocity of a gas molecule....

We assume that in 1 cc of gas at 0° at a pressure of 760 mm of mercury, there are 21 trillion molecules, so that the oxygen molecule has a weight of 7- and the nitrogen molecule a weight of 6 hundred-thousand-trillionth gram. The former moves with an average velocity of 461 meters/sec, the latter 492 meters/sec, so that the kinetic energy ($\frac{1}{2}mv^2$) is equally large for each.

The gas molecules behave like completely elastic spheres in their mutual interactions. Their collisions with the mote-particles may or may not be elastic likewise....[4]

The particle acquires after such a collision, if it is perfectly elastic, a velocity $2av/(a+b)$, where v is the velocity of the gas molecule, a = mass of gas molecule, b = mass of particle. Since the smallest fission fungus still weighs about 300 million times as much as a gas molecule, it acquires a velocity of only 0.002 mm/sec. If the collisions are not elastic, the effect is still smaller.

The motion which a sun-mote, and on the whole any particle found in the air, can acquire by the collisions of an individual gas molecule or a multitude of such molecules is therefore so extraordinarily small, and the number of simultaneous collisions against the particle from all sides is so extraordinarily large, that the particle behaves just as if it were completely at rest.

As for Brownian movement, Wiener and Exner showed

that its cause is to be sought in the liquid and is to be ascribed to the internal motions peculiar to the liquid state. But should this be understood in the sense that it is the collisions themselves of the liquid molecules moving in different directions, and not the molecular forces which cause the particles visible in the microscope to dance, then such an assumption would be no better founded than the analogous supposition for the dance of sun-motes.

It is not possible, said Nägeli, to give an exact value of the molecular velocity in liquids, but it must be less than in gases, so that the same arguments apply even more strongly. Nägeli concluded that the cause of particle motion must lie not in thermal molecular motions but in attractive and repulsive forces.

§15.3]   CRITICISMS OF MOLECULAR-IMPACT THEORY   669

The British chemist William Ramsay, writing three years after Nägeli but with no apparent knowledge of his paper, used similar arguments against the idea that Brownian movement could result from the impacts of individual molecules. The active particles have masses at least 125 million times as great as that of a water molecule, according to Ramsay's estimate.

> If molecules do not coalesce and move as a whole, then they would appear to have no possible power of giving motion to a mass so much larger than themselves. But that molecules have arrangement is probable, owing to the power which some liquids possess of rotating the plane of polarized light.
>
> Clerk-Maxwell supposed for some time that the attraction of two molecules varies inversely as the fifth power of the distance. If attraction at distance 2 is 1, attraction at distance 1 would be 64. Why do not all molecules therefore coalesce? Probably, because their own proper motion, of which heat represents the higher harmonics, causes them to fly apart again. The wavelength of that motion is not so minute, and although we possess no means of ascertaining the amplitude of such vibrations, still their rate is so prodigious as to give rise to an almost incredible impact.[5]

The last paragraph quoted reveals how little most scientists really understood of the kinetic theory of gases in the 1880's. For example, Maxwell had assumed that molecules *repel* each other with an inverse fifth power force, not attract (§12.3). On the other hand, Ramsay was going far beyond any established results of kinetic theory when he suggested here, and again in 1892, that water molecules might move together "in complex groups of considerable mass, and of some stability" which could produce Brownian movement.[6]

The same idea was brought forward a few years later by the French physicist Léon Gouy. He admitted that while one could not explain Brownian movement simply by invoking "the uncoordinated movements of the molecules, which one often regards as constituting thermal movement," "one may conceive that molecular movements in liquids are partly coordinated for spaces comparable to 1 micron.... The existence of Brownian movement seems to show that in reality something similar to this takes place."[7] This might seem like circular reasoning, but Gouy was fortunate enough to be addressing physicists, in the language of physics, at a time when they were ready to pay attention, and thus Gouy got a certain amount of credit for "discovering" the cause of Brownian movement.[8] (Some of this he deserved for his experimental work, in which he showed that external influences such as strong magnetic fields have no effect on the movements, so that the cause must be sought in factors internal to the liquid.) Of more historical significance than his qualitative discussion of the cause of Brownian movement, is his suggestion that it might offer an exception to the Second Law of Thermodynamics:

> Whatever idea one may have as to the cause that produces [the movement], it is no less certain that work is expended on these particles, and one can conceive a mechanism by which a portion of this work might become available. Imagine, for example, that one of these solid particles is suspended by a thread of diameter very small compared to its own, from a rachet wheel; impulses in a certain direction make the wheel turn, and we can recover the work. This mechanism is clearly unrealisable, but there is no theoretical reason to prevent it from functioning. Work could be produced at the expense of the heat of the surrounding medium, in opposition to Carnot's principle. It appears that one can then make precise the meaning of Helmholtz's reservations about this principle, in the case of living tissues; this principle would then be exact only for the gross mechanisms that we know how to make, and it would cease to be applicable when the *receptor* organ has dimensions comparable to 1 micron.[9]

Gouy's remarks found a responsive reader in Henri Poincaré, who pointed out their significance to his audience at the Congress of Arts and Science in St. Louis (1904):

> The biologist, armed with his microscope, long ago noticed in his preparations disorderly movements of little particles in suspension: this is the Brownian movement; he first thought this was a vital phenomenon, but he soon saw that the inanimate bodies danced with no less ardor than the others; then he turned the matter over to the physicists. Unhappily, the physicists remained long uninterested in this question; the light is focused to illuminate the microscopic preparation, thought they; with light goes heat; hence inequalities of temperature and interior currents produce the movements in the liquid of which we speak.
>
> M. Gouy, however, looked more closely, and he saw, or thought he saw, that this explanation is untenable, that the movements become more brisk as the particles are smaller, but that they are not influenced by the mode of illumination.
>
> If, then, these movements never cease, or rather are reborn

without ceasing, without borrowing anything from an external source of energy, what ought we to believe? To be sure, we should not renounce our belief in the conservation of energy, but we see under our eyes now motion transformed into heat by friction, now heat changed inversely into motion, and that without loss since the movement lasts forever. This is the contrary of the principle of Carnot.

If this be so, to see the world return backward, we no longer have need of the infinitely subtle eye of Maxwell's demon; our microscope suffices us. Bodies too large, those, for example, which are a tenth of a millimeter, are hit from all sides by moving atoms, but they do not budge, because these shocks are very numerous and the law of chance makes them compensate each other: but the smaller particles receive too few shocks for this compensation to take place with certainty and are incessantly knocked about.[10]

Poincaré's interest in Brownian movement presents us with a new historical problem: Why did he not go ahead and work out the quantitative theory of it himself? The only evidence I have found that throws any light on this question is contained in a thesis by L. Bachelier, presented to the Faculté des Sciences de Paris in 1900. The thesis is dedicated to Poincaré, who was also one of the examiners, and is mainly concerned with the theory of speculation on the Paris stock market. With this particular application apparently foremost in his mind, Bachelier developed a theory of stochastic processes similar to the modern theory of Brownian movement (the "Wiener process"). At the end of the thesis there is printed in one paragraph the "Seconde Thèse: Propositions données par la Faculté" which reads as follows:

Résistance d'une masse liquide indéfinie pourvue de frottements intérieurs, régis par les formules de Navier, aux petits mouvements variés de translation d'une sphère solide, immergée dans cette masse et adhérente à la couche fluide qui la touche.[11]

But there is no hint in Bachelier's thesis that the problem of stock-market speculation has anything to do with the problem of the resistance of a viscous fluid to the movement of a solid sphere.

Notes for §15.3

1. Karl Nägeli, *Mün. Ber.* **9**, 389 (1879).
2. K. Nägeli and S. Schwendener, *Das Mikroskop* (Leipzig, 1867; 2. Aufl. 1877),

English trans. *The Microscope* (New York, 1892). Nägeli, *Mechanisch-physiologisch Theorie der Abstammungslehre* (München und Leipzig, 1884). (There is a section on pp. 729–35 of the last-named book dealing with heat and molecular motion.) For a modern evaluation of Nägeli's research see J. S. Wilkie, *Ann. Sci.* **17**, 27 (1961) and earlier papers. His failure to appreciate the significance of Mendel's ideas is attributed to the influence of Romantic philosophy by Bentley Glass, in *Studies in Intellectual History* (Baltimore: Johns Hopkins Press, 1953), p. 148.
3. See Loschmidt's paper cited in §1.8, note 4.
4. Nägeli, *op. cit.* (note 1), my translation.
5. William Ramsay, *Proceedings of the Bristol Naturalists' Society* **3**, 299 (1882).
6. Ramsay, *Chemical News* **65**, 90 (1892); *Proceedings of the Chemical Society* **8**, 17 (1894).
7. See §15.2, note 14; also the reprint with notes by J. Thirion in *Revue des Questions Scientifiques* [3] **15**, 251 (1909) which includes a shorter paper on the same subject by Gouy.
8. J. Perrin, *C.R. Paris* **146**, 967 (1908); G. L. de Haas-Lorentz, *Die Brownsche Bewegung und einige verwandte Erscheinungen* (Braunschweig: Vieweg, 1913), p. 9.
9. Gouy cites Helmholtz, *Sur la thermodynamique des théorèmes chimiques*, (Academie de Berlin, 1882) Traduit par M. G. Chaperon dans le Journal de Physique, 1884. William Thomson, who first proposed the generalized "Dissipation of energy" version of the Second Law, suggested that organic life might be excepted from its domain of validity:

"Any restoration of mechanical energy, without more than an equivalent of dissipation, is impossible in inanimate material processes, and is probably never effected by means of organized matter, either endowed with vegetable life or subjected to the will of an animated creature." [*Phil. Mag.* **4**, 306 (1852)].

In 1862, he agreed with his brother James Thomson that plants and animals might possess a "vital principle" which could reverse the dissipation of energy; see James Thomson's *Collected Papers in Physics and Engineering*, ed. J. Larmor (Cambridge: Cambridge University Press, 1912), p. 1v.
10. Jules Henri Poincaré, in *Congress of Arts and Science, Universal Exposition, St. Louis, 1904* (Boston and New York: Houghton, Mifflin and Co., 1905), I, 604.
11. L. Bachelier, Thèses présentées a la Faculté des Sciences de Paris pour obtenir le Grade de Docteur es Sciences Mathematiques (Paris: Gauthier-Villars, 1900). The Première Thèse, "Theorie de la Speculation," was also published in *Ann. Sci. École Normale Superieure* [3] **17**, 21–86 (1900). For a recent discussion of the relation between Bachelier's work and Brownian movement, see M. F. M. Osborne, *Operations Research* **7**, 807 (1959).

15.4 Einstein's theory of Brownian movement

In his "Autobiographical Notes," written for the Schilpp collection *Albert Einstein: Philosopher–Scientist* (1949), Einstein indicates the motivation for his work on the theory of Brownian movement and its relation to the state of physics at the beginning of this century:

> Not acquainted with the earlier investigations of Boltzmann and Gibbs, which had appeared earlier and actually exhausted the subject, I developed the statistical mechanics and the molecular-kinetic theory of thermodynamics which was based on the former. My major aim in this was to find facts which would guarantee as much as possible the existence of atoms of definite finite size. In the midst of this I discovered that, according to atomistic theory, there would have to be a movement of suspended microscopic particles open to observation, without knowing that observations concerning the Brownian motion were already long familiar.

Since there is some doubt about the accuracy of these statements, it should be noted that Einstein himself recognized at the beginning of this essay that "Every reminiscence is colored by today's being what it is, and therefore by a deceptive point of view."

After summarizing his theory of Brownian movement, Einstein continued his reminiscences with the following remarks:

> The agreement of these considerations with experience together with Planck's determination of the true molecular size from the law of radiation (for high temperatures) convinced the sceptics, who were quite numerous at that time (Ostwald, Mach) of the reality of atoms. The antipathy of these scholars towards atomic theory can indubitably be traced back to their positivistic philosophical attitude. This is an interesting example of the fact that even scholars of audacious spirit and fine instinct can be obstructed in the interpretation of facts by philosophical prejudices. The prejudice—which has by no means died out in the meantime—consists in the faith that facts by themselves can and should yield scientific knowledge without free conceptual construction. Such a misconception is possible only because one does not easily become aware of the free choice of such concepts, which, through verification and long usage, appear to be immediately connected with the empirical material.[1]

Not only did Einstein's theory provide a decisive breakthrough in the understanding of the phenomenon of Brownian motion; it also, in the opinion of Max Born, did "more than any other work to convince physicists of the reality of atoms and molecules, of the kinetic theory of heat, and of the fundamental part of probability in the natural laws."[2] It will therefore be worthwhile to digress somewhat from the history of Brownian movement itself to recall some of the background of late 19th-century theoretical physics underlying Einstein's theory.

Kinetic theories of matter, identifying heat qualitatively or quantitatively with molecular motion, had been frequently proposed since the 17th century (§1.3). At the time of Robert Brown's work, vague qualitative notions about thermal molecular motion were common enough, but Herapath's enthusiastic though inaccurate attempts to advance to something like the modern kinetic theory of gases had been rebuffed by the Royal Society in 1821 (§2.3). The kinetic theory was revived after the general acceptance of the law of conservation of energy in the middle of the 19th century; kinetic theory was then thought of as the molecular extension of the "mechanical theory of heat" or thermodynamics. The theory was developed extensively by Clausius, Maxwell, and Boltzmann. Einstein, contrary to his statement at the beginning of the above quotation, was familiar with Boltzmann's treatise, *Vorlesungen über Gastheorie*; he cites it on the fourth page of his own first paper on kinetic theory in 1902.[3] Two aspects of Boltzmann's work had a special attraction for Einstein: first, the formulation of "statistical mechanics" (as we now call it, following Gibbs) by means of generalized Lagrange–Hamilton dynamics; and second, the emphasis on statistical fluctuation phenomena. Einstein's purpose, as he says, was not only to connect thermodynamics with general mechanics, but also to deal with phenomena explicitly involving the atomic structure of matter. Moreover, he was very much interested in putting Planck's quantum theory into the language of statistical mechanics, and the papers on Brownian movement contain several references to the radiation distribution law. Perhaps the best way to characterize Einstein's work on kinetic theory is to say that he wanted to remove the restriction to *gases*, which seemed to limit the applicability of most of the work of Clausius, Maxwell, and Boltzmann, and develop a theory sufficiently general to deal with liquids, solids, and radiation.[4]

Einstein was certainly aware of the attacks on kinetic theory which had been made by Mach, Ostwald, and their followers in the 1890's. In its simplest form, the issue was: Why should we base our theories of matter on atomic hypotheses, when a phenomenological description such as thermodynamics contains all the necessary information about a physical system and also avoids the various paradoxes and inconsistencies that plague atomic theory? Einstein spoke directly to this point in his first paper on Brownian movement when he contrasted the predictions of classical thermodynamics and of the molecular-kinetic theory of heat. According to the former, small suspended particles in a liquid will not exert any force on a semipermeable membrane placed in the liquid, although it is known that molecules of a dissolved nonelectrolyte will exert an osmotic pressure in such a situation. On the other hand, the molecular-kinetic theory maintains that "a dissolved molecule is differentiated from a suspended body *solely* by its dimensions" and a certain number of suspended particles must produce the same osmotic pressure as the same number of molecules in solution. Thus, in Einstein's view, classical thermodynamics introduces an artificial distinction between suspended particles and dissolved molecules, based on the fact that the latter are too small for us to see and so we do not recognize them as being the same kind of entity.

It is obvious that any theory of Brownian movement must deal with the motion of particles in liquids. Unfortunately, at the time Einstein first began to work on this problem, there was no adequate quantitative theory of liquids developed from the molecular viewpoint comparable to the kinetic theory of gases. In fact, the kinetic theory of liquids is such a difficult subject that even today it has not advanced as far as gas theory had gone by 1905. On the other hand, there was in existence, late in the 19th century, a sizable body of theoretical research on solids and liquids treated as continua. Much of this research had originally been done for the purpose of developing theories of propagation of light through the ether, but the substantive mathematical results turned out to be more relevant to the properties of "real" liquids and solids. In particular, there was a well-established formula—the "Stokes formula"—for the force resisting the motion of a sphere through a liquid.

The origin of the Stokes formula itself is worth noting. In 1845, Stokes had derived his general equations for the flow of fluids, taking account of "internal friction" (viscosity).[5] The original application of these equations was to the computation of air-resistance corrections to results obtained with swinging pendulums. This subject was discussed in another paper in 1850; he remarked that

> On account of the inconvenience and expense attending experiments in a vacuum apparatus, the observations are usually made in air, and then it becomes necessary to apply a small correction, in order to reduce the observed result to what would have been observed had the pendulum been swung in a vacuum.[6]

Stokes quoted some experiments of Colonel Sabine (1829), which according to Sabine might indicate "an inherent property in the elastic fluids, analogous to that of viscidity in liquids, of resistance to the motion of bodies passing through them...."[7] However, Stokes argued that previous theories of the motion of solids through viscous fluids had erroneously assumed that the fluid simply glides past the surface of the solid, ignoring the tangential action between the surface and the fluid. As evidence for such tangential action–*i.e.* for the assumption that the layer of fluid in contact with the surface does not move past it–Stokes quoted an experiment of Sir James South:

> ... on attaching a piece of gold leaf to the bottom of a pendulum, so as to stick out in a direction perpendicular to the surface, and then setting the pendulum in motion, Sir James South found that the gold leaf retained its perpendicular position just as if the pendulum had been at rest; and it was not until the gold leaf carried by the pendulum had been removed to some distance from the surface, that it began to lag behind. This experiment shews clearly the existence of a tangential action between the pendulum and the air, and between one layer of air and another.[8]

Stokes was able to solve his hydrodynamic equations in the special case of a sphere moving uniformly in a fluid; he found that, provided terms involving the square of the velocity may be neglected, a force K imparts to a sphere of radius P a velocity $K/6\pi kP$, where k is the viscosity of the fluid. (We are now following Einstein's notation.) The surprising feature of this result to Stokes was that the resistance to an object moving with a given speed is proportional to its radius rather than to its surface area. One consequence of this fact is that "the resistance to a minute globule of water falling through the air with its terminal velocity depends almost wholly on the internal friction of air.... The terminal velocity thus obtained is so small in the case of small globules such as those of which we may conceive a cloud to be composed, that the apparent suspension of the clouds does not seem to present any difficulty."[9]

Even before his long memoir containing these results had appeared, Stokes had already published a shorter paper "On the Constitution of the Luminiferous ether" in which he assumed that the ether close to the surface of the earth is at rest relative to it, and then applied the hydrodynamical theory which he had developed for this case.[10] Throughout the 19th century, this hypothesis was elaborated and criticized by theoretical physicists in discussions of the ether.[11] It is therefore nor surprising that Einstein was familiar with the Stokes formula; although the only reference he gives is to Kirchhoff's *Vorlesungen über Mechanik*,[12] he may have been led to that source by

§15.4]   EINSTEIN'S THEORY OF BROWNIAN MOVEMENT   677

his reading of Lorentz's papers.[13] The formula was also used indirectly by J. J. Thomson in his determination of the electrical charge on ions produced by Röntgen rays, in 1898.[14]

The other cornerstone of Einstein's theory was J. H. van't Hoff's theory of osmotic pressure in solutions.[15] But this theory applied (or was thought to apply) only to dissolved molecules which are about the same size as the molecules of the liquid. Nevertheless, van't Hoff's formula for osmotic pressure was one of the very few quantitative results known to be valid for liquids.

Here were two theories dealing with particles in liquids: the hydrodynamic theory of Stokes, based on the assumption that the liquid is a continuous medium which sticks to a solid surface moving through it with not too high a speed; and the osmotic theory of van't Hoff, based on the assumption that the particle is itself a molecule mixed in with a molecular liquid. To put it another way: the equations of hydrodynamics are valid in situations where solid boundaries or suspended particles are acted on by a steady force originating in the liquid, and turbulence or random molecular motion in the liquid has no significant effect on the motion of the particle. (Osborne Reynolds and others in the 1880's had already shown the need to avoid this restriction, and to take account of surface effects.)[16] The osmotic theory, on the other hand, is valid in situations where *all* the motion of the dissolved particle must be attributed to random molecular motion. Thus the two theories seemed to have mutually exclusive regions of validity; it required the reckless genius of an Einstein to ignore that difficulty and boldly combine them. How much that move upset Einstein's contemporaries is well illustrated by the voluminous appendix which was published with the reprint of Einstein's papers on Brownian movement in Ostwald's *Klassiker*.[17]

Since the details of Einstein's theory are easily accessible in a paperback reprint of the English translation of the *Klassiker* edition,[18] I will give here only a summary. Einstein argued that the suspended particles visible in the microscope should exert an osmotic pressure against a semipermeable membrane, just as the molecules of a dissolved substance do. In fact, the pressure is just that which would be exerted by an ideal gas of the same number of point-atoms in the same space, provided that one can ignore interactions between the particles. To justify this assertion, Einstein carried out a short calculation with the help of his own statistical mechanical formulation. This calculation probably obscured the argument for most readers at the time, since the use of what we now call the "partition function" or "phase integral" had not yet become familiar. What the formalism does, or should do, in such an investigation, is to free one from the need for calculating pressures and thermodynamic functions by considering collisions of particles with the wall of the container, and computing the time for a particle with a certain speed to go back and forth between two opposite walls. Clearly the microscopic particles do not bounce back and forth between parallel walls, moving in straight lines in the way that one thinks of gas molecules doing in elementary kinetic-theory derivations. The point is that these particles are nevertheless in statistical thermal equilibrium with the molecules of the fluid, and this is what determines their average pressure and energy.

Einstein could therefore use the ideal gas equation for the osmotic pressure

$$p = \frac{RT}{V^*}\frac{n}{N} = \frac{RT}{N}\nu \tag{1}$$

where T = absolute (Kelvin) temperature, n = number of suspended particles in a volume V^* (which is partitioned out of a larger volume V), N = Avogadro's number, and the concentration is $\nu = n/V$.

The basis of Einstein's description of Brownian movement is the notion that the suspended particles are "diffusing" through the liquid, in such a way that dynamical equilibrium is always maintained between the osmotic force originating in a concentration gradient, and the viscous force which must retard the motion of a particle according to hydrodynamics. The osmotic force tends to push the particles from regions of high concentration to regions of low concentration; the magnitude of the force is the same as the pressure-gradient in the direction of motion. One way to develop the argument would be to write down the equation relating force to pressure-gradient,

$$K\nu = \frac{\partial p}{\partial x} \tag{2}$$

and then substitute the value for pressure given by van't Hoff's ideal gas law (1) on the right-hand side, thereby arriving at the equation

$$K\nu = \frac{RT}{N}\frac{\partial \nu}{\partial x} \tag{3}$$

We could then use the Stokes formula for the velocity of a particle moving through a viscous medium,

$$v = \frac{K}{6\pi k P} \tag{4}$$

§15.4] EINSTEIN'S THEORY OF BROWNIAN MOVEMENT 679

to eliminate the force K from the left-hand side of eq. (3), and obtain

$$6\pi k P v \nu = \frac{RT}{N} \frac{\partial \nu}{\partial x} \tag{5}$$

We would then have a relation between several quantities that presumably can be measured: the radius of the particle (P), its velocity (v), the concentration of particles (ν), the concentration gradient $\partial \nu / \partial x$, Avogadro's number (N), the gas constant (R), and the temperature (T).

However, Einstein did not propose eq. (5) as the basic prediction of the theory to be tested by experiment. In fact, he did not even follow the sequence indicated above, deriving eq. (3) from (2), although as Fürth suggested in his notes[17] this would have been the most direct route. Instead, Einstein gave a derivation of (3) which superficially seems to be independent of any formula for the pressure, such as (1)–actually, both eqs. (3) and (1) are consequences of the same statistical mechanical formulation–and proceeded instead from expressions for energy and entropy. Again, the effect was probably to obscure the argument for readers who did not have Einstein's own familiarity with statistical mechanics.

Another reason why Einstein did not propose eq. (5), even though it is perfectly consistent with his development up to this point, is that he wanted to use a mathematical description of "diffusion" which avoided ascribing a well-defined instantaneous velocity to a particle. Perhaps this is why he did not even introduce a symbol for velocity as we have done in eq. (4), but immediately translated that equation into an expression for the number of particles passing unit area per unit of time, $\nu K/6\pi k P$. He could then equate this expression to the conventional formula for the rate of diffusion, namely, $-D(\partial \nu / \partial x)$, where D is the "coefficient of diffusion." The velocity of an individual particle is never mentioned in the paper after this paragraph.

Einstein's first two basic equations are then the relation between osmotic force and concentration gradient,

$$K\nu = \frac{RT}{N} \frac{\partial \nu}{\partial x} \tag{3}$$

and the relation between diffusion rate and the flow of particles through a viscous medium resulting from this same force,

$$\frac{K\nu}{6\pi k P} = D \frac{\partial \nu}{\partial x} \tag{6}$$

It is perhaps a further sign of Einstein's preference for abstract formulations that he writes both of these equations with two terms on the left-hand side equated to zero on the right-hand side.

By combining eqs. (3) and (6) Einstein obtained the following expression for the diffusion constant

$$D = \frac{RT}{N} \frac{1}{6\pi k P} \tag{7}$$

Einstein now turned to a "closer consideration of the irregular movements which arise from thermal molecular movement,"[19] or what we would now call the mathematical description of a certain type of stochastic process. Hardly any of the mathematics is original with Einstein; what is new is the attempt to describe particle motions that are in principle still deterministic (on the molecular level) by a certain mode of probabilistic analysis.

Let us assume, Einstein said, that the motion of each particle is independent of the others, and moreover that "the movements of one and the same particle after different intervals of time [are] mutually independent processes, so long as we think of these intervals of time as being chosen not too small." To be more specific, assume that the movements of a particle in two consecutive intervals of time τ are independent. Let the number of particles which experience in the time interval τ, a displacement which lies between Δ and $\Delta + d\Delta$, be expressed in the form

$$dn = n\phi(\Delta)\, d\Delta \tag{8}$$

(That is, $\phi(\Delta)$ is *defined* by eq. (8), in the traditional indirect manner of theoretical physics.) Further, let the concentration ν be regarded as a function of space and time, $f(x, t)$. The values of this function after the time-interval τ has elapsed can be computed in terms of the distribution function for displacements, $\phi(\Delta)$; that is, $f(x, t + \tau)$ will depend on the values of $f(x + \Delta, t)$ for all possible values of Δ, weighted by $\phi(\Delta)$. If τ is very small (note that this contradicts the original assumption about τ), and if only small values of Δ need be taken into account, then one can derive a differential equation which relates the time-variations of f to its space-variations

$$\frac{\partial f}{\partial t} = D \frac{\partial^2 f}{\partial x^2} \tag{9}$$

where

$$D = \frac{1}{\tau} \int_{-\infty}^{+\infty} \frac{\Delta^2}{2} \phi(\Delta)\, d\Delta \tag{10}$$

§15.4] EINSTEIN'S THEORY OF BROWNIAN MOVEMENT

(This formula comes from an expansion of $f(x + \Delta, t)$ in powers of Δ.)

In solving eq. (9), we may as well (as Einstein remarks) take advantage of our assumption that the particles move independently of each other, and therefore interpret x not as the actual space coordinate of a particle but rather as the distance it has moved from its starting place at $t = 0$. The solution of (9) is then

$$f(x, t) = \frac{n}{\sqrt{4\pi D}} \frac{e^{-x^2/4Dt}}{\sqrt{t}} \tag{11}$$

Einstein could now obtain immediately the root-mean-square displacement of a particle in the x-direction,

$$\lambda_x = \sqrt{\overline{x^2}} = \sqrt{2Dt} \tag{12}$$

Thus, *the mean displacement is proportional to the square root of the time.*

All that is left is to substitute into eq. (12) the expression previously found for the diffusion constant [(eq. (7)]

$$\lambda_x = \sqrt{t}\sqrt{\frac{RT}{N}\frac{1}{3\pi kP}} \tag{13}$$

This is Einstein's final result in this paper, and he noted that it can be used either to predict the mean displacement if one assumes values for N, R, T, K, and P, or else to calculate N if everything else is known.

In this first paper, Einstein did not emphasize very strongly the significance of his result that λ_x is proportional to the square root of the time, and in fact it is quite probable that most early readers of the paper gave up in bewilderment before they got to this result. Einstein wrote two more papers, published in the *Zeitschrift für Elektrochemie* in 1907 and 1908, calling his results to the attention of experimentalists and attempting to explain them more simply.[20] In the first of these papers he pointed out that according to the molecular theory of heat (*i.e.* the Waterston–Maxwell equipartition theorem) the mean-square velocity of a suspended particle should be determined by the equation

$$\frac{m}{2}\overline{v^2} = \frac{3}{2}\frac{RT}{N} \tag{14}$$

For the colloidal platinum solutions investigated by Svedberg,[21] in which the mass of the particles is about 2.5×10^{-15} g, this equation indicates a root-mean-square velocity of 8.6 cm/sec. However, Einstein showed that there is no possibility of observing this velocity, because of the very rapid viscous damping, which one can calculate from the Stokes formula. The velocity of such a particle would drop to $\frac{1}{10}$ of its initial value in about 3.3×10^{-7} seconds. "But, at the same time," said Einstein, "we must assume that the particle gets new impulses to movement during this time by some process that is the inverse of viscosity, so that it retains a velocity which on an average is equal to $\sqrt{\overline{v^2}}$. But since we must imagine that the direction and magnitude of these impulses are (approximately) independent of the original direction of motion and velocity of the particle, we must conclude that the velocity and direction of motion of the particle will be already very greatly altered in the extraordinarily short time $Q[=3.3 \times 10^{-7}$ seconds], and, indeed, in a totally irregular manner. It is therefore impossible–at least for ultramicroscopic particles–to ascertain $\sqrt{\overline{v^2}}$ by observation."[22]

According to Einstein's theory, the mean velocity in an interval τ will be inversely proportional to $\sqrt{\tau}$; that is, it increases without limit as the time interval becomes smaller. Hence any attempt to measure the "instantaneous" velocity of particles in Brownian movement will give erratic and meaningless results. It is for just this reason that all the efforts of experimentalists, who knew nothing more of kinetic theory than the equipartition theorem, had failed to lead to any definite conclusion about the average speeds of suspended particles. They were simply measuring the wrong thing until Einstein pointed out that only the ratio of mean-square displacement to time could be expected to have any theoretical significance. One can hardly find a better example in the history of science of the complete failure of experiment and observation, unguided (until 1905) by theory, to unearth the simple laws governing a phenomenon.

The peculiar nature of random motion governed by a diffusion equation had reared its head in physics once before. In 1854, William Thomson (later Lord Kelvin) had applied the diffusion equation (*i.e.* Fourier's equation for heat conduction) in his theoretical studies of the motion of electricity in telegraph lines.[23] After going through almost exactly the same mathematical analysis that Einstein was to make 50 years later, Thomson wrote:

> We may infer that the retardations of signals are proportional to the squares of the distances, and not to the distances simply; and hence different observers, believing they have found a "velocity of electric propagation," may well have obtained widely discrepant results; and the apparent velocity would, *caeteris paribus*, be the less, the greater the length of wire used in the observation.[24]

§15.4]   EINSTEIN'S THEORY OF BROWNIAN MOVEMENT   683

In an article on the "Velocity of electricity" published in *Nichol's Cyclopedia* in 1856, Thomson quoted nine different results for the "velocity of electricity" ranging from 1430 to 288 000 miles per second, and pointed out that the diversity of values measured under different conditions can probably be attributed to the fact that the actual time required for an electric impulse to get from one place to another is proportional to the square of the distance.[25] The validity of Thomson's "law of squares" was of considerable economic importance at the time because the technical problems involved in laying the Atlantic Cable were just then being thrashed out. Thomson had to defend his theoretical prediction against experiments that appeared to contradict it; those engineers who were enthusiastic about pushing ahead with the cable did not like the law of squares. Thomson argued that the transmission of messages would be very slow at large distances unless the cable is made very thick, and said:

> Capitalists ought to require a very "matter-of-fact" proof of the attainability of a sufficient rapidity in the communication of actual messages, by whatever cable may be proposed, before sinking so large an amount of property in the Atlantic, as would be involved in any cable of ordinary or of extraordinary great lateral dimensions to form an electric communication between Britain and America.[26]

Apparently the scientists who attempted to measure the velocity of particles in Browian movement later in the 19th century had not followed the dispute about Thomson's law of squares in the electric telegraph problem, and they obtained a similar collection of wildly varying results, none of them in agreement with the equipartition theorem.[27]

Einstein was also well aware that his formula for the mean-square displacement could not be applicable to very short time intervals. The formula (12) seems to imply an infinite instantaneous velocity. But, as Einstein noted in a second paper written in 1905, "we have implicitly assumed in our development that the events during the time t are to be looked upon as phenomena independent of the events in the time immediately preceding. But this assumption becomes harder to justify the smaller the time t is chosen."[28] One can ignore this objection and retain the assumption that events at any time are completely independent of those at any other time; one then has what might be called a mathematical idealization of Brownian movement, sometimes called a "Wiener process."[29] The most common method for modifying the assumptions at short times to give a physically reasonable result leads to the "Ornstein–Uhlenbeck process."[30] Both processes are still of considerable interest in mathematical physics and other applications, and there is really no need to choose between them except for a particular purpose.

Another important result obtained by Einstein in his second paper is a formula for the probability distribution of the vertical distance of a particle of density ρ and volume v from the bottom of a container

$$dW = \text{const.} \exp\left[-(N/RT)v(\rho - \rho_0)gx\right] dx \qquad (15)$$

Since this formula follows directly from the Maxwell–Boltzmann distribution law,[31] we will not discuss its derivation. The formula was used by Perrin in some of his experimental work,[32] to determine N, and thus played an important role in establishing the "real existence" of the atom (see §15.6).

Notes for §15.4

1. *Albert Einstein Philosopher-Scientist*, ed. P. A. Schilpp (New York: Library of Living Philosophers, 1949; reprinted by Harper, 1959), p. 1 (quoted from pp. 46–49).
2. Max Born, in *Albert Einstein Philosopher-Scientist*, p. 161.
3. A. Einstein, *Ann. Phys.* [4] **17**, 549 (1905); see publications cited in notes 17 and 18, below.
4. On Einstein's early work in statistical mechanics see M. J. Klein, *Natural Philosopher* **2**, 59 (1963); *Science* **157**, 509 (1967). J. Mehra, *Physica* **79A**, 447 (1975).
5. G. G. Stokes, *Trans. Camb. Phil. Soc.* **8**, 287 (1845, pub. 1849); *Mathematical and Physical Papers* (Cambridge, 1880–1905; New York: Johnson Reprint, 1966), **1**, 75. See also his review of earlier work of Challis, Green, Airy, Barré de Saint-Venant, Navier, Poisson, and others, in *B.A. Rep.* **16**, 1 (1846); *Papers* **1**, 157.

   It should be noted that in the work discussed here Stokes used an equation less general than what is now called the "Navier–Stokes equation," in that he assumed that the terms involving the dependence of pressure on the rate of change of density could be dropped. This is tantamount to ignoring what is now called the "bulk viscosity" as an independent fluid parameter.
6. G. G. Stokes, *Trans. Camb. Phil. Soc.* **9**, [8]–[106] (1850, pub. 1856); *Papers* **3**, 1.
7. *Papers* **3**, 3, quoted from p. 232 of Sabine's paper in *Phil. Trans.* **119**, 207 (1829). On the misleading conclusions derived by Stokes from Sabine's experiment see §5.2, note 8.
8. *Papers*, **3**, 7. The *Dictionary of National Biography* has an interesting article on Sir James South.
9. *Papers*, **3**, 10.
10. G. G. Stokes, *Phil. Mag.* [3] **34**, 343 (1848); *Papers* **2**, 8.
11. E. T. Whittaker, *A History of the Theories of Aether and Electricity* (London: Nelson, 1951). K. F. Schaffner, *Nineteenth-Century Aether Theories* (New York: Pergamon Press, 1972).

12. G. R. Kirchhoff, *Vorlesungen über Mechanik* [*Vorlesungen über Mathematische Physik*, erster Band] (Leipzig, 4. Aufl. 1897), Sechsundzwanzigste Vorlesung, p. 380.
13. See for example H. A. Lorentz, *Abhandlungen über theoretische Physik* (Leipzig and Berlin: Teubner, 1907), p. 23, based on a paper first published in 1896. Lorentz cited Kirchhoff's *Mechanik* for the derivation of the Stokes formula. On Einstein's reading of Lorentz see R. S. Shankland, *Am. J. Phys.* **41**, 895 (1973). The formula is also derived in August Föppl's *Vorlesungen über technischen Mechanik* (Leipzig: Teubner, 1910), **6**, §71, a work which Einstein may have consulted; cf. G. Holton, *Thematic Origins of Scientific Thought* (Cambridge, Mass.: Harvard University Press, 1973), p. 207.
14. J. J. Thomson, *Phil. Mag.* [5] **46**, 528 (1898); this work is mentioned in Lorentz's paper cited in note 13.
15. J. H. van't Hoff, *Kongliga Svenska Vetenskaps-Academiens Handlingar*, Stockholm, **21** (17), 1 (1884); for further details see J. R. Partington, *A History of Chemistry* (London: Macmillan, 1964), **4**, 654f.
16. O. Reynolds, *Phil. Trans.* **174**, 935 (1884). H. Rouse and S. Ince, *History of Hydraulics* (Iowa City: Iowa Institute of Hydraulic Research, 1957; New York: Dover Pubs., 1963).
17. Albert Einstein, *Untersuchungen über die Theorie der Brownschen Bewegungen*, hrsg. R. Fürth (Leipzig: Akademische Verlagsgesellschaft, 1922).
18. Albert Einstein, *Investigations on the Theory of the Brownian Movement*, trans. A. D. Cowper (London: Methuen, 1926; New York: Dover Pubs., 1956).
19. *Ibid.*, p. 12.
20. *Ibid.*, pp. 63–67, 68–85. (The translation is marred by several misprints which have not been corrected in the reprint.)
21. T. Svedberg, *Z. Elektrochemie* **12**, 853, 909 (1906).
22. A. Einstein, *op. cit.* (note 18), p. 66.
23. William Thomson, *Proc. R.S. London* **7**, 382 (1855), reprinted in his *Mathematical and Physical Papers*, (Cambridge, 1884), **2**, 61.
24. Thomson, *Papers* 65.
25. *Ibid.*, pp. 131–37.
26. *Ibid.*, pp. 92–102 (reprinted from *Athenaeum*, 1856).
27. See for example the table of results of Regnauld, Wiener, Ramsay, R. Exner, and Zsigmondy, in T. Svedberg, *Ion* **1**, 373 (1909).
28. Einstein, *Investigations*, p. 34.
29. Norbert Wiener, *Proc. Nat. Acad. Sci. U.S.A.* **7**, 253, 294 (1921) and many other papers. The modern theory is summarized by E. B. Dynkin, *Markov Processes*, trans. from Russian (Berlin: Springer-Verlag, 1965). On the connection with quantum statistical mechanics see S. G. Brush, *Reviews of Modern Physics* **33**, 79 (1961).
30. G. E. Uhlenbeck and L. S. Ornstein, *Phys. Rev.* [2] **36**, 823 (1930).
31. L. Boltzmann, *Wien. Ber.* **58**, 517 (1868), J. C. Maxwell, *Nature* **8**, 537 (1873); see our discussion in §§5.3, 6.1 and 10.6. On the justification of the $-\rho_0$ term (buoyancy correction) in the exponential see R. Baierlein, *Am. J. Phys.* **37**, 315 (1969).
32. J. Perrin, *C.R. Paris* **146**, 967 (1908), reprinted in *Oeuvres Scientifiques de Jean Perrin* (Paris: Centre National de la Recherche Scientifique, 1950). The use of this experiment to measure Boltzmann's constant has been described by M. Horne, P. Farago, and J. Oliver, *Am. J. Phys.* **41**, 344 (1973).

15.5 Smoluchowski's theory of Brownian movement

Shortly after the publication of Einstein's first paper on Brownian movement in 1905, there appeared a paper by the Polish physicist Marian von Smoluchowski on the same subject.[1] Smoluchowski begins by citing Einstein's work, and says that Einstein's results "agree completely with those that I obtained several years ago by following a completely different line of thought." However, he thinks that his own method is "more direct, simpler, and more convincing than Einstein's." This judgment is of course a matter of taste, but it is probably true that most physicists and chemists at that time found it easier to follow Smoluchowski's arguments, based on combinatorics and the mean-free-path approximation of kinetic theory, than Einstein's, based on abstract statistical mechanics and the diffusion equation. Smoluchowski also made a much greater effort to review the experimental results that had some bearing on the theory, as well as to justify his theoretical assumptions.

According to Smoluchowski, observations show that the motion becomes more rapid, the smaller the diameter of the particles, though few reliable measurements have been made. Different observers have reported contradictory results concerning the influence of the medium, but it is clear that the motion is liveliest in fluids of the smallest viscosity. The motion is almost entirely independent of external influences. Smoluchowski then cites and criticizes several theoretical explanations of Brownian movement that have been proposed:

> ... there follows immediately from [the independence of external influences] the untenability of any theories based on an external energy source; and especially any supposition that one has here convection streams originating from temperature inequalities. The inadmissibility of these latter explanations follows moreover from simple considerations of another kind. The motions would have to stop completely in water at a temperature of 4°, whereas actually they continue down to the freezing point with scarcely diminished strength (Meade Bache). The reduction of the thickness of the liquid layer to a small fraction of a millimeter by placing it on a cover glass would be expected to diminish the mobility greatly, whereas there is no evidence of this effect. Calculations show that in this case a temperature drop of the order of magnitude of 100 000° in 1 cm would be necessary to create a convection stream of the observed velocity. It is known that such streams occur in

containers of larger dimensions, but these collective motions of a larger number of particles are completely different from the irregular vibrating Brownian motions.

(Here Smoluchowski attacks one of the most common misconceptions about the molecular-kinetic explanation of Brownian movement: the idea that a large number of molecules near the suspended particle must move in unison in order to produce the observed motion. Smoluchowski shows, below, that such cooperative motions need not be postulated arbitrarily, but rather that equivalent *fluctuations* must be expected as a natural consequence of the randomness of molecular motions.)

It may be noted [he continues] that the maximum temperature difference produced in the neighborhood of a spherical completely black particle exposed to direct sunlight is $ca/k = (\frac{1}{300})°$ (assuming: radiation intensity $c_3 = \frac{1}{30}$, radius $a = 10^{-4}$ cm, thermal conductivity $k = 10^{-3}$ (water)). This is in agreement with the previous remarks about the impossibility of Regnauld's explanation on the basis of the origin of convection streams in the neighborhood of a particle as a result of absorption of radiation at its surface.

The independence of the Brownian phenomenon of the intensity of illumination also contradicts Koláček's and Quincke's theories, which find in it an analogy to radiometer motions, or to the various phenomena of periodic capillary motions investigated by Quincke, respectively. It seems very difficult to understand how a continuous radiation can give rise to the periodic expansion of warmer over colder fluid layers at the surface of each particle, assumed by Quincke, and how there can be any connection between the extraordinary phenomenon of periodic capillary motion that occurs in certain cases (oil in soap solution, alcohol in salt solution, etc.) and the very general phenomenon of Brownian motion which is independent of the substance. It is indeed very probable that a sufficiently strong radiation can give rise to motions, but these would be completely different from Brownian motions.

There remain therefore only the theories that assume internal energy sources. We must reject at the outset the hypotheses of the existence of mutual repulsive forces (Meade Bache) and of electrical forces of a similar kind (Jevons, Raehlmann), since these could effect only a certain arrangement of the particles but not a progressive motion, and in particular since their nature presents a new problem.

Also the opinion that one has here to do with phenomena of capillary energy is untenable. Maltezos assumes that small impurities are the basic cause which disturbs the capillary equilibrium, while Mensbrugghe refers to the example of bits of camphor dancing on water. However, this does not explain the fact that intentional impurities have no effect, that completely insoluble substances (diamond, graphite) move, and especially that the motions do not stop at the time when everything should be balanced. The microscopically small gas bubbles enclosed in minerals must certainly have reached the capillary equilibrium state, yet they still move.

Having gone through the ritual of demolishing all other explanations of the phenomenon (a procedure which Einstein did not consider necessary), Smoluchowski turns to the kinetic theory which he clearly had already decided to use. At this point it should be noted that Smoluchowski had already done some substantial research in the kinetic theory of rarefied gases–in particular, the problem of the temperature-discontinuity at a solid surface–and had just published a paper extending the work of Jeans on the "persistence of velocities" in collisions of gas molecules.[2] Jeans had been trying to improve on the elementary mean-free-path calculations for kinetic theory, in which it was usually assumed that after a collision, a molecule "forgets" its previous motion and simply assumes the average velocity and direction of motion characteristic of the place in the gas where the collision occurred. That assumption had been the basis of Maxwell's original calculation of the viscosity of a gas in 1859, though Maxwell later abandoned it when he developed his more general theory based on transfer equations.[3] Nevertheless, the mean-free-path theory was easier to apply in many calculations, since Maxwell's equations could only be solved in the special case of inverse fifth power repulsive forces. Smoluchowski was thus carrying on a type of research previously pursued by Clausius, Maxwell, O. E. Meyer, Tait, and Jeans, in which one tries to describe the effect of collisions on the path of a molecule and thus on the properties of a gas (§12.2). Einstein, on the other hand, was working along lines first opened up by Boltzmann, Maxwell (in his later papers), and Gibbs; there, the objective was to deduce more general results from a postulated probability distribution for configurations of the entire system of molecules (described by

§15.5] SMOLUCHOWSKI'S THEORY OF BROWNIAN MOVEMENT 689

generalized coordinates), without making specific assumptions about the intermolecular forces and collisions that determine transitions from one configuration to another. The two methodologies–kinetic theory and statistical mechanics–lead to similar results in one region of application, the equilibrium properties of gases, but kinetic theory could be extended to the calculation of transport properties of gases, whereas statistical mechanics could be extended to the calculation of equilibrium properties of liquids and solids. The explanation of Brownian motion involves a subject which neither kinetic theory nor statistical mechanics could claim to have conquered: transport properties and fluctuations in liquids. It was therefore valuable to have both viewpoints brought to bear on the problem.

Smoluchowski recognized the reason why previous attempts to apply kinetic theory to Brownian movement had failed: they had been based either on a naive invocation of the equipartition theorem, or on a consideration of individual collisions of the molecules with the suspended particles. He mentions Nägeli,[4] who thought that he could refute the kinetic explanation "by indicating the smallness of the velocity produced by a collision. Thus a molecule of water, colliding with a particle of diameter 10^{-4} cm (and of density 1), would impart to it a velocity of only 3×10^{-6} cm/sec, which is much less than the order of magnitude of Brownian motion. In actuality the successive impulses would combine with each other, but Nägeli thought that on the average they must cancel out, since they act in all directions of space, and that the end result could not be noticeably greater."[5] Smoluchowski shows the fallacy of this argument, and at the same time gives us a glimpse of an alternative derivation of Einstein's result that the mean displacement is proportional to the square root of the time interval, by a simple combinatorial calculation. Suppose a gambling game consists of a sequence of random events–for example, throws of dice–in which there is an equal probability of winning or losing each time. Then the probability that in n throws there will be m favorable and $n-m$ unfavorable ones (hence a net gain of $2m - n$) is

$$p_{n,m} = \frac{n!}{2^n m!(n-m)!} = \frac{1}{2^n} \binom{n}{m} \tag{1}$$

The average positive or negative deviation from the value zero (*i.e.* the average of the absolute value of the net total gain or loss in n throws) is

$$\nu = 2 \sum_{m=n/2}^{n} (2m-n)p_{n,m} = \frac{n}{2^n} \binom{n}{n/2} \tag{2}$$

For large n, this reduces to

$$\nu = \sqrt{\frac{2n}{\pi}} \tag{3}$$

According to eq. (3), the *velocity* acquired by a suspended particle as a result of random impacts of molecules will be proportional to the square root of the number of impacts. Thus, if there are 10^{16} impacts per second, each of which transfers a velocity component in the X direction of $\pm 10^{-6}$ cm/sec, then each particle will acquire a velocity of 100 cm/sec after one second.

While this calculation shows the error in Nägeli's argument, it does not by itself lead to the correct result, according to Smoluchowski. It is not true, for example, that each collision changes the velocity of the particle by the same amount on the average; instead, the probability of increases in the velocity becomes less, the greater the velocity itself. One expects that in the equilibrium state the particles will have an average velocity given by the equipartition theorem,

$$C = c\sqrt{\frac{m}{M}} \tag{4}$$

(c = average velocity of the molecules, m = mass of molecules, M = mass of particles.) However, the velocity calculated from this formula is generally much larger than what is actually observed (1000 times larger, in one case). This paradox can easily be explained if we remember that according to kinetic theory the particle will be changing its direction of motion as a result of molecular impacts more than 10^{16} times per second, so that we cannot observe the instantaneous velocity but only the displacement over a large number of segments of a zigzag path. This is of course just Einstein's explanation, but it gains concreteness with the help of the kinetic-theory picture.

In constructing his quantitative theory, Smoluchowski argues that the magnitude of the velocity of the particle will always fluctuate around its equilibrium value given by (4), but its direction will change by a small amount at each impact of a molecule. The average change in direction is assumed to be $3mc/4M$, according to "the laws of collisions of elastic spheres."[6] Therefore he assumes that the magnitude of the velocity of the particle is always constant, but its direction changes at each impact by an amount $3mc/4MC$ in a random direction. He also assumes that the molecular impacts occur at equal time intervals, so that the path of the particle is a chain made up of segments of equal lengths.

§15.5] SMOLUCHOWSKI'S THEORY OF BROWNIAN MOVEMENT

In all this it is perfectly clear that Smoluchowski is simply adapting the mean-free-path description of the path of a gas molecule, with the single difference that here the persistence of motion after a collision is almost complete, whereas in the case of a gas molecule moving among other gas molecules there is relatively little persistence of motion.

Smoluchowski thus reduces the problem of Brownian movement to the mathematical problem: find the mean-square end-to-end distance of a chain composed of n segments, each of length l, each rotated in a randomly chosen direction from the direction of the preceding one by a small angle ϵ. The general solution, which Smoluchowski obtains by solving a recursion equation involving an integral over n trigonometric functions, is

$$\Delta_n^2 = l^2 \left\{ \frac{2n}{\delta} + 1 - n - 2\frac{(1-\delta)^2 - (1-\delta)^{n+2}}{\delta^2} \right\} \quad (5)$$

where $\delta = 1 - \cos\epsilon \approx \epsilon^2/2$. In the limit when $n\delta$ is small, this reduces to

$$\Delta = nl\left(1 - \frac{n\delta}{6}\right) \quad (6)$$

(This is almost the same as straight-line motion, with a small correction for curvature.) When n is large, the root-mean-square displacement reduces to

$$\Delta = l\sqrt{\frac{2n}{\delta}} \quad (7)$$

Substituting $\epsilon = 3mc/4MC$ ($= 3C/4c$ according to eq. (4)) and $l = C/n$ for the length of each segment (if n segments are traversed in one second), Smoluchowski arrives at the result

$$\Delta = \frac{8}{3}\frac{c}{\sqrt{n}} \quad (8)$$

In order to put his result into a form directly comparable with Einstein's—that is, to express the average displacement in terms of the size of the particle and the viscosity and temperature of the medium—Smoluchowski uses another result from the kinetic theory of gases. The number of collisions per second experienced by a molecule of radius R, at rest in a gas containing N point molecules in unit volume moving with average velocity c, is[6]

$$n = NR^2\pi c \quad (9)$$

The average change in the velocity C caused by each collision is

$2mC/3M$. Hence the resisting force of the medium is

$$S = \frac{2\pi}{3} R^2 \rho c = \frac{2}{3} mn \tag{10}$$

So the number of collisions, n, can be related to the resisting force, S, by the formula

$$n = \frac{3}{2} \frac{S}{m} \tag{11}$$

If we now assume that the Stokes formula for the resisting force is applicable,

$$S = 6\pi\mu R$$

we can substitute $n = 9\pi\mu R/m$ into equation (5.8) and obtain the result

$$\Delta = \frac{8}{9\sqrt{\pi}} \frac{c\sqrt{m}}{\sqrt{\mu R}} \tag{12}$$

Recalling that Einstein's λ_x in eq. (13) [§15.4] is the average component of displacement in the x-direction, and thus corresponds to Smoluchowski's Δ divided by $\sqrt{3}$, and remembering that $RT/N = \overline{mc^2}/3$, we see that Smoluchowski's result is smaller than Einstein's by a factor of $\sqrt{\frac{27}{64}}$ but is otherwise identical.

(It should be noted that Smoluchowski did not write down explicitly the step corresponding to our eq. (11), and that his eq. (23) has a misprint, ρ in place of S, so his derivation is somewhat puzzling at first sight.)

The slight discrepancy in the numerical factor is perhaps not surprising in view of the various approximations used by Einstein and Smoluchowski; as Fürth points out in his notes to Smoluchowski's paper in the *Klassiker* reprint, Smoluchowski himself adopted Einstein's formula in his later papers.[7]

Notes for §15.5

1. M. R. von Smolan Smoluchowski, *Rozprawy Kraków* **A46**, 257 (1906): German trans. in *Ann. Phys.* [4] **21**, 756 (1906), reprinted in *Abhandlungen über die Brownsche Bewegung und verwandte Erscheinigen* (Leipzig: Akademische Verlagsgesellschaft m. b. H., 1923). The quotations in the text are my translation from the German version. A. Teske, *The History of Physics and the Philosophy of Science: Selected Essays* (Wroclaw: Ossolineum, 1972), p. 34.
2. M. R. von Smolan Smoluchowski, *Rozprawy Kraków* **A46**, 129 (1906); French trans.

reprinted in *Pisma Marjana Smoluchowskiego* (Cracovie: Academic Polonaise des Sciences et des Lettres, 1924–27), **1**, 479.
3. See §12.3.
4. See §15.3, note 1.
5. Smoluchowski, *Abhandlungen über die Brownsche Bewegung*, p. 7.
6. *Ibid.*, p. 9. In his attempt to reproduce this result, Fürth obtains the result 0.806 mc/M (*Ibid.*, pp. 116–17).
7. *Ibid.*, pp. 118–19.

15.6 Perrin's experiments and the reality of atoms

The French physicist Jean Perrin (1870–1942) occupies a pivotal position in the history of Brownian movement. He is generally credited with having established the Einstein–Smoluchowski theory by his experiments, but just as important was his role as propagandist for atomism. In addition, his emphasis on the analogy of Brownian-movement paths with non-differentiable functions in mathematics seems to have stimulated some of the later research on functional integrals, especially that of Wiener.[1]

Perrin's earlier work had been in the field of cathode rays and X-rays. In his first paper (1895), he found evidence that cathode rays are negatively charged particles, and in 1896 he was awarded the Joule prize of the Royal Society of London for his experiments on cathode rays. After an interval of several years, during which he was occupied with organizing a course in physical chemistry at the Sorbonne, he returned to the laboratory in 1903 and started to work on contact electricity and colloidal solutions.[2]

Two works published during this pedagogical interlude give some indication of his familiarity with contemporary theoretical physics. In a short contribution to a symposium on molecular hypotheses at Paris in 1901, he proposed a "nucleo-planetary" model of atomic structure, with a positively-charged "sun" surrounded by many smaller negatively charged "planets." The periods of rotation of these planets might correspond to different wavelengths in the emission spectrum.[3] In 1903, he published the first volume of a textbook on physical chemistry. In the preface to this book, he reviewed the status of molecular hypotheses. While conceding that science should not base itself on atomism if that meant simply reducing the visible to the invisible or unknowable, he maintained that atomic hypotheses could be legitimate if they dealt with sensations that were at least possible even if they had not yet been realized. He suggested, as an analogy, that the germ theory of disease might have been developed and successfully tested before the invention of the microscope; the microbes would have been hypothetical entities, yet, as we know now, they could eventually be observed. Perrin was therefore receptive to atomistic theories, all the more so because he seemed to think that the alternative, "energetics," had degenerated into a pseudo-religious cult.[4]

Perrin's interest in Brownian movement was first clearly shown in a lecture he gave to the Société de Philosophie in 1906. In this lecture he discussed the meaning and limits of validity of the Second Law of Thermodynamics, and raised the question of its compatibility with molecular hypotheses.[5] Boltzmann had maintained that the Second Law is perfectly consistent with kinetic theory as long as one demands only that the law be *statistically* valid; fluctuations that correspond to entropy decreases can occur, but so rarely that it is extremely improbable that they would be observed in any actual experiment. Critics of the kinetic theory, such as Zermelo, argued that this "merely" statistical validity was not good enough; irreversibility is a fundamental property of natural processes, and any molecular hypothesis – or perhaps all conceivable molecular hypotheses based on Newtonian mechanics – that permits any exceptions must be wrong (§14.7). Referring to this criticism (which he attributes to Lippmann) Perrin said:

> If one recalls all the beautiful discoveries that we owe to molecular hypotheses, he would hesitate to support this radical opinion. But since one has no direct proof of the existence of molecules, he can only go by esthetic reasons if he has not succeeded in proving by experimental arguments that the second law does not have the character of absolute rigor, in the name of which one would sacrifice the molecular theories. We are going to try to show that such arguments exist.
>
> Briefly, we are going to show that sufficiently careful observation reveals that at every instant, in a mass of fluid, there is an irregular spontaneous agitation which cannot be reconciled with Carnot's principle except just on the condition of admitting that his principle has the probabilistic character suggested to us by molecular hypotheses.[6]

He then discussed the phenomena of Brownian movement, together with a qualitative kinetic explanation, and then proposes a method for violating the Second Law. Suppose one starts with a liquid containing particles only in its lower layers. As a result of Brownian movement, these particles will tend to move upwards. By inserting in the liquid a piston made of a membrane permeable to the liquid but not to the particles, one could obtain mechanical work from this upward motion.

This is just the concept of "osmotic pressure" of the particles in Brownian movement that Einstein had used in his theoretical paper the previous year, though at this point Perrin did not mention Einstein.

In another popularization, written at about the time for the *Revue du Mois*, Perrin drew an analogy between the physical discontinuity of matter and the mathematical properties of curves without tangents.[7] He pointed out that in teaching the concept of limit, we usually draw a curve and calculate the average velocity of a point moving on the curve from the quotient of its finite displacement and corresponding finite time interval. Then we say to the student: "You understand, don't you, that when this distance tends to zero, the average velocity tends to a limit." If the student is not too bright or too critical he will be intimidated by this demonstration and agree that of course the curve has an instantaneous velocity or tangent at any point.

But, said Perrin, the mathematicians of the preceding century have finally recognized that it is futile to attempt to prove rigorously, by such geometric arguments, that every continuous function has a derivative.

> But they still thought the only interesting functions were the ones that can be differentiated. Now, however, an important school, developing with rigor the notion of continuity, has created a new mathematics, within which the old theory of functions is only the study (profound, to be sure) of a group of singular cases. It is curves with derivatives that are now the exceptions; or, if one prefers the geometrical language, curves with no tangent at any point become the rule, while the familiar regular curves become some kind of curiosities, doubtless interesting, but still very special.

There are still those, he noted, who consider themselves men of good sense, who would say that these developments, while interesting to mathematicians, have nothing to do with the real world. But they are quite wrong, Perrin asserted, for one has only to look in the microscope to see that ordinary matter, which appeared perfectly continuous, homogeneous, and static to the naked eye, is really discontinuous, heterogeneous, and dynamic. In short, we have no reason to think that the properties of matter vary in a regular way as we go from one point of space to another. The ultramicroscope, recently developed by Siedentopf, Zsygmondy, Cotton, and Mouton, reveals the fine structure of matter more clearly than was ever possible before. Again, Perrin gave a qualitative discussion of Brownian movement, and suggested that such phenomena support the kinetic theory, but now he suggests that his interest in the subject is more than that of an expositor and commentator on the state of other people's work:

> I will say only, announcing the results of reasonings that will be detailed in a subsequent article of M. Langevin, that the properties of fluids imply that molecules have diameters of about ten-millionth of a millimeter, and masses so small that a hundred billion molecules of water weigh scarcely a thousandth of a milligram.[8]

Langevin's simplified version of the Einstein theory was presented to the Académie des Sciences in Paris on 9 March, 1908.[9] Perrin wrote later, "ever since I became, through M. Langevin, acquainted with the theory, it has been my aim to apply to it the test of experiment."[10] In the previous year, Seddig[11] and Svedberg[12] had attempted to verify Einstein's formula (eq. 13, §15.4), but Perrin did not think their results were conclusive.[13]

In his first series of experiments, Perrin attempted to verify the formula (15) in §15.4 for the equilibrium distribution of the particles, which he wrote in the form

$$2.3 \log \frac{n_0}{n} = \frac{1}{k} mgh \left(1 - \frac{1}{\rho}\right) \qquad (1)$$

It will be recalled that this equation can be derived directly from the Maxwell–Boltzmann distribution law; the derivation does not depend on the more dubious assumptions which Einstein used in obtaining his displacement formula, such as the validity of Stokes' law for very small particles. However, Perrin did use the Stokes law to determine the mass of the particles in an associated experiment. He found that by taking the value of Avogadro's number that could then be estimated from kinetic theory, *i.e.* 7×10^{23}, he could fit his data very well; the agreement was even better if he chose $N = 6.7 \times 10^{23}$. He concluded:

> The average kinetic energy of a granule of the colloid is therefore equal to that of a molecule. This is, established by experiment, the hypothesis that Einstein and Langevin have indicated as equivalent to that of M. Gouy (theorem of equipartition of kinetic energies). At the same time, the kinetic theory of fluids seems to gain some support, and molecules become a little more tangible.[14]

At the meeting of the Académie des Sciences on 18 May 1908, one week after the presentation of Perrin's first report, Victor Henri reported a cinematographic study of Brownian movement, conducted in order to check Einstein's displacement formula.[15] Svedberg[12] had found that the displacements are six or seven times larger than those calculated from Einstein's formula. Henri's experimental results were four times larger than the theoretical prediction, although the proportionality of Δ^2 to t was confirmed. Henri suggested that perhaps Stokes' law does not apply to such small particles.

Since the use of Stokes' law appeared to be the weakest link in Einstein's chain of deductions, Perrin next carried out a direct experimental test of the law for small particles of gamboge. He concluded that Stokes' law is valid, at least for the average displacement of a particle in a short time, for particles as small as $\frac{1}{10}$ of a micron.[16]

Returning to the experiment on distribution of particles, Perrin presented more data and suggested that Brownian movement might offer a new and more precise way of determining Avogadro's number.[17] He followed up this suggestion in a note presented on 5 October 1908, in which he reviewed four methods of determining N: (1) from kinetic theory; (2) from electrolysis, combined with measurements of the electronic charge; (3) by Planck and Lorentz, from Planck's "beautiful electromagnetic theory of black-body radiation;" (4) from Brownian movement. Perrin's best value of N is now 71×10^{22}; the corresponding value of the electronic charge would be 4.1×10^{-10} esu.[18]

On 30 November 1908, Chaudesaiges, a student working in Perrin's laboratory, reported on a new test of the Einstein displacement formula, using spherical grains of gamboge whose radius had been precisely measured.[19] Chaudesaiges found that Einstein's formula is completely exact, not only with respect to the proportionality constant, provided that one takes $N = 64 \times 10^{22}$. The distribution of displacements was stated to follow the law of errors, as expected.

Perrin conducted some further experiments, to verify Einstein's formula[20] for rotational motion,[21] and to obtain more accurate values of N and atomic parameters such as the electronic charge.[22] In 1909, he was awarded the Prix Gaston Planté for his work on cathode rays, X-rays, and especially Brownian movement.[2] For the next few years, he seems to have devoted much of his time to popularizing the significance of his work on Brownian movement, in particular the idea that atoms have now been proved to exist.[23] In this he was surprisingly successful. In fact, the willingness of scientists to believe in the "reality" of atoms after 1908, in contrast to previous insistence on their "hypothetical" character, is quite amazing.

The evidence provided by the Brownian-movement experiments of Perrin and others seems rather flimsy, compared to what was already available from other sources. The fact that one could determine

Avogadro's number and the charge on the electron by one more method seems hardly sufficient to justify such profound metaphysical conclusions. Several independent methods of determining these parameters had been known since 1870 or before, to say nothing of the many successes of kinetic theory in predicting the properties of gases. Perhaps it was the novelty of Einstein's deduction of the displacement formula, tying together seemingly unrelated properties of liquids, that startled scientists into conceding that the evidence for atomism had now become irrefutable. But the evidence for the quantitative validity of the displacement formula was not yet very good. Perrin had to explain away the discordant results of Victor Henri in order to be able to claim that Chaudesaige's measurements constituted a verification of the theory:

> It is necessary to admit that some unknown complication or some systematic source of error has falsified the results of Victor Henri, for the measurements that I am going to summarize leave no doubt of the rigorous exactitude of the formula proposed by Einstein.[24]

The following statements are typical of the reactions of scientists to Perrin's work on Brownian movement. Nernst, in the 6th edition of his *Theoretische Chemie* in 1909, added a new section on kinetic theory and heat, in which he discussed Brownian movement in connection with the limits of validity of the Second Law. He concluded the section with the remark:

> In view of the *ocular* confirmation of the picture which the kinetic theory provides us of the world of molecules, one must admit that this theory begins to lose its hypothetical character.[25]

Arrhenius, in a lecture in Paris on 13 March 1911, reviewed the work of Perrin and Svedberg and said:

> After this, it does not seem possible to doubt that the molecular theory entertained by the philosophers of antiquity, Leucippus and Democritos, has attained the truth, at least in essentials.[26]

The most dramatic reaction was that of Ostwald, because he had been an outspoken critic of atomism as late as 1906. In his Ingersoll lecture at Harvard, he had said

> ... as I have been maintaining for the last ten years, the matter-and-motion theory (or scientific materialism) has outgrown itself and must be replaced by another theory, to which the name *Energetics* has been given.... The question as to the identity or non-identity of

the different portions of water is without meaning, since there is no means of singling out the individual parts of the water and identifying them . . . atoms are only hypothetical things . . . .[27]

In the preface of the 4th edition of his *Grundriss der allgemeinen Chemie*, written in 1909, Ostwald completely reversed himself:

> *I have convinced myself that we have recently come into possession of experimental proof of the discrete or grainy nature of matter, for which the atomic hypothesis had vainly sought for centuries, even millenia.* The isolation and counting of gas ions on the one hand–which the exhaustive and excellent work of J. J. Thomson has crowned with complete success–and the agreement of Brownian movements with the predictions of the kinetic hypothesis on the other hand, which has been shown by a series of researchers, most completely by J. Perrin–this evidence now justifies even the most cautious scientist in speaking of the *experimental* proof of the atomistic nature of space-filling matter. What has up to now been called the atomistic hypothesis is thereby raised to the level of a well-founded theory, which therefore deserves its place in any textbook intended as an introduction to the scientific subject of general chemistry.[28]

There was only one major dissent from the scientific consensus on the reality of atoms after 1908. Ernst Mach, in 1909, reprinted his essay on the principle of conservation of work, in which he wrote, reviewing the history of the subject:

> . . . it was concluded that, if heat can be transformed into mechanical work, heat consists in mechanical processes–in motion. This conclusion, which has spread over the whole cultivated world like wildfire, had, as an effect, a huge mass of literature on this subject, and now people are everywhere eagerly bent on explaining heat by means of motions; they determine the velocities, the average distances, and the paths of the molecules, and there is hardly a single problem which could not, people say, be completely solved in this way by means of sufficiently long calculations and of different hypotheses. No wonder that in all this clamour the voice of one of the most eminent, that of the great founder of the mechanical theory of heat, J. R. Mayer, is unheard: "Just as little as, from the connexion between the tendency to fall (*Fallkraft*) and motion, we can conclude that the essence of this tendency is motion, just so little does this conclusion hold for heat. Rather might we conclude

the opposite, that, in order to become heat, motion–whether simple or vibrating, like light or radiant heat–must cease to be motion." (*Mechanik der Wärme*, Stuttgart, 1867, p. 9.)..."[29]

If, then, we are astonished at the discovery that heat is motion, we are astonished at something which has never been discovered. It is quite irrelevant for scientific purposes whether we think of heat as a substance or not.[30]

Albert Einstein received a copy of this essay and wrote to Mach as follows (9.8.1909); enclosing copies of some of his own works:

Especially I ask you to glance at the paper on Brownian movement, since here there is a motion that one must believe to be "heat motion."[31]

But Mach was not convinced by this argument, and continued to assert that atomism was merely a hypothesis, though perhaps a useful one (§8.7).

It took a long time for Brownian movement to work itself into the mainstream of physical science. But when it was finally recognized as a phenomenon worthy of serious study, the consequences were striking; and the vindication of the kinetic-molecular theory was only the beginning.

Notes for §15.6

1. Norbert Wiener, *I am a Mathematician* (Garden City, N.Y.: Doubleday, 1956), p. 33f.
2. For biographical information see Mary Jo Nye, *Molecular Reality: A Perspective on the Scientific Work of Jean Perrin* (New York: American Elsevier, 1972); Ferdinand Lot, *Jean Perrin et les Atomes* (Paris: Seghers, 1963); Albert Ranc, *Jean Perrin, un grand Savant au Service du Socialisme* (Paris: Editions de la Liberté, 1945); Louis de Broglie, Hommage National à Jean Perrin (Paris: Institut de France, Academie des Sciences, 1962). "Prix Gaston Planté," *C.R. Paris* **149**, 1207 (1909). V. V. Raman, *Physics Teacher* **8**, 380 (1970).

    A selection of Perrin's papers was reprinted in *Oeuvres Scientifiques de Jean Perrin* (Paris: Centre National de la Recherche Scientifique, 1950), hereafter cited as *Oeuvres*.
3. J. Perrin, *Revue Scientifique* **15**, 449 (1901); *Oeuvres*, p. 165.
4. J. Perrin, *Traité de Chimie Physique. Les Principes* (Paris: Gauthier-Villars, 1903).
5. J. Perrin, *Bull. Soc. Philosophie* **6**, 81 (1906); *Oeuvres*, p. 57.
6. J. Perrin, *Oeuvres*, p. 68.
7. J. Perrin, *Revue du Mois* **1**, 323 (1906).
8. *Ibid.*, p. 339.
9. P. Langevin, *C.R. Paris* **146**, 530 (1908), reprinted in *Oeuvres Scientifiques de Paul Langevin* (Paris: Centre National de la Recherche Scientifique, 1950).

§15.6] PERRIN'S EXPERIMENTS 701

10. J. Perrin, *Atoms*, trans. by D. L. Hammick (London: Constable, 2nd ed., 1923), p. 114. According to Nye (*op. cit.*, note 2, p. 97) Perrin had already begun his experiments before learning of Einstein's theory.
11. Max Seddig, *Sitzungsberichte der Gesellschaft zur Beförderung der gesammten Naturwissenschaften (Marburg)*, p. 182 (1907); *Nat. Rund.* **23**, 377 (1908).
12. Theodor Svedberg, *Studien zur Lehre von der Kolloiden Lösungen* (Upsala, 1907). See also *Die Existenz der Moleküle: Experimentelle Studien* (Leipzig: Akademische Verlagsgesselschaft m.b.H., 1912); *Jahrbuch der Radioactivität* **10**, 467 (1913).
13. J. Perrin, *Atoms*, p. 120, note 1.
14. See §15.4, note 32.
15. Victor Henri, *C.R. Paris* **146**, 1024 (1908).
16. J. Perrin, *C.R. Paris* **147**, 475 (7 Sept. 1908).
17. J. Perrin, *C.R. Paris* **147**, 530 (21 Sept. 1908).
18. J. Perrin, *C.R. Paris* **147**, 594 (1908).
19. Chaudesaiges, *C.R. Paris* **147**, 1044 (30 Nov. 1908).
20. A. Einstein, *Investigations on the Theory of the Brownian Movement*, p. 33 [from *Ann. Phys.*, 1906].
21. J. Perrin, *C.R. Paris* **149**, 549 (1909).
22. Perrin and Dabrowski, *C.R. Paris* **149**, 477 (1909).
23. See the bibliography in *Oeuvres*, pp. vii–xii.
24. J. Perrin, *Ann. Chim. Phys.* [8] **18**, 1 (1909); *Oeuvres*, p. 171 (quotation from p. 214). See also *Atoms*, p. 121.
25. W. Nernst, *Theoretische Chemie* (Stuttgart: Enke, 6. Aufl. 1909), p. 212. E. Rutherford, *Science* **30**, 291 (1909).
26. S. Arrhénius, *Conférences sur quelques thèmes choisis de la chimie physique pure et appliquee, faites à l'Université de Paris du 6 au 13 Mars 1911* (Paris: Hermann, 1912), quotation from p. 12.
27. Wilhelm Ostwald, *Individuality and Immortality* (Boston: Houghton, Mifflin & Co., 1906); quotations from pp. 7, 40–41.
28. Ostwald, *Grundriss der allgemeinen Chemie* (Leipzig: Engelmann, 4. Aufl. 1909), quotation from the "Vorbericht."
29. E. Mach, *Die Geschichte und die Wurzel des Satzes von der Erhaltung der Arbeit*, Vortrag . . . 2. unveränderter Abdruck nach der in Prag 1872 ersch. 1. Aufl. (Leipzig: Barth, 1909); English trans., *History and Root of the Principle of the Conservation of Energy* [sic] (Chicago" Open Court, 1911), quotation from p. 37 of the English trans.
30. *Ibid.*, p. 47.
31. Friedrich Herneck, *Forschungen und Fortschritte* **37**, 239 (1963).

PART D

Bibliography

CHAPTER 16

The Literature of Kinetic Theory

16.1 Quantitative aspects of the history of kinetic theory

In monographs on topics in the history of science it is customary to provide a systematic bibliography of the sources used. It seemed to me that it would not be desirable to load the end of this book with redundant or more complete citations of the thousands of minor works mentioned in the notes to the text. On the other hand since I had already compiled a fairly complete bibliography of *primary* sources for my own use, and I wanted to give full titles and annotations for a number of these works which were not discussed in the text, I decided to give here a list that might serve some other purposes. In §16.3 will be found citations of 539 publications on kinetic theory during the 19th century. In my judgment this list includes 99% of the papers and books that contain either original contributions or specific criticisms of the theory, excluding biographical works, philosophical discussions and other "secondary" sources. Reviews and textbook expositions have been included insofar as they contain substantive (though not necessarily original) material on kinetic theory. In addition I have given birth and death dates of authors and the places where they worked (when this information was available to me), and citations of abstracts, reprints and translations of the papers. (The first citation is the first *full* publication of the paper, even if this came after an abstract published separately.)

With this population of research publications one can follow the growth of the literature in a well-defined subfield of science from its beginning (taking account of the fact that the first publication was

Bernoulli's *Hydrodynamica* in 1738) up to the time when it split into several other subfields (following the introduction of quantum theory, Gibbsian statistical mechanics, Brownian movement theory and other seminal publications in the early 1900's). I hope that this collection (which is of course based on my personal judgment as to which publications to include) will be useful to people who are interested in studying the gross statistical features of the history and sociology of science.

From the bibliography one can easily prepare tables such as table 16.1-1 showing the distribution of dates of publication. The figures do not fit very well the usual "exponential growth" model of science, but can be understood only with some further knowledge of particular events. Thus the great expansion of the literature in the 1870's reflects the entrance of several new scientists into the field, attracted by the pathbreaking works of Clausius, Maxwell, Boltzmann, and van der Waals which suggested new research problems; the considerable interest in the radiometer reinforced this expansion in the latter part of the decade. But the significant decline in the number of publications in

Table 16.1-1. Distribution of dates of publication

Time period	Number of publications	Number of important publications*
Before 1801	1	1
1801–15	0	0
1816–20	1	0
1821–25	14	2
1826–30	0	0
1831–35	1	0
1836–40	2	1
1841–45	2	1
1846–50	5	3
1851–55	2	0
1856–60	12	4
1801–65	15	1
1866–70	21	3
1871–75	55	6
1876–80	79	6
1881–85	49	1
1886–90	79	1
1891–95	113	2
1896–1900	89	7
Total	540	39

* See table 1.1-1.

the 1880's suggests that many of these new scientists were not really competent or persistent enough to solve the problems arising in the kinetic theory, so they moved on to other fields. The second expansion, in the 1890's, seems to be attributable to the fact that certain key problems such as the ratio of specific heats and irreversibility could be discussed without a great deal of elaborate calculation and thus held the interest of several scientists who were not prepared to do full-time research in kinetic theory.

The weakness of all quantitative research in history is the fact that some events, people, and publications are qualitatively more important than others; counting all the papers published on a subject can give a very misleading picture of the actual progress being made in the field. But since one cannot avoid subjective judgments in writing the history of the field, why not go a step further and decide, subjectively, what were the most important publications? I have done this and have given the list in table 1.1-1. There are 39 "important" publications between 1738 and 1900, all but one of them after 1815. Their distribution is shown in table 16.1-1; it indicates that (if you accept my judgments) the first peak of activity, in the 1870's, included substantial progress in the theory (12 important publications), while the second did not (only two important publications during 1891–95 out of a total of 113).

Having defined important publications (by listing them), one can also define "major authors" – those who wrote ten or more publications and/or one or more "important" publications. They are listed in table 16.1-2, along with their annual rates of production for the period when they were active. (This figure is somewhat misleading for people like Herapath and Kelvin who were very active at the beginning and end of the time period but turned their attention to other matters in between.) These 25 major authors accounted for 302 of the publications up to 1900; the other 115 authors accounted for the other 238.

The geographical distribution of kinetic theory research is shown in tables 16.1-3 and 16.1-4. I have guessed at the nationalities of the authors, using their location and the place and language of their major publications; the figures in table 16.1-3 are probably not very accurate but may give some idea of where most of the work, and most of the important work, was going on. It is somewhat easier to pin down the place of publication of a journal or book than the nationality of its author (table 16.1-4). Putting these two tables together one can conclude that the English-speaking and German-speaking countries, with Scotland and Austria making substantial contributions in proportion to their population; there were more minor authors in Germany, but more minor publications in England.

Table 16.1-2. Kinetic theory publications by selected authors, to 1900 (number of articles and books by each person who wrote ten or more publications, and/or one or more "important" publications, as listed in table 1.1-1)

	Dates of first and last publications	Number of publications	Annual rate	Important papers
Bernoulli	1738	1	1	1
Boltzmann	1866–1900	57	1.6	8
Bryan	1891–1900	20	2.0	0
Burbury	1876–1900	25	1.0	1
Clausius	1857–88	24	0.75	3
Herapath	1816–47	12	0.4	4
Joule	1847–48	3	1.5	1
Kelvin	1874–1900	6	0.2	2
Krönig	1856	1	1.0	1
Loschmidt	1865–76	5	0.4	1
Maxwell	1860–79	15	0.75	6
Meyer	1861–91	10	0.9	1
Natanson	1887–97	19	1.7	0
Poincaré	1889–96	7	0.9	1
Preston	1877–91	13	0.9	0
Rayleigh	1873–1900	9	0.3	1
Reynolds	1874–97	11	0.5	1
Smoluchowski	1898–1900	4	1.3	1
Stefan	1863–72	5	0.5	1
Stoney	1858–1900	15	0.3	0
Sutherland	1893–97	6	1.2	1
Tait	1875–96	17	0.8	0
van der Waals	1873–99	9	0.3	1
Waterston	1843–59	5	0.3	2
Zermelo	1896–1900	3	0.6	1

D. S. L. Cardwell presents a graph showing the numbers of papers relating to heat and thermodynamics published in British, French and German periodicals, 1851–1900 [*The Organisation of Science in England* (London: Heinemann, rev. ed. 1972), p. 189]. The British decline from their peak around 1850; the French overtake them in the 1860's and the Germans in the 1880's.

Finally, one can compare the growth of the literature in kinetic theory with that in physics, chemistry, and mathematics as a whole during this period. The figures in table 16.1-5 show that neither fits the exponential model very well, though the *totals* for physical science disciplines might be found to increase more regularly than the relatively small sample used here.

Table 16.1-3. Nationalities of authors

Country	Total number of authors	Major authors as listed in table 16.1-2
England	26	7
Scotland	9	4
Ireland	4	1
U.S.A.	9	0
Australia	1	1
Germany	35	3
Austria	11	3
France	9	1
Holland	10	1
Poland	7	3
Italy	8	0
Russia	6	0
Belgium	2	0
Hungary	1	0
Switzerland	1	1
Czechoslovakia	1	0

Table 16.1-4. Distribution of place of publication and major journals

	Number of publications (total = 540)	Important publications (total = 39)
England		
Nature	73	2
Philosophical Magazine	62	2
British Association Reports	17	0
Royal Society Proceedings and *Philosophical Transactions*	12	5
Annals of Philosophy	11	2
Other journals	14	4
Books	6	2
Total	195	17
Scotland		
Royal Society of Edinburgh Proceedings and Transactions	17	1
Other journals	1	0
Books	1	1
Total	19	2

Table 16.1-4. (cont.)

	Number of publications	Important publications
Ireland		
Journals	6	0
Total	6	0
Germany		
Annalen der Physik	63	5
Other journals	33	1
Books	9	3
Total	105	9
Austria		
Akademie der Wissenschaften, Wien, Sitzungsberichte	60	8
Books	1	0
Total	61	8
France		
Comptes Rendus, Académie des Sciences, Paris	19	0
Other journals	10	0
Books	8	0
Total	37	0
Holland		
Verslagen, K. Akademie van Wetenschappen, Amsterdam	18	0
Archives Néerlandaises	5	0
Other journals	1	0
Books	1	1
Total	25	1
Russia		
Zhurnal Russkago Fiziko-Khimicheskago Obshchestva, Fiz. Chast.	17	0
Other journals	4	0
Total	21	0
Poland		
Journals	18	0
Books	3	1
Total	21	1
Italy		
Journals	14	0
Books	3	9
Total	17	0
United States of America		
Journals	13	0
Total	13	0

Table 16.1-4. (cont.)

	Number of publications	Important publications
Switzerland		
Journals	6	9
Total	6	0
Belgium		
Journals	5	0
Total	5	0
Hungary		
Journals	3	0
Total	3	0
Sweden		
Journals	1	1
Total	1	1
Estonia		
Books	1	0
Total	1	1

Table 16.1-5. Distribution of dates of publication for publications in kinetic theory compared with publications in physics, chemistry, and mathematics*

Decade	Publications in kinetic theory by decade, as percentage of all kinetic-theory publications in period 1801–1900	Publications in physics, chemistry, and mathematics by decade, as percentage of all such publications in period 1801–1900
1801–10	0.0	1.9
1811–20	0.2	2.3
1821–30	2.6	2.8
1831–40	0.5	4.6
1841–50	1.3	6.4
1851–60	2.6	6.9
1861–70	6.7	9.4
1871–80	24.8	15.6
1881–90	23.7	23.2
1891–1900	37.4	27.1

* Based on a random sample of 3768 papers selected from the Royal Society *Catalogue of Scientific Papers, 1800–1900* and classified by Mrs. Lorelei Krakowski. The total number of physics/chemistry/mathematics papers in the sample was 1015; the total number of publications in kinetic theory for this period was 539. (Mrs. Krakowski's work was done as part of a project supported by Contract No. NAS 5-21293 with the National Aeronautics and Space Administration.)

16.2 Journals and abbreviations*

Acta Math.
   Acta Mathematica, Stockholm, 1882– .
Am. J. Math.
   American Journal of Mathematics. Baltimore: Johns Hopkins University, 1878– .
Am. J. Phys.
   American Journal of Physics. New York: American Institute of Physics, 1933– .
Am. J. Sci.
   American Journal of Science, New Haven, Conn., 1818– . Also known as *Silliman's Journal of Science*, 1820–79.
Amsterdam Proc.
   Proceedings of the Section of Sciences, Koninklijke Akademie van Wetenschappen te Amsterdam. (English translation of *Amsterdam Verslagen*, beginning with meeting of 28 May 1898, i.e. *Verslagen* [4] 7.)
Amsterdam Verh.
   Verhandelingen der Koninklijke Akademie van Wetenschappen, Amsterdam, 1854– ; in two sections after 1892.
Amsterdam Verslagen
   Verslagen en Mededeelingen der Koninklijke Akademie van Wetenschappen, Afdeeling Natuurkunde, 1853–65. Tweede Reeks [2] 1866–84. Derde Reeks [3] 1884–92.
Amsterdam Verslagen [4]
   Verslagen van der Zittingen van de Wis- en Natuurkundige Afdeeling der Koninklijke Akademie van Wetenschappen, Amsterdam, 1892– . After 1896 the title changed to: *Verslagen van de Gewone Vergaderingen der Wis- en Natuurkundige Afdeeling*. English translations may be found in *Amsterdam Proc.*, beginning in 1898.
Ann. Chem. Pharm.
   Annalen der Chemie und Pharmacie, Leipzig, 1832– ; also known as *Justus Liebig's Annalen der Chemie und Pharmacie*.
Ann. Chim. Phys.
   Annales de Chimie et de Physique, Paris, 1789– ; [2] 1816–40; [3] 1841–63; [4] 1864–73; [5] 1874–83; [6] 1884–93; [7] 1894–1903; [8] 1904–13; also known as *Annales de Chimie*.
Ann. école norm.
   Annales Scientifique de l'École Normale Supérieure, Paris, 1864– ; [2] 1872–83; [3] 1884– .
Ann. Math.
   Annals of Mathematics, Princeton, 1884– ; [2] 1900– .
Ann. Phil.
   Annals of Philosophy; or, Magazine of Chemistry, Mineralogy, Mechanics, Natural History, Agriculture and the Arts, London, 1813–26; [2] 18 –26; merged into *Phil. Mag.*
Ann. Phys.
   Annalen der Physik und Chemie, 1799– ; [2] 1824–76, also known as *Poggen-*

---

* Information about publication is given for identification purposes but is not intended to be complete; see the Royal Society *Catalogue* or *Union List of Serials* for further details.

dorff's Annalen; [3] 1877–99, also known as *Wiedemann's Annalen*; [4] 1900–28, also known as *Drude's Annalen.*
Ann. Phys. Beibl.
Beiblätter zu den Annalen der Physik und Chemie, 1877–1919; united with *Fortschr.* and *Halbmonatliches Literaturverzeichnis* to form *Physikalische Berichte.*
Ann. Phys. Erg.
Annalen der Physik, Ergänzungsband, 1842–78.
Ann. Sci.
Annals of Science, An International Quarterly Review of the History of Science and Technology since the Renaissance. London: Taylor & Francis, 1936–
Ap. J.
Astrophysical Journal. Chicago, 1895– .
Arch. Hist. Exact Sci.
Archive for History of Exact Sciences. Berlin, Heidelberg, and New York: Springer-Verlag, 1962– .
Arch. Math.
Archiv der Mathematik und Physik, Greifswald, Leipzig, and Dresden, 1841– ; [2] 1884–1900; [3] 1901–20.
Arch. Mus. Teyler
Archives du Musée Teyler, Haarlem, 1867– ; [2] 1881–1911; [3] 1912– .
Arch. Néerl.
Archives Néerlandaises des Sciences Exactes et Naturelles, publiées par La Société Hollandaises des Sciences à Harlem, 1866– ; [2] 1898–1911; [3a] and [3b], 1911– .
Arch. Rat. Mech. Anal.
Archive for Rational Mechanics and Analysis. Berlin, Heidelberg, and New York: Springer-Verlag, 1957– .
Arch. Sci. Phys.
Bibliothèque Universelle, Archives des Sciences Physiques et Naturelles, Genève, 1846–57 [2] 1858–78; [3] 1878–95; [4] 1896–1918; [5] 1919– .
Atti Roma
Atti della R. Accademia dei Lincei, Roma, 1870– ; [2] 1873–76; [3] (with "*Transunti*" added to title) 1876–84; [4] (with "*Rendiconti*" added to title) 1884–91.
Atti Roma [5]
Atti della Reale Accademia dei Lincei, Roma, Rendiconti, Classe di Scienze Fisiche, Matematiche e Naturali, Serie Quinta [5] 1892–1924.
Austr. Ass. Rep.
Report of the... Meeting of the Australasian Association for the Advancement of Science, Sydney, 1888–1926.
B.A. Rep.
Report of the... Meeting of the British Association for the Advancement of Science, 1831– .
Ber. D. Chem. Ges.
Berichte der Deutschen Chemischen Gesellschaft, Berlin, 1868– .
Berlin. Ber.
Sitzungsberichte der [Königlich] Preussischen Akademie der Wissenschaften, Physikalisch-mathematische Klass, Berlin, 1882–1921. (Supersedes *Berlin. Monatsber...*,).
Berlin Monatsber.
Monatsberichte der Königlich Preussischen Akademie der Wissenschaften zu Berlin, 1856–81. (Supersedes *Bericht der...*, and superseded by *Berlin. Ber.*)

Bibl. Univ., see *Arch. Sci. Phys.*
Bonn. Ber.
  Sitzungsberichte der Niederrheinischen Gesellschaft fuer Natur- und Heilkunde zu Bonn, 1851–1905.
Brit. J. Hist. Sci.
  The British Journal for the History of Science. London: The British Society for the History of Science, 1964– .
Bull. Acad. Sci. Bruxelles
  Bulletin de l'Academie Royale des Sciences, des lettres et des Beaux-Arts de Belgique, 1832– ; [2] 1857–80; [3] 1881–98.
Bull. Acad. Sci. Bruxelles [4]
  Bulletins de la Classe des Sciences, Académie Royale de Belgique, Bruxelles, 1899– ; [5] 1911– .
Bull. Acad. Sci. Cracovie
  Bulletin International de l'Academie des Sciences de Cracovie, Comptes rendus des séances, 1889– . (After 1900, *Classes des Sciences Mathématiques et Naturelles*.) The Academy is also known as: Akademija Umiejętności, Kraków.
Bull. Am. Math. Soc.
  Bulletin of the American Mathematical Society, A Historical and Critical Review of Mathematical Science, New York, 1894– .
Bull. Soc. Vaud.
  Bulletin de la Société Vaudoise des Sciences Naturelles, Lausanne, 1842– .
Chem. Cent.
  Chemisches Centralblatt, Deutsche Chemische Gesellschaft, Leipzig and Berlin, 1856–1906 (supersedes *Pharmaceutisches Centralblatt*, 1830–49, and *Chemisch-Pharmaceutisches Central-blatt*, 1850–55). Also known as *Chemisches Zentralblatt*.
Chem. News
  Chemical News and Journal of Physical Science, 1859–1932.
C. R. Paris
  Comptes Rendus hebdomadaires des Séances de l'Academie des Sciences, Paris, 1835– .
Carl's Rep.
  Repertorium für experimental-Physik (ed. Carl), München and Leipzig, 1865–82. Also known as *Repertorium für physikalische Technik*, 1865–67; superseded by *Exner's Rep.*, 1883–91.
Dinglers polyt. J.
  Dinglers polytechnisches Journal, Berlin and Stuttgart, 1820–1931. Also known as *Polytechnisches Journal*, 1820–74.
Edinburgh J. Sci.
  Edinburgh Journal of Science. Edinburgh, 1824–29; [2] 1829–32.
Edinburgh New Phil. J.
  Edinburgh New Philosophical Journal, 1826–64; [2] 1855–64. Superseded by the *Quarterly Journal of Science*, later *Journal of Science*.
Enc. math. Wiss.
  Encyklopädie der mathematischen Wissenschaften. Teubner, Leipzig.
Enc. Sci. Math.
  Encyclopédie des Sciences Mathématiques Pures et Appliquees, Gauthier-Villars, Paris, 1915. (French edition of *Enc. math. Wiss.*)
Exner's Rep.

Repertorium der Physik (ed. F. Exner), München and Leipzig, 1883–91 (supersedes *Carl's Rep.*).

Fortschr.
  Fortschritte der Physik, Deutsche Physikalische Gesellschaft, Berlin, 1845–1918. United with *Halbmonatliches Literaturverzeichnis* and *Ann. Phys. Beibl.* to form *Physikalische Berichte.*

Freiburg Ber.
  Berichte der naturforschenden Gesellschaft zu Freiburg i B., 1886– . (Supersedes *Berichte über die Verhandlungen*..., 1855–85.)

Gerland's Beit. Geophysik
  Beiträge zur Geophysik, Zeitschrift für physikalische Erdkunde (ed. G. Gerland), Stuttgart and Leipzig, 1887– .

Giorn. Mat.
  Giornale di matematiche (Battaglini), Naples, 1863– .

Göttingen Abh.
  Abhandlungen der Königliche Gesellschaft der Wissenschaften zu Göttingen, 1838–95. (Supersedes its *Commentationes*, and superseded by: *Mathematisch-Physikalische Klasse, Abhandlungen.*)

Göttingen Nachr.
  Nachrichten von der Königl. Gesellschaft der Wissenschaften und der Georg-Augusts-Universität zu Göttingen, 1884–93; *Nachrichten von der Königl. Gesellschaft der Wissenschaften zu Göttingen, Mathematisch-Physikalische Klasse*, 1894–1900.

Grad. J.
  The Graduate Journal. Austin, Texas: The University of Texas.

Hand. Ned. Nat. Gen. Cong.
  Handelingen van het Nederlandsch Natuur- en Geneeskundig Congress, Haarlem, 1887– .

Hist. Sci.
  History of Science. Cambridge: Heffer & Sons, 1962– (distributed in U.S.A. by Science History Publications, New York).

Hist. Stud. Phys. Sci.
  Historical Studies in the Physical Sciences. Philadelphia: University of Pennsylvania Press, 1969– .

L'Institut
  L'Institut, Journal Universel des Sciences et des Sociétés savants en France et à l'étranger. Premiere Section, Sciences mathématiques, physiques et naturelles, 1833–76.

Ion
  Ion, A Journal of Electronics, Atomistics, Ionology, Radioactivity etc.

Isis
  Isis, An International Review devoted to the History of Science and its Cultural Influences. Washington, D.C.: The History of Science Society, 1912–

J. Am. Chem. Soc.
  Journal of the American Chemical Society. 1879– .

J. Chem. Ed.
  Journal of Chemical Education

J. École Polyt.
  Journal de l'Ecole Polytechnique, Paris, 1795–1894; [2] 1895– .

J. Franklin Inst.
Journal of the Franklin Institute, Philadelphia, 1826– .

J. Math.
Liouville's Journal de Mathématiques pures et appliquees, Paris, 1836– ; [2] 1856–74; [3] 1875–84; [4] 1885–94; [5] 1895–1904; [6] 1905–14; [7] 1915–18; [8] 1919–21.

J. Phys.
Journal de physique théorique et appliquée, Paris, 1872– ; [2] 1882–91; [3] 1892–1901; [4] 1902–10; [5] 1911–19.

J. Phys. Chem.
The Journal of Physical Chemistry, Ithaca and Baltimore, 1896–

J. phys. chim. hist. nat.
Journal de physique, de Chimie, et de l'Histoire Naturelle, Paris, 1793–1823. (Previously known as *Observations sur la physique*....)

J. r. ang. Math.
Journal für die reine und angewandte Mathematik (ed. Crelle), Berlin, 1826–

Jahresber. D. Math. Ver.
Jahresbericht der Deutschen Mathematiker-Vereinigung, Berlin and Leipzig, 1892– .

Kosmos
Kosmos, Czasopismo Polskiego Towarzystwo Przyrodników Imienia Kopernika, Lemberg, 1875–1927.

Leiden Comm. Supp.
Rijksuniversiteit, Leyden, *Communications of the Physical Laboratory, Supplement*, 1899–

Leipzig. Ber.
Berichte über die Verhandlungen, K. Sächsischen Gesellschaft der Wissenschaften, Mathematische-Physikalische Klasse, Leipzig, 1849– .

Lumière Electrique
La Lumière Electrique, Journal Universel d'Électricité, Paris, 1879– .

Manchester Mem.
Memoirs of the Manchester Literary and Philosophical Society, 1785– ; [2] 1805–60; [3] 1862–87; [4] 1888–96; 1897– without series number.

Manchester Proc.
Proceedings of the Literary and Philosophical Society of Manchester, 1857–87. (Combined with *Manchester Mem.* 1888–).

Mat. Ert.
Mathematikae Ertekezések, Budapest.

Mat. Sb.
Matematicheskii Sbornik, Moskovskoe Matematicheskoe Obshchestvo, Moscow, 1866– .

Math. Ann.
Mathematische Annalen, Leipzig and Berlin, 1869– .

Mem. Acad. Sci. Bruxelles
Mémoires de l'Academie Royale des Sciences, des lettres et des Beaux-Arts de Belgique, Bruxelles, 1820– .

Mem. Accad. Lincei
R. Accademia dei Lincei, Roma. *Atti. Classe di scienze fisiche, matematiche e naturali, Memoire*, [3] 1876–84; [4] 1884–90; [5] 1894–1924.

Mem. Am. Acad.
　Memoirs of the American Academy of Arts and Sciences, Cambridge, Mass.
Meteorol. Z.
　Meteorologische Zeitschrift, herausgegeben im Auftrage der Österreichischen Gesellschaft für Meteorologie und der Deutschen Meteorologischen Gesellschaft; zugleich Zeitschrift der Österreichischen Gesellschaft für Meteorologie, Wien, 1884– (first two vols. pub. by the D.M.G. only).
Mem. Soc. Phys. Geneve
　Memoires de la Société de Physique et d'Histoire Naturelle de Geneve, 1821– .
Mondes
　Les Mondes, Revue hebdomadaire des Sciences et de leurs applications aux arts et à l'industrie, Paris. Also known as *Cosmos,* 1863–1901.
Monthly Notices R.A.S.
　Monthly Notices of the Royal Astronomical Society. 1827– .
Monthly Weather Rev.
　Monthly Weather Review, U.S. Dept. of Agriculture, Washington, D.C. 1873– .
Müeg. Lap.
　Müegyetemi Lapok; havi folyöirat a mathematika, Budapest, 1876–78.
Mün. Ber.
　Sitzungsberichte der mathematisch-physikalischen Classe der Königlichen Bayerischen Akademie der Wissenschaften zu München, 1860– .
N. Cim.
　Il Nuovo Cimento, Giornale di Fisica, di Chimica e delle loro Applicazioni alla medicina, alla Farmacia ed alle Arti Industriali, Pisa, 1851–68; [2] 1869–76; [3] 1877–94; [4] 1895–1900; [5] 1901–10.
Nat. Hist. Rev.
　Natural History Review, a Quarterly Journal of Biological Science, Natural History Society of Dublin, 1854–65; [2] 1861–65.
Nat. Rund.
　Naturwissenschaftliche Rundschau, Braunschweig, 1886–1912.
Nat. Woch.
　Naturwissenschaftliche Wochenschrift, Berlin and Jena, 1887–1922.
Naturwiss.
　Die Naturwissenschaften, Berlin, 1913– . (Supersedes *Naturwissenschaftliche Rundschau.*)
Nieuw Arch. Wisk.
　Nieuw Archief voor Wiskunde, Amsterdam, 1875–93; [2] 1895– .
Pam. Tow. Nauk
　Pamiętnik Towarzystwa Nauk Scisłych w Paryzu (Paris).
Phil. Mag.
　The London, Edinburgh, and Dublin Philosophical Magazine and Journal of Science, 1798–1826; [2] 1827–32; [3] 1832–50; [4] 1851–75; [5] 1876–1900; [6] 1901–25.
Phil. Trans.
　Philosophical Transactions of the Royal Society of London, 1665–1886; Series A, containing papers of a mathematical and physical character, 1887– .
Phys. Abst.
　Physics Abstracts (*Science Abstracts, Section A,* Institution of Electrical Engineers, London, 1898–　).
Phys. Rev.
　The Physical Review, a Journal of Experimental and Theoretical Physics, Cornell

University, 1894–1912; [2] by the American Physical Society, Lancaster, Pa., and Ithaca, N.Y., 1913–

Phys. Soc. Abst.
(Section of abstracts in *Proceedings of the Physical Society*, London; superseded by *Phys. Abst.*)

Phys. Z.
Physikalische Zeitschrift, 1899–

Pop. Astr.
Popular Astronomy, A critical review of Astronomy and Allied Sciences, Carleton College, Northfield, Minnesota, 1894–

Pop. Sci. Mon.
Popular Science Monthly, New York, 1872–

Prace Mat.-Fiz.
Prace Matematyczno-Fizyczne, Warszawa, 1888–1952.

Presse Sci. deux mondes.
Presse Scientifique des deux mondes, Paris, 1860–67.

Proc. A.A.A.S.
Proceedings of the American Association for the Advancement of Science, Washington, 1848– .

Proc. Am. Acad.
Proceedings of the American Academy of Arts and Sciences, Boston, 1846– .

Proc. Am. Phil. Soc.
Proceedings of the American Philosophical Society, Philadelphia, 1840– .

Proc. Birmingham Phil. Soc.
Proceedings of the Birmingham Natural History and Philosophical Society, 1876– (known as Birmingham Philosophical Society, 1876–93).

Proc. Cambridge Phil. Soc.
Proceedings of the Cambridge Philosophical Society, Cambridge, Eng. 1843– .

Proc. Glasgow Phil. Soc.
Proceedings of the Philosophical Society of Glasgow, 1841– .

Proc. Inst. Mech. Eng.
Institution of Mechanical Engineers, London. *Proceedings*, 1847– .

Proc. London Math. Soc.
London Mathematical Society, *Proceedings*, 1865– ; [2] 1903– .

Proc. Nat. Acad. Sci. U.S.A.
Proceedings of the National Academy of Sciences of the U.S.A., Baltimore, 1915– .

Proc. Phys. Soc. London
Proceedings of the Physical Society, London, 1874– .

Proc. Roy. Inst.
Proceedings of the Royal Institution of Great Britain, London, 1851– .

Proc. R. Irish Acad.
Proceedings of the Royal Irish Academy, Dublin, 1836– ; [2] 1869–88 ("Science" added to title); [3] 1889–1901.

Proc. R.S. Dublin
Scientific Proceedings of the Royal Society of Dublin, 1877– .

Proc. R.S. Edinburgh
Proceedings of the Royal Society of Edinburgh, 1832– .

Proc. R.S. London
Proceedings of the Royal Society of London, 1800– . (Also known as *Abstracts of*

the papers printed in the *Philosophical Transactions*, 1800–43; *Abstracts of the papers communicated to the Royal Society*, 1843–54.) Series A, 1905– .

Q. J. Math.
   Quarterly Journal of Mathematics (known as *Quarterly Journal of Pure and Applied Mathematics*, 1857–1927).

Radium
   Radium; *la radioactivité et les radiations, les sciences qui s'y rattachent et leurs applications*, Paris, 1904–19. (United with *J. Phys.* to form *Journal de physique et le Radium*.)

Rend. Acc. Lincei
   R. Accademia dei Lincei. Atti. Rendiconti. [4] 1884–91. *Classe di scienze fisiche, matematiche e naturali. Rendiconti.* [5] 1892–1924.

Rev. Gen. Sci.
   Revue Generale des Sciences pures et appliquées, Paris, 1890–1940.

Rev. Met.
   Revue de Metaphysique et de Morale, Paris, 1893– .

Riv. Sci.-Ind.
   Rivista Scientifico-Industriale delle principali scoperte ed invenzioni fatte nelle scienze e nelle industrie, Firenze, 1869– .

Rozprawy Kraków
   Rozprawy i Sprawozdania z Posiedzeń Wydziału Matematyczno-Przyrodniczego Akademii Umiejetnósci, Kraków, 1874–18; [2] 1891–1902; [3] 1901– .

Sci. Proc. Ohio Mech. Inst.
   Scientific Proceedings of the Ohio Mechanics Institute, Cincinnati, 1882–83.

Sci. Trans. Dublin
   Scientific Transactions of the Royal Dublin Society, [2] 1877–1909.

Science
   Science, Cambridge, Mass. 1883–94; [2] 1895– ; includes official proceedings and papers of the American Association for the Advancement of Science, 1901– .

Séances Soc. Fr. Phys.
   Séances de la Société Française de Physique, Paris, 1873–1901. *Bulletin des Séances*, 1901–10. Merged into *J. Phys.*

Smiths. Rep.
   Annual Report of the Board of Regents of the Smithsonian Institution, Washington, 1850– .

Spr. Tow. Nauk
   Sprawozdania Towarzystwa Naukowego Warszawskiego, Warsaw, 1908–12. *Wydzial III Nauk matematycznych i przyredniczych*, 1913–18.

Taylor's Sci. Mem.
   Scientific Memoirs, selected from the Transactions of Foreign Academies of Science and Learned Societies, and from Foreign Journals. London, 1837–52; [2] "Natural Philosophy" 1852–53.

Tokyo Sūg. Buts.
   Tōkyō Sūgaku Butsurigaku Kwai Kiji Gaiyō.

Trans. Camb. Phil. Soc.
   Transactions of the Cambridge Philosophical Society, 1822– .

Trans. R.S. Edinburgh
   Transactions of the Royal Society of Edinburgh, 1788– .

Trudy Varsh.
  Trudy Varshavskago Obshestva Estestvoispytatelei, Protokoly Otdeleniya Fiziki i khimii, 1889– .
Uch. Zap. Mos. Un.
  Moscow, Universitet. Fiziko-mekhanicheskii matematicheskii fakul'tet. *Uchenya Zapiski,* 1880–1916.
Utrecht Vet. Coll. Comm.
  Communications from the Laboratory of Physics and Physical Chemistry of the Veterinary College of Utrecht.
Verh. D. phys. Ges.
  Verhandlungen der Deutschen physikalischen Gesellschaft, Berlin, 1882– ; [2] 1899–1919.
Verh. Ges. D. Naturf. Aerzte
  Verhandlungen der Gesellschaft Deutscher Naturforscher und Aerzte, 1822– . Early volumes have title: *Amtlicher Bericht über die... Versammlung Deutscher Naturforscher und Ärzte....*
Vest. Opytn. Fiziki
  Vestnik Opytnoi Fiziki i Elementarnoi Matematiki, Odessa, 1886–1908. (Also known as: *Messager de physique experimental et de mathematique élémentaire,* Odessa.)
Vid. Selsk. Overs.
  Oversigt over K. Danske Videnskabernes selskab forhandlinger, Copenhagen, 1814– ; after 1917 with subtitle of series, *Mathematisk-fysiske meddelelser.*
Vid. Selsk. Skr.
  Det K. Danske Videnskabernes Selskabs Skrivter, Copenhagen, 1824– .
Vierteljahrsber. Wiener Ver. Förd. phys. chem. Unt.
  Vierteljahrsberischte des Wiener Vereines zur Förderung des physikalischen und chemischen Unterrichtes (zugleich Organ der Chemisch-Physikalischen Gessellschaft, Wien).
Vopr. Fiz.
  Voprosy Fiziki, St. Petersburg.
Werd.-Gymn. Festschr.
  Festschrift des Friedrich-Werder'schen Gymnasium, Berlin.
Wiad. mat.
  Wiadomości matematiyczne, Warsaw, 1877– .
Wien. Alm.
  K. Akademie der Wissenschaften, Wien. *Almanach,* 1851– .
Wien. Anz.
  K. Akademie der Wissenschaften, Wien. Mathematisch-Naturwissenschaftliche Klasse. *Anzieger,* 1864– .
Wien. Ber.
  K. Akademie der Wissenschaften, Wien. Mathematisch-Naturwissenschaftliche Klasse. *Sitzungsberichte,* 1848– . (Teil II or IIa) Changed to Österreichische Akademie..., 1946– .
Wien. Denk.
  Kaiserliche Akademie der Wissenschaften in Wien. Mathematisch-Naturwissenschaftlich Klasse. *Denkschriften,* 1850– .
Wk. Polsk. Ucz.
  Wklad polskich uczonych do fizyki statystyczno-molekularnej, ed. T. Piech. (Zrodla do dziejow Nauki i techniki, III) Wroclaw-Warszawa-Kraków: Zaklad Narodowy

§16.3] BIBLIOGRAPHY OF RESEARCH PUBLICATIONS, 1801-1900

Imienia Ossolinskich Wydawnictwo Polskiej Akademii Nauk, 1962. (Reprints and translations of papers by L. Bodaszewski, E. and W. Natanson, and M. Smoluchowski.)
Z. anorg. Chem.
  Zeitschrift für anorganische Chemie, Hamburg and Leipzig, 1892– . (Since 1915, title is Zeitschrift für anorganische und allgemeine Chemie.)
Z. Elektrochemie
  Zeitschrift für Elektrochemie und angewandte physikalische Chemie (Deutsche Bunsen-Gessellschaft für angewandte physikalische Chemie), Halle a. S. 1894– .
Z. ges. Naturw.
  Zeitschrift für die gesammten Naturwissenschaften, Berlin/Halle, 1853–81. Changed to Zeitschrift für Naturwissenschaften, Leipzig, 1882– .
Z. Math. Phys.
  Zeitschrift für Mathematik und Physik (ed. Schlömilch), Leipzig, 1856–1917.
Z. Phys.
  Zeitschrift für Physik, herausgegeben von der Deutschen Physikalischen Gesellschaft als Ergänzung zu ihren "Verhandlungen", 1920– (supersedes Ber. D. Phys. Ges.).
Z. phys. Chem.
  Zeitschrift für physikalische Chemie, Stöchiometrie und Verwandtschaftlehre, Leipzig, 1887– .
Z. phys. chem. Unt.
  Zeitschrift für den physikalischen und chemischen Unterricht, 1887– .
Zap. Mat. Otd. Nor. Obsh. Est.
  Zapiski Matematicheskago Otdeleniya Novorossiiskago Obshchestva Estestvoispitatelei, Odessa, 1872– .
Zhur. Russ. Fiz.-Khim. Obsh.
  Zhurnal Russkago Fiziko-Khimicheskago Obshchestva pri Imperatorskom S.-Petersburgskom Universitet, 1869–1906. . . . Chast Fizicheskaya, 1908–1930.

16.3 Bibliography of research publications on the kinetic theory of gases, 1801-1900

Gerrit Bakker
1896. Ueber die potentielle Energie und das Virial der Molekularkräfte usw.
  Z. phys. Chem. 21, 497 (1896).

Charles Edward Basevi
1895. Argon and the kinetic theory.
  Nature 52, 221 (1895).
    Ratio of specific heats.
1895a. Clausius' virial theorem.
  Nature 52, 413 (1895).
    Theorem said to be untrue. See Burbury (1895a) and Baynes (1895).

Robert Edward Baynes (1849–1921)
Oxford
1895. Clausius' virial theorem.
  Nature 52, 568 (1895).
    Criticizes Basevi (1895a).

Hans Benndorff
1896. Weiterfuhrung der Annäherungsrechnung in der Maxwell'sche Gastheorie.
Wien. Ber. **105**, 646 (1896).
Calc. of collision integral by method of spherical harmonics.

Joseph Louis Francois Bertrand (1822–1900)
Paris
1889. *Calcul des probabilités.*
Gauthier-Villars, Paris, 1889, pp. 29–32.
Criticism of Maxwell's proof of velocity-distribution.
1896. Sur la théorie des gaz.
C.R. Paris **122**, 1083 (1896); *Phys. Soc. Abst.* **2**, 398 (1896).
Criticizes both of Maxwell's proofs of velocity distribution.
1896a. [Sur la théorie des gaz. Réponse à M. Boltzmann].
C.R. Paris **122**, 1174, 1314 (1896); *Phys. Soc. Abst.* **2**, 398 (1896).
Reply to Boltzmann (1896a).

Pietro Blaserna (1836–1918)
Palermo, Rome
1869. Sur la vitesse moyenne du mouvement de translation des molécules dans les gaz imparfait.
C.R. Paris **69**, 134 (1869).
1895. Sulla teoria cinetica dei gas.
Atti Roma [5] **4**, 315 (1895).
Remarks on equation of state.

Böhnert
1891. Beseitigung einer Fehlerquelle in den Grundgleichungen der kinetischen Gastheorie.
Nat. Woch. **6**, 319, 346 (1891).
See O. E. Meyer (1891). New method of calculating avg. velocity; ratio of specific heats.

Ludwig Boltzmann (1844–1906)
Vienna, Graz, Munich, Leipzig
Wissenschaftliche Abhandlungen
J. A. Barth, Leipzig, 1909, 3 vols. Reprinted by Chelsea Pub. Co., New York, 1968.
Hereafter cited as *Abh.*
1866. Über die mechanische Bedeutung des zweiten Hauptsatzes der Wärmetheorie.
Wien. Ber. **53**, 195 (1866); *Wien. Anz.* **3**, 36 (1866); *Abh.* **1**, 9.
1867. Über die Anzahl der Atome in den Gasmolekülen und die innere Arbeit in Gasen.
Wien. Ber. **56**, 682 (1867); *Wien. Anz.* **4**, 235 (1867); *Abh.* **1**, 34.
Ratio of specific heats.
1868. Studien über das Gleichgewicht der lebendigen Kraft zwischen bewegten materiellen Punkten.
Wien. Ber. **58**, 517 (1868); *Wien. Anz.* **5**, 196 (1868); *Abh.* **1**, 49.
Effect of forces on molecular distribution ("Boltzmann factor").
1868a. Lösung eines mechanischen Problems.
Wien. Ber. **58**, 1035 (1868); *Wien. Anz.* **5**, 257 (1868); *Abh.* **1**, 97.
Orbits of two particles interacting with a particular force law.

[Boltzmann, cont.]
1871. Boiling-points of organic bodies.
Phil. Mag. [4] **42**, 393 (1871); *Abh.* **1**, 199.
Dependence of velocity of gas molecules on temperature.
1871a. Zur Priorität der Auffindung der Beziehung zwischen dem zweiten Hauptsatze der mechanischen Wärmetheorie und dem Prinzip der kleinsten Wirkung.
Ann. Phys. [2] **143**, 211 (1871); *Abh.* **1**, 228.
See Clausius (1870a, 1872, 1872a).
1871b. Über das Wärmegleichgewicht zwischen mehratomigen Gasmolekülen.
Wien. Ber. **63**, 397 (1871); *Wien. Anz.* **8**, 46 (1871); *Abh.* **1**, 237.
1871c. Einige allgemeine Sätze über Wärmegleichgewicht.
Wien. Ber. **63**, 679 (1871); *Wien. Anz.* **8**, 55 (1871); *Abh.* **1**, 259.
Jacobi's principle of last multipliers.
1871d. Analytischer Beweis des zweiten Hauptsatzes der mechanischen Wärmetheorie aus den Sätzen über das Gleichgewicht der lebendigen Kraft.
Wien. Ber. **63**, 712 (1871); *Wien. Anz.* **8**, 92 (1871); *Abh.* **1**, 288.
1872. Über das Wirkungsgesetz der Molekularkräfte.
Wien. Ber. **66**, 213 (1872); *Wien. Anz.* **9**, 134 (1872); *Abh.* **1**, 309.
Variation of effective diameter with temperature.
1872a. Weitere Studien über das Wärmegleichgewicht unter Gasmolekülen.
Wien. Ber. **66**, 275 (1872); *Wien. Anz.* **9**, 23 (1872); *Abh.* **1**, 316. ET* in Brush, *Kinetic Theory* **2**, 88.
H-theorem, simple case of transport equation. Criticism of Maxwell's second proof of his vel.-dist.-law.
1875. Über das Wärmegleichgewicht von Gasen, auf welche äussere Krafte wirken.
Wien. Ber. **72**, 427 (1875); *Wien. Anz.* **12**, 174 (1875); *Phil. Mag.* [4] **50**, 495 (1875); *Abh.* **2**, 1.
H-theorem.
1875a. Bemerkungen über die Wärmeleitung der Gase.
Wien. Ber. **72**, 458 (1876); *Wien. Anz.* **12**, 174 (1875); *Ann. Phys.* [2] **157**, 457 (1876); *Phil. Mag.* [4] **50**, 495 (1875). *Abh.* **2**, 31.
1876. Über die Aufstellung und Integration der Gleichungen, welche die Molecularbewegung in Gasen bestimmen.
Wien. Ber. **74**, 503 (1876); *Wien. Anz.* **13**, 204 (1876); *Abh.* **2**, 55.
Reply to Loschmidt's paradoxes, extension of (1875); case of visible motion of the gas.
1876a. Über die Natur der Gasmolecüle.
Wien. Ber. **74**, 553 (1877); *Wien. Anz.* **13**, 204 (1876); *Ann. Phys.* [2] **160**, 175 (1877); *Phil. Mag.* [5] **3**, 320 (1877); *Abh.* **2**, 103.
Ratio of specific heats.
1877. Bemerkungen über einige Probleme der mechanischen Wärmetheorie.
Wien. Ber. **75**, 62 (1877); *Wien. Anz.* **14**, 9 (1877); *Abh.* **2**, 112.
(1) Specific heats of liquids; (2) relation between Stosszahlansatz and H-theorem (reply to Loschmidt); (3) mechanical meaning of 2nd law. ET of section (2) in Brush, *Kinetic Theory* **2**, 188.
1877a. Über die Beziehung zwischen dem zweiten Hauptsatze der mechanischen Wärmetheorie und der Wahrscheinlichkeitsrechnung, respective den Sätzen über das Wärmegleichgewicht.

---

* ET: English translation.

[Boltzmann, cont.]
        Wien. Ber. **76**, 373 (1877); Wien. Anz. **14**, 196 (1877); Ann. Phys. Beibl. **3**, 166 (1879); Phil. Mag. [5] **6**, 236 (1878); Abh. **2**, 164.
        Maxwell's velocity distribution is the most probable. Relation between entropy and probability.
1878. Weitere Bemerkungen über einige Probleme der mechanischen Wärmetheorie.
        Wien. Ber. **78**, 7 (1878); Wien. Anz. **15**, 115 (1878); Abh. **2**, 250; Phil. Mag. [5] **6**, 236 (1878).
        Probability + 2nd law; equilibrium of heavy gas; diffusion.
1878a. Über die Beziehung der Diffusionsphänomene zum zweiten Hauptsatze der mechanischen Wärmetheorie.
        Wien. Ber. **78**, 733 (1878); Wien. Anz. **15**, 177 (1878); Phil. Mag. [5] **6**, 236 (1878); Abh. **2**, 289.
        Reply to Preston (1878).
1880. Erwiderung auf die Notiz des Herrn. O. E. Meyer: "Ueber eine veranderte Form usw."
        Ann. Phys. [3] **11**, 529 (1880); Abh. **2**, 358.
1880a. Zur Theorie der Gasreibung.
        Wien. Ber. **81**, 117 (1880), **84**, 40, 1230 (1882); Wien Anz. **17**, 11, 213 (1880), **18**, 272 (1881); Abh. **2**, 388, 431, 523.
1881. Über einige das Wärmegleichgewicht betreffende Sätze.
        Wien. Ber. **84**, 136 (1881); Wien. Anz. **18**, 148 (1881); Abh. **2**, 572.
        Number of ways of distributing energy among molecules.
1881a. Referat über die Abhandlung von J. C. Maxwell "Über Boltzmanns Theorem betreffend die mittlere Verteilung der lebendigen Kraft in einem System materieller Punkte."
        Ann. Phys. Beibl. **5**, 403 (1881); Phil. Mag. [5] **14**, 299 (1882); Abh. **2**, 582.
1882. Zur Theorie der Gasdiffusion.
        Wien. Ber. **86**, 63 (1882), **88**, 835 (1883); Wien. Anz. **19**, 128 (1882), 183 (1883); Abh. **3**, 3, 38.
1883. Zu K. Strecker's Abhandlungen: Die specifische Wärme der gasförmigen zweiatomigen Verbindungen von Chlor, Brom, Jod usw.
        Ann. Phys. [3] **18**, 309 (1883); Abh. **3**, 64.
1883a. Über die Arbeitsquantum, welches bei chemischen Verbindungen gewonnen werden kann.
        Wien. Ber. **88**, 861 (1883); Wien. Anz. **20**, 204 (1883); Ann. Phys. [3] **22**, 39 (1884); Abh. **3**, 66.
1884. *Ueber die Möglichkeit der Begründung einer kinetischen Gastheorie auf anziehende Kräfte allein.*
        Wien. Ber. **89**, 714 (1884); Wien. Anz. **21**, 100 (1884); Ann. Phys. [3] **24**, 37 (1885); Exner's Rep. **21**, 1 (1885); Abh. **3**, 101.
1884a. Ueber die Eigenschaften monocyclischer und anderer damit verwandter Systeme.
        Wien. Ber. **90**, 231 (1884); J. r. ang. Math. **98**, 68 (1885); Wien. Anz. **21**, 153, 171 (1884); Abh. **3**, 122.
        Ergoden.
1885. Über einige Fälle, wo die lebendige Kraft nicht integrierende Nenner des Differentials der zugeführten Energie ist.
        Wien. Ber. **92**, 853 (1885); Exner's Rep. **22**, 135 (1886); Abh. **3**, 153.
1886. Neuer Beweis eines von Helmholtz aufgestellten Theorme betreffend die

[Boltzmann, cont.]
Eigenschaften monocyclischer Systeme.
Göttingen Nachr. 209 (1886); *Abh.* 3, 176.
1886a. Ueber die zum theoretischen Beweise des Avogadroschen Gesetzes erforderlichen Voraussetzungen.
Wien. Ber. **94**, 613 (1887); *Wien. Anz.* **23**, 174 (1886); *Phil. Mag.* [5] **23**, 305 (1887); *Abh.* 3, 225.

Equipartition, H-theorem, comments on paper of Tait (1886). See Tait's reply (1887a). There is an appendix which appears only in the English version, reprinted in *Abh.* 3, 255.

1887. Ueber die mechanischen Analogien des zweiten Hauptsatzes der Thermodynamik.
J. r. ang. Math. **100**, 201 (1887); *Abh.* 3, 258.

Ergodic hypothesis.

1887a. Neuer Beweis zweier Sätze über das Wärmegleichgewicht unter mehratomigen Gasmolekülen.
Wien. Ber. **95**, 153 (1887); *Wien. Anz.* **24**, 25 (1887); *Abh.* 3, 272.

Takes account of the objection of Lorentz (1887) to his previous (1872a) proof; possible nonexistence of inverse collisions.

1887b. Ueber einige Fragen der kinetischen Gastheorie.
Wien. Ber. **96**, 891 (1887); *Wien. Anz.* **24**, 228 (1887); *Phil. Mag.* [5] **25**, 81 (1888); *Abh.* 3, 293.

Reply to Tait (1886) and criticisms of his definition of mean free path; proof of Maxwell velocity distribution; equilibrium of a gas under external forces; viscosity.

1888. Ueber das Gleichgewicht der lebendigen Kraft zwischen progressiver und Rotations-Bewegung bei Gasmolekulen.
Berlin. Ber. 1395 (1888); *Abh.* 3, 366.

Comments on Burnside's paper (1887).

1892. III Teil der Studien über Gleichgewicht der lebendigen Kraft. *Mün. Ber.* **22** (3) 329 (1892); *Phil. Mag.* [5] **35**, 153 (1893); *Abh.* 3, 428.

Case when kinetic energy is not a sum of squares; considers a test-case of Kelvin.

1894. On the application of the determinantal relation to the kinetic theory of polyatomic gases.
B.A. Rep. **64**, 102 (1894); *Abh.* 3, 520.

Appendix to Bryan's report (1894).

1894a. On Maxwell's method of deriving the equations of hydrodynamics from the kinetic theory of gases.
B.A. Rep. **64**, 579 (1894); *Abh.* 3, 526.

Suggests that Maxwell may have left an unpublished manuscript on the application of spherical harmonics to gas theory. See Maxwell's (1879) notes in square brackets.

1894b. Über den Beweis des Maxwellschen Geschwindigkeitsverteilungsgesetzes unter Gasmolekülen.
Mün. Ber. **24** (3) 207 (1894); *Ann. Phys.* [3] **53**, 955 (1894); *Phys. Soc. Abst.* **1**, 96 (1895); *Abh.* 3, 528.

Statistical independence of velocities and coordinates of two molecules before but not after they collide. See Planck's comment (1895) and Boltzmann's reply (1895a).

1894c. See G. H. Bryan (1894a).

[Boltzmann, cont.]
1895.   On certain questions of the theory of gases.
   Nature **51**, 413, 581 (1895).
   "(1) Is the theory of gases a true physical theory as valuable as any other physical theory? (2) What can we demand from any physical theory?"
1895a. Nochmals das Maxwell'sche Vertheilungsgesetz der Geschwindigkeiten.
   Ann. Phys. [3] **55**, 223 (1895); *Mün. Ber.* **25**, 25 (1896); *Abh.* **3**, 532.
   Reply to Planck (1895).
1895b. On the minimum theorem in the theory of gases.
   Nature **52**, 221 (1895); *Abh.* **3**, 546.
   Burbury's assumption of external disturbances is not necessary to prove H-theorem, provided mean free path is large.
1896. Über die Berechnung der Abweichungen der Gas vom Boyle–Charleschen Gesetz und der Dissoziation derselben.
   Wien. Ber. **105**, 695 (1896); *Wien. Anz.* **33**, 200 (1896); *Abh.* **3**, 547.
   3rd virial coefficient for hard spheres (cf. Jäger 1896) and 2nd coefficient for Maxwellian molecules.
1896a. Sur la théorie des gaz.
   C.R. Paris **122**, 1173, 1314 (1896); *Phys. Soc. Abst.* **2**, 398 (1896); *Abh.* **3**, 564, 566.
   Discussion with Bertrand on Maxwell's proofs of velocity distribution.
1896b. *Vorlesungen über Gastheorie.* I. Teil.
   J. A. Barth, Leipzig, 1896; reprinted 1910.

   Leçons sur la Théorie des Gaz, traduites par A. Galloti, avec une introduction et des notes de M. Brillouin. Premiere Partie.
   Gauthier-Villars, Paris, 1902.

   Lektsii po Teorii Gazov. Translated, with notes, edited by B. I. Davidov.
   Gosudarst. Izd.-vo tekhniko-teoret. Literatury, 1956.

   Lectures on Gas Theory. Translated, with introduction and notes, by S. G. Brush.
   University of California Press, Berkeley, 1964.
   General principles; transport theory for low density gases. Elastic spheres and centers of force.
1896c. Entgegnung auf die Wärmetheoretischen Betrachtungen des Hrn. E. Zermelo.
   Ann. Phys. [3] **57**, 773 (1896); *Phys. Soc. Abst.* **2**, 245 (1896); *Abh.* **3**, 567. ET in Brush, *Kinetic Theory* **2**, 218.
1897. Zu Hrn. Zermelo's Abhandlung "Ueber die mechanische Erklärung irreversibler Vorgänge."
   Ann. Phys. [2] **60**, 392, 776 (1897); *Phys. Soc. Abst.* **3**, 211 (1897); *Abh.* **3**, 579. ET in Brush, *Kinetic Theory* **2**, 238.
1897a. Ueber einen mechanischen Satz Poincaré's.
   Wien. Ber. **106**, 12 (1897); *Wien. Anz.* **34**, 3 (1897); *Phys. Soc. Abst.* **3**, 350 (1897); *Abh.* **3**, 587.
1898. *Vorlesungen über Gastheorie.* II. Teil.
   J. A. Barth, Leipzig, 1898; reprinted 1912.

   Leçons sur la Théorie des Gaz, traduites par A. Galloti et H. Bénard. Seconde Partie. Gauthier-Villars, Paris, 1905.
   (The Russian and English translations have the second part bound together with the first: see 1896b.)
   Van der Waals' theory; gases with compound molecules; gas dissociation;

[Boltzmann, cont.]
principles of general mechanics; Ergoden; further discussion of the H-theorem.
1898a. Ueber die sogenannte H-curve.
Math. Ann. **50**, 325 (1898); *Abh.* **3**, 629.
1898b. Ueber die kinetische Ableitung der Formeln für den Druck des gesättigen Dampfes, für den Dissociationsgrad von Gasen und für die Entropie eines van der Waals'sche Gesetz befolgenden Gases.
Verh. Ges. D. Naturf. Aerzte. Th. **2**, 74 (1898); *Abh.* **3**, 642.
1898c. Sur le rapport des deux chaleurs spécifiques des gaz.
C.R. Paris **127**, 1009 (1898); *Abh.* **3**, 645.
1899. Ueber eine Modification der van der Waals'schen Zustandsgleichung. (with H. Mache)
Ann. Phys. [3] **68**, 350 (1899); *Wien. Anz.* **36**, 87 (1899); *Abh.* **3**, 651.
1899a. Ueber die Bedeutung der Constante b des van der Waals'schen Gesetzes (with H. Mache).
Trans. Camb. Phil. Soc. **18**, 91 (1900); *Abh.* **3**, 654.
Equation of state of dissociating molecules.
1899b. Ueber die Zustandsgleichung van der Waals.
Amsterdam Verslagen [4] **7**, 477 (1899); *Amsterdam Proc.* **1**, 398 (1899); *Abh.* **3**, 658.
Fourth virial coefficient for hard spheres; includes a letter to van der Waals inviting his comments on the discrepancy between this result and that of van Laar (1899).
1900. Notiz über die Formel für die Drucke der Gase.
Arch. Néerl. [2] **5**, 76 (1900); *Abh.* **3**, 671.

R. H. M. Bosanquet
Oxford
1877. Notes on the theory of sound.
Phil. Mag. [5] **3**, 271 (1877).
Ratio of sp. heats is 1.4 for smooth hard solid of revolution. (Cf. Boltzmann 1876a)

[Louis] Marcel Brillouin (1854–1948)
Paris
1899. Théorie de la diffusion des gaz sans paroi poreuse. Propagation du son dans les mélanges.
Ann. Chim. [7] **18**, 433 (1899); *Phys. Abst.* **3**, 298 (1900).
1900. Théorie moleculaire des gaz. Diffusion du mouvement et de l'énergie. *Ann. Chim.* [7] **20**, 440 (1900).
See Chapman and Cowling, *Math. Theory of Nonuniform Gases* (Cambridge, 1952) p. 383 and Ikenberry and Truesdell, *J. Rat. Mech. Anal.* **5**, 41 (1956). General expression for velocity distribution in nonuniform gas.
1900a. La Diffusion des Gaz sans paroi poreuse dépend-elle de la concentration?
Cong. Int. Phys. **1**, 512 (1900).

Alexander Crum Brown (1838–1922)
Edinburgh
1885. Difficulties connected with the Dynamical Theory of Gases. *Nature* **32**, 352, 533 (1885).
Equipartition, specific heats, effect of temperature on chemical reactions.

George Hartley Bryan (1864–1928)
Cambridge, Bangor

1891. Researches related to the connection of the Second Law with Dynamical Principles.
> B.A. Rep. **61**, 85 (1891).
>> Report of committee (J. Larmor and G. H. Bryan) on the present state of our knowledge of thermodynamics, specially with regard to the Second Law, Part I. (See Bryan 1894 for Part II.) Work of Rankine, Clausius, Szily, Helmholtz, J. J. Thomson, Boltzmann, Burbury, and Nichols. Erratum in Part II. See Rayleigh (1902); Watson (1892).

1893. The moon's atmosphere and the kinetic theory of gases.
> B.A. Rep. **63**, 682 (1893).

1893a. The atmospheres of the Moon, Planets and Sun.
> Science **22**, 311 (1893).

1893b. The second law of thermodynamics
> Nature **49**, 197 (1893).
>> Further remarks on the subject of 1891 paper.

1894. The laws of distribution of energy and their limitations.
> B.A. Rep. **64**, 64 (1894).
>> Equipartition in the kinetic theory. See Boltzmann (1894). "The results stated in the first twelve lines of Part I (1891), Section III 44 are now known to be erroneous."

1894a. Ueber die mechanische Analogie des Wärmegleichgewichtes zweier sich beruhrender Körper.
> (with L. Boltzmann)
> Wien. Ber. **103**, 1125 (1894); Nature **51**, 454 (1895); Phys. Soc. Abst. **13**, 485 (1895).
>> Model for thermal equilibrium between two gases not requiring any diffusion or transfer of energy by solids; used by Boltzmann (1896b, §19).

1894b. Prof. Boltzmann and the kinetic theory of gases.
> Nature **51**, 31 (1894).
>> Complains that Boltzmann's verbal remarks at the British Association meeting (Oxford) on ratio of specific heats are being misquoted. The discrepancy does not make kinetic theory useless.

1894c. The kinetic theory of gases.
> Nature **51**, 152 (1894).
>> Transfer of energy between gas molecules and the ether.

1894d. The kinetic theory of gases.
> Nature **51**, 176 (1894).
>> Assumptions needed for H-theorem.

1894e. A simple test case of Maxwell's law of partition of energy.
> Proc. Camb. Phil. Soc. **8**, 250 (1894).
>> Specific heats for molecules with axial symmetry.

1894f. On Maxwell's law of partition of energy.
> Proc. London Math. Soc. **26**, 57 (1894) (title only); Nature **51**, 262 (1895).
>> Connection between time and ensemble averages.

1895. The kinetic theory of gases.
> Nature **51**, 319 (1895).
>> Conditions for proof of equipartition; may be inapplicable to liquids and solids.

[Bryan, cont.]
1895a. The assumptions in Boltzmann's minimum theorem.
Nature **52**, 29 (1895).
H-theorem cannot apply to reversed motion because molecules are not statistically independent after a collision.
1895b. The kinetic theory of gases.
Nature **52**, 244 (1895).
Opinions of Burbury and Boltzmann on H-theorem can be reconciled; restrictions on validity of the theorem.
1895c. Note on a simple graphic illustration of the determinental relation of dynamics.
Phil. Mag. [5] **39**, 531 (1895); *Proc. Phys. Soc. London* **13**, 481 (1895).
1896. On some difficulties connected with the kinetic theory of gases.
B.A. Rep. **66**, 721 (1896).
Suggests that for liquids and solids it may be necessary to deduce kinetic theory from thermodynamics rather than vice versa.
1897. On certain applications of the theory of probability to natural phenomena.
Am. J. Math. **19**, 283 (1897).
Possible use of probability theory in deriving Maxwell–Boltzmann distribution.
1899. On the permanence of certain gases in the atmospheres of planets.
B.A. Rep. **69**, 634 (1899).
1900. Energy accelerations. A study in energy partition and irreversibility.
Arch. Néerl. [2] **5**, 279 (1900); *B.A. Rep.* **70**, 634 (1900).
1900a. The kinetic theory of planetary atmospheres.
Nature **62**, 126 (1900); *Phil. Trans.* **196**, 1 (1901).

John Young Buchanan (1844–1925)
Edinburgh
1888. On a law of distribution of molecular velocities amongst the molecules of a fluid.
Phil. Mag. [5] **25**, 165 (1888).

Samuel Hawksley Burbury (1831–1911)
London
1876. On the Second Law of Thermodynamics in connection with the Kinetic Theory of Gases.
Phil. Mag. [5] **1**, 61 (1876).
1886. The Foundations of the Kinetic Theory of Gases. Note on Professor Tait's paper (No. 131, p. 343).
Phil. Mag. [5] **21**, 481 (1886).
Collisions tend to bring about Maxwell's distribution; Tait has postulated more than necessary.
1887. On the diffusion of gases – A simple case of diffusion.
Phil. Mag. [5] **24**, 471 (1887).
Criticism of Tait's treatment (1887) of a diffusion problem. See Tait's reply (1888b).
1888. On the diffusion of gases; a reply to Professor Tait.
Phil. Mag. [5] **25**, 128 (1888).
See Tait (1888b).
1890. On some problems in the kinetic theory of gases.
Phil. Mag. [5] **30**, 298 (1890).
H-theorem, transport properties of hard spheres.

[Burbury, cont.]
1892. Prof. Burnside's paper on the partition of energy, R.S.E. July 1887.
Nature **45**, 533 (1892).
1892a. See Watson.
1892b. On the collisions of elastic bodies.
Phil. Trans. **183A**, 407 (1892).
See Bryan (1894), p. 83ff. Discussion of test cases proposed by Thomson and Burnside. Use of H-theorem. Rate of decay of disturbances.
1893. The second law of thermodynamics.
Nature **49**, 150 (1893), 246 (1894); Phil. Mag. [5] **37**, 574 (1894).
Tries to derive Maxwell's distribution from the condition that dQ/T must be a complete differential.
1894. The ratio of the specific heats of gases.
Nature **51**, 127 (1894).
1894a. Boltzmann's minimum function.
Nature **51**, 78, 320 (1894), **52**, 104 (1895).
Reply to Culverwell (1894); H diminishes in a real system because of external disturbances, introducing randomness.
1894b. The kinetic theory of gases.
Nature **51**, 175 (1894), **52**, 316 (1895).
H-theorem for molecules with short-range forces.
1894c. On the law of distribution of energy.
Phil. Mag. [5] **37**, 143 (1894).
Probability theory used to prove Maxwell–Boltzmann distribution.
1895. An extension of Boltzmann's minimum theorem.
Proc. London Math. Soc. **26**, 431 (1895).
Calculates entropy for colliding spheres with correlated velocities.
1895a. Clausius' virial theorem.
Nature **52**, 568 (1895).
Criticism of Basevi's paper (1895).
1896. On the application of the kinetic theory to dense gases.
Phil. Trans. **A187**, 1 (1896).
Correlation of velocities.
1896a. On Boltzmann's law of the equality of mean kinetic energy for each degree of freedom.
Proc. London Math. Soc. **27**, 214 (1896), erratum **29**, 225 (1898).
1896b. On the stationary motion of a system of equal elastic spheres in a field of no forces when their aggregate volume is *not* infinitely small compared with the space in which they move.
B.A. Rep. **66**, 716 (1896).
Velocities of nearby spheres are correlated.
1897. On the stationary motion of a system of equal elastic spheres of finite diameter.
Proc. London Math. Soc. **28**, 331 (1897).
Velocities of nearby spheres are correlated.
1898. On the general theory of stationary motion in an infinite system of molecules.
Proc. London Math. Soc. **29**, 225 (1898).
Effect of correlated velocities on gas properties.
1899. *A treatise on the kinetic theory of gases.*
Cambridge University Press, 1899.
Exposition based on assumption of correlated velocities; consequences for equipartition and H-theorem.

[Burbury, cont.]
1900. The law of partition of kinetic energy.
   Phil. Mag. [5] **49**, 226 (1900).
   Comments on Rayleigh's paper (1900).
1900a. On certain supposed irreversible processes.
   Phil. Mag. [5] **49**, 475 (1900).
   H-theorem.
1900b. On the law of partition of energy.
   Phil. Mag. [5] **50**, 584 (1900).
   Assumptions needed to prove equipartition.
1900c. Ueber die Grundhypothesen der kinetischen Gastheorie.
   Ann. Phys. [4] **3**, 355 (1900); **4**, 646 (1901).
   Reply to Zemplen (1900).

William Burnside
1887. On the partition of energy between the translatory and rotational motions of a set of nonhomogeneous elastic spheres.
   Trans. R.S. Edinburgh **33**, 501 (1887).
1888. On a simplified proof of Maxwell's theorem.
   Proc. R.S. Edinburgh **15**, 106 (1888).
   Average energy exchanged at an impact.
1892. Prof. Burnside's paper on the Partition of Energy, R.S.E., July 1887.
   Nature **45**, 533 (1892).
   Reply to Watson (1892) and Boltzmann (1888).

Christoph Hendrik Diederik Buys-Ballot (1817–90)
Utrecht
1858. Ueber die Art von Bewegung, welche wir Wärme und Electricität nennen.
   Ann. Phys. [2] **103**, 240 (1858).
   If kinetic theory is correct, why is gas diffusion so slow? See Clausius (1858) for reply.

"C" (Anonymous)
1821. Observations on Mr. Herapath's Theory.
   Ann. Phil. [2] **2**, 418 (1821).
1821a. Observations upon D's answer to C's remarks upon Mr. Herapath's Theory.
   Ann. Phil. [2] **4**, 197 (1822).
   Collisions of "hard" vs. "elastic" bodies.

Charles Cellérier (1818–89)
Geneva
1882. Note sur la répartition des vitesses moléculaires dans le gaz.
   Arch. Sci. Phys. **6**, 337 (1881); *Phil. Mag.* [5] **13**, 47 (1882).
1891. Lois des chocs moléculaires.
   J. Math. [4] **7**, 141 (1891).
   Theory of collisions of molecules of various shapes.

Rudolf Julius Emmanuel Clausius (1822–88)
Zurich, Würzburg, Bonn

1857. Ueber die Art der Bewegung, welche wir Wärme nennen.
Ann. Phys. [2] **100**, 353 (1857); Phil. Mag. [4] **14**, 108 (1857); Fortschr. **1857**, 282 (1859); Z. Math. Phys. **2**, 170 (1857); Ann. Chim. [3] **50**, 497 (1857); Arch. Sci. Phys. **36**, 293 (1857); N. Cim. **6**, 435 (1857); ET reprinted in Brush, Kinetic Theory **1**, 111.

1858. Ueber die mittlere Länge der Wege, welche bei der Molecularbewegung gasförmigen Körper von den einzelnen Molecülen zurückgelegt werden, nebst einigen anderen Bemerkungen über die mechanischen Wärmetheorie.
Ann. Phys. [2] **105**, 239 (1858); Phil. Mag. [4] **17**, 81 (1859); Arch. Sci. Phys. **4**, 341 (1859); ET reprinted in Brush, Kinetic Theory **1**, 135.

1860. On the dynamical theory of gases.
Phil. Mag. [4] **19**, 434 (1860).
Factor of $\tfrac{4}{3}$ vs. $\sqrt{2}$ in mean-free-path formula.

1862. Ueber die Wärmeleitung gasförmiger Körper.
Ann. Phys. [2] **115**, 1 (1862); Phil. Mag. [4] **23**, 417, 512 (1862); Presse Sci. Deux Mondes, p. 24 (1862) (2); Arch. Sci. Phys. **16**, 134 (186); Z. ges. Naturw. **20**, 216 (186).

1863. Über die Molecularbewegungen in gasförmigen Körpern.
Wien. Ber. **46**, 402 (1863).
Comment on Puschl (1862).

1864. Ueber den Einfluss der Schwere auf die Bewegungen der Gasmoleküle.
Z. Math. Phys. **9**, 375 (1864).

1870. Ueber einen auf. die Wärme anwendbaren mechanischen Satz.
Bonn. Ber., p. 114 (1870); Ann. Phys. [2] **141**, 124 (1870); Carl's Rep. **6**, 197 (1870); Phil. Mag. [4] **40**, 122 (1870); Z. Math. Phys. **17**, 82 (1872). ET reprinted in Brush, Kinetic Theory **1**, 172.
Virial theorem.

1870a. Ueber die Zurückführung des zweiten Hauptsatzes der mechanischen Wärmetheorie auf allgemeine mechanische Principien.
Bonn. Ber. **167** (1870); Carl's Rep. **7**, 27 (1871); Ann. Phys. [2] **142**, 433 (1871); Phil. Mag. [4] **42**, 161 (1871).
Relation between virial theorem and thermodynamics.

1871. Ueber die Anwendung einer von ihm aufgestellten mechanischen Gleichung auf die Bewegung eines materiellen Punctes um ein festes Anziehungscentrum und zweier materiellen Puncte um einander.
Göttingen Nachr., p. 245 (1871); Math. Ann. **4**, 231 (1871); Phil. Mag. [4] **42**, 321 (1871).
Virial for interacting particles.

1872. Bemerkungen zu der Prioritätsreclamation des Hrn. Boltzmann.
Ann. Phys. [2] **144**, 265 (1872).
Mechanical proof of Second Law of Thermodynamics; see Boltzmann (1871a).

1872a. Ueber den Zusammenhang des zweiten Hauptsatzes der mechanischen Wärmetheorie met dem Hamilton'sche Princip.
Ann. Phys. [2] **146**, 585 (1872); Phil. Mag. [4] **44**, 365 (1872).
Comments on Szily's work (1872).

[Clausius, cont.]
1872b. Ueber die Beziehungen zwischen den bei Centralbewegungen vorkommenden characteristischen Grossen.
Göttingen Nachr. 600 (1872); *Math. Ann.* **6**, 390 (1873); *Phil. Mag.* [4] **46**, 1 (1873).
1873. Ueber einen neuen mechanischen Satz in Bezug auf stationäre Bewegungen.
Ann. Phys. [2], **150**, 106 (1873); *J. Math.* **19**, 193 (1874); *Bonn. Ber.* **30**, 136 (1873); *Phil. Mag.* [4] **46**, 236, 266 (1873).
Virial theorem, Lagrange's equations, Hamilton's principle, effect of external forces.
1874. Ueber verschiedene Formen des Virials.
Ann. Phys. [2] *Jubelband*, p. 411 (1874); *Phil. Mag.* [4] **48**, 1 (1874).
1874a. Sur une équation mécanique qui correspond à l'équation $\int dQ/T = 0$.
C.R. Paris **78**, 461 (1874).
1874b. Sur un cas spécial du viriel.
C.R. Paris **78**, 1731 (1874).
1874c. Ueber den Satz vom mittleren Ergal und seine Anwendung auf die Molecularbewegungen der Gase.
Bonn. Ber., p. 183 (1874); *Ann. Phys.* [2] *Ergänz.* **7**, 215 (1876); *Phil. Mag.* [4] **50**, 26, 101, 191 (1875).
1878. Ueber die Beziehung der durch Diffusion geleisteten Arbeit zum zweiten Hauptsatze der mechanische Wärmetheorie.
Ann. Phys. [3] **4**, 341 (1878); *Phil. Mag.* [5] **6**, 237 (1878).
Refutes Preston's (1878) exception to Second Law.
1880. Ueber das Verhalten der Kohlensäure in Bezug auf Druck, Volumen und Temperatur.
Ann. Phys. [3] **9**, 337 (1880); *Phil. Mag.* [5] **9**, 393 (1880); *Ann. Chim.* [4] **30**, 358 (1883).
Modification of van der Waals' equation.
1880a. Ueber einige neue Untersuchungen über die mittlere Weglänge der Gasmolecule.
Ann. Phys. [3] **10**, 92 (1880).
1884. Ueber die zur Erklarung des zweiten Hauptsatzes der mechanischen Wärmetheorie dienenden mechanischen Gleichungen.
Berlin. Ber., p. 663 (1884).
1885. Sur les dimensions des molecules et leurs distances relatives.
Lumière Electrique **17**, 241 (1885).
Agrees with Maxwell that mean-free-path formula should have a factor $\sqrt{2}$ instead of $\frac{4}{3}$.
1886. Examen des objections faites par M. Hirn á la théorie cinétique des gaz.
Bull. Acad. Sci. Bruxelles [3] **11**, 173 (1886).
1888. Die Kinetische Theorie der Gase.
F. Vieweg und Sohn, Braunschweig, 1889–91.
Published as vol. 3 of the 2nd ed. of *Abhandlungen über die mechanische Wärmetheorie*.

S. R. Cook
Cornell
1899. On the escape of gases from the planets according to the kinetic theory.
Proc. A.A.A.S., p. 120 (1899).
1900. On the escape of gases from planetary atmospheres.
Ap. J. **11**, 36 (1900); *Nature* **62**, 501 (1900), **62**, 54 (1900).
See Stoney (1900, 1900a, 1900b).

Hans Cornelius (1863–1947)
Frankfurt
1893. Notiz über das Verhältnis der Energien der fortschreitenden Bewegungder Moleküle und der inneren Molekularbewegung der Gase.
Z. phys. Chem. **11**, 403 (1893).
Ratio of specific heats.

Severinus Corrigan
St. Paul (Minn.)
1895. *The Constitution and Functions of Gases, The Nature of Radiance, and the Law of Radiation; a treatise demonstrating the existence of some heretofore unknown properties and functions of the atmospheric, and other, gases; embodying a determination of the pressure, density, temperature and probable nature of the luminiferous aether, and of the effective temperature of the sun; and indicating the probable origin of all thermal, electric and magnetic forces.*
The Pioneer Press Co., Saint Paul, 1895.
Kinetic theory of polyatomic gases; proposed a model for molecules with internal motion of atoms; ratio of sp. heats and spectral lines, molecular constants.

Edward Parnall Culverwell (1855–1931)
Dublin
1890. Note on Boltzmann's Kinetic Theory of Gases, and on Sir W. Thomson's Address to Section A, British Association, 1884.
Phil. Mag. [5] **30**, 95 (1890).
Counterexample to equipartition; reversibility paradox; temperature equilibrium and irreversibility are due to interactions with aether.
1890a. Possibility of irreversible molecular motions.
B.A. Rep. **60**, 744 (1890).
Molecular motions may actually be irreversible or may deviate slightly from Newton's laws individually while obeying them *en masse*.
1892. Lord Kelvin's test-case on the Maxwell–Boltzmann law.
Nature **46**, 76 (1892).
1894. Dr. Watson's proof of Boltzmann's theorem on permanence of distributions.
Nature **50**, 617 (1894).
What is the H-theorem supposed to prove?
1894a. The kinetic theory of gases.
Nature **51**, 78 (1894).
Disagrees with Bryan's (1894b) account of Boltzmann's views.
1894b. Boltzmann's minimum theorem.
Nature **51**, 105, 246 (1894).
In order to prove H-theorem one has to assume the "average state" has already been attained; there seems to be some confusion about what the theorem means anyway.
1895. Prof. Boltzmann's letter on the kinetic theory of gases.
Nature **51**, 581 (1895).
Reply to Boltzmann (1895). Reversibility paradox.
1895a. Boltzmann's minimum theorem.
Nature **52**, 149 (1895).
Criticizes proof of H-theorem.

§16.3] BIBLIOGRAPHY OF RESEARCH PUBLICATIONS, 1801–1900   735

Paul Czermak
Graz
1884. Der Werth der Integrale A_1 und A_2 der Maxwell'schen Gastheorie unter Zugrundelegung eines Kraftgesetzes $-K/r^5$.
   Wien. Ber. **89**, 722 (1884).
   See Maxwell (1866), Boltzmann (1884).

"D" (Anonymous)
1821. Reply to C's observations on Mr. Herapath's theory.
   Ann. Phil. [2] **3**, 290, 357 (1821).
1822. D's second reply to C on Mr. Herapath's theory.
   Phil. Mag. **60**, 285 (1822).
   Criticizes Herapath's theory of collisions of hard bodies.

James Dewar (1842–1923)
Edinburgh, Cambridge
1875. Charcoal vacua (with P. G. Tait).
   Nature **12**, 217 (1875).
   Report of a paper read to the Royal Society, Edinburgh. Radiometer action attributed to long mean free path in rarefied gases.

Conrad Heinrich Dieterici (1858–1929)
Hannover, Kiel
1898. Kinetische Theorie der Flüssigkeiten.
   Ann. Phys. [3] **66**, 826 (1898).
1899. Ueber den kritischen Zustand.
   Ann. Phys. [3] **69**, 685 (1899).

Pietro Donnini
1879. Sull' equivalente meccanico di calore, la teoria cinetica ed il calore antomico dei gas.
   N. Cim. [3] **5**, 97 (1879).
   Ratio of specific heats.

Eugene Karl Dühring (1833–1921)
1878. *Neue Grundgesetze zur rationellen Physik und Chemie.*
   Fues's Verlag (R. Reisland) 1878; part 2 (pub. 1886) with Ulrich Dühring.
   Comments on Mayer, cons. of force. Chap. II: gases. Relation between compressibility and excluded volume of molecules.

H. Ebert
1889. Zur Anwendung des Doppler'schen Principes auf leuchtende Gasmolecule.
   Ann. Phys. [3] **36**, 466 (1889).
   Theory of spectral line width.

Henry Turner Eddy (1844–1921)
Cincinnati
1883. An extension of the theorem of the virial and its application to the kinetic theory of gases.

[Eddy, cont.]
> *Sci. Proc. Ohio Mech. Inst.* **2**, 26 (1883); *J. Franklin Inst.* [3] **115**, 339, 409 (1883).
>> Rotational motion of molecules, ratio of specific heats.

1883a. Developments in the kinetic theory of solids, liquids, and gases.
> *Sci. Proc. Ohio Mech. Inst.* **2**, 82, 121 (1883).
>> Virial theorem applied to vibrations of atoms in molecules.

Rudolph Heinrich Finkener (1834–1902)
Berlin
1876. Ueber das Radiometer von Crookes.
> *Ann. Phys.* [2] **158**, 572 (1876).

George Francis Fitzgerald (1851–1901)
Dublin
> *Scientific Writings of the late George Francis Fitzgerald.*
> Hodges, Figgis and Company, Dublin, 1902.
>> Edited with an introduction by J. Larmor. Cited as *Writings*.

1878. On the mechanical theory of Crookes' Force.
> *Sci. Trans. Dublin* **1**, 57 (1878); *Phil. Mag.* [5] **7**, 15 (1879); *Writings*, p. 18.
>> Uses Clausius' (1862) theory of heat conductivity, supports Stoney's theory of the radiometer.

1881. On Professor Osborne Reynold's paper "On certain dimensional properties of matter in the gaseous state."
> *Phil. Mag.* [5] **11**, 103 (1881).

1895. The kinetic theory of gases.
> *Nature* **51**, 221 (1895); *Writings*, p. 321.
>> Why does it not apply to the solar system? Exchange of energy with the ether should be included in a complete theory.

1895a. On some considerations showing that Maxwell's theorem of the equal partition of energy among the degrees of freedom of atoms is not inconsistent with the various internal movements exhibited by the spectra of gases.
> *Proc. R.S. London* **57**, 312 (1895); *Nature* **51**, 453; *Writings*, p. 338.
>> Since electrons in different atoms are connected with each other through interactions with ether, one should not have to add more than a few degrees of freedom to take account of several thousand electrons.

1900. On the size at which heat movements are manifested in matter.
> *Nature* **61**, 612 (1900).
>> Equipartition requires heating the ether as well as the atoms.

George Carey Foster (1835–1919)
London
1877. The radiometer and its lessons.
> *Nature* **17**, 5, 43, 80, 142 (1877).
>> Criticism of Reynolds' theory; heat conductivity increases at very low density; supports Stoney.

P. C. F. Frowein
1879. Eene bekende formule van Clausius.
> *Nieuw Arch. Wisk.* **5**, 191 (1879).
>> Mean-free-path formula.

Boris B. Galitzin(1816–1916)
1895. Zur Theorie der Verbreiterung der Spectrallinien.
Ann. Phys. [3] **56**, 78 (1895).

Josiah Willard Gibbs (1839–1903)
Yale
The Collected Works of J. Willard Gibbs, 2 vols., 2nd ed. Longmans, Green and Co., London and New York, 1928; reprinted by Yale University Press, New Haven, 1948, and by Dover Pubs., Inc., New York, 1960.
Hereafter cited as *Works*. The 1st edition, *The Scientific Papers of J. Willard Gibbs* (Longmans, Green and Co., London and New York, 1906) does not include *Statistical Mechanics* (1902).
1884. On the fundamental formula of statistical mechanics, with applications to astronomy and thermodynamics (Abstract).
Proc. A.A.A.S. **33**, 57 (1884); *Works* **2**, part 2, 16.

George Gore (1826–1908)
Birmingham
1894. Mechanical energy of molecules of Gases.
Phil. Mag. [5] **37**, 340 (1894).
"Rediscovery" of equipartition theorem. See Lodge (1894).

Władisław Gosiewski (1844–1911)
Paris, Warsaw
1899. O rozdziale predkosci w ukladzie dynamicznym, ozwionym ruchem umiejscowionym.
Prace Mat.-Fiz. **10**, 16 (1899).
Distribution of velocities in a dynamical system with stationary motion.

Gilberto Govi (1826–89)
1876. Sur la cause des mouvements dans le radiometre de M. Crookes.
C.R. Paris **82**, 1410 (1876).

G. Gross
Dresden
1890. Zur Diffusion der Gase.
Ann. Phys. [3] **40**, 424 (1890).

Frederick Guthrie
Graaf Reinet College
1873. Kinetic theory of gases.
Nature **8**, 67, 486 (1873).
Gravity should make the top of a column of gas hotter than bottom. See Maxwell (1873a).
1874. Molecular motion.
Nature **10**, 123 (1874).
Objections to Maxwell's (1873a) assumptions about the number of molecules which collide with specified velocities.

Gustav Hansemann (1829–1902)
Berlin
1871. Die Atome und ihre Bewegungen; ein Versuch zur Verallgemeinerung der Krönig-Clausius'schen Theorie der Gase.
E. H. Mayer, Coln u. Leipzig, 1871.
See Boltzmann, *Fortschr.* **26**, 470 (1870); Maxwell, "Atom" in 9th ed. of *Encyclopedia Britannica* (1878) or *Papers* **2**, 445.
1871a. Ueber die innere Beschaffenheit der Gase.
Ann. Phys. [2] **144**, 82 (1871).
Numerical factor in Krönig's derivation of gas law.
1872. Druck und elastischer Stoss.
Ann. Phys. [2] **146**, 620 (1872).
Reply to Sellmeier (1872).
1874. Ueber den Einfluss der Anziehung auf die Temperatur der Weltkörper.
Ann. Phys. Erg. **6**, 417 (1874).
Effect of gravity on temperature-equilibrium in the atmosphere.

Hermann Ludwig Ferdinand von Helmholtz (1821–94)
Berlin
Wissenschaftliche Abhandlungen, 3 vols.
Barth, Leipzig, 1882–95. Cited as *Abh.*
1884. Studien zur Statik monocyklischer Systeme.
Berlin. Ber. **159**, 311, 755 (1884); *Abh.* **3**, 119, 163, 173.
1884a. Principien der Statik monocyklischer Systeme.
J. r. ang. Math. **97**, 111, 317 (1884); *Abh.* **3**, 142, 179.

John Herapath (1790–1868)
Bristol
1816. On the physical properties of gases.
Ann. Phil. **8**, 56 (1816).
Preliminary announcement of kinetic theory.
1821. A mathematical inquiry into the causes, laws, and principal phenomena of heat, gases, gravitation, etc.
Ann. Phil. [2] **1**, 273, 340, 401 (1921). Reprinted with 1847 book (see below).
Collisions of hard bodies, deduction of ideal gas laws, theory of specific heats and latent heats, diffusion, velocity of sound, effect of finite size of atoms.
1821a. Tables of temperature, and a mathematical development of the causes and laws of phenomena which have been adduced in support of hypotheses of "Calorific Capacity," "Latent Heat," etc.
Ann. Phil. [2] **2**, 50, 89, 201, 257, 363, 435, **19**, 16 (1821).
The "true temperature" is proportional to the molecular velocity. Temperature of mixtures. Theory of evaporation and melting.
1821b. Reply to Mr. Tredgold.
Ann. Phil. [2] **2**, 462 (1821).
1822. Reply to X.
Ann. Phil. [2] **3**, 29 (1822).
1822a. Remarks on Dr. Thomson's paper, "On the influence of humidity in modifying the specific gravity of gases."
Ann. Phil. [2] **3**, 419 (1822).
Elasticity of steam not simply proportional to its density.

[Herapath, cont.]
1822b. On the hypothesis of gaseous repulsion.
Phil. Mag. **60**, 18 (1822).
The hypothesis is not consistent with most properties of gases.
1823. Observations on M. Laplace's Communication to the Royal Academy of Sciences, "Sur l'Attraction des Sphères, et sur la Repulsion des Fluides Elastiques."
Phil. Mag. **62**, 61, 136 (1823).
Objections to Laplace's derivation of the gas laws from caloric theory.
1832. On the velocity of sound (abstract).
B.A. Rep. **2**, 557 (1832).
(See 1836 for text.)
1836. Exact calculation of the velocity of sound.
Railway Magazine **1**, 22 (1836).
1836a. On the physical constitution of the world.
Railway Magazine **1**, 104 (1836).
(Letter written 6 November 1832, and "read before the Bristol Literary and Philosophical Society about four years ago.")
1847. *Mathematical Physics*, 2 vols.
Whittaker and Co., London, 1847. Reprinted, with 1821 paper, in *Mathematical Physics and Selected Papers by John Herapath*, edited and with an Introduction and Bibliography by Stephen G. Brush (New York: Johnson Reprint Corp., 1972).
Kinetic theory, specific and latent heats, pp. 212–236 (vol. 1).

Alexander Steward Herschel (1836–1907)
Newcastle
1878. On the use of the virial in thermodynamics.
Nature **18**, 39 (1878).
Comments on paper by Preston (1877c).

Heinrich Rudolf Hertz (1857–1894)
Berlin
1882. Ueber die Verdunstung der Flüssigkeiten, insbesondere des Quecksilbers, im luftleeren Raume.
Ann. Phys. [3] **17**, 177 (1882); *Chem. News* **47**, 64 (1883).
Formula for number of molecules of a vapor striking a surface.

William Mitchinson Hicks (1850–1934)
Cambridge, Sheffield
1877. On some effects of dissociation on the physical properties of gases.
Phil. Mag. [5] **3**, 401, **4**, 80, 174 (1877).
Uses Maxwell distribution to calculate effect of dissociation on pressure.
1885. On Boltzmann's theorem.
B.A. Rep. **55**, 905 (1885).
Equipartition cannot apply to internal degrees of freedom since they cannot have more than a finite amount of energy.

Gustav Adolph Hirn (1815–90)
Colmar
1868. *Analyse Élémentaire de l'Universe.*
Gauthier-Villars, Paris, 1868.
Criticism of kinetic theory.

[Hirn, cont.]
1881. Recherches expérimentales sur la relation qui existe entre la résistance de l'air et sa température. Consequences physiques et philosophiques qui découlent de ces expériences.
Mem. Acad. Sci. Bruxelles **43**, (1881); published separately by Gauthier-Villars, Paris.
    Consequences of his experiments for kinetic theory, p. 52–91. See Sandrucci (1886–88, 1893).
1884. Recherches experimentales et analytiques sur les lois de l'ecoulement et du choc des gaz en fonction de la temperature.
Mem. Acad. Sci. Bruxelles **46** (1886); *Ann. Chim.* [6] **7**, 289 (1886); *Recherches expérimentales sur la limite de la vitesse que prend un gaz quand il passe d'une pression à une autre plus faible.*
Gauthier-Villars, Paris, 1886 (this is a reprint of the first part which is largely experimental).
    Attack on kinetic theory, pp. 97–154.
1886. Réflexions générales au sujet des rapports de MM. les Commissaires Examinateurs de ce Memoir.
Mem. Acad. Sci. Bruxelles **46** (1886) (printed after the above paper, beginning p. 155).
    Further discussion of kinetic theory, reply to Folie.
1886a. La cinétique moderne et le dynamisme de l'avenir. Reponse à diverses critiques faites par M. Clausius, aux conclusions de mes travaux précédentes.
Mem. Acad. Sci. Bruxelles **46** (1886).
1886b. Réflexions sur une critique de M. Hugoniot, relative aux lois d'écoulement des gaz.
C.R. Paris **103**, 109 (1886).
    See also **103**, 371, 1232. Hirn repeats his view that his experiments disprove the kinetic theory, while Hugoniot is interested only in explaining them by hydrodynamics.
1888. Réflexions relatives à la Note précédente de M. Ladislas Natanson.
C.R. Paris **107**, 166 (1888).
    Natanson has explained how molecules escaping from a container can have a higher mean velocity than those inside by invoking statistical arguments which do not please Hirn.

Nathaniel Dana Carlile Hodges (1852–1927)
Harvard
1880. On the mean free path of a molecule.
Am. J. Sci. [3] **19**, 222 (1880); *Phil. Mag.* [5] **9**, 177 (1880).
    Mean path in regions of varying density, *e.g.* liquid–gas interface.
1900. Note on the law of distribution of velocities among gas molecules.
Phys. Rev. **10**, 253 (1900).
    Maxwell's law derived from principle of least action.

Jan Leendert Hoorweg (1841–1919)
Utrecht
1876. Sur la propagation du son d'après la nouvelle théorie des gaz.
Arch. Néerl. **11**, 131 (1876).

R. Hoppe
1858. Ueber Bewegung und Beschaffenheit der Atome.
> *Ann. Phys.* [2] **104**, 279 (1858).
> Comments on Krönig, Clausius papers; comparison of kinetic and vibration theories, and Redtenbacher's "Dynamidensystem."
1860. Erwiderung auf ein Artikel von Clausius, nebst einer Bemerkung zur Erklärung der Erdwärme.
> *Ann. Phys.* [2] **110**, 598 (1860).
> Effect of gravity on equilibrium; comments on Clausius theory of mean free path.

Louis Aimé-Charles Houllevigue (1863–   )
Caen (France)
1895. Sur la theorie cinétique des fluides pesants.
> *J. Phys.* [3] **4**, 301 (1895).
> Effect of gravity on thermal equilibrium in a gas.

W. H. Howard
Adrian (Michigan)
1893. The Moon's atmosphere.
> *Science* **21**, 233 (1893).
> Examples of Mars and Jupiter not consistent with kinetic theory explanation. Molecules should be treated as "interdependent centers of activity" not independent masses.

Gustav Jäger (1865–1938)
Vienna
1891. Zur Theorie der Dissociation der Gase.
> *Wien. Ber.* **100**, 1182 (1891); **104**, 671 (1895); *Ann. Phys. Beibl.* (1892); *Phil. Mag.* [5] **34**, 143 (1892).
> Uses Maxwell distribution.
1892. Ueber die Art der Kräfte, welche Gasmolekeln auf einander ausüben.
> *Wien. Ber.* **101**, 1520 (1892).
> Relation between pressure and intermolecular force, using series in inverse powers of volume. Cf. Kamerlingh Onnes (1881).
1892a. Ueber die Temperaturfunction der Zustandsgleichung der Gase.
> *Wien. Ber.* **101**, 1675 (1892).
> Theories based on dissociation.
1893. Über die kinetische Theorie der inneren Reibung der Flüssigkeiten.
> *Wien. Ber.* **102**, 253 (1893).
> Mean-free-path formula corrected for molecular size.
1895. Sur le chemin moyen des molécules gazeuses.
> *Arch. Sci. Phys.* **34**, 376 (1895).
1896. Die Gasdruckformel mit Berucksichtigung des Molecularvolumens.
> *Wien. Ber.* **105**, 15 (1896).
> Third virial coefficient for hard spheres; proposes closed form for equation of state which is consistent with virial coefficients.
1896a. Über den Einfluss des Molecularvolumens auf die mittlere Weglänge der Gasmolekeln.
> *Wien. Ber.* **105**, 97 (1896).

[Jäger, cont.]
1896b. Zu Theorie der Zustandsgleichung der Gase.
Wien. Ber. **105**, 791 (1896).
Maxwell's distribution used to calculate frequency of various kinds of collisions in a system of dissociating molecules.
1899. Über den Einfluss des Molecularvolumens auf die innere Reibung der Gase.
Wien. Ber. **108**, 447 (1899); **109**, 74 (1900).

Emil Carl Gustav Georg Jochmann (1833–1871)
Berlin
1859. Ueber die Molecularconstitution der Gas.
Ann. Phys. [2] **108**, 153 (1859).
Criticism of Krönig–Clausius hypothesis. Temperature-differences would be equalized too rapidly if the kinetic theory were true.

James Prescott Joule (1818–89)
Manchester
The Scientific Papers of James Prescott Joule.
Physical Society of London, 1884.
Cited as *Papers.*
1847. On Matter, Living Force, and Heat.
Manchester Courier, May 5 and 12 (1847); *Papers*, p. 265.
1848. On the mechanical equivalent of heat, and on the constitution of elastic fluids.
B.A. Rep. **18** (2) 21 (1848). *Papers*, p. 288.
Pressure of a gas proportional to *vis viva* of its particles, according to Davy or Herapath; velocity of hydrogen molecule and calculated specific heats.
1848a. Some remarks on heat, and the constitution of elastic fluids.
Manchester Mem. [2] **9**, 107 (1851) (read 1848); *Phil. Mag.* [4] **14**, 211 (1857); *Ann. Chim.* [3] **50**, 381 (1857); *Arch. Sci. Phys.* **36**, 349 (1857); *Papers*, p. 290; *Am. J. Phys.* **17**, 63 (1949).
Expanded version of the preceding paper.

Sir William Thomson, Baron Kelvin (1824–1907)
Glasgow
Mathematical and Physical Papers.
Cambridge University Press.
The 2nd ed. was arranged and revised with notes by J. Larmor; hereafter cited as *Papers.*
1874. The kinetic theory of the dissipation of energy.
Proc. R.S. Edinburgh **8**, 325 (1874); *Nature* **9**, 441 (1874); *Phil. Mag.* [5] **33**, 291 (1892); *Papers* **5**, 11. Brush, *Kinetic Theory* **2**, 176.
Calculates probability that all the oxygen molecules of air are on one side of container and all nitrogen molecules on the other. See Natanson (1892).
1884. On the distribution of energy among colliding groups of molecules.
Proc. R.S. Edinburgh **13**, 2 (1884) (title only).
1891. On some test-cases for the Maxwell–Boltzmann doctrine regarding distribution of energy.
Proc. R.S. London **50**, 79 (1891); *Nature* **44**, 355 (1891); *Papers* **4**, 484.
Hollow spherical shell containing a solid sphere. See Rayleigh (1892).
1891a. On periodic motion of a finite conservative system.

[Kelvin, cont.]
Phil. Mag. [5] **32**, 375, 555 (1891); *Papers* **4**, 497.
Cf. Poincaré (1890).
1892. On a decisive test-case disproving the Maxwell–Boltzmann doctrine regarding distribution of kinetic energy.
Phil. Mag. [5] **33**, 466 (1892); *Proc. R.S. London* **51**, 397 (1892); *Papers* **4**, 495.
1900. Nineteenth century clouds over the dynamical theory of heat and light.
Proc. Roy. Inst. **16**, 363 (1900); *Phil. Mag.* [6] **2**, 1 (1901); *Baltimore Lectures on Molecular Dynamics and the Wave Theory of Light* (C. J. Clay and Sons, London, 1904), p. 486.
Equipartition, specific heats, test cases.

Gustav Robert Kirchhoff (1824–87)
Berlin
1894. *Vorlesungen über die Theorie der Warme.*
B. G. Teubner, Leipzig, 1894.
Discussion of Maxwell's transport theory, p. 134ff.

C. J. Kool
1892. Sur la correction qu'exige l'équation $\Sigma \frac{1}{2}mv^2 = \frac{3}{2}PV$ en vertu de l'etendu des molécules.
Bull. Soc. Vaud. **28**, 271 (1892); *Arch. Sci. Phys.* **28**, 72 (1892).
Effect of molecular volume on equation of state.
1892a. Sur la longueur exacte du chemin parcouru en moyenne dans un gaz par les molécules entre deux collisions successives.
Bull. Soc. Vaud. **28**, 211 (1892); *Arch. Sci. Phys.* **28**, 74 (1892).
1894. De la correction qu'exige l'équation $\Sigma \frac{1}{2}mv^2 = \frac{3}{2}PV$ en vertu de l'attraction qui existe entre les molécules des gaz.
Bull. Soc. Vaud. **30**, 209 (1894).
1899. Sur la longueur exacte du chemin parcouru en moyenne dans un gaz par les molécules entre deux collisions successives.
Bull. Soc. Vaud. **35**, 383 (1899).

Diederik Johannes Korteweg
1876. Over de Berekening van den Gemmiddelden Botsingsafstand der Gasmoleculen, met in achtneming van al hunne afmetingen.
Amsterdam Verslagen [2] **10**, 349 (1876); *Arch. Néerl.* **12**, 241 (1877).
1876a. Berekening van de Vermeerding welke de Spanning van een gas Tengevolge van de Botsingen der Moleculen Ondergaat.
Amsterdam Verslagen [2] **10**, 363 (1876); *Arch. Néerl.* **12**, 254 (1877).
1881. Ueber den Einfluss der räumlichen Ausdehnung der Moleküle auf den Druck eines Gases.
Ann. Phys. [3] **12**, 136 (1881).
Discussion of van der Waals' equation and that of Clausius (1880), comparison with experiments of Regnault.
1892. On van der Waals's Isothermal Equation.
Nature **45**, 152 (1892).
Comments on Tait (1891, 1891a).
1892a. On van der Waals's Isothermal Equation.
Nature **45**, 277 (1892).

August Karl Krönig (1822–79)
Berlin
1856. Grundzüge einer Theorie der Gase.
Ann. Phys. [2] **33**, 315 (1856); *N. Cim.* **6**, 435 (1857).

August Adolph Eduard Eberhardt Kundt (1839–94)
Strasbourg
1875. Ueber Reibung und Wärmeleitung verdünnter Gase. (with E. Warburg)
Ann. Phys. [2] **155**, 337 (1875); *Phil. Mag.* [4] **50**, 53 (1875); *Berlin. Monatsber.*, p. 160 (1875).

Johannes Jacobus van Laar (1860–19??)
Amsterdam
1893. Die theoretische Berechnung der Dampfdrucke gesättigter Dämpfe.
Z. phys. Chem. **11**, 433 (1893).
Equations of state of Clausius and van der Waals.
1899. Berekening der tweede correctie op de grootheid b der toestands-vergelijking van van der Waals.
Amsterdam Verslagen [4] **7**, 350 (1899); *Amsterdam Proc.* **1**, 273 (1899); *Arch. Mus. Teyler* **6**, 237 (1900).
Calculation of third virial coefficient for hard spheres by a method suggested by van der Waals (1899). Cf. Boltzmann (1899c).

Victor Edler von Lang (1838–1921)
Vienna
1871. Zur dynamischen Theorie der Gase.
Wien. Ber. **64**, 485 (1871), **65**, 415 (1872); *Ann. Phys.* [2] **145**, 290, **147**, 157 (1872).
Mean-free-path derivations of kinetic theory formulae for pressure, viscosity, heat conductivity, with a factor of $\frac{1}{12}$ in the latter case where Clausius obtained $\frac{5}{24}$. (Subsequently revised to $\frac{1}{4}$.)

Joseph Larmor (1857–1942)
Cambridge
Mathematical and Physical Papers.
Cambridge University Press, 1929.
Cited as *Papers*.
1894. The kinetic theory of gases.
Nature **51**, 152 (1894).
Recurrence and reversibility paradoxes.
1895. A dynamical theory of the electric and luminiferous medium, Part III. Relations with material media.
Phil. Trans. **A190**, 205 (1898).
(For Parts I and II see *Phil. Trans.* **A185**, 719 (1895), **A186**, 695 (1896).)
Suggests reason why internal motions of molecules are not much affected by temperature–specific heat paradox.

John Le Conte (1818–1921)
Columbia (S. Car.)
1864. On the adequecy of LaPlace's explanation to account for the discrepancy between the computed and the observed velocity of sound in air and gases.
Phil. Mag. [4] **27**, 1 (1864).
Refers to Herapath's comments on velocity of sound.

Alfred Constant Hector Ledieu (1830–91)
Brest
1882. Considérations sur la theorie cinetique des gaz et sur l'état vibratoire de la matiére.
C.R. Paris **94**, 691 (1882).

Amand Jean Leray (1828–?)
Angers
1869. *Constitution de la matiere et ses mouvements, nature et cause de la pesanteur.*
Au bureau du journal Les Mondes, Paris, 1869.
1885. *Essai sur la synthese des forces physiques.*
Gauthier-Villars, Paris, 1885.
1891. La théorie cinetique des gaz.
Séances Soc. Fr. Phys., p. 168 (1891).
1892. *Complément de l'essai sur la synthèses des forces physiques.*
Paris, 1892.
Chaleur et pesanteur... Théories cinétiques... Cohesion et affinité.

Ferdinand Franz Lippich (1838–1913)
Graz
1870. Über die Breite der Spectrallinien.
Ann. Phys. [2] **139**, 465 (1870).

Gabriel Lippmann (1845–1921)
Paris
1900. La théorie cinétique des gaz et le principe de Carnot.
Cong. Int. Phys. **1**, 546 (1900).
Proposed perpetual motion of second kind using alternating induction currents generated by electric charges in molecules.

George Downing Liveing (1827–1924)
Cambridge
1885. Kinetic theory of gases.
Nature **32**, 533 (1885).
Equipartition.

Oliver Joseph Lodge (1851–1940)
London, Birmingham
1894. Molecular energy of gases.
Phil. Mag. [5] **37**, 419, 507 (1894).
Gore (1894) has merely rediscovered the equipartition theorem by trivial series of calculations.

Hendrik Antoon Lorentz (1853–1928)
Leiden
> *Collected Papers*
> Martinus Nijhoff, The Hague, 1934–39, 9 vols.
> > Edited by P. Zeeman and A. D. Fooker; cited as *Papers*.
> *Abhandlungen über Theoretische Physik.*
> B. G. Teubner, Leipzig, 1907.
> > Cited as *Abh.*

1880. De bewegingsvergelijkingen der gassen en de voortplanting van het geluid volgens de kinetische gastheorie.
> *Amsterdam Verslagen* [2] **15**, 350 (1880); *Arch. Néerl.* **16**, 1 (1881); *Ann. Phys. Beibl.* **5**, 174 (1881); *Abh.*, p. 72 (1907); *Papers* **6**, 1 (1938).
> > "The equations of motion of a gas, and the propagation of sound according to the kinetic theory of gases." Series solution of Boltzmann equation.

1881. Über die Anwendung des Satzes vom Virial in der kinetischen Theorie der Gase.
> *Ann. Phys.* [3] **12**, 127, 660 (1881); *Abh.*, p. 114; *Papers* **6**, 40.
> > Derivation of second virial coefficient for hard spheres.

1882. Over de bewegingen die onden den invloed der zwaartekracht, ten gevolge van temperatuurverschillen, in eene gasmassa optreden. *Amsterdam Verslagen* [2] **17**, 179 (1882); *Arch. Néerl.* **17**, 193 (1882); *Papers* **6**, 51.
> "On the motions produced in a mass of gas, under the influence of gravity, as a consequence of temperature differences."

1887. Über das Gleichgewicht der lebendigen Kraft unter Gasmolekülen. *Wien. Ber.* **95**, 115 (1887); *Abh.*, p. 124; *Papers* **6**, 74.
> Criticism of Boltzmann's proof of H-theorem for polyatomic gases; nonexistence of inverse collisions of certain cases.

1896. Over de entropie eener gasmassa.
> *Amsterdam Verslagen* [4] **5**, 252 (1896); *Fortschr.* 228 (1896); *Abh.*, p. 164; *Papers* **6**, 134.
> > H-theorem, identification of H with gas entropy.

Josef Loschmidt (1821–95)
Vienna

1865. Zur Grösse der Luftmolecule.
> *Wien. Ber.* **52**, 395 (1865); *Wien. Anz.* **2**, 162 (1865); *Z. Math. Phys.* **10**, 511 (1865).
> > Uses mean-free-path formula to estimate size of molecule.

1866. Zur Theorie der Gase.
> *Wien. Ber.* **54**, 646 (1866).
> > Discusses validity of kinetic theory.

1867. Theorie des Gleichgewichtes und der Bewegung eines Systems von Punkten. *Wien Ber.* **55**, 523 (1867).
> Poinsot's theorem.

1869. Der zweite Satz der mechanischen Wärmetheorie.
> *Wien Ber.* **59**, 385 (1869).

1876. Über den Zustand des Wärmegleichgewichtes eines Systems von Korpern mit Rucksicht auf die Schwerkraft.
> *Wien Ber.* **73**, 128, 366 (1876), **75**, 287, **76**, 209 (1877).
> > Objection to H-theorem, reversibility paradox.

Gustav Lubeck (1847–)
Berlin
1882. Die Bewegung eines Kugelförmigen Atoms in einem idealen Gas.
Werd.-Gymn. Festschr. 295 (1881); *Ann. Phys. Beibl.* **6**, 451 (1882); *Phil. Mag.* [5] **14**, 157 (1882).
Derivation of Maxwell distribution by variational method.

Carlo del Lungo
1896. Sopra la teoria cinetica dei gas.
Atti Roma [5] **5** (1) 467 (1896); *Phys. Soc. Abst.* **2**, 399 (1896).
Comments on Bertrand (1896, 1896a) vs. Boltzmann (1896a); conditions for validity of Maxwell distribution law.

Heinrich Mache (1876–1954)
Vienna
1899. See L. Boltzmann (1899, 1899a).

Felice Marco (1836–)
1875. *L'Unité dynamique des forces et des phenomenes de la nature ou l'atome tourbillon.*
Gauthier-Villars, Paris, 1875.
1877. La cause del moto di rotazione del molinello del radiometro di Crookes.
Atti Roma [3] **1**, 16 (1877).

James Clerk Maxwell (1831–79)
Aberdeen, London, Cambridge
The Scientific Papers of James Clerk Maxwell.
Cambridge University Press, 1890, 2 vol.; reprinted by Hermann, Paris, 1927; reprinted by Dover, New York, 1952, 1965. Edited by W. D. Niven. Hereafter cited as *Papers*.
1860. Illustrations of the dynamical theory of gases.
Phil. Mag. [4], **19**, 19, **20**, 21, 33 (1860); *B.A. Rep.* **29** (2) 9 (1859); *Athenaeum* 468 (1859); *L'Institut*, p. 364 (1859); *Papers* **1**, 377. Brush, *Kinetic Theory* **1**, 148.
I. On the motions and collisions of perfectly elastic spheres.
II. On the process of diffusion of two or more kinds of moving particles among one another.
III. On the collision of perfectly elastic bodies of any form.
1861. On the results of Bernoulli's theory of gases as applied to their internal friction, their diffusion, and their conductivity for heat.
B.A. Rep. **30**, 15 (1861).
1867. On the dynamical theory of gases.
Phil. Trans. **157**, 49 (1867); *Phil. Mag.* [4] **32**, 390 (1866), **35**, 129, 185 (1868); *Proc. R.S. London* **15**, 146 (1867); *Papers* **2**, 26. Brush, *Kinetic Theory* **2**, 23.
Transport theory, special case of inverse fifth power forces.
1871. *Theory of Heat.* Longmans, Green, and Co., London, 1871, 2nd and 3rd eds., 1872, 4th ed., 1875, 5th ed., 1877, 6th ed., 1880, 7th ed., 1883; new edition with corrections and additions by Rayleigh, 1897, 1902.

Teoriya teploty v elementarnoi obrabotky. Translated from the 7th English ed. by A. L. Korol'kov. Tip. I. N. Kumnerev, Kiev, 1888.

Theorie der Wärme. Nach der 4. Aufl. des Originals in's Deutsche übertragen von Dr. F. Auerbach. Maruschke & Berendt, Breslau, 1877.
See p. 328 for "Maxwell's demon."

[Maxwell, cont.]
1873. Clerk-Maxwell's kinetic theory of gases.
Nature **8**, 84 (1873).
Reply to Guthrie (1873).
1873a. On Loschmidt's experiments on diffusion in relation to the kinetic theory of gases.
Nature **8**, 298 (1873); *Mondes* **32**, 164 (1873); *Papers* **2**, 343.
1873b. On the equilibrium of temperature of a gaseous column subjected to gravity.
Nature **8**, 527 (1873).
Another reply to Guthrie.
1873c. On the final state of a system of molecules in motion subject to forces of any kind.
Nature **8**, 537 (1873); *B.A. Rep.* **43**, 29 (1873); *Papers* **2**, 351.
Reply to Guthrie (1873); effect of gravity on temperature of a column of gas; derivation of "Maxwell–Boltzmann" distribution.
1874. Molecular motion
Nature **10**, 123 (1874).
Reply to Guthrie (1874).
1874a. Van der Waals on the continuity of the gaseous and liquid states.
Nature **10**, 477 (1874); *Papers* **2**, 407.
Generally favorable account but criticized van der Waals' use of virial theorem and proposed alternative form for equation of state.
1875. On the dynamical evidence of the molecular constitution of bodies.
Nature **11**, 357, 374 (1875); *J. Chem. Soc.* **13**, 493 (1875); *Gazz. Chem. Ital.* **5**, 190 (1875); *Papers* **2**, 418.
Derivation of equal areas rule for two-phase region of gas–liquid system.
1877. The kinetic theory of gases.
Nature **16**, 242 (1877).
Review of Watson's book (1876); criticism of Boltzmann (1876a) proposal to explain specific heats of diatomic molecules.
1878. Diffusion.
Encyclopedia Britannica (9th ed.) **7**, 214 (1878); *Papers* **2**, 625.
1879. On stresses in rarified gases arising from inequalities of temperature.
Phil. Trans. **170**, 231 (1880); *Proc. R.S. London* **27**, 304 (1878); *Papers* **2**, 681.
Radiometer theory.
1879a. On Boltzmann's theorem on the average distribution of energy in a system of material points.
Trans. Camb. Phil. Soc. **12**, 547 (1879); *Ann. Phys. Beibl.* **5**, 403 (1881); *Phil. Mag.* [5] **14**, 299 (1882); *Papers* **2**, 713.

Rudolf Adriaan Mees (1844–86)
Groningen
1878. Over de theorie van den radiometer.
Amsterdam Verslagen [2] **13**, 265 (1878); *Arch. Néerl.* **14**, 97 (1879); *Ann. Phys. Beibl.* **4**, 522 (1880).
Objections to theories of the radiometer proposed by Reynolds, Zöllner, Neesen, and Meyer; proposes his own theory.
1880. De voortplanting van vlakke geluidsgolven in gassen, volgens de kinetische gastheorie.
Amsterdam Verslagen [2] **15**, 394 (1880); *Ann. Phys. Beibl.* **5**, 244 (1881).
"The propagation of plane sound-waves in gases, according to the kinetic theory of gases."

Oskar Emil Meyer (1834–1915)
Göttingen, Breslau
1861. Ueber die Reibung der Flüssigkeiten.
Ann. Phys. [2] **113**, 55, 193, 383 (1861); J. r. ang. Math. **59**, 229 (1861); **62**, 201 (1863).
1865. Ueber die innere Reibung der Gase.
Ann. Phys. [2] **125**, 177, 401, 564 (1865).
1866. Ueber die Reibung der Gase.
Ann. Phys. [2] **127**, 253, 353 (1866).
1866. *De Gasorum Theoria.*
Maelzer, Vratislaviae, 1866.
Discussion of Maxwell and Clausius mean free path theories.
1874. Zur Theorie der inneren Reibung.
J. r. ang. Math. **78**, 130 (1874), **80**, 315 (1875).
1877. *Die Kinetische Theorie der Gase.*
Maruschke and Berendt, Breslau, 1877; 2nd ed., 1899.
The Kinetic Theory of Gases.
Longmans, Green and Co., London, 1899. Translated from the 2nd rev. ed. by R. E. Baynes.
1879. Ueber einen Beweis des Maxwell'schen Gesetzes für das Gleichgewicht von Gasmolecülen.
Ann. Phys. [3] **7**, 317 (1879).
1880. Ueber eine veränderte Form meines Beweises für das Maxwell'sche Gesetz der Energievertheilung.
Ann. Phys. [3] **10**, 296 (1880).
See Boltzmann 1877a, 1880.
1888. Bemerkungen über einen Punkt aus der kinetischen Theorie der Gase.
Z. phys. Chem. **2**, 340 (1888).
Diffusion. Reply to Ostwald (1888).
1891. Ueber die Grundgleichung der kinetischen Gastheorie.
Nat. Woch. **6**, 346 (1891).

Albert Abraham Michelson (1852–1931)
Worcester (Mass.), Chicago
1892. On the application of interference methods to spectroscopic measurements.
Phil. Mag. [5] **34**, 280 (1892).
Width of spectral lines predicted from Maxwell distribution.
1895. On the broadening of spectral lines.
Ap. J. **2**, 251 (1895).
Extends work of Rayleigh (1889).

Vladimir Aleksandrovich Michelson
1886. Prostiishii byvod vtorago zakona termodinamiki iz' nachal' analiticheskoi mekhaniki.
Mat. Sb. **13**, 229 (1886).
"Simple derivation of the second law of thermodynamics based on the principles of analytical mechanics."

William Muir
1879. Mr. Preston on general temperature-equilibrium.
Nature **20**, 6 (1879).
See Preston (1879, 1897a).

Eduard Mulder (1832–1924)
Utrecht
1870. Geschwindigkeit der Molecularbewegung und des Schalles in Gasen.
Ann. Phys. [2] **140**, 288 (1870).

Johann Heinrich Jacob Müller (1846–1875)
Zürich
1874. Ueber ein aus der Hamilton'schen Theorie der Bewegung hervorgehendes mechanisches Princip.
Ann. Phys. [2] **152**, 105 (1874); *Phil. Mag.* [4] **48**, 274 (1874).
Attempts to derive second law of thermodynamics.

Joseph John Murphy (1827–)
Dunmurry (Ireland)
1875. Equilibrium in gases.
Nature **12**, 26 (1875).
Nichols' statement (1875) that a column of gas is cooler at the top is contrary to "well-known meteorological facts."

Karl W. von Nägeli (1827–91)
1879. Ueber die Bewegungen kleinster Körperchen.
Mün. Ber. **9**, 389 (1879).
Criticism of kinetic theory explanation of Brownian motion.

Władysław [Ladislaus] Natanson (1864–1937)
Dorpat, Warsaw, Kraków
1887. *Ueber die kinetische Theorie unvollkommener Gase.*
Dissertation, Dorpat; Druck von C. Mattiesen, 1887.
Equation of state; association and dissociation of gas molecules.
1888. Teorya cynetyczna gazów niedoskonałych.
Kosmos **13**, 58 (1888); *Ann. Phys.* [3] **33**, 683 (1888).
"Kinetic theory of imperfect gases." Summary of his dissertation. See Pirogov (1889).
1888a. Sur l'explication d'une expérience de Joule, d'après la théorie cinétique des gaz.
C.R. Paris **107**, 164 (1888); *Prace Mat.-Fiz.* **2**, 75 (1890).
Hirn's experiments (1881, 1884) can be explained by taking account of the spread of velocities; the molecules which come out of the hole have higher mean velocity than those inside. See Hirn's reply (1888).
1888b. Przyczynek do deoryi dyssocyacyi.
Kosmos **13**, 222 (1888).
"Contribution to the theory of dissociation."
1888c. Ueber die Geschwindigkeit, mit welcher Gase den Maxwell'schen Zustand erreichen.
Ann. Phys. [3] **34**, 970 (1888).
1888d. Uwagi nad drugiém prawem mechanicznej teoryi ciepła.
Kosmos **13**, 256 (1888).
"Notes on the second law of the mechanical theory of heat." Discusses Maxwell's demon.

[Natanson, cont.]
1888e. O zadaniu Tait'a.
Prace Mat.-Fiz. **1**, 26 (1888).
Velocity distribution, H-theorem, relaxation to equilibrium.
1889. Ueber die Warmeerscheinungen bei der Ausdehnung der Gase.
Ann. Phys. [3] **37**, 341 (1889).
1889a. Ueber die kinetische Theorie der Dissociationserscheinungen in Gasen.
Ann. Phys. [3] **38**, 288 (1889); *Phil. Mag.* [5] **29**, 18 (1889).
1889b. O wpływie spotkán, w których uczestniczy dowolna liczba czasteczek, na równowage cieplna gazów.
Kosmos **14**, 47 (1889).
"On the effect of collisions, in which an arbitrary number of particles participate, on the thermal equilibrium of gases."
1890. *Wstep do fiziki teoretycznej.*
Warsaw, 1890.
1891. O rozpraszaniu energii.
Kosmos **16**, 30 (1891).
"On the dissipation of energy." Reply to Olearski (1890).
1892. Dynamical illustration of the isothermal formula.
Phil. Mag. [5] **33**, 301 (1892); *Arch. Sci. Phys.* [3] **28**, 112 (1892).
Equation of state. Association of molecules near critical point.
1892a. On the probabilities of molecular configurations.
Phil. Mag. [5] **34**, 51 (1892).
Comments on Kelvin's paper (1874).
1893. O znaczeniu kinetycznem funkcyi dysypacyjnej.
Rozprawy Kraków **9**, 171 (1895); *Bull. Acad. Sci. Cracovie,* p. 348 (1893); *C.R. Paris* **117**, 539 (1893); *Phil. Mag.* [5] **39**, 455 (1895); *Z. phys. Chem.* **13**, 437 (1894); *Phys. Soc. Abst.* **1**, 6 (1895). *Wk. Polsk. Ucz.*
"On the kinetic interpretation of the dissipation function." Rayleigh's dissipation function expressed in terms of molecular magnitudes and related to theory of viscosity.
1894. Uwaga termodynamiczna o prawie Maxwell'a.
Prace Mat.-Fiz. **5**, 118 (1894); *Z. phys. Chem.* **14**, 151 (1894).
"Thermodynamic interpretation of Maxwell's law." (See Bryan 1894, pp. 84–85.)
1895. O energii kinetycznej ruchu ciepla i o funckcyi dysypacyjnen odpowiedniej.
Rozprawy Kraków. **7**, 273 (1895); *Bull. Acad. Sci. Cracovie* **2**, 295 (1894); *Z. phys. Chem.* **16**, 289 (1895); *Phil. Mag.* [5] **39**, 501 (1895); *Phys. Soc. Abst.* **1**, 267 (1895).
"On the kinetic energy of the motion of heat and the corresponding dissipation function." Heat conduction, relaxation.
1896. O prawach zjawisk nieodwracalnych.
Rozprawy Kraków **10**, 309 (1896); *Bull. Acad. Sci. Cracovie,* p. 117 (1896); *Phil. Mag.* [5] **41**, 385 (1896). *Wk. Polsk. Ucz.*
"On the laws of irreversible phenomena." Extension of Maxwell's (1866) relaxation theory.
1897. O teoryi kinetycznej ruchu wirowego.
Rozprawy Kraków **13**, 154 (1898); *Bull. Acad. Sci. Cracovie,* p. 155 (1897);
Equations of hydrodynamics derived from kinetic theory. "On the kinetic theory of turbulent motion."

Alexander Nicolaus Franz Naumann (1837–1922)
Tübingen
1867. Ueber specifische Wärme der Gase für gleiche Volume bei constantem Drucke.
    Ann. Chem. Pharm. **142**, 265 (1867); *Phil. Mag.* [4] **34**, 205 (1867).
1867a. Ueber die Geschwindigkeit der Bewegungen der Atome.
    Ann. Chem. Pharm. **142**, 284 (1867); *Phil. Mag.* [4] **34**, 373 (1867).
        Motion of atoms inside a molecule.
1869. Das Avogadro'sche Gesetz abgeleitet aus der Grundvorstellung der mechanischen Gastheorie.
    Ann. Chem. Pharm. Suppl. **7**, 339 (1870); *Phil. Mag.* [4] **39**, 317 (1870); *J. Franklin Inst.* **59**, 353 (1870); *Ber. D. Chem. Ges.* **2**, 690 (1869).

R. C. Nichols
1875. On the dynamical evidence of molecular constitution.
    Nature **11**, 486 (1875).
        Various objections to kinetic theory. See Murphy (1875).
1876. On the proof of the Second Law of Thermodynamics.
    Phil. Mag. [5] **1**, 369 (1876).
        Proof using virial theorem; criticism of Szily's proofs. See Bryan (1894), p. 96.

William Augustus Norton (1810–1883)
Yale
1879. On the force of effective molecular action.
    Am. J. Sci. **17**, 346, 433 (1879).
        Claims that kinetic theory is inferior to his own theory [*ibid.* **38**, 61, 207, (1864), **39**, 237, **40**, 61 (1865), **41**, 61, 196 (1866), **46**, 167 (1868), **49**, 24 (1870), [2] **3**, 327, 440, **4**, 8 (1872), **5**, 186 (1873), **17**, 183 (1879), *Proc. A.A.A.S.* **29**, 222 (1880)].

Kazimierz Olearski (1855–1936)
Lemberg
1890. (Review of Natanson's *Wstep do fisiki teoretyczncy*.)
    Kosmos **15**, 522 (1890).
        See Natanson (1891).

Heike Kamerlingh Onnes (1853–1926)
Groningen, Delft, Leiden
1881. Algemeene theorie der vloeistoffen.
    Amsterdam Verslagen [2] **16**, 241 (1881); *Amsterdam Verh.* **21**, No. 4 (1881); *Ann. Phys. Beibl.* **5**, 718 (1881); *Arch. Néerl.* **30**, 113 (1897).
        "General theory of fluids." Introduction of virial series for equation of state; vapor–liquid equilibrium; similarity of the isotherms is related to similarity of molecular movements. See van der Waals (1895).

Wilhelm Ostwald (1853–1932)
Leipzig
1888. Bemerkungen über einen Punkt aus der kinetischen Theorie der Gase.
    Z. phys. Chem. **2**, 81, 342 (1888).
        The law that diffusion is proportional to $m^{\frac{1}{2}}$ can be derived without using the kinetic theory.

§16.3] BIBLIOGRAPHY OF RESEARCH PUBLICATIONS, 1801–1900   753

Leopold von Pfaundler (1839–1920)
Innsbruck
1877. Über die Anwendung des Doppler'schen Princips auf die fortschreitende Bewegnung lauchtender Gasmoleküle.
Wien. Ber. **76**, 852 (1878); *Wien. Anz.* **14**, 238 (1877).
Width of spectral lines.

N. N. Pirogov
1885. Neskol'ko dopolnenii k' kineticheskoi teorii gazov'.
Zhur. Russ. Fiz.-Khim. Obsh. **17**, 114, 281 (1885); *Fortschr.* (2), p. 237 (1886).
"Some contributions to the kinetic theory of gases." On the Maxwell–Boltzmann distribution.
1886. Novoe analiticheskoe dokazatelstvo 2 nachala termodinamika.
Zhur. Russ. Fiz.-Khim. Obsh. **18**, 307 (1886); *Fortschr.* (2), p. 238 (1886).
"New analytical proof of the second law of thermodynamics"; discusses H-theorem.
1886a. Predel'nyya skorosti b' gazakh'.
Zhur. Russ. Fiz.-Khim. Obsh. **18**, 93 (1886); *Fortschr.* (2), p. 238 (1886).
"Limiting velocities in gasses and Watson's theory of the rotational motion of molecules."
1886b. Osnovaniya kineticheskoi teorii mnogoetomnykh' gazov'.
Zhur. Russ. Fiz.-Khim. Obsh. **18**, *Suppl.* (1886); *Fortschr.* (2), p. 238 (1886).
"Foundations of the kinetic theory of polyatomic gases."
1887. Poyasneniya k' "zametk" G. Stankevicha.
Zhur. Russ. Fiz.-Khim. Obsh. **19**, 133 (1887).
"Reply to G. Stankevich's 'Note' (on limiting velocity of gases and Watson's theory)."
1888. O virial' sil'.
Zhur. Russ. Fiz.-Khim. Obsh. **20**, 1 (1888); **21**, 219 (1889); **23**, 127 (1891); *Fortschr.*, p. 207 (1889), p. 248 (1891); *Z. Math. Phys.* **37**, 257 (1891).
"On the virial of force." Derives a new equation of state for hard spheres and discusses effect of interatomic forces in polyatomic gases.
1889. O nesovershennykh' gazakh'.
Zhur. Russ. Fiz.-Khim. Obsh. **21**, 44 (1889); *Ann. Phys. Beibl.* **14**, 259 (1890); *Fortschr.*, p. 209 (1889).
"On imperfect gases." Criticizes Natanson (1888, 1888c).
1889a. O zakon' Maxwell'ya.
Zhur. Russ. Fiz.-Khim. Obsh. **21**, 76 (1889).
"On Maxwell's Law."
1890. O zakon' Boltzmann'a.
Zhur. Russ. Fiz.-Khim. Obsh. **22**, 44 (1890); *Exner's Rep.* **27**, 515 (1891).
"On Boltzmann's law." New derivation of Maxwell–Boltzmann distribution.

Max Karl Ernst Ludwig Planck (1858–1947)
Munich, Berlin
Physikalische Abhandlungen und Vorträge, 3 vol.
F. Vieweg & Sohn, Braunschweig, 1958.
Hereafter cited as *Abh.*
1895. Ueber den Beweis des Maxwell'schen Geschwindigkeitsvertheilungsgesetzes unter Gasmolekülen.
Mün. Ber. **24**, 391 (1895); *Ann. Phys.* [3] **55**, 220 (1895); *Abh.* **1**, 442.

Jules Henri Poincaré (1854–1912)
Paris
    Oeuvres, 10 vols.
    Gauthier-Villars, Paris, 1951–54.
        Edited by G. Darboux. Cited as *Oeuvres.*
1889. Sur les tentatives d'explication mécanique des principles de la thermodynamique.
    C.R. Paris **108**, 550 (1889); *Oeuvres* **10**, 231.
        Helmholtz's theory (1884, 1884a) not applicable to irreversible processes.
1890. Sur le problème des trois corps et les équations de la dynamique.
    Acta Math. **13**, 1 (1890); *Oeuvres* **7**, 262.
        Recurrence of dynamical systems. See Zermelo (1896), Boltzmann (1897a).
        Partial ET in Brush, *Kinetic Theory* **2**, 194.
1893. Sur une objection à la théorie cinétique des gaz.
    C.R. Paris **116**, 1017 (1893); *Oeuvres* **10**, 240.
        Maxwell (1866) failed to take account of vibrations; his relation between heat conductivity and viscosity is wrong.
1893a. Sur la théorie cinétique des gaz.
    C.R. Paris **116**, 1165 (1893); *Oeuvres* **10**, 244.
        Vibrational energy.
1893b. Le mécanisme et l'expérience.
    Rev. Met. **1**, 535 (1893). ET in Brush, *Kinetic Theory* **2**, 203.
        Recurrence paradox.
1894. Sur la théorie cinétique des gaz.
    Rev. Gen. Sci. **5**, 513 (1894), *Oeuvres* **10**, 246.
        Kelvin's objections to equipartition.
1896. *Calcul des probabilités.*
    Gauthier-Villars, Paris, 1896.
        Page 21: criticism of Maxwell's proof of distribution law.

Samuel Tolver Preston (1844–?)
Heatherfield, Eastcliff, Bournemouth, Hamburg
1877. Mode of the propagation of sound, and the physical condition determining its velocity on the basis of the kinetic theory of gases.
    Phil. Mag. [5] **3**, 441 (1877).
        According to Waterston (1859) a wave can be propagated by collisions of spheres; speed of sound should be proportional to molecular speed; Maxwell, to whom this paper was sent, showed ratio is $\sqrt{5/3}$.
1877a. On the equilibrium of pressure in gases.
    Phil. Mag. [5] **4**, 77 (1877).
        Further remarks on propagation of sound.
1877b. On the nature of what is commonly termed a "vacuum."
    Phil. Mag. [5] **4**, 110 (1877).
        Still enough molecules to make radiometer work. See Stoney (1877).
1877c. On the diffusion of matter in relation to the Second Law of Thermodynamics.
    Nature **17**, 31 (1877); *Ann. Phys. Beibl.* **2**, 248 (1878).
        Gas diffusion an exception to Second Law? Cf. Herschel (1878).
1878. On a means for converting the heat-motion possessed by matter at normal temperature into work.
    Nature **17**, 202 (1878).
        See Clausius (1878).

[Preston, cont.]
1878a. On the availability of normal-temperature heat-energy.
Nature **18**, 92 (1878).
1878b. On the view of the propagation of sound demanded by the acceptance of the kinetic theory of gases.
Nature **18**, 253 (1878).
1878c. A question raised by the observed absence of an atmosphere in the moon.
Nature **19**, 3 (1878).
    According to kinetic theory some molecules have very high velocities and can thus overcome pull of gravity and escape.
1879. On the possibility of explaining the continuance of life in the universe consistent with the tendency to temperature-equilibrium.
Nature **19**, 460, 555 (1879).
    See Muir (1879).
1879a. Temperature equilibrium in the universe in relation to the kinetic theory.
Nature **20**, 28 (1879).
1879b. On the possibility of accounting for the continuance of recurring changes in the universe, consistently with the tendency to temperature-equilibrium.
Phil. Mag. [5] **8**, 152 (1879); *Wien. Ber.* **87**, 806 (1883).
1880. A question regarding one of the physical premises upon which the finality of universal change is based.
Phil. Mag. [5] **10**, 338 (1880).
    "Heat death" of the universe.
1891. Some remarks on the kinetic theory of gases.
Phil. Mag. [5] **31**, 441 (1891); *Science* **22**, 191 (1893).
    Large molecular velocities prevented by friction in moving through the ether; loss of vibrational energy.

Ernst Pringsheim (1859–1917)
Breslau
1883. Ueber das Radiometer.
Ann. Phys. [3] **18**, 1 (1883).

Johann Puluj (1845–1918)
Vienna
1878. Ueber die Reibung der Dampf.
Wien. Ber. **78**, 279 (1878); *Phil. Mag.* [5] **6**, 157 (1878); *Wien. Anz.* **15**, 140 (1878); *Carl's Rep.* **14**, 573 (1878), **15**, 427 (1879).
    Temperature-variation of viscosity attributed to variation of molecular size (suggested by Stefan).

Karl Puschl (1825–1912)
Melk, Seitenstetten
1862. Über den Wärmezustand der Gase.
Wien. Ber. **45**, 357 (1862).
    Kinetic derivation of ideal gas laws, effect of ether. See Clausius (1863).
1863. Notiz über die Molecularbewegung in Gasen.
Wien. Ber. **48**, 35 (1863).
    Numerical factor in pressure formulae of Krönig (1856), Clausius (1857).

[Puschl, cont.]
1875. Ueber eine Modification der herrschenden Gastheorie.
  Wien. Ber. **70**, 413 (1875); *Carl's Rep.* **11**, 42 (1875).
    Specific heats.

William John Macquorn Rankine (1820–72)
Edinburgh, Glasgow
1852. On the reconcentration of the mechanical energy of the universe.
  B.A. Rep **22** (2), 12 (1852); *Phil. Mag.* [4] **4**, 358 (1852); *Edinburgh New Phil. J.* **54**, 88 (1853).
1864. On the use of mechanical hypotheses in science, and especially in the theory of heat.
  Proc. Glasgow Phil. Soc. **5**, 126 (1864).
    Comparison of theories of heat based on molecular motion.
1871. On the hypothesis of molecular motions in thermodynamics.
  Phil. Mag. [4] **41**, 62 (1871).

John William Strutt, Third Baron (Lord) Rayleigh (1842–1919)
Cambridge, London
  Scientific Papers, 6 vols.
  Cambridge University Press, 1899–1920.
    Cited as *Papers*.
1873. Note on a natural limit to the sharpness of spectral lines.
  Nature **8**, 474 (1873); *Papers* **1**, 183.
1889. On the limit to interference when light is radiated from moving molecules.
  Phil. Mag. [5] **27**, 298 (1889); *Papers* **3**, 258.
    Width of spectral lines calculated using Maxwell distribution.
1891a. Dynamical problems in illustration of the theory of gases.
  Phil. Mag. [5] **32**, 424 (1891); *Papers* **3**, 473.
    Approach to equilibrium; one-dimensional problems.
1891. On van der Waals' treatment of Laplace's pressure in the virial equation: a letter to Prof. Tait.
  Nature **44**, 499, 597 (1891); *Papers* **3**, 465.
1892. On the virial of a system of hard colliding bodies.
  Nature **45**, 80 (1892); *Papers* **3**, 469.
    Defends van der Waals' equation, criticizes that of Maxwell (1874). Pressure of hard spheres always prop. to kinetic energy at const. volume.
1892a. Remarks on Maxwell's investigation respecting Boltzmann's theorem.
  Phil. Mag. [5] **33**, 356 (1892); *Papers* **3**, 554.
    Equipartition, ergodic hypothesis; comments on Kelvin (1891), Bryan (1891).
1900. The law of partition of kinetic energy.
  Phil. Mag. [5] **49**, 98 (1900); *Papers* **4**, 433.
    Discusses Kelvin's test-cases. See Burbury (1900).
1900a. On the viscosity of argon as affected by temperature.
  Proc. R.S. London **66**, 68 (1900); *Papers* **4**, 452.
    Temperature-dependence of viscosity deduced by dimensional argument.
1900b. On a theorem analogous to the virial theorem.
  Phil. Mag. [5] **50**, 210 (1900); *Papers* **4**, 491.
    Relation between x-coordinate and y-component of force, etc.

Maximilian Reinganum (1876–1914)
Göttingen, Munster, Freiberg
1899. Theorie und Aufstellung einer Zustandsgleichung.
Dissertation. Druck der Univ.-Buchdruckerei von E. A. Huth, Göttingen, 1899.
Equation of state for hard spheres, inverse power forces.
1900. Über die molekulare Anziehung in schwach comprimierten Gasen.
Arch. Néerl. [2] 5, 574 (1900).
Equation of state for hard spheres with attractive forces.

Osborne Reynolds (1842–1912)
Manchester
Papers on Mechanical and Physical Subjects, 3 vols.
Cambridge University Press, 1900–1903.
Cited as Papers.
1874. On the forces caused by evaporation from and condensation at a surface.
Proc. R.S. London 22, 401 (1874); Chem. News 30, 11 (1874).
Repulsion of hot and attraction of cold bodies.
1874a. On the surface-forces caused by the communication of heat.
Phil. Mag. [4] 48, 389 (1874).
Surface-forces will be important at low density when convection currents diminish. Explanation of Crookes' experiments.
1875. The attraction and repulsion caused by the radiation of heat.
Nature 12, 6 (1875).
Further remarks on radiometer theory.
1876. On the force caused by the communication of heat between a surface and a gas, and on a new photometer.
Proc. R.S. London 24, 388 (1876); Phil. Mag. [5] 2, 231 (1876); Phil. Trans. 166, 725 (1877); Papers 1, 170.
(Mainly discussion of radiometer experiments.)
1878. The radiometer and its lessons.
Nature 17, 220 (1878).
Reply to Stoney (1878) and criticism of his theory.
1879. On Mr. G. F. Fitzgerald's paper "On the mechanical theory of Crookes' force."
Phil. Mag. [5] 7, 179 (1879).
Criticism of Stoney–Fitzgerald theory. The force acts between surface and gas, not between two surfaces.
1879a. On certain dimensional properties of matter in the gaseous state. Part I: Experimental researches on thermal transpiration of gases through porous plates, and on the laws of transpiration and impulsion, including an experimental proof that a gas is not a continuous plenum. Part II: On an extension of the dynamical theory of gas which includes the stresses, tangential and normal, caused by a varying condition of the gas, and affords an explanation of the phenomena of transpiration and impulsion.
Proc. R.S. London 28, 304 (1879); Phil. Trans. 170, 727 (1880); Nature 19, 435 (1879); Papers 1, 257.
Dependence of radiometer effect on the size of vanes. General theory of rarefied gases. Cf. Maxwell (1879).

[Reynolds, cont.]
1880. Note on thermal transpiration.
    Proc. R.S. London **30**, 300 (1880); *Papers* **1**, 391.
    Reply to Maxwell (1879); see also Stokes (1880).
1881. Certain dimensional properties of matter in the gaseous state. An answer to Mr. George Francis Fitzgerald.
    Phil. Mag. [5] **11**, 335 (1881).
    Thermal transpiration, radiometer theory.
1883. On the equations of motion and the boundary conditions for viscous fluids.
    B.A. Rep. (1883); *Papers* **2**, 132.
    Attempts to derive corrections to hydrodynamic equations from kinetic theory.
1897. Thermal transpiration and radiometer motion.
    Phil. Mag. [5] **43**, 142 (1897).
    Reply to Sutherland (1896).

Franz Richarz (1860–1920)
Bonn
1891. Zur kinetische Theorie mehratomiger Gase.
    Ann. Phys. [3] **48**, 467 (1893). *Verh. D. phys. Ges.* **10**, 73 (1891).
    Virial theorem.

Karl Robida (1804–77)
Klagenfurt
1864. Zur Theorie der Gase.
    Z. Math. Phys. **9**, 218 (1864).
    Effect of gravity.

Eugenie Aleksandrovich Rogovsky (1855–?)
St. Pokrowskaya
1882. O stroenii zemnoi atmosfery.
    Zhur. Russ. Fiz.-Khim. Obsh. **14**, 276 (1882).
    "On the structure of the earth's atmosphere."
1884. Ob atmosferakh planet.
    Zhur. Russ. Fiz.-Khim. Obsh. **16**, 76 (1884).
    "On atmospheres of planets."
1884a. O stroenii zemnoi atmosfery i obshikh zakonakh teorii gazov.
    Zhur. Russ. Fiz.-Khim. Obsh. **16**, 25, 185 (1884).
    "On the structure of the earth's atmosphere and the general laws of the theory of gases."
1884b. Otvet na "zametku" G. Stankevicha po povody stat'o "O Stroenii zemnoi atmosfery" i t.d.
    Zhur. Russ. Fiz.-Khim. Obsh. **16**, 314 (1884).
    "Reply to G. Stankevich's 'Note' on the memoir 'On the structure of the earth's atmosphere.'"

Antonio Roiti (1843–1921)
Florence
1876. La velocità teorica del suono e la velocità molecolare die gas.
    Atti Roma [3] **1**, 31 (1876) (title only); *N. Cim.* **16**, 218 (1876); *Mem. Accad. Lincei* **1**, 39 (1877).

[Roiti, cont.]
1877. Intorno ai rapporti che passano tra la velocità molecolare dei gas e la velocità teorica del suono [Risposta al prof. Brusotti].
Atti. Roma [3] **1**, 171 (1877).
1877a. Sulla propagazione del suono nella odierna teoria degli aeriformi.
Mem. Accad. Lincei **1**, 762 (1877); N. Cim. [3] **2**, 42 (1877); Ann. Phys. Beibl. **2**, 113 (1878).

Jacques Émil Joseph Ronkar
1884. Sur la conductibilite des corps gazeux pour la chaleur.
Bull. Acad. Sci. Bruxelles [3] **8**, 204 (1884).
Modification of Clausius theory to explain deviations from $\sqrt{T}$ law.

Arthur William Rücker (1848–1916)
Leeds
1880. On a suggestion as to the constitution of chlorine, offered by the dynamical theory of gases.
Phil. Mag. [5] **9**, 35 (1880); Proc. Phys. Soc. London **3**, 163 (1880).
Specific heat ratio is 1.4 for rigid surface of revolution.

Alessandro Atride Lorenzo Giuseppe Sandrucci (1861–?)
Pisa
1886. Sopra une obbiezone mossa da G. A. Hirn alla teoria cinetica dei gas.
N. Cim. [3] **20**, 193 (1886); Ann. Phys. Beibl. **11**, 686 (1887).
Weisbach formula for gas efflux.
1886a. Consequenze analitiche di una formula indicante la velocità molecolare totale di un corpo qualunque.
Riv. Sci.-Ind., nos. 13–14 (1886).
1887. Su l'accordo della teoria cinetica dei gas colla termodinamica, e sopra un principio della cinetica ammesso finora come vero.
Atti Roma [4] **3** (1) 205 (1887); Ann. Phys. Beibl. **11**, 686 (1887).
1887a. Sopra la constante R nell'isoterma dei gas perfetti.
Giorn. Mat. **25**, 73 (1887).
Molecular velocities.
1888. Sopra l'inesattezza di un principio ritenuto giusto nella teoria cinetica dei gas.
Atti Roma [4] **4**, 461 (1888).
Further discussion of Hirn's objections to kinetic theory.
1893. Sulla recenti esperienze di G. A. Hirn e sulle leggi dell'efflusso dei gas.
Atti Roma [5] **2**, 209 (1893).
1895. *La teorie su l'efflusso dei gas e gli esperimenti di G. A. Hirn.*
Firenze, 1895.

August von Schmidt (1840–1929)
Stuttgart
1900. Das Warmegleichgewicht der Atmosphäre nach den Vorstellungen der kinetischen Gastheorie.
Gerland's Beit. Geophysik **4**, 1 (1900).

Gustav Johann Leopold Schmidt (1826–83)
Pribaum, Prague
1860. Ein Beitrag zur Mechanik der Gase.
   Wien. Ber. **39**, 41 (1860).
   Discusses gas theories of Clausius (1857) and Redtenbacher (*Dynamidensystem*, 1857).

Arthur Schuster (1851–1934)
Manchester
1877. The radiometer and its lessons.
   Nature **17**, 143 (1877).
   Discusses theories of Reynolds, Stoney.
1895. The kinetic theory of gases.
   Nature **51**, 293 (1895).
   Molecular electrons could produce many spectral lines without having more than a few degrees of freedom.

W. Sellmeier
1872. Druck und elastischer Stoss.
   Ann. Phys. [2] **145**, 162 (1872).
   Comment on Hansemann (1871a); Sellmeier supports Clausius.

Charles Marie Etienne Theophile Simon (1825–80)
Paris
1876. Sur le rapport des deux chaleurs spécifiques d'un gaz.
   C.R. Paris **83**, 726 (1876); *Phil. Mag.* [5] **2**, 478 (1876).
   Uses formula of Yvon-Villarceau (1876) for internal motions of molecule.

Marian Ritter von Smolan Smoluchowski (1872–1917)
Vienna, Lemberg, Kraków
   Pisma Marjana Smoluchowskiego, 3 vols.
   Drukarnia Uniwersytetu Jagiellónskiego, Krakowie, 1924–27.
      Cited as *Pisma*.
1898. Ueber den Temperatursprung bei Wärmeleitung in Gasen.
   Wien. Ber. **107**, 304 (1898), **108**, 5 (1899); *Pisma* **1**, 113, 199: Polish trans. in *Wk. Polsk. Ucz.*
1898a. Ueber Wärmeleitung in verdunnten Gasen.
   Ann. Phys. [3] **64**, 101 (1898); *Phil. Mag.* [5] **46**, 192 (1898); *Pisma* **1**, 139.
   Criticism and extension of Maxwell's work (1879); experiments.
1898b. O przewodnictwie cieplnem gazów. Według dotychczasowych teoryj i dóswiadczén.
   Prace Mat.-Fiz. **10**, 33 (1898); *Pisma* **1**, 165.
   "On the thermal conductivity of gases according to theory and experiment."
   Solutions of Boltzmann equation.
1900. O Atmosferze ziemi i planet.
   Ksiega Uniw. lwowski, p. 1; *Phys. Z.* **2**, 307 (1900); *Pisma* **1**, 217, 263.

Nikolai Yakovlevich Sonin (1849–1915)
St. Petersburg, Warsaw
1889. O tak' nazyvaemom' fizicheskom' zakone fan' der' Vaal' sa.
   Trudy Varsh. **1** (5), 9 (1889).
   "On the so-called physical law of van der Waals." Remarks on $[1-(b/v)]^{-1}$ vs. $[1+(b/v)]$; relation between a/v^2 and the force law.

[Sonin, cont.]
1889a. O tak' nazyvaemom' fizicheskom' zakone fan' der' Vaal'sa i evo vidoizmeneniyakh'.
Trudy Varsh. **1**(6) 1 (1889).
"On the so-called physical law of van der Waals and its modification."
1889b. O primenenii uravneniya viriala k' kineticheskoi teorii gazov'.
Trudy Varsh. **1** (7) 1 (1889).
"On the application of the virial equation to the kinetic theory of gases."

Hermann Christian Otto Staigmuller (1857–1908)
Stuttgart
1898. Beitrage zur kinetischen Theorie mehratomiger Gase.
Ann. Phys. [3] **65**, 655 (1898); *Phys. Abst.* **1**, 696 (1898).
Specific heats.
1898a. Versuch einer theoretischen Ableitung der Constanten des Gestezes von Dulong und Petit.
Ann. Phys. [3] **65**, 670 (1898).

Boris Vyacheslavovich Stankevich (1860–?)
Warsaw
1884. Zametka na stat'yu Rogovskogo "O stroenii zemnoi atmosfery".
Zhur. Russ. Fiz.-Khim. Obsh. **16**, 311 (1884).
"Note on Rogovsky's paper 'On the structure of the earth's atmosphere.'"
1884a. Otvet' G. Rogovskomu ("O stroenii zemnoi atmosfery i t.d.").
Zhur. Russ. Fiz.-Khim. Obsh. **16**, 493 (1884).
"Reply to Rogovsky's paper ('On the structure of the earth's atmosphere')."
1885a. Kineticheskaya teoriya gazov' v' matematicheskom' izloshenii.
Uch. Zap. Mos. Un. **6**, 79 (1885).
"The kinetic theory of gases mathematically stated."
1886. Zur dynamischen Gastheorie.
Ann. Phys. [3] **29**, 153 (1886).
Binary collisions, Jacobian for transformation of coordinates; see Boltzmann (1886a).
1886a. O raspredelenii energii vrashatel'nago dvisheniya na molekuly gaza.
Mat. Sb. **13**, 129 (1886).
"On the distribution of rotational energy among the molecules of a gas."
1887. Zametka na stat'i G. Pirogova.
Zhur. Russ. Fiz.-Khim. Obsh. **19**, 32 (1887).
"Note on Pirogov's papers" (1886a).
1888. Etyudy po kineticheskoi teorii stroeniya tel'.
Zap. Mat. Otd. Nor. Obsh. Est. **8**, i, 1 (1888); *Fortschr.* (2), p. 234 (1888).
"Studies in the kinetic theory of the structure of bodies." Modification of van der Waals' theory.

Josef Stefan (1835–1893)
Vienna
1863. Bemerkungen zur Theorie der Gase.
Wien. Ber. **47**, 81 (1863); *Ann. Phys.* [2] **118**, 492 (1863).
Heat conduction, comments on Clausius' (1862) theory.
1863a. Ueber die Fortpflanzung der Wärme.
Wien. Ber. **47**, 326 (1863); *Ann. Phys.* [2] **125**, 257 (1865).
Kinetic theory of heat conduction.

[Stefan, cont.]
1836b. Ueber die Fortpflanzungsgeschwindigkeit des Schalles in gasförmigen Körpern.
Ann. Phys. [2] **118**, 494 (1863).
By applying Krönig's theory it is possible to derive Newton's value for the speed of sound, $\sqrt{p/\rho}$. A more complete theory (perhaps taking account of the internal motions of molecules, as suggested by Clausius) would have to give Laplace's correction.
1871. Ueber das Gleichgewicht und die Bewegung, insbesondere die Diffusion von Gasmengen.
Wien. Ber. **63**, 63 (1871).
Theory of gas diffusion; comments on Maxwell's theory.
1872. Über die dynamische Theorie der Diffusion der Gase.
Wien. Ber. **65**, 323 (1872).
Kinetic theory of diffusion and viscosity.

George Johnstone Stoney (1826–1911)
Dublin
1858. Notes on the molecular constitution of matter.
Proc. R. Irish Acad. **7**, 37 (1858); *Nat. Hist. Rev.* **5**, 210 (1858).
Reviews static gas theories and finds them unsatisfactory; suggests investigation of kinetic theory might be useful.
1867. On the physical constitution of the sun and stars.
Phil. Mag. [4] **34**, 304 (1867); *Proc. R.S. London* **16**, 25 (1868); **17**, 1 (1868).
Kinetic theory indicates separation of lighter molecules in an atmosphere of varying temperature.
1876. On Crookes' Radiometer.
Phil. Mag. [5] **1**, 177, 305 (1876); Nature **13**, 420 (1876).
1877. On the penetration of heat across layers of gas.
Phil. Mag. [5] **4**, 424 (1877); *Sci. Trans. Dublin* **1**, 13 (1877); *Proc. R.S. Dublin* **1**, 51 (1878).
Discusses "Crookes' layer" in rarefied gas.
1878. The radiometer and its lessons.
Nature **17**, 181, 261 (1878).
See Reynolds (1878), Change in thermal conductivity of gas at low pressures is important.
1878a. On the mechanical theory of Crookes' (or polarization) stress in gases.
Phil. Mag. [5] **6**, 401 (1878); *Sci. Trans. Dublin* **1**, 39 (1878).
Application of Clausius' (1862) theory to heat conduction in rarefied gases.
1879. On complete expansions for the conduction of heat and the polarization stress in gases.
B.A. Rep. **49**, 256 (1879).
Polarization stress is proportional to square of heat flow plus higher powers. Cf. Maxwell (1879).
1887. Curious consequences of a well-known dynamical theorem.
Proc. R.S. Dublin **5**, 448 (1887); *Phil. Mag.* [5] **23**, 544 (1887).
Second Law of Thermodynamics is not a "true dynamical law" since it is not invariant under time-reversal. However, time does not exist apart from events in the universe.
1893. Suggestion as to a possible source of the energy required for the life of Bacilli, and as to the cause of their small size.

[Stoney, cont.]
> *Proc. R.S. Dublin* **8**, 154 (1893); *Phil. Mag.* [5] **34**, 389 (1893).
> Statistical nature of Second Law and possible exceptions to it.

1895. Note on the motions of and within molecules; and on the significance of the ratio of the two specific heats in gases.
> *Proc. R.S. London* **58**, 177 (1895); *Nature* **52**, 286 (1895).

1895a. On the kinetic theory of gas, regarded as illustrating nature.
> *Proc. R.S. Dublin* **8**, 351 (1895); *Phil. Mag.* [5] **40**, 362 (1895).
> Ergodic hypothesis; specific heat paradox; motions of electrons in molecules; effect of ether.

1897. Of atmospheres upon planets and satellites.
> *Sci. Trans. Dublin* **6**, 305 (1897); *Ap. J.* **7**, 25 (1898).

1900. Escape of gases from planetary atmospheres.
> *Nature* **61**, 515, **62**, 78 (1900).
> Criticizes use of Maxwell distribution by Cook (1900); implications of presence of small amount of helium in earth's atmosphere.

1900a. On the escape of gases from planetary atmospheres according to the kinetic theory.
> *Ap. J.* **11**, 251, 357 (1900).

1900b. Note on inquiries as to the escape of gases from atmospheres.
> *Proc. R.S. London* **67**, 286 (1900).

Gavriil Konstantinovich Suslov (1857–?)
St. Petersburg

1886. Opyt' prilosheniya kineticheskoi teorii gazov'k' vybodu zakonov' soprotivleniya.
> *Zhur. Russ. Fiz.-Khim. Obsh.* **18**, 79 (1886).
> "Application of the kinetic theory of gases to derivation of the laws of resistance."

William Sutherland (1859–1912)
Melbourne

1893. The laws of molecular force.
> *Phil. Mag.* [5] **35**, 211, **36**, 150 (1893); *Proc. Phys. Soc. London* **12**, 30 (1894).
> Virial theorem used to test fourth power law; remarks on Waterston's work; Brownian motion.

1893a. The viscosity of gases and molecular forces.
> *Phil. Mag.* [5] **35**, 211, **36**, 150 (1893).
> Collisions of attracting spheres, explanation of apparent change of effective diameter with temperature, new formula for viscosity.

1894. The attraction of unlike molecules. I. The diffusion of gases.
> *Phil. Mag.* [5] **38**, 1 (1894).

1895. The viscosity of mixed gases.
> *Phil. Mag.* [5] **40**, 421 (1895).

1896. Thermal transpiration and radiometer motion.
> *Phil. Mag.* [5] **42**, 373, 476 (1896); **44**, 52 (1897).
> Criticizes methods of Reynolds (1879), applies those of Clausius (1862) instead. See Reynolds (1897).

1897. Boyle's law at very low pressures.
> *Phil. Mag.* [5] **43**, 11 (1897).
> Theory of surface-condensation and its possible effects on pressure.

Coloman (Kálmán) Szily von Nagy-Szigeth (1838–1924)
Budapest
1872. Das Hamilton'sche Princip und der zweite Hauptsatz der mechanische Wärmetheorie.
Ann. Phys. [2] **145**, 295 (1872); Phil. Mag. [4] **43**, 339 (1872); N. Cim. **2**, 259 (1877).
See Clausius (1872a).
1873. [Das dynamische Princip von Hamilton in der Thermodynamik.]
Ann. Phys. [2] **149**, 74 (1873); Phil. Mag. [4] **46**, 426 (1873).
Translation; original source not verified.
1875. [Der zweite Hauptsatz der mechanischen Wärmetheorie abgeleitet aus dem ersten.]
Ann. Phys. Erg. **7**, 154 (1876); Phil. Mag. [5] **1**, 22 (1876).
Translation; original source not verified. See Nichols (1876); Szily (1876).
1876. [On the dynamical signification of the quantities occurring in the mechanical theory of heat.]
Phil. Mag. [5] **2**, 254 (1876); Ann. Phys. [2] **160**, 435 (1877); Carl's Rep. **13**, 97 (1877).
Translation; original source not verified. Admits that Nichols' (1876) criticism of his previous paper is just; tries again to derive Second Law.

Peter Guthrie Tait (1831–1901)
Edinburgh
Scientific Papers, 2 vols.
Cambridge University Press, 1890–1900.
Cited as *Papers*.
1875. See Dewar (1875).
1884. Note on a theorem of Clerk-Maxwell.
Proc. R.S. Edinburgh **13**, 21 (1884).
New proof of velocity distribution; doubts validity of Boltzmann's generalization to all degrees of freedom.
1886. On the foundations of the kinetic theory of gases.
Trans. R.S. Edinburgh **33**, 65 (1886); Phil. Mag. [5] **21**, 343 (1886); Papers **2**, 124.*
Equipartition theorem; mean-free-path definition; rate of approach to equilibrium; effects of gravity, Maxwell–Boltzmann distribution. See Burbury (1886); Natanson (1888).
1886a. On the partition of energy among groups of colliding spheres.
Proc. R.S. Edinburgh **13**, 537 (1886).
1887. On the foundations of the kinetic theory of gases. II.
Trans. R.S. Edinburgh **33**, 251 (1887); Phil. Mag. [5] **23**, 141, 433 (1887); Papers **2**, 153.
Theory of viscosity and diffusion.
1887a. The assumptions required for the proof of Avogadro's law.
Phil. Mag. [5] **23**, 433 (1887).
Reply to Boltzmann (1886a).

---

* "Translated into Russian by Captain J. Gerebiateffe and published with annotations expanding Tait's mathematical processes in the Russian *Review of Artillery* (1894)"–C. G. Knott, *Life and Scientific Work of Peter Guthrie Tait* (Cambridge University Press, 1911), p. 25.

[Tait, cont.]
1888. On some questions in the kinetic theory of gases. Reply to Prof. Boltzmann.
Phil. Mag. [5] **25**, 172 (1888).
Effect of collisions in producing equilibrium; definition of mean path.
1888a. Reply to Prof. Boltzmann.
Proc. R.S. Edinburgh **15**, 140 (1888).
Proof of velocity distribution; viscosity theory.
1888b. Note on the motion of a gas "in mass."
Phil. Mag. [5] **25**, 38 (1888).
Reply to Burbury (1887).
1888c. Numerical and other additions to his paper, read on 6 December 1886, on the foundations of the kinetic theory of gases.
Proc. R.S. Edinburgh **14**, 46 (1888).
1888d. On the foundations of the kinetic theory of gases. III.
Trans. R.S. Edinburgh **35**, 1029 (1890, read 1888); *Papers* **2**, 179.
Analysis of collisions; mean free path; viscosity; thermal cond.
1889. On the mean free path and the average number of collisions per particle per second in a group of equal spheres.
Proc. R.S. Edinburgh **15**, 225 (1889).
1891. On van der Waals's treatment of Laplace's pressure in the virial equation: in answer to Lord Rayleigh.
Nature **44**, 546 (1891).
1891a. On van der Waals's treatment of Laplace's pressure in the virial equation: in answer to Lord Rayleigh.
Nature **44**, 627 (1891).
Criticism of van der Waals (1873).
1891b. On the virial equation for gases and vapors.
Nature **45**, 199 (1891).
Reply to Korteweg (1892).
1891c. On the foundations of the kinetic theory of gases. IV.
Trans. R.S. Edinburgh **36**, 257 (1892) (read 1889 and 1891); *Papers* **2**, 192 .
Equations of state of van der Waals and Clausius; virial equation for attracting spheres; relation between kinetic energy and temperature.
1896. Note on Clerk-Maxwell's law of distribution of velocity in a group of equal colliding spheres.
Proc. R.S. Edinburgh **21**, 123 (1897, read 1896); *Papers* **2**, 427.
Defends Maxwell's proof against criticisms of Bertrand (1896) and Boltzmann; extension of his previous work (1886, 1887). Quotes letter from Boltzmann saying flaws in Maxwell's proof are trivial.

Joseph John Thomson (1856–1940)
Cambridge
1887. Some applications of dynamical principles to physical phenomena. Part II.
Phil. Trans. **178**, 471 (1887).
Comments on relation of Second Law to mechanics.
1888. *Applications of Dynamics to Physics and Chemistry.*
Macmillan, London; reprinted by Dawsons of Pall Mall, London, 1968.
Chapter VI, proof that mean translatory energies of two substances in contact tend to become equal. Chapter XI, theory of evaporation. Chapter XII, properties of dilute solutions. Chapter XIII, dissociation.

[Thomson, J. J., cont.]
1900. On a view of the constitution of a luminous gas suggested by Lorentz's theory of dispersion.
Arch. Néerl. [2] **5**, 642 (1900).
Specific heats; relation between spectral lines and number of degrees of freedom in molecule.

William Thomson: see Lord Kelvin

E. Töpler
1886. Zur Ermittelung des Luftwiderstandes nach der kinetischen Theorie. Gerold, Wien, 1886.
Exner's Rep. **23**, 162 (1887); Ann. Phys. Beibl. **11**, 747 (1887).

Thomas Tredgold (1788–1829)
London
1821. A refutation of Mr Herapath's Mathematical Inquiry into the causes, laws, etc. of heat, gases, gravitation, etc.
Phil. Mag. **58**, 130, 260 (1821).

Aroldo Violi
1883. Sulla relazione di alcune proprietà fisiche degli aeriforme, col rapporto die calori specifici a pressione costante ed a volume costante.
N. Cim. [3] **14**, 183, 207 (1883).
Ratio of sp. heats, number of degrees of freedom in molecules.
1884. La velocità molecolari degli aeriformi.
Atti Roma [3] **8**, 22 (1884).
1888. L'isotherma die gas.
Atti Roma [4] **4**, 285, 316, 462, 513 (1888); Phil. Mag. [5] **27**, 527 (1889); Ann. Phys. Beibl. **13**, 66.
Derives new equation of state.

Woldemar Voigt (1850–1919)
Göttingen
1885. Die Erwärmung eines Gases durch Compression nach der kinetischen Gastheorie.
Göttingen Nachr., p. 228 (1885).
1896. Fluorescenz und kinetische Theorie.
Göttingen Nachr., p. 184 (1896); Phys. Soc. Abst. **2**, 342 (1896).

Johannes Diderik van der Waals (1837–1923)
Leiden, Amsterdam
1873. Over de continuiteit van den gas- en vloeistoftoestand.
Dissertation, Leiden, 1873.
Ueber den Uebergangs-Zustand zwischen Gas und Flüssigkeit.
A. W. Sijthoff, Leiden, 1873.
Die Continuität des Gasförmigen und Flussigen Zustandes. Aus dem Hollandischen Übersetzt und mit Zusatzen Versehen, von Dr. F. Roth. J. A. Barth, Leipzig, 1881.
Physical Memoirs **1**, Part 3. English trans. by R. Threlfall and J. F. Adair (published by Taylor and Francis for the Physical Society, London, 1890).
Ann. Phys. Beibl. **1**, 10 (1877).
See Maxwell's review (1874).

[Van der Waals, cont.]
1876. Over het betrekkelijk aantal botsingen, dat een molekuul ondergaat, wanneer het zich beweegt door bewegende molekulen of door molekulen, die men onderstelt stil te staan; alsmede over den invloed van de afmetingen der molekulen volgens de richting der relative beweging op het aantal dier botsingen.
Amsterdam Verslagen [2] **10**, 321 (1876); *Arch. Néerl.* **12**, 201 (1877).
"On the relative number of collisions experienced by a molecule as it moves through a medium of molecules in motion or a medium or molecules assumed at rest; and on the influence which the dimensions of the molecules in the direction of relative motion exert on the number of collisions."

1876a. Over het aantalbotsingen en den gemiddelden botsings-afstand in gasmengsels.
Amsterdam Verslagen [2] **10**, 337 (1876); *Arch. Néerl.* **12**, 217 (1877).
"On the number of collisions and the mean collision-distance in gas mixtures."

1880. De betrekking tusschen spanning, volumen en temperatur bij dissociatie. *Amsterdam Verslagen* [2] **15**, 199 (1880).
"On the pressure–volume–temperature relation in the case of dissociation." Refers to Gibbs' theory (*Trans. Conn. Acad.* **3**, Part I).

1889. Molekulair-theorie voor een mengsel van twee stoffen.
Amsterdam Verslagen [3] **6**, 163 (1889) (abstract); *Arch. Néerl.* **24**, 1 (1891), *Z. phys. Chem.* **5**, 133 (1890).
"Molecular theory of binary mixtures" (equation of state).

1895. Over de kinetische beteekenis van den thermodynamischen potentiaal.
Amsterdam Verslagen [4] **3**, 205 (1895); *Arch. Néerl.* **30**, 137 (1897).
"On the kinetic meaning of the thermodynamic potential." Cf. Onnes (1881).

1896. Eene bijdrage tot de kennis der toestandsvergelijking.
Amsterdam Verslagen [4] **5**, 150 (1896); *Fortschr.* (2), p. 199 (1896).
"A contribution to the knowledge of the equation of state." Third virial coefficient for hard spheres; cf. Boltzmann (1896); Jäger (1896).

1899. Eenvoudige afleiding van de toestandsvergelijking voor stoffen met uitgebreide en samengestelde molekulen.
Amsterdam Verslagen [4] **7**, 160 (1899); *Amsterdam Proc.* **1**, 138 (1899).
"Simple deduction of the equation of state for substances with extended and composite molecules." Method for calculating virial coefficients, used by van Laar (1899).

1899a. Over de afleiding der toestandsvergelijking. Discussie met Prof. Boltzmann.
Amsterdam Verslagen [4] **7**, 537 (1899); *Amsterdam Proc.* **1**, 468 (1899); *Arch. Néerl.* [2] **4**, 299 (1901).
"On the deduction of the equation of state. Discussion with Prof. Boltzmann."

Emil Gabrial Warburg (1846–1931)
Strasbourg
1875. See Kundt (1875).

John James Waterston (1811–83)
Bombay, Edinburgh
The Collected Scientific Papers of John James Waterston. Edited, with a biography, by J. S. Haldane. Oliver and Boyd, Edinburgh, 1928.
Cited as *Papers*.
1843. *Thoughts on the mental functions; being an attempt to treat metaphysics as a branch of the physiology of the nervous system.*

[Waterston, cont.]
>Oliver and Boyd, Edinburgh, 1843; *Papers* 3, 183, 167.
>>Contains his first formulation of the kinetic theory, including equipartition theorem (stated incorrectly).

1845. An account of a mathematical theory of gases.
>Circulated privately; *Papers*, p. 320.
>>Long abstract of 1846 paper.

1846. On the physics of media that are composed of free and perfectly elastic molecules in a state of motion.
>*Proc. R.S. London* **5**, 604 (1846) (abstract); *Phil. Trans.* **183A**, 5 (1893); *Papers*, p. 207.

1851. On a general theory of gases.
>*B.A. Rep.* **21**, 6 (1851); *Papers*, p. 318.
>>States results of his kinetic theory.

1859. On the theory of sound.
>*Phil. Mag.* [4] **16**, 481 (1859); *Papers*, p. 345. Reprinted in R. B. Lindsay, ed., *Acoustics* (Stroudsberg, Pa.: Dowden, Hutchinson & Ross, 1973), p. 308.
>>Cf. Preston (1877).

Henry William Watson (1827–1903)
Coventry

1876. *A Treatise on the Kinetic Theory of Gases.*
>Clarendon Press, Oxford, 1876; 2nd ed., 1893.
>>Rederives some of Boltzmann's results on equilibrium distribution of energy and Second Law of Thermodynamics, using generalized coordinates.

1884. The propagation of sound in gas on the kinetic theory.
>*Proc. Birmingham Phil. Soc.* **4**, 242 (1884).
>>Ratio of specific heats.

1892. On the Boltzmann–Maxwell law of partition of kinetic energy.
>*Nature* **45**, 512 (1892).
>>Comments on Bryan (1891); Burnside (1887).

1892a. On a proposition in the kinetic theory of gases.
>*Nature* **46**, 29 (1892).
>>Proof of velocity distribution; comments on Rayleigh (1892a).

1892b. Maxwell's law of distribution of energy (with S. H. Burbury).
>*Nature* **46**, 100 (1892).

1894. Boltzmann's minimum theorem.
>*Nature* **51**, 105 (1894).
>>Reply to Culverwell (1894). Reversibility paradox does not contradict H-theorem.

1895. The kinetic theory of gases.
>*Nature* **51**, 222 (1895).
>>Equipartition theorem implies discontinuous change in mean square velocity when billiard balls become slightly nonspherical.

Max Bernhard Weinstein (1852–1918)
Berlin

1895. Ueber die Zustandsgleichung der Körper und die absolute Temperatur.
>*Ann. Phys.* [3] **54**, 544 (1895).
>>Mathematical transformations of virial equation, calculations for colliding molecules.

Ernst Eilhard Gustav Wiedemann (1852–1928)
Leipzig
1878. Untersuchungen uber die Natur der Spectra.
Ann. Phys. [3] **5**, 500 (1878); *Phil. Mag.* [5] **7**, 77 (1879).
Pressure-broadening of spectral lines.

Cornelis Harm Wind (1867–1911)
Gröningen
1897. Über den dem Liouville'schen Satze entsprechenden Satz der Gastheorie.
Wien. Ber. **106**, 21 (1897).
Criticizes Boltzmann's assumption about inverse collisions.

"X" (Anonymous)
1821. Remarks on Mr. Herapath's theory.
Ann. Phil. [2] **2**, 223 (1821).
Does not understand proof that pressure is prop. to *square* of velocity.
1821a. Further remarks on Mr. Herapath's theory.
Ann. Phil. [2] **2**, 390 (1821).
Doubts existence of absolute cold; gas would have zero volume.

Antoine Joseph François Yvon-Villarceau (1813–83)
Paris
1872. Sur un nouveau théorème de mécanique générale.
C.R. Paris **75**, 232, 377, 990 (1872).
Theorem similar to virial theorem.
1876. Note sur les déterminations théorique et experimentale du rapport des deux chaleurs spécifiques, dans les gaz parfaits dont les molécules seraient monoatomiques.
C.R. Paris. **82**, 1127, 1175 (1876).

Gyözö Zemplen (1879–1916)
Budapest
1900. Ueber die Grundhypothesen der kinetischen Gastheorie.
Ann. Phys. [4] **2**, 404 (1900).
Criticism of Burbury's book (1899).
1900a. Ueber die Grundhypothesen der kinetischen Gastheorie.
Ann. Phys. [4] **3**, 761 (1900).
Reply to Burbury (1900c).

Ernst Friedrich Ferdinand Zermelo (1871–1953)
Berlin, Göttingen
1896. Ueber einen Satze der Dynamik und die mechanische Wärmetheorie.
Ann. Phys. [3] **57**, 485 (1896). ET in Brush, *Kinetic Theory* 2, 208.
Recurrence paradox – see Poincaré (1890); Boltzmann (1896c, 1897, 1897a); Dugas (1959, p. 206).
1896a. Ueber mechanische Erklärungen irreversibler Vorgänge. Eine Antwort auf Hrn. Boltzmann's "Entgegnung." ET in Brush, *Kinetic Theory* 2, 229.
Ann. Phys. [3] **59**, 793 (1896).
1900. Ueber die Anwendung der Wahrscheinlichkeitsrechnung auf dynamische Systeme.
Phys. Z. **1**, 317 (1900).

Index to Books 1 and 2*

ABEL, N. H., 447
ABNEY, W. DE W., 512, 533
absolute zero (*see also* temperature), 23–26, 27
absorption of sound, 86–87
accommodation coefficient, 230
action at a distance, 21, 48, 388
ADDAMS, R., 323
adiabatic compression and expansion, 31, 34, 88, 151, 540
AEPINUS, F. U. T., 138, 144
aether, *see* ether
AGASSIZ, L., 557, 559
agnosticism, 58
AICHI, K., 442
air, heat transfer, 471–73, 478–80; thermal conductivity, 491, 494, 496, 500–1; viscosity, 435–41
air pressure, 9–13, 17, 123–24
AIRY, G. B., 684
AKASOFU, S.-I., 456
ALDER, B., 271, 415
ALEMBERT, J. LE R. D', 563
ALIOTTA, A., 68
ALLEN, Z., 332, 595
AMAGAT, E. H., 203
AMANO, K., 523
American science, 88, 157, 709–10

AMONTONS, G., 23
ANDERSON, D. L., 208
ANDLER, C., 637
ANDRADE, E. N. DA C., 34
ANDREWS, F. C., 281
ANDREWS, J. P., 189
ANDREWS, T., 77, 257–58, 394, 485, 489, 520
ÅNGSTRÖM, A., 524
ÅNGSTRÖM, K., 511, 533
ARAGO, D. F. J., 42, 123
ARCHER, C. T., 504, 534
argon, 354, 431; heat conduction, 503
Aristotelian science, 12, 36–39, 41
ARNOTT, N., 323
ARONSON, S., 22, 51
ARRHENIUS, S., 299, 508–9, 582, 698
art, 52
ASCHKINASS, E., 533
astronomy, 37–38, 41, 387, 545–48, 584
ATKINS, K. R., 18
Atlantic cable, 682
atmosphere, 23, 168, 464, 526; chemical composition, 45, 50, 138, 199; escape, 138; heat/radiation transfer, 88, 472, 487, 508, 555; pressure variation with height, 75, 119, 166; resistance to railway train, 123–24,

* *The Kind of Motion We Call Heat* is published in two books: page numbers 1–300 in the index refer to Book 1 and page numbers 301–770 refer to Book 2.

atmosphere (*Cont'd*)
  126; temperature variation with height, 70, 71, 74, 119, 122–23, 153–54, 195–6, 349–50, 588–89
atom (*see also* molecule), 21; atmospheric model, 45, 53, 92, 138, 153, 204–5, 277, 283, 296–97, 319; billiard ball, 3, 21, 54, 73–74, 153, 337, 355, 401–2, 436; electrostatic models, 322; forces between, 13–14, 30, 35, 45, 46, 54, 64, 74, 92, 138, 170, 173, 177–78, 265, 277–78, 386–95, 417, 441, 504; hard, 14, 63, 110, 131, 354; size, 45–47, 75–78, 83, 138, 199, 278; springy, 13; vortex 92, 207–8, 388
atomic weight, 45, 93, 278
atomism and anti-atomism, 8, 21, 42, 55, 56, 60, 63–65, 90–102, 197, 208–9, 263–64, 274–99, 528, 641–42, 656, 673, 674, 693–94, 698–700
AUERBACH, F., 582
AUERBACH, L., 296
AVENARIUS, R. H. L., 61, 291
AVOGADRO, A., 46, 47
Avogadro's hypothesis, 45–48, 50, 113, 139, 147, 167, 174, 196–98
Avogadro's number, 76, 83, 101, 696–97
AYRES, C., 67

BABBAGE, C., 322, 595
BABINET, J., 323, 595
BACHE, R. M., 664, 686, 687
BACHELIER, L., 101, 671
BACON, F., 329, 608
BADASH, L., 548
BAER, K. E. VON, 613
BAIERLEIN, R., 685
BAILEY, C., 15
BAILEY, E., 564
BAILLIE, J., 637
BAILLY, C., 320
BAILLY, J.-S., 552, 555, 563
BAILY, F., 437, 442
BAIRE, R. L., 382, 384
BAKER, R., 49
BAKEWELL, R., 662
BAKKER, G., 721
BAKLAEV, B. G., 377
BALARD, 332

BALAZS, N. L., 581
BANCROFT, W. D., 102
BANKS, J., 658
BANNAWITZ, E., 503
BARBER, E., 331
BARKER, J. A., 506
BARNETT, M. K., 26, 27, 32, 562, 579
BARR, E. S., 313, 477, 489, 533
BARRÉ DE SAINT-VENANT, A. J. C., 684
BARTHOLOMEW, M., 565
BARTOLI, A., 244, 518–19, 539–40
BARTON, J., 322
BARUS, C., 441
BARZUN, J., 582
BASEVI, C. E., 721
BATEMAN, H., 532
BAUER, E., 522, 652
BAUMGARTNER, A., 322
BAYNES, R. E., 721
BEAUFORT, 136
BECKER, O., 638
BECQUEREL, A. [C.], 324
BECQUEREL, [A.] E., 482
BELLONE, E., 312
BENNDORFF, H., 722
BÉRARD, J. E., 163
BERGER, G. K., 640
BERGMAN, T., 28
BERGSON, H., 68
BERINGER, C. C., 563
BERNAL, J. D., 512
BERNHARDT, H., 549, 613, 614, 615
BERNKOPF, M., 447
BERNOULLI, D., 6, 39, 150, 391, 489; kinetic theory, 20, 22, 69, 70, 110, 131, 171, 209, 264, 401, 706, 708
BERNOULLI, J., 492
BERNSTEIN, H. T., 132, 431
BERRY, A. J., 90, 528
BERTHELOT, [P. E.] M., 512
BERTHOLLET, C. L., 29, 477
BERTRAND, J. L. F., 597, 722
BERZELIUS, J. J., 47, 55, 320
BESSEL, F. W., 437
bibliography, 5–6, 721–69
BICKERTON, A. W., 582
BIEL, J., 549
biology, 40, 52–53, 55, 57–58, 657, 663

Book 1, pp. 1–300; Book 2, pp. 301–770.

INDEX   xvii

BIOT, J. B., 42, 155, 313, 477, 482, 552, 556
BIRD, G., 324
BIRD, R. B., 506
BIRKS, T. R., 200
BLACK, J., 28, 33, 256, 291, 292, 305, 312, 476
BLACKMORE, J., 68, 284, 294, 295, 296, 299, 549
BLAIR, 22
BLASERNA, P., 722
BLÜH, O., 296
BLUMENTHAL, 449
BOAS, M. (*see also* M. B. HALL), 16, 396, 550
BOBETIC, M. V., 506
BOCK, A. M., 22
BÖHNERT, 722
BOGUSLAWSKI, S., 449
BOHM, D., 640
BOHR, N., 91, 294, 299
BOLTZMANN, L. (*see also* H-theorem), 7, 203, 231–32, 243–44, 350, 593, 618–19, 673, 674, 706, 708, 722–27; atomism, 94, 97–98, 244–47, 292–93; Brownian movement, 654; diatomic molecules, 79, 86, 195, 354–55, 360; dynamism, 277–78, 633; equation of state, 269, 270, 414–17; ergodic hypothesis, 79, 363–77, 384; Gibbs, 633, 639; H-curve, 622–24, 627, 633–35; heat conduction, 497–98, 502; infinity and infinitesimals, 371, 610, 635, 645–46; Mach, 287, 289–93, 615, 636; Maxwell, 233–35, 242, 243, 347, 370, 446, 588, 601, 611; Meyer, 615; molecular disorder, 621, 625, 626–27, 648–51; monocyclic systems, 367, 617–18; Planck, 616, 642–48, 654; radiation theory, 88, 244, 306, 518–19, 522, 540–42, 645–47; recurrence paradox, 239–40, 289, 625, 632–36, 645, 650; reversibility paradox, 239, 588, 606–7; statistical interpretation of Second Law, 63, 80–84, 94–96, 99, 236, 239, 279, 287, 598–602, 606–12, 616, 622–25, 632–37, 694; statistical method, 3, 170, 183, 415, 587, 597,

606–12, 636; suicide, 246–47, 293; Tait, 357–58, 362, 431; thermal equilibrium, 235–36, 241, 351, 357–58, 366, 419, 614; time direction, 289–90, 635–36, 650; tombstone, 84, 609; transport theory, 74, 80–81, 84–85, 88, 180, 236–37, 422, 432, 442–45, 449–50, 458, 467; velocity distribution for finite system, 188, 371; van der Waals, 246, 414–16, 418
Boltzmann constant, 74, 234, 685
Boltzmann equation, 80, 236–37, 443–44, 447–49, 456–59, 462, 600
Boltzmann factor (*see also* Maxwell-Boltzmann distribution), 74–75, 101, 196, 233, 348, 420
BOLYAI, J., 289
BOLZA, H., 449
BOLZANO, B., 378
BORDEAUX, A., 89
BOREL, E., 380, 383, 384–85, 626
BORK, A. M., 548, 654
BORN, M., 49, 279, 449, 673
BOSANQUET, R. H. M., 86, 195, 356, 727
BOSCOVICH, R., 53, 54, 92, 96, 277, 359, 388, 392–93, 397, 584, 594, 639
BOSTOCK, J., 313
BOTTO, J. D., 666
BOTTOMLEY, J. T., 520, 523, 527
BOTZEN, A., 532
BOUSSINESQ, J., 103, 532
BOUTAN, A., 512
BOWDEN, R. C., 363
BOWDITCH, N., 365
BOX, T., 511
BOYER, C. B., 26, 66, 381
BOYLE, R., 3, 16, 90, 388, 396, 545, 563; pneumatics, 1–12; theory of air pressure, 13, 15, 35, 149
Boyle's law, 11–12, 17, 19, 30, 70, 77, 162, 199, 389; deviations from, 127–28, 258, 398–400
BRACE, D. B., 482
BRAHE, T., 37–38, 41
BRANDES, H. W., 322
BRAUN, J. A., 26
BREITENBACH, P., 432
BREWSTER, D., 307, 318: Brownian movement, 659–60

Book 1, pp. 1–300; Book 2, pp. 301–770.

BRIDGMAN, P. W., 63, 582
BRILLOUIN, M., 450, 462, 727
BRINTON, C., 36
British Association for the Advancement of Science, 95, 140, 157, 360, 561, 617
British science, 84, 88, 94, 157–58, 559, 708–9
BROCK, W. H., 93, 198, 281, 488
BRODA, E., 248
BRODIE, B., 93, 196
BROGLIE, L. DE, 299, 626, 700
BRONSTEIN, M., 639
BROUGHAM, H., 122
BROUWER, L. E. J., 80, 242, 380, 383–84
BROWN, A., 332
BROWN, A. C., 356, 727
BROWN, R., 488, 655, 657–62
BROWN, S. C., 30, 33, 34, 308, 476, 488
Brownian movement, 91, 101, 247, 298, 530, 649, 654–700; as violation of Second Law, 616, 656, 670–71, 694
BRUCKNER, A., 446
BRÜCKE, E., 320
BRUNHES, B., 543, 549, 639
BRUNO, G., 38
BRUNOLD, C., 549
BRUSH, C. F., 22, 527–28
BRUSH, S. G., 8, 18, 34, 49, 69, 90, 133, 164, 188, 194, 198, 248, 270, 273, 281, 385, 397, 425, 462, 498, 507, 532, 548, 582, 626, 654, 685
BRYAN, G. H., 95, 281, 360–61, 377, 617–22, 626, 638, 708, 728
BUCHANAN, J. Y., 729
BUCHHOLZ, H., 248
BUCKLE, H. T., 333
BUCKLEY, H., 159
BUCKLEY, J. H., 582
BUDDE, E., 666
BÜCHEL, W., 549, 639
BUFF, H., 489
BUFFON, G.-L. LECLERC, COMTE DE, 471, 552–53, 557, 657, 659, 661
BULLARD, E. C., 563
BUNGE, M., 299
BURBURY, S. H., 7, 95, 336, 356, 357, 359–60, 372, 614, 616, 619, 625, 626, 651, 708, 729–31; correlation of

Book 1, pp. 1–300; Book 2, pp. 301–770.

colliding molecules, 360, 620–21, 624
BURCHFIELD, J. D., 566
BUREAU, F., 312
BURKE, J. G., 114, 538
BURNETT, D., 467
BURNSIDE, W., 358–59, 731
BURR, A. C., 474–75, 477, 484, 505
BURSTYN, H. L., 180, 489
BURTT, E. A., 16, 396, 550
BUYS-BALLOT, C. H. D., 50, 72, 177, 180, 324, 492, 494, 585, 731
BYWATER, J., 661, 662–63

CAGNIARD DE LA TOUR, C., 77, 257, 394
CAJORI, F., 26, 30, 32, 308, 396
CALLENDAR, G. S., 511
caloric theory, 8, 27–32, 47, 112, 151, 160, 203, 208, 209, 304–12, 316, 318–21, 330–31, 394–95, 398, 529, 567, 663
Cambridge University Library, 229
CAMP, W. J., 418
CAMPBELL, L., 185, 201
CANNIZZARO, S., 48, 93, 175, 197–98, 203
CANNON, W. F., 565
CANTONI, G., 665
CANTOR, G., 79, 242, 378–80
capillarity, 30, 75, 76, 142, 264, 283, 395, 687–88
ČAPEK, M., 581, 595, 637
CARATHEODORY, C., 297, 631
carbon dioxide (CO_2), 77, 78, 163, 257–58, 267, 400, 480, 483, 485, 500; absorption of radiation, 508, 511
carbon monoxide (CO), 78
CARDWELL, D. S. L., 109, 113, 149, 159, 164, 318, 333, 476, 477, 482, 579, 708
CARMAN, E. H., 506
CARNOT, L., 567, 568
CARNOT, [N. L.] S. (see also thermodynamics, second law), motive power of heat, 31, 34, 540, 551, 567–69; nature of heat, 311, 325, 331, 579
CARO, E. M., 590
CARPENTER, E., 68
CARPENTER, W. B., 213–14, 663
CARUS, P., 68

INDEX

CASSIRER, E., 67, 247, 584, 654
cathode rays, 181, 227, 693
Catholic Church, 38–39
CAUCHY, A.-L., 199, 244, 378
CAUSEY, R. L., 51
CAVENDISH, H., 33, 388
CAYLEY, A., 358
CAZIN, A., 205, 332
CELLÉRIER, C., 731
centripetal acceleration, 15
CHALLIS, J., 51, 120, 155, 204, 322, 332, 684
CHALMERS, A. F., 198, 356, 404
CHALMERS, T. W., 333
CHAMBERLIN, T. C., 511
CHAMBERS, R., 323
CHAMBERS, W., 323
CHARLES, J. A. C., 23; gas law, 23
CHAPMAN, S., 455, 463–65, 468; thermal diffusion, 466; transport theory, 85, 180, 193, 237, 425, 432, 442, 448–56, 462, 499, 502, 506
Chaudesaiges, 697
chemical atomic theory, 45–47, 93, 175, 196–97, 278–79, 286, 292, 297
chemical reactions, 29, 34, 46
chemistry, 36, 53, 55
Cherbuliez, 22, 155
CHMELKA, F., 295
CHRISTIANSEN, C., 504, 517
CHRYSTAL, G., 226
CHURCH, W. C., 512
CHWOLSON, O., 512, 523
CLAPEYRON, B.-P.-E., 331, 542, 568
CLARK, X., 68
CLARKE, S., 546, 550
classicism, 52, 55
CLAUSIUS, R., 6, 7, 142, 168, 486, 585, 596, 732–33; dielectric constant, 78; equation of state, 181–82, 269, 414, 416; equipartition, 174, 339–40; heat conduction, 73, 181, 192, 436, 492–94, 499; Herapath, 130; internal motion of molecules, 172, 174–75, 339; irreversibility, 575–79, 583; Joule, 165; kinetic theory, 71–72, 113, 141, 155, 157, 160, 164, 168–82, 264–65, 330, 401, 422, 446, 706, 708; Maxwell, 181–182, 427, 492–94, 498; mean free path, 72, 177–81, 408, 426, 431, 438, 462, 493–94, 585–86; phase changes, 173–74; reception of his kinetic theory, 35, 48, 131, 175, 197, 202–3, 205, 208, 209, 283, 435; specific heats, 148, 175, 340, 354; statistical method, 168–69, 178, 182, 492–93, 583, 585–86; thermodynamics, 35, 232, 279, 419, 569–71, 573–79, 581, 593, 617, 641–42; virial theorem, 75, 181, 208, 404–6
Clausius–Mossotti formula, 78
CLEGHORN, W., 28–29
CLEMENS, F. A., 323
CLÉMENT, N., 27, 118, 149, 152
CLIFFORD, W. K., 201, 208–9
clockwork universe, 40, 396, 545–46, 553–54
COBB, G. C., 18
COCKE, W. J., 639
Coggia's comet, 215
COHEN, I. B., 18, 281, 296, 550
COLBURN, Z., 35
COLDING, L., 164, 203, 318, 325, 326–27, 562, 595
COLE, T. M., JR., 322
COLERIDGE, S. T., 53
COLLINGWOOD, R. G., 49
collisions or impacts, 41, 110–11, 127, 131, 135, 173, 178, 186, 341, 343–45, 610–11, 688–90; effect of attractive forces, 428–29, 433–34; notation, 457; rate, 413–14, 461–62; reversibility, 237, 239, 444, 613, 618
COMBE, J., 125
comets' tails, 34, 215
COMSTOCK, J. L., 320
COMTE, A., 60, 67
CONANT, J. B., 16, 17, 18
condensation coefficient, 75, 78
CONDON, E. U., 87–88, 151
CONDORCET, A. N., MARQUIS DE, 563, 584, 594
CONGREVE, R., 67
convection, 128, 471, 473–75, 486, 488, 490, 515, 524–26, 536–39; word, 488
COOK, S. R., 733
COOK, W. R., 432

Book 1, pp. 1–300; Book 2, pp. 301–770.

COOKE, J. P., 323
cooling law, 25, 470–81, 515–16, 535
Copenhagen philosophy, 651
COPERNICUS, N., 37, 41
CORNELIUS, H., 734
CORNELL, E. S., 313, 477
corresponding states, law of, 254, 268
CORRIGAN, S., 734
COSTABEL, P., 27, 314
COTTON, A., 695
COUCHE, 323
COUES, S. E., 323
COULOMB, C. A., 417, 437–38, 529
COULSON, C. A., 281
COULSON, T., 324
COUPER, A. S., 55
COUPER, P., 322
COURNOT, A.-A., 183, 549, 595
COWLING, T. G., 432, 449, 468, 499, 506, 566
CRAWFORD, A., 26, 112
CRESCAS, H., 381
critical point, liquid–gas, 77, 257–63, 266–67, 270–71, 394, 531; solid–liquid, 271–72
CROLL, J., 68, 204
CROMBIE, A. C., 188
CROOKES, W., 85, 88, 143, 214, 294, 323, 476, 526–27: radiometer, 211–21, 228, 323, 340
CROSLAND, M., 15, 89, 312
CROUNE, W., 12
CROVA, A., 512
CROWLEY, M. E., 16
CULVERWELL, E. P., 95, 361, 616–18, 620–24, 643, 734
CURRY, C. E., 248
CURTISS, C. F., 506
cyclic history, 553–54, 565, 577, 627–29
CZERMAK, P., 735

DABROWSKI, 701
DAGUIN, P. A., 332
DAHL, P. F., 164, 333
DAKO, M., 549
DALLAS, D. M., 198
DALTON, J., 24, 26, 119, 153, 165, 392, 477, 478, 480, 483, 485, 642; atomic theory, 29, 45–47, 93, 292

DAMPIER, W. C., 159
DANCER, J. B., 665
DANIELL, J., 324, 480, 488
DARMSTAEDTER, L., 159
DARWIN, C., 40, 55, 561, 604; Brown, 661–62
DARWIN, G. H., 565
DASCOLA, G., 313
DAUB, E. E., 89, 141, 145, 167, 168, 176, 549, 582, 583, 597, 598, 614, 615
DAUBEN, J. W., 381
DAUVILLIER, A., 641
DAVIES, C. H., 522
DAVIES, G., 553
DAVIES, P. C. W., 549
DAVY, H., 161, 480, 485, 548, 565; dynamism, 53, 54; electrochemical theory, 55; heat, 26, 28, 29, 31–33, 114, 118, 120, 305, 306, 307, 308–9, 313, 330–32, 334, 475; Herapath, 115–19, 122, 126, 158; retirement from Presidency of Royal Society, 119–21, 126
DAVY, J., 120, 121
DAY, A. L., 523
DE BROGLIE, L., 299, 626, 700
DEBYE, P., 91, 411, 523
degeneration, 548, 562
DE GROOT, S. R., vii, 256
DE HAAS-LORENTZ, G. L., 672
DE LA BECHE, H., 580
DE LA PROVOSTAYE, F. H., 313, 480
DE LA RIVE, A., 27, 163, 318
DE LA ROCHE, F., 163, 482
DELEVSKY, J., 637
DELISLE, J. N., 49
DELLINGSHAUSEN, N., 595
DELSAULX, J., 666
DELUC, J. A., 21, 26, 48, 112, 117, 171, 477
DEMILT, C., 51
DEMOCRITUS, 91, 131, 698
DE MORGAN, A., 550
DENBIGH, K. G., 639
density, 73, 75, 78
DESAINS, P., 313, 480, 501, 512, 516
DESCARTES, R., 3, 13, 21, 38, 41, 110, 113, 127, 135, 207, 388, 396, 545
DESORMES, C. B., 27, 118, 149, 152

Book 1, pp. 1–300; Book 2, pp. 301–770.

DESPRETZ, C., 324, 482, 516
determinism, 40, 62, 96, 241, 545, 583–84, 591, 630, 634, 648–49, 654
DEVIENNE, F. M., 533
DEWAR, J., 216, 735
diathermancy, 487–88, 508–9, 527
DICKINS, B. G., 503, 504
DIETERICI, C. H., 735
diffusion, 69, 72, 73, 78, 84–85, 110, 127, 153, 166–67, 177, 192–93, 203, 464, 592, 615, 643, 678–80; thermal, 85, 180, 423, 450, 466
DIJKSTERHUIS, E. J., 16, 550
dimensional analysis, 221, 430, 498, 499
dimensionality (of set of points), 378–80, 383–84
DINGLE, H., 159
DIRAC, P. A. M., 299
disorder, molecular (see also randomness), 82–84, 95, 100, 584, 616, 621, 634, 640, 643–44
dispersion of sound, 86–87, 208
dissipation of energy, 80, 236, 551, 561–62, 602–3, 629–30, 643
DIXON, E. T., 68
DONKIN, W. F., 587
DONNINI, P., 735
DOOTSON, F. W., 85, 466
DORLING, J., 198
DORN, E., 78
DORSMAN, 534
DORTOUS DE MAIRAN, J. J., 551–52, 563
DRAKE, S., 68
DRAPER, J. W., 307, 324, 480–81, 482, 514, 516
DREHER, E., 103
DRUMMOND, J., 661
DRYDEN, H. L., 532
DU BOIS, H., 415
DU BOIS-REYMOND,•E., 55, 58, 582, 584
DUCARLA-BONIFAS, M., 473
DÜHRING, E. K., 735
DUFOUR, L., 227
DUHAMEL, J.-M.-C., 484
DUHEM, P., 26, 33, 61, 68, 94, 245, 280, 419, 638
DULONG, P. L., 313, 515–16; specific heats, 25, 27, 153, 478, 506; speed of sound, 123
Dulong–Petit cooling law, 25, 84, 478–81, 489, 501–2, 508–10, 515–17, 520–21
DUMAS, J.-B.-A., 174, 197
DUNOYER, L., 22, 533
DUPRÉ, A., 76
DUPRÉ, P., 89
Dutch science, 88, 94, 254, 419, 709–10
DUTTA, M., 581
dynamism, 40, 45, 53, 74, 92, 96, 204, 277, 297, 629, 633
DYNKIN, E. B., 685

EARNSHAW, S., 322
earth, action on moon, 109, 110; age, 40, 552, 557–58, 561, 565; internal temperature and cooling, 126, 471, 551–65; rotation, 631
EBERT, H., 735
ECKERLEIN, P. A., 503, 504
economy of thought, 61–62, 288, 293
EDDINGTON, A. S., 244, 639
EDDY, H. T., 244, 518, 735–36
EDLER, J., 520
EDWARDS, L. P., 36, 38
EDWARDS, [H.] MILNE, 659
effusion, 127, 129, 225
EGGARTER, T. P., 627
EHRENBERG, C. G., 89, 138
EHRENFEST, P., 79, 238, 242, 280, 281, 306, 364–65, 369, 372–77, 384, 415, 596, 605, 627, 639, 640, 641, 650
EHRENFEST, T., 79, 238, 242, 280, 281, 306, 364–65, 369, 372–77, 384, 596, 605, 627, 639, 640, 650
EICHHORN, W., 503, 504
EINSTEIN, A., 44–45; Boltzmann, 673, 674; Brownian movement, 91, 101, 298, 649, 656–57, 672–84, 686, 695–98, 700; Mach, 275, 276, 294, 296, 700; quantum theory, 3, 91, 317, 542; relativity theory, 244, 274, 299, 544; statistical thermodynamics, 100, 372
EISENMAN, H. J., 533
EISENSCHITZ, R., 418
elasticity, 13, 15, 63–65, 110–11, 168, 243, 245, 355, 545
electrical discharges, 88

Book 1, pp. 1–300; Book 2, pp. 301–770.

electrical forces between atoms, 47–48
electricity, contact, 76, 293; motion in wires, 682–83
electromagnetic waves, 32, 45, 275, 306, 644–48
electromagnetism, 41, 44, 51, 243–44, 277, 544, 644
electron, 90, 199, 697
elements, 249
ELIADE, M., 637
ELIOT, G., 667
ELKANA, Y., 66, 164, 248
ELLIS, C. H., 125
ELLIS, R. L., 584, 596
ELSTER, J., 294
EMMETT, J. B., 34
empiricism, 53
empiriocriticism, 55, 57, 61–63, 280, 297–98
energetics, 55, 61–62, 97, 245, 280, 287, 297, 544, 642, 656, 694, 698
Energy Conservation Law, 9, 35, 42, 44, 51, 54, 58, 61, 66, 110–11, 129, 164, 234, 285, 325–28, 546, 569
ENFIELD, W., 566
ensemble, 79, 242, 368–69, 419–20, 611
ENSKOG, D., 465–68; Hilbert, 457, 460; thermal diffusion, 450; transport theory, 85, 180, 193, 237, 414, 425, 432, 445, 450, 456–59, 462, 498, 499, 502, 506, 525, 530
entropy, 80, 92, 95, 100, 238–40, 287, 368, 576–79, 582, 592–93, 604–5, 641, 679; disorder, 83–84, 236, 241, 291, 351, 593, 607–12; radiation, 540–41
EPICURUS, 62, 91
EPSTEIN, P. S., 49, 614
equation of state, 172, 181, 400, 461; effect of intermolecular forces, 264–66, 270, 398–99, 416–18; hard spheres, 408–15; van der Waals, 266, 410
equilibrium, thermal, 80, 82, 111, 113, 139, 147, 236, 316, 445, 554, 606–7
equipartition theorem, 79, 86, 101, 158, 356–62, 366, 618–19, 633, 656, 681–83, 689–90, 696; internal motions of molecules, 175, 194–95, 345; mixture of gases, 71, 113, 139, 146–47, 174, 196, 337, 344; radiation, 305–6, 324–25

Ergoden, 241–42, 364, 367–70
ergodic hypothesis, 79–80, 97, 241–42, 335, 338, 363–85, 611
ERICSSON, J., 509, 512, 516
ERIKSSON, G., 65
errors, law of, 72, 170, 184–85, 587
ERXLEBEN, J. C. P., 476
ether, 3, 11, 20, 32, 43, 44, 95, 120, 149, 171, 205, 277, 303–5, 337, 355–56, 361–62, 472, 475, 481, 547, 617–19, 675–76
ethylene ["olefiant gas"], 439, 480
ETTINGSHAUSEN, A. R. VON, 277, 282, 284
EUCKEN, A., 502–3, 506
EULER, L., 39, 391, 473, 547; $F = ma$, 14; gas theory, 19, 150, 171; sound, 150, 403
Eulerian approach to history of science, 8
evaporation, 112, 173, 660–61
EVERITT, C. W. F., vii, 145, 156, 183, 185, 188, 210, 229, 350, 522, 532, 613
evolution, 36, 40, 55, 58, 61, 63, 548, 562, 604
EXLEY, T., 322
EXNER, F., 78
EXNER, R., 668, 685
expansion, free, of gases, 31, 34–35, 164, 166
expansion, thermal, 23–24
explanation in science, 4

FARADAY, M., 134, 145, 313; Brownian movement, 659; dynamism, 53–54, 277, 397; electromagnetism, 43–45; heat, 325, 327
FARAGO, P., 685
FARBER, E., 483
FARMER, J. B., 662
FAYE, H., 512
FECHNER, G., 55, 202, 278, 283, 291
FEDDERSEN, W., 226–27
FEDEROV, E., 103
FENYES, I., 626
FERGUSON, A., 229
FERGUSON, E. S., 512
FERREL, W., 88, 520–21

Book 1, pp. 1–300; Book 2, pp. 301–770.

INDEX xxiii

FESTING, E. R., 512, 533
FEUER, L., 247, 276
FICHTE, I. H. VON, 278, 284, 397
FICHTE, J. G., 52
FICK, A., 209, 582
field theory, 43, 45
FIERZ, M., 421
FINKENER, R. H., 736
FINN, B. S., 155, 312
FISCHER, E. S., 311
FISCHER, J. C., 313–14
FISHER, M. E., 273
FISHER, S. J., 22
FITZGERALD, G. F., 198, 217, 227, 228, 361, 619, 736
FIZEAU, A.-H.-L., 512
FLAMM, W., vii, 248
FLECK, G. M., 102
FLEMING, D., 482
FLEMING, R. S., 397
FLOURENS, P., 662
fluctuations, 62–63, 83, 96, 174, 240, 666, 674, 687, 689, 694; fluctuation-dissipation theorem, 423
FÖPPL, A., 685
FOGLE, B., 456
FONG, P., 581
FORBES, J. D., 146, 305, 313, 318, 329, 330, 333, 482, 584, 587
forces (*see also* atom), 39–40, 44–45, 277, 387
FORMAN, P., 65, 654
FOSTER, G. C., 736
FOUILLEE, A., 68
FOURCROY, A. F. DE, 29
FOURIER, J. B. J., 113, 378; heat conduction theory, 31, 42, 97, 308, 315–16, 471, 477, 524, 555–58, 560, 564, 584, 588; nature of heat, 312, 320, 331, 481
Fourier components of radiation, 646–48
FOWLE, F. E., 512
FOWLER, R. H., 608, 615
FOX, R., 26, 27, 32, 33, 49, 149, 312, 333, 397, 482
FRANCK, E. U., 507
FRANK, P., 66, 293–94, 296
FRANKFURT, U. I., 164
FRANKLAND, E., 55

FRANKLIN, B., 473
FRANKLIN, W. S., 562, 639
FRANZ, R., 487, 511
FRÉCHET, M. R., 382
FREDHOLM, I., 447, 467
FRENCH, A. P., 89
French science, 84, 89, 320, 321, 709–10
FRESNEL, A., 32, 42–43, 213, 305, 310, 323
FRIEDMAN, F. L., 22
FRISCH, H. L., 416
FROWEIN, P. C. F., 736
FÜRTH, R., 679, 692–93
FULLMER, J. Z., 120, 121
FULTON, J. F., 16
funiculus, 11

GADOLIN, J., 26
GALILEI, G., 9, 16, 38, 39, 41, 90, 379
GALITZIN, B. B., 737
GAL-OR, B., 549
GALTON, F., 56, 170, 229
GANOT, A., 332
GARBER, E. W., vii, 176, 182, 183, 188, 229, 498, 499, 532, 596
GARDNER, M., 549
GARNETT, W., 185, 201, 215, 226
gases (*see also* pressure and other properties; plasma), 15; mixtures, 45–46, 84–85; solidification, 128; surface interactions, 84, 85, 217–25, 222–24, 340, 485–86, 528; word, 183
gases, dense, 460–62, 530
gases, rarefied, 84, 85, 88, 216–21, 227; heat conduction, 501, 526–28; sound propagation, 11
GAUDIN, M. A., 322
GAUSS, C. F., 289
GAY-LUSSAC, J. L., 23, 25, 27, 42, 47, 50, 113, 149, 152, 197, 311, 314; free expansion of gases, 31, 34; gas law, 23, 30, 77, 153, 166, 170, 173, 234, 576; law of combining volumes, 46, 50, 153
GEITEL, H., 294
geology, 40; uniformitarian, 558
geophysics, 463–65, 545, 551–62
GEORGESCU-ROEGEN, N., 582

Book 1, pp. 1–300; Book 2, pp. 301–770.

GERHARDT, C. F., 174
GERLAND, E., 17
German/Austrian science, 84, 88, 89, 565, 708–10
GERSON, E. F., 614
GIBBS, J. W., 5, 88, 100, 170, 596–87, 639, 737; atomism, 94, 98, 279; ensembles and statistical mechanics, 99, 241–42, 369, 419–20, 608, 611, 651, 673; entropy increase, 280, 604, 633, 634, 650; thermodynamics, 279
Gibbs paradox, 592, 604
GILBERT, D., 107, 109, 115–17, 119, 121
GILLISPIE, C. C., 8, 159, 183, 342, 565, 587, 594
GIRDLESTONE, A. G., 200
GLASS, B., 672
GLEICHEN, F. W. VON, 661
GMELIN, L., 322
God, 545–47, 583
GOETHE, J. W. VON, 397
GOLD, T., 549
GOLDBERG, S., 653
GOODFIELD, J., 563
GOODMAN, D. C., 662
GOODWIN, G., 125
GOPAL, E. S. R., 263
GORDON, L., 568
GORE, G., 737
GORMAN, M., 533
GORNSHTEIN, T. N., 333
GOSIEWSKI, W., 93, 155, 207, 737
GOUGH, J. B., 15
GOUY, L., 100, 616, 667, 669–70, 696
GOVI, G., 737
GOWER, B., 65
GRAD, H., 449, 522
GRAETZ, L., 503, 504, 517, 520, 523, 524
GRAHAM, T., 70, 89, 127, 129, 144, 147, 153, 167, 191, 193, 200, 203, 225, 324, 436, 438–40
GRANT, R. E., 662
GRANT, S., 661
GRASSMANN, H. G., 50
GRATTAN-GUINNESS, I., 312, 564
gravity, 387, 547; effect of temperature on, 110; effect on temperature of air, 70, 71, 74, 83, 123, 154, 195–96, 349–50, 588–89, 602; kinetic theory, 21,
22, 48, 135, 336, 388, 396
GRAY, P. L., 523
GREEN, G., 43, 684
GREENAWAY, F., 50
GREENE, J. C., 654
GREENOUGH, G., 558–59
GREGORY, G., 34
GREGORY, H., 504, 534
GREGORY, J. W., 566
GREGORY, O. G., 123
GREISZ, C. G., 332
GRIFFITH, J. W., 663
GROENEVELD, J., 416
GROSS, G., 737
GROVE, W. R., 320, 323, 325, 326, 485, 520
GRÜNBAUM, A., 549
GRUND, F. J., 320
GRUNWALD, M., 248
GUARESCHI, I., 666
GUERICKE, O. VON, 10, 16, 216, 389
GUERLAC, H., 22, 67
GUGGENHEIM, E. A., 273
GUILLEMIN, A., 333
GURNEY, G., 122
GUTH, E., 449
GUTHRIE, F., 196, 349–50, 737
GUYTON DE MORVEAU, L. B., 29

H-theorem, 80, 82, 95, 100, 187, 238–40, 297, 351–52, 443–47, 584, 600–1, 612–13, 618, 620–24, 649–50; H replaces E, 619, 626; quantum, 636; reversibility objection, 618, 620–24, 643
HAAR, D. TER, vii, 159, 549, 614, 640
HAAS, A. E., 164
HABER, F. C., vii, 49, 563, 565
HADFIELD, E., 125
HAHN, R., viii, 584, 594
HALDANE, J. S., 139, 143, 144
HALL, A. R., 36, 49, 550
HALL, M. B. (*see also* BOAS, M.), 16, 18, 550
HAMILTON, W., 189
HAMILTON, W. R. (*see also* Lagrange-Hamilton), 277
HAMTIL, C. N., 581
HANCOCK, H., 442

Book 1, pp. 1–300; Book 2, pp. 301–770.

HANKINS, T. L., 18
HANLEY, H. J. M., 506
HANSEMANN, G., 738
HAPPEL, H., 415
HARCOURT, W. V., 33
HARDY, W. B., 396, 397, 418
HARE, R., 34, 114, 322
HARTLEY, H., vii, 144
HARTNETT, J. P., 533
HASSE, H. R., 432
HAURWITZ, B., 456
HAWKINS, D., 18
HAWKINS, T., 638
HAWTHORNE, R. M., 89
HAYCRAFT, W. T., 153
HAYEK, F. A., 67
heat (*see also* radiation; convection), mechanical equivalent, 34, 148, 163, 305, 307, 325; nature of, 8–9, 27–32, 35, 42, 61, 170–71, 204–5, 285, 303–34, 585; repulsive force, 317, 319–20, 391; wave theory, 32, 146, 171, 303–34; weight, 29, 34, 312
heat conduction, 31, 40, 42, 315–16, 481, 517, 524, 555–57; in gases, 29, 73, 78, 84, 128, 181, 192, 436, 453, 473–76, 481, 483–508, 524–31
heat death, 240, 286, 551–52, 555, 578, 582, 607, 612, 630–32, 635
heat transfer (different modes), 469–76, 481, 483–88, 515–16, 519, 524–39, 569
HEATHCOTE, N. H. DE V., 33
HEAVISIDE, O., 45
HECKE, E., 449
HEILBRON, J., 144
HEIMANN, P. M., 362
HEINE, 628
HEINE, H., 511
HEINTZ, W., 34
HEISENBERG, W., 91, 294, 299
helium, heat conduction, 503; specific heats ratio, 79, 354; superfluidity, 4, 532
HELL, B., 164
HELLINGER, E., 449
HELM, G., 61, 68, 96, 245, 280, 638
HELMHOLTZ, H. VON, 55, 243, 277, 533, 598, 626, 652, 738; energy conservation, 35, 325, 556; kinetic theory, 202; monocyclic systems, 233, 367–68, 593, 617, 628; slip, 218, 230; thermodynamics, 279, 633, 670; vortex theory, 92, 206; Waterston and Joule, 141, 209; wave theory of heat, 326
HEMMER, P. C., 418
HENFREY, A., 663
HENLEY, E. M., 549
HENRI, V., 696–97, 698
HENRY, J., 130, 320, 324, 330
HENRY, W., 33, 34, 312, 477
HENRY, W. C., 47, 120
HENSEL, F., 507
HERACLITUS, 28
HERAPATH, J., 6, 21, 48, 74, 107–9, 113, 121–25, 195, 391, 588, 738–39; absolute zero, 118, 128; Avogadro's hypothesis, 48; Boyle's law deviations, 127–28, 264, 401; diffusion, 127; heat transfer, 128, 489, 498; kinetic theory, 69–70, 86, 110–13, 154, 422, 585, 707–8; Laplace, 30, 118; Railway Magazine, 122–25; reception of theory, 24, 35, 54, 115–18, 129–32, 161, 203, 206; speed of molecule, 123; speed of sound, 123, 152, 155, 203, 401; temperature scale, 25, 111–12
HERAPATH, W., 107, 132
HERCUS, E. O., 503, 504, 581
HERING, E., 291
HERIVEL, J., 15, 19, 312, 321
HERMANN, A., 34, 65, 102
HERMANN, J., 19, 22, 492
HERO, 288
HERSCHEL, A. S., 739
HERSCHEL, J., 184–85, 201, 263, 342, 394, 482, 488, 587, 597, 613
HERSCHEL, W., radiant heat, 305, 308, 475
HERTZ, H., 66, 244, 739
HERTZ, P., 372
HERZFELD, K. F., 90, 247, 614
HESKETH, R., 363, 626
HESSE, M. B., 66, 188
HICKOK, L. P., 332
HICKS, W. M., 739

Book 1, pp. 1–300; Book 2, pp. 301–770.

HIEBERT, E. N., vii, 66, 68, 281, 282, 284, 296, 548, 549, 639, 643, 651–52, 653
HIGGINS, B., 28, 33
HIGGINS, L. D., 442
HILBERT, D., 447–48, 457, 458, 460
HILTS, V. L., 176
HIRN, G.-A., 94, 188, 325, 577, 739–40
HIRSCHFELDER, J. O., 506
HIRTH, G., 582
historiography and historical method, 5, 8, 19, 36, 91, 102, 544, 628, 657; quantitative/bibliographical, 705–11
HO, C. Y., 507
HOBBES, T., 17
HODGES, N. D. C., 740
HÖPPLER, F., 424
HÖNIGSWALD, R., 247
HOGG, J., 332
HOLLAND, H., 50
HOLLAND, P. W., 34
HOLMES, A., 566
HOLT, N. R., 102
HOLTON, G., vii, 685
HOLTZMANN, K., 325
HOOKE, R., 10, 12, 19, 595
HOORWEG, J. L., 740
HOOVER, W. G., 271, 416
HOOYKAAS, R., 155
HOPKINS, W., 559–62, 572
HOPLEY, I. B., 188
HOPPE, B., 65
HOPPE, E., 581
HOPPE, R., 494, 741
HORNE, M., 685
HORNE, R. A., 15
horror vacui, 9–11
HOULLEVIGUE, L. A. C., 741
HOWARD, W. H., 741
HOWARTH, H. E., 18
HOYLE, F., 67
HUDSON, H., 333
HULL, G. F., 211
HUMBOLDT, A. VON, 50, 661
HUME, D., 299
HUNT, R., 318–19, 322
HUNTER, J., 554
HUTTON, J., 553–54, 558, 563
HUXLEY, T. H., 58–59, 67, 582, 663

HUYGENS, C., 13, 17, 41, 90
HWANG, S.-T., 549
hydrodynamics, 39, 81, 86–87, 92, 227, 424, 432–33, 454, 675–77
hydrogen, 71, 73, 77, 78, 161, 174, 399, 480, 483, 485–87; heat conduction, 494, 497, 500, 503
hypothetico-deductive method, 194–95, 387, 389–90

idealism, 55–57, 66
IHDE, A. J., 50, 51
IKENBERRY, E., 499
immortality, 132
INCE, S., 22, 425, 685
indeterminism, 95, 647–48 (*see also* randomness)
INFELD, L., 44–45
integral equations, 444, 447–48, 456–59, 467
INTEMANN, H., 506
ionized gases, *see* plasma
irreversibility, 40, 42, 80, 82–84, 95, 100, 128, 182, 236, 239–40, 291, 352, 543–654, 694, 707
IRVINE, W., 26
ISENKRAHE, C., 22
"Ising model", 18
IVORY, J., 597

JACOBI, C. G. J., 406
JÄGER, G., 248, 414, 416, 425, 431, 460–62, 525, 741–42
JAKI, S. L., 637
JAKOB, M., 532, 534
JAMES, C. G. F., 432
JAMES, W. S., 18, 26
JAMIN, J., 332, 512
JAMMER, M., 396, 522, 653–54
JANOSSY, L., 549
JEANS, J. H., 158, 256, 280, 321, 365, 418, 427, 442, 533, 640, 651, 688
JEFFREYS, J., 273
JELLETT, J. H., 50
JEVONS, W. S., 665, 687
JOANNIS, 333
JOCHMANN, E. [C. G. G.], 188, 492, 494, 742

Book 1, pp. 1–300; Book 2, pp. 301–770.

JOHN, V., 595
JOHNSTON, J., 322, 324
JONSON, B., 423
JOULE, J. P., 6, 127, 131, 170, 333, 742; heat and energy conservation, 35, 44, 325, 327, 329, 330–32, 569; Herapath, 70, 118, 123, 130, 161; kinetic theory, 160–65, 167, 330, 585, 708; mechanical equivalent of heat, 148, 149, 307, 570; speed of molecule, 69, 71, 123, 131, 141
Joule-Thomson experiment, 35, 140, 164, 265, 403, 417, 441, 501, 578
journals, scientific, 709–21
Jupiter, 552

KAC, M., 383, 418, 614
KAGAL'NIKOVA, I. I., 22
KAISER, W., 102
KAMERLINGH ONNES, H., 256, 534, 752
KANE, R., 323
KANGRO, H., vii, 513, 523, 532, 640, 652
KANNULUIK, W. G., 503, 504, 506
KANT, I., 31, 53, 277, 547
KAPLAN, L. D., 511
KARGON, R. H., 15, 35, 323
Karlsruhe Congress, 51, 93, 175
KÁRMÁN, T. VON, 449
KÄRRE, K., 468
KAUFMAN, W. A., 637
KEESOM, W. H., 270, 416
KEKULE, A., 55
KELHAM, B. B., 33
KELLAND, P., 34, 140, 319, 322–23, 331, 482, 560
KELVIN, LORD (William Thomson) (*see also* Joule-Thomson experiment), 7, 43, 224, 225, 277, 523, 544, 559, 565, 585, 589, 596, 742–43; atoms, 56, 68, 76–77, 142, 201; Boscovich theory, 392–93; cooling of earth, 560–61; dissipation of energy, 80, 286, 561–62, 569, 572, 577, 629, 672; equipartition, 79, 359–60, 371, 372, 544, 618; evolution, 49, 561, 604; field theory, 55; free expansion experiment, 35, 140, 164; gas-surface interactions, 219–20, 226; heat theories, 306, 330–34; Herapath, 130; kinetic theory, 206, 707–8; kineticism, 44, 50; mechanical models, 580; motion of electricity, 682–83; radiation pressure, 518; reversibility paradox, 83, 239, 352, 602–4; specific heats, 152; sun's heat, 140; thermodynamics, 35, 279, 307, 568–73, 580–81; vortex atom, 92, 161, 206–7, 277, 396
KEMP, G., 12
KENNARD, E. H., 228, 431, 533
KENNELLY, A. E., 532
KEPLER, J., 14, 39, 41
KERKER, M., 579
KESTIN, J., 506
KEYNES, J. M., 651
KHINCHIN, A. I., 615
KILPATRICK, J. E., 416
kineticism or kinetic world-view, 40, 44, 48, 50, 54, 92, 306
KIRCHHOFF, G. R., 99–100, 442, 625, 639, 643, 676, 743; theory of sound, 87, 88
KIRKWOOD, J. G., 415
KIRWAN, R., 553, 563
KLAUS, A., 7, 159
KLEIN, M. J., viii, 103, 248, 269, 281, 306, 371–72, 377, 397, 421, 523, 549, 582, 597, 613, 614, 626, 640, 652, 653, 684
KNIGHT, D. M., 50, 65, 93, 120, 198, 281
KNOTT, C. G., 597, 602
KNUDSEN, M., 230, 455, 528–29
KNUDSEN, O., 132
KOCH, 503
KOCH, R., 55
KÖHLER, W., 549, 640
KOENIG, F. O., 579, 580
KOHLER, M., 534
KOHNSTAMM, P. A., 414, 415
KOLÁČEK, F., 687
KOLBE, H., 55
KOOL, C. J., 743
KORTEWEG, D. J., 413, 414, 416, 743
KOYRÉ, A., 396, 550
KRAKOWSKI, L., 711
KRAMERS, H. A., 411
KRAUSSOLD, H., 532
KRÖNIG, A. K., 35, 71, 131, 141, 142, 154, 157, 165–67, 171–72, 197, 202, 209, 283, 340, 585, 708, 744

Book 1, pp. 1–300; Book 2, pp. 301–770.

KRONSTADT, B., 89, 129
krypton, 354
KUBRUN, D., 500
KUHN, T. S., vii, 65, 66, 155, 164, 294, 312, 389; Carnot's theory, 34; disorder, 627; energy conservation, 325, 328, 333, 579; paradigms, 12, 526; scientific revolutions, 36
KUNDT, A. [A. E. E.], 744; rarefied gases, 217–18, 501, 517, 528; specific heats of mercury, 79, 354
KURLBAUM, F., 523
KUTTA, W., 504
KUZNETSOV, B. G., 164
KUZNETSOV, P. G., 582

LAAR, J. J. VAN, 256, 414, 416, 744
LABY, T. H., 503, 504
LADENBURG, R., 511
LAGRANGE, J. L., 150–51, 547, 554
Lagrange–Hamilton dynamics, 99, 633, 674
Lagrangian approach to history of science, 8
LAMB, H., 87, 225
LAMBERT, J. H., 26, 473
LAMÉ, G., 312, 318, 321
LAMPA, A., 295, 296
LANCHESTER, 523
LANDAU, L., 639
LANDOLT, H., 34
LANDSBERG, H., 511
LANDSBERG, P. T., 549
LANE, J. H., 512
LANG, V. VON, 498, 744
LANGEVIN, P., 450, 696
LANGLEY, S. P., 22, 32, 88, 307, 313, 477, 510, 512, 526
LANGMUIR, I., 482, 530, 532
LAPLACE, P. S. DE, 26, 34, 118, 183, 547, 554; Cagniard de la Tour experiment, 394–95; caloric theory, 30–31, 118, 207, 305, 312, 320; capillarity, 77, 264; determinism, 62, 545, 584, 594; force law, 265, 394–95; light, 42–43; mechanistic physics, 49, 53, 310; speed of sound, 31, 85–87, 151–54, 308, 395
LA PROVOSTAYE, F. H. DE, 313, 480

LARDNER, D., 318
LARMOR, J., 95, 214, 225, 455, 617, 622, 744
LASAREFF, P., 533
LASSWITZ, K., 15
latent heat, 28, 33, 75, 112, 142, 153, 161, 174, 256–57, 291, 322–23
LAURENT, A., 174
LAVOISIER, A. L., 29, 39, 305, 312
LEAR, J., 456
LEVEDEV, P. N., 211
LEBESGUE, H., 80, 380, 631
LEBOWITZ, J. L., viii
LE CHATELIER, H., 523
LECHER, E., 511, 517
LECONTE, J., 203–4, 745
LEDIEU, A. C. H., 103, 745
LEENDERTZ, W., 256
LEES, C. H., 532
LEEUWENHOEK, A. VAN, 661
LEIBNIZ, G. W., 114, 396, 546, 551
LEICESTER, H. M., 21
LEIDENFROST, W., 531
LENARD, P., 181
LENIN, V. I., 275
LENNARD-JONES, J. E., 270, 278, 425
LENZ, W., 18
LENZEN, V. G., 549
LERAY, A. J., 745
LESAGE, G. L., 21, 48, 131, 171, 396, 492
LESLIE, J., 135, 305, 475–76, 480, 483, 485, 489, 534–39, 566
LEUCIPPUS, 698
level of complexity, 272–73
LEVENSMA, T. P., 256
LEVENSPUT, O., 549
LEVY, M., 103
LEWIS, R. M., 383
LIDDELL, H. G., 582
light, 28, 31–32, 41, 171; dispersion, 76, 86; electromagnetic theory, 55; particle theory, 305, 309, 317, 319, 542; pressure, 211, 215, 229, 518; refraction, 78, 146; speed, 13, 18; wave theory, 42–43, 201, 211, 213, 305, 309–10, 675
LILEY, P. E., viii, 507, 532
LILLEY, S., 33, 307
LINDBERGH, C. A., 66

Book 1, pp. 1–300; Book 2, pp. 301–770.

LINDSAY, R. B., 11, 158
LINDSAY, T. M., 15
LINUS, F., 11, 17, 606
Liouville's theorem, 634
LIPPICH, F. F., 745
LIPPMANN, E. O. VON, 188
LIPPMANN, G., 543, 694, 745
LIPSCHITZ, R., 406
liquids, ideal model, 271; mixtures, 111–12, 116; molecular motions, 660–61, 668–70, 675; specific heat, 112; statistical equilibrium, 624; surface tension, 75; transition to gas, 77, 112, 173–74, 256–57, 266, 392, 394; transport processes, 530–31
Lissajous figures, 365, 369–70, 373
literature, 52, 628
LITTRÉ, E., 60
LIVEING, G. D., 745
LOBACHEVSKI, N., 289
LODGE, O. [J.], 68, 356, 582, 745
LOEB, L. B., 158, 228, 431, 533
LOMONOSOV, M. V., 21, 595
LONDON, F., 418, 532
LORENTZ, H. A., 242, 450, 499, 613, 645, 677, 685, 697, 746; density and refractive index, 78; equation of state, 411–13; sound, 86, 90
LORENZ, L. V., 76, 78, 89, 517, 525
LOSCHMIDT, J., 6, 84, 708, 746; effect of gravity on temperature, 196, 605; reversibility paradox, 83, 239, 352, 588, 602, 605–6; size of molecules, 75–76, 201, 202, 231–32, 264
Loschmidt's number, 76, 77, 89, 197
LOT, F., 700
LOVEJOY, A. O., 52
LOVELL, D. J., 313
LOVITT, W. V., 449
LUBBOCK, J. W., 140, 145, 330
LUBECK, G., 747
LUCRETIUS, 8, 28, 91, 131, 583, 591
LUDWIG, C., 55
LÜROTH, J., 382
LUMMER, O., 99, 521, 523
LUNGO, C. DEL, 747
LUNN, A. C., 450
LUNN, F., 482, 488
LUNNON, R. G., 90
LYELL, C., 558–59

MACDONALD, D. K. C., 159
MACDONALD, I. G., 89
MACH, E., 26, 63, 66, 68, 246, 274, 419, 476, 638; atomism, 60, 91, 94, 245, 274–99, 673, 674, 699–700; Boltzmann, 287, 289–93, 295; empiriocriticism, 61, 280, 544, 548–49; energetics, 287, 290; entropy and time, 287, 289, 636; Fechner, 283–84, 291; kinetic theory, 203, 283, 285, 290; Mayer, 285; opinions on his views, 274–76, 298–99; Ostwald, 290, 298; Stallo, 280, 292, 293, 298
MACHE, H., 747
MACIVER, I., 125, 126
MACKAYE, J., 22
MADDEN, E. H., 68
MADDISON, R. E. W., 16
MAGIE, W. F., 18, 654
MAGNUS, G., 168, 475, 486–87, 509
MAGRINI, L., 203
MAIRAN, J. J. DORTOUS DE, 551–52, 563
MAJUMDAR, R., 415
MALTEZOS, C., 688
MANABE, S., 513
MANDELBAUM, M., 49, 67
MANN, W. B., 504
MANUEL, F. E., 49
MARALDI, 49
MARCET, F., 163
MARCO, F., 747
MARGENAU, H., 418
MARIÉ-DAVY, 332
"Mariotte's Law", 18, 127, 153, 166, 170, 173, 576
MARKOVIĆ, Z., 66
MARTIN, L. H., 503, 504
MARTINE, G., 26, 476
MARVIN, F. S., 67
MARX, C. M., 662
MASON, E. A., viii, 89, 129, 193, 506, 532, 542, 626
materialism, 8, 39, 44, 53, 55–61, 67, 92, 591, 602, 628–30; 698; reaction against, 55–57, 63, 93, 245
mathematics, 42, 335–36, 372, 378–80, 446, 695
MATHIAS, O. B., 579, 580, 626
MATSON, F. W., 67

Book 1, pp. 1–300; Book 2, pp. 301–770.

MATTEUCCI, C., 318
Matthew effect, 623
MAXWELL, J. C., 6, 7, 21, 22, 50, 51, 79, 88, 90, 98, 164, 205, 226, 277, 528, 559, 604, 642, 706, 708, 747–48; atomic magnitudes, 77, 197, 201, 207; Boltzmann, 355, 445–46, 610–11; Clausius, 131, 181, 190, 196, 431, 491–95; Crookes, 214–15, 217, 219, 221; diffusion, 85, 192–93; dog, 215; electromagnetic theory, 43, 45, 55, 78, 99, 214, 243, 244, 306, 518, 541, 644; equal-areas rule, 269; equipartition, 72–73, 79, 113, 194–96, 491; ergodic hypothesis, 79, 363–77; ether, 51; gravity, 48; effect of gravity on temperature, 123, 195–96, 349–50, 605; heat conduction, 84, 192, 195, 402, 476, 484, 489–501, 531; Herapath, 131–32, 190; mean free path, 73, 411, 425–26, 489–91, 495; Meyer, 439–40; poem on molecules, 593; radiant heat, 307, 330; radiometer, 85, 214–24, 518; Rankine, 580; reception of kinetic theory, 35, 48, 201–3; reversibility, 602; Reynolds, 216, 218–27; Second Law of Thermodynamics, 279, 650; specific heats, 148, 352, 354–56, 402; speed of sound, 86; statistical method, 3, 69, 170–71, 183–87, 344, 587–93, 620; thermal transpiration, 218–19, 222–25; William Thomson, 224–25; transport theory, 81, 180, 191–92, 214, 237, 402, 423, 432–34, 445, 446, 449–51, 467; velocity distribution, 72, 185–88, 196, 233–35, 342–43, 345–47, 444–45, 587–89; viscosity, 11, 73–74, 189–91, 201, 264, 402, 424–25, 434–42, 494, 499, 688; vortex atom, 92, 206, 277; van der Waals, 251, 268–69, 417; Waterston, 141, 142, 155
Maxwell–Boltzmann distribution, 182, 196, 234, 240, 241, 348–49, 352, 416, 449, 619, 624, 684, 696
Maxwell demon, 65, 82, 170, 279, 297, 589–90, 603, 631–32, 650, 671
Maxwellian molecules, 74, 84, 85, 191, 237, 416–17, 428, 440, 449, 453, 456, 496, 504, 669
Maxwell number, 500
MAYER, J. E., 614
MAYER, J. R., 34, 35, 54, 61, 114, 164, 285, 320, 325, 327–28, 331–32, 626
MCADAMS, W. H., 532
MCCARTY, R. D., 506
MCCORMMACH, R., 388, 548, 565, 653
MCKIE, D., 33
MCLAUGHLIN, E., 534
MEADOWS, J. J., 511
mean free path, 70, 72–73, 75, 76, 78, 84–85, 88, 138, 177–81, 202, 216, 221, 264, 408–10, 414, 426–30, 438, 449, 461, 493–94, 498, 500, 525–27, 530, 585–86, 624, 688–90
measure (of set of points), 379–80, 383–84, 630–31
mechanical philosophers, 3, 9, 15, 54, 388
mechanics, 58, 387; celestial, 107, 545–47; Second Law of Thermodynamics, 232–33, 279, 291, 367–68, 593, 617–18, 632
mechanism, 40, 48, 51, 55–56, 94, 96, 115, 239, 245, 278–79, 290–91, 543, 628, 630, 631–33
MEES, R. A., 748
MEHRA, J., 460, 549, 684
MEITNER, L., 654
MELDRUM, A. N., 198
MELLONI, M., 146, 171, 305, 308–9, 313, 318, 320, 324, 325, 327, 330, 333, 487, 489
melting, 112, 271–72
MENDEL, G., 672
MENDELEEV, D. I., 93
MENDENHALL, C. E., 521, 522
MENDOZA, E., viii, 108, 113, 132, 164, 312, 314, 568, 595
MENSBRUGGHE, G. VAN DER, 688
mercury, 9–10, 79, 354; heat conduction, 503; specific heat, 502, 506
MEREDITH, F. M., 126
MERTON, R., 331, 627
MERZ, J. T., 7, 34–35, 51, 159, 177, 583, 595
meteorology, *see* atmosphere
MEWES, R., 333
MEYER, F. I. F., 662
MEYER, L., 175

Book 1, pp. 1–300; Book 2, pp. 301–770.

MEYER, O. E., 7, 78, 84, 202, 245, 403, 502, 506, 615, 708, 749; diffusion, 84–85, 193; Maxwell, 436, 438; viscosity, 191, 202, 403, 428, 435–41, 506
MEYER, S., 294–95
MEYERSON, E., 595, 637
MICHELS, A., viii, 256, 532
MICHELSON, A. A., 88, 749
MICHELSON, V. A., 749
Michelson–Morley experiment, 544
microstates, 241, 607–10
MIDDLETON, W. E. K., 16, 22, 26, 473, 476
MILL, J. S., 60
MILLER, D. G., 102
MILLER-HAUENFELS, A. R. V., 333
MILLIKAN, R. A., 103, 582
MILLS, E. J., 264
MILNE, E. A., 582
MILNE–EDWARDS, H., 659
MILVERTON, S. W., 504
MINTER, C. C., 534
MITCHELL, A. C., 476, 532
MITCHELL, J. M., JR., 511
MÖLLER, F., 511
MOHLER, N. M., 550
MOHR, [C.] F., 320, 324, 325–26, 499, 595
molecular disorder or chaos, 82, 620–25, 643, 648–51
molecules (*see also* atom), compression, 414; early use of word, 657; number of atoms, 46–47, 73, 139, 147, 167, 174; size, 75–78, 89, 138, 142, 202, 423, 696; speed, 69, 71, 123, 138, 150, 174, 177, 668
MOLL, G., 123
momentum, 14, 34, 41, 111, 166
MOMIGLIANO, A. D., 637
monocyclic systems, 233, 367–68
MONROE, E., 415
MONTROLL, E., vii
moon, 109, 110, 464, 552, 559
MOORE, E. H., 382
MORGAN, C. L., 68
MORRIS, R. J., JR., 32, 333
MORRISON, P., 281, 549
MORVEAU, L. B. GUYTON DE, 29
MOSER, L., 324

MOSSOTTI, O. F., 34, 78, 138, 144, 281, 319, 322
MOTTE, A., 396
MOTT-SMITH, M., 579
MOUTON, 695
MÜLLER, E., 504
MÜLLER, J., 323, 324, 332
MÜLLER, J. H. J., 750
MUIR, W., 582, 749
MULDER, E., 750
MUNCKE, G. W., 125, 324, 482, 662
MURNAGHAN, F. D., 532
MURPHY, J. J., 614, 750
MURRAY, J., 312, 477, 554–55
MURRAY, J. A. H., 67
MURRAY, R. L., 18
music and mathematics, 446

NABL, J., 247, 614, 627, 649
NÄGELI, K., 667–68, 672, 689–90, 750
NAGAOKA, H., 91
NAGEL, E., 608
NAIRN, J. H., 416
NARR, F., 499–500
NASH, L. K., 33, 50, 51
NATANSON, L., 188, 708, 750–51
Naturphilosophie, 31, 51–54, 62, 115, 277, 319
NAUMANN, A. [N. F.], 202, 752
NAVIER, C. L. M. H., 199, 424, 684
Navier–Stokes equations, 81, 432–33, 671, 684
NEEDHAM, J., 657, 659, 661
neon, heat conduction, 503
neo-romanticism, 55, 59, 62
NERNST, W., 91, 698
NEUMANN, C., 226
NEUMANN, F., 114, 565
NEVILLE, R. G., 16, 18
NEWCOMB, S., 203, 333, 565
NEWCOMEN, T., 256
NEWTON, I., 3, 18, 38, 90, 97, 99, 107, 126, 288, 315, 396, 480; cooling law, 470–71, 478, 508–10, 525, 535, 556, 564–65; forces between atoms, 14, 18, 24, 111, 207, 386–88; gas pressure, 13–14, 20, 30, 35, 39, 93, 138–39, 153, 189, 207, 389–91; hard

Book 1, pp. 1–300; Book 2, pp. 301–770.

NEWTON, I. (Cont'd)
  atoms, 114, 391, 546; heat, 307, 309, 471–72; irreversibility, 135, 545–47, 550; sound, 149–50; viscosity, 423–24
  Newton's Laws of Motion, 4, 14, 39; reversibility of, 62, 95, 236, 556, 617–18
  Newtonian science/worldview, 35, 39–40, 51, 274, 553, 583–84, 655
NICHOL, J. P., 566
NICHOLS, E. F., 211
NICHOLS, R. C., 614, 752
NICOLSON, M., 550
NIETZSCHE, F., 628–30
NISBET, R. A., 49
nitrogen, 73, 163, 174, 500
nitrous oxide (N_2O), 78
NIVEN, W. D., 431
NOBILI, L., 308
NORTON, W. A., 93, 205, 332, 752
numeratom, 111
NYE, M. J., 103, 700, 701

OBENDORF, D. L., 533
OBERMAYER, A. VON, 84, 432, 441, 506
OERSTED, H. C., 31, 44, 54, 135, 144, 277, 323
ÖPIK, E. J., 637
OESPER, R. E., 324
O'HARA, J. G., 208
OLEARSKI, K., 752
olefiant gas, see ethylene
OLIVER, J., 685
OLMSTED, D., 34, 312, 333
OLSON, R. G., viii, 397, 475, 532, 538
ONNES, H. K., 256, 534, 752
ORNSTEIN, L. S., 372, 411, 685
Ornstein–Uhlenbeck process, 684
OSBORNE, M. F. M., 672
osmosis and osmotic pressure, 665, 675, 677–78, 695
OSTWALD, W., 68, 246, 295, 677, 752; atomism, 60, 93, 94, 101, 281, 298, 673, 674, 698–99; Boltzmann, 246, 290, 642; energetics, 61–62, 96–97, 245, 280, 640, 698
oxygen, 73, 163, 174

PACEY, A. J., 20, 129, 329
PAMBOUR, F. M. G. DE, 124
PANNEKOEK, A., 550, 654
PAPIN, D., 256
PARENT, A., 492
PARRY, W. E., 123
PARS, L. A., 580
PARTINGTON, J. R., 26, 50, 51, 89, 90, 114, 149, 165, 193, 194, 258, 263, 356, 432, 476, 506, 523, 583, 608, 615, 685
PASCAL, B., 9, 10, 16, 389
PASCHEN, F., 512, 521, 523
PASTEUR, L., 55
PAULI, E., 504
PAULI, W., 636
PAULY, P. J., 210
PAYEN, J., 27
Peano curve, 372, 380, 385
PEARSON, K., 68, 280
PECLET, E., 324, 482, 489
PECQUET, J., 13, 17
pedesis, 665
PEEL, R., 120
PEIERLS, R. E., 549
PEIRCE, C. S., 62, 549
PEKERIS, C. L., 455
PERIER, F., 10, 16
periodic table, 93
PERNTER, J., 511
PERRIN, J., 693–94; atomism, 693–99; Brownian movement, 91, 101–2, 298, 656, 666, 672, 684, 694–98
PERRY, J., 583
PERRY, W., 107
PERSON, C. C., 27, 322
PERSOZ, J., 324
PETERSON, H., 67
PETIT, A. L., 515: cooling law, see Dulong; specific heats, 25, 27, 153, 478, 506
PETRIE, W., 323
PETRUCCIOLI, S., 548
PETZOLDT, J., 292, 296
PEYRE, J.-M.-M., 31
PFAUNDLER, L., 562, 753
PFEFFER, R., 629
phase transitions, 77, 78, 110, 112, 128, 161, 173, 250, 256–63, 266–68, 271–73, 392–94, 413–14, 415

Book 1, pp. 1–300; Book 2, pp. 301–770.

philosophy of science, 91, 194, 272–73, 274–76, 286, 389
physiology, 55, 59
PICTET, M. A., 308, 309
PIDDUCK, F. B., 450
PIERSON, S., 314
PIETSCHMANN, H., 65
PIHL, M., 89
PINAUD, A., 320
PIOTROWSKI, G. VON, 229, 230
PIROGOV, N. N., 753
PLANCHEREL, M., 80, 242, 383–84
PLANCK, M., 164, 548, 632, 753; atomism and kinetic theory, 641–42, 652–53; Boltzmann, 235, 297, 372, 625, 640–48; Clausius, 176, 642; energetics, 640, 642–43; entropy, 641–43, 647; indeterminacy, 647–48; irreversibility, 640–41, 643–48; Kirchhoff's lectures, 643; Mach, 275, 295–96, 297, 299; quantum theory, 3, 5, 88, 99, 234, 306, 372, 469, 512, 519, 544, 610, 640, 648–49, 673, 697; radiation theory, 306, 469, 641, 644–48, 673; rule in history of science, 94, 640; statistical thermodynamics, 99–100, 608, 640; Waterston, 141, 145; Zermelo, 644, 653
PLANK, J., 506
plasma, 250, 271–72, 417, 529
PLASS, G. N., 511
platinum, 480–81, 485, 508, 513
platonism, 38
PLAYFAIR, J., 553–55, 558, 564
PLAYFAIR, L., 570, 580
PLEDGE, H. T., 159
POE, E. A., 637
POGGENDORFF, J. C., 322, 486
POINCARÉ, [J.] H., 7, 68, 94, 101, 360, 372, 447, 499, 543, 583, 638, 646, 708, 754; Brownian movement, 616, 670–71; recurrence theorem, 95, 239, 352, 384, 628, 630–32, 638
POINCARÉ, L., 103
POISEUILLE, J. L. M., 424, 438
POISSON, S. D., 30, 31, 34, 42–43, 183, 398, 416, 482, 528, 547, 564, 628, 630, 684; bright spot, 43, 49, 310; caloric

theory, 118, 318, 320, 481; speed of sound, 151, 308, 395
POLLOCK, J. A., 506
POLTZ, H., 534
POMPE, A., 506
POPPER, K., 389, 640
positivism, 55, 57, 59–60, 67, 275, 315, 528, 673
POST, H. R., 102
potential, 43
POTTER, R., 209
POUILLET, C., 320, 323, 508
POWELL, B., 140, 144, 313, 323, 330, 566
POWELL, R. W., 507
POWER, H., 12, 15, 38, 389
Power–Townley Law, 12, 19
Prandtl number, 506
precursoritis, 19, 400
pressure, air, 9–11, 389; effect of interatomic forces, 398–99, 407, 416–18; kinetic theory, 14–15, 20, 69, 111, 162, 164, 166, 265; repulsive theory, 14, 390–91
PRESTON, S. T., 22, 86, 141, 145, 155, 210, 582, 708, 754–55
PRESTON, T., 482, 489, 511
PREVOST, P., 131, 308, 479, 492, 521
PRICE, D. J., 22, 51
PRIESTLEY, J., 484
PRIGOGINE, I., 281
PRINGSHEIM, E., 99, 228, 521, 755
probability, 185, 340, 587, 608; molecular states, 74, 83–84, 606–10
process, 40, 655
PROUT, W., 47, 488
PROVOSTAYE, F. DE LA, 313, 480, 501, 516
PRZIBRAM, K., 295
psychology, 55, 61, 276
PTOLEMY, C., 36, 37, 154
PULUJ, J., 431, 441, 755
PUSCHL, K., 332, 755

qualitative vs. quantitative, 12–13, 18, 102, 707
quantum theory, 3, 4, 90–91, 235, 270, 271, 299, 362, 419, 420, 610, 648–49
QUETELET, A., 55, 170, 183–84, 587
QUINCKE, G. H., 687

Book 1, pp. 1–300; Book 2, pp. 301–770.

QUINTUS ICILIUS, G. VON, 332

RABINOVICH, N. L., 381
radiation, 88; absorption by gases, 487, 489, 508–9, 525, 531; black body, 4, 58, 88, 99–100, 303, 324–25, 518–21, 539–42, 641, 644–48; Brownian movement, 663–65, 686–87; cooling law, 84, 479, 508–10, 513–21; exchanges law, 479, 521; frequency distribution, 481, 487, 513, 519, 524, 641, 647; heat, 20, 28, 30–32, 34, 41, 87, 128, 146, 171, 244, 303, 305–32, 472–77, 481, 524, 573; pressure, 211, 214–15, 518, 539–41; quantum theory, 3
radioactivity, 90, 92, 649
radiometer, 85, 88, 143, 210–30, 340, 498, 518, 687, 707
RÁDL, F., 663
RADZIYEVSKII, V. V., 22, 51
RAE, J., 549
RAEHLMANN, F., 687
railways, 122–27, 135–36
RAMAN, V. V., 579, 583, 700
RAMSAUER, C., 181
RAMSAY, W., 79, 246, 299, 354, 665, 669, 685
RANC, A., 700
RAND, W., 8
RANDALL, W. W., 26
randomness, 40, 49, 62–63, 95, 100, 182, 183, 187, 352, 366, 544–46, 583–99, 610, 612, 619–22, 636–37, 648, 650–51; random or stochastic process, 178, 655, 671, 680–84, 689
RANKINE, W. J. M., 295, 581, 756; heat and thermodynamics, 35, 279, 280, 307, 328, 570, 578, 583; Herapath–Waterston theory, 140, 141, 339; molecular vortex, 92, 140, 161, 206, 578, 580; reconcentration of energy, 573–74, 577
RANSOME, A., 204
RASOOL, S. I., 511
RASPAIL, F., 662
RAVETZ, J. R., 312, 564
RAYLEIGH, LORD (J. W. Strutt), 7, 89, 140, 307, 375, 498, 708, 756; black body radiation, 88, 321, 324–25, 641; equation of state, 413; kinetic theory, 201, 372, 625; specific heats, 354; sound, 87–88, 90; viscosity, 191, 430–31, 455; Waterston, 156–58, 330
Rayleigh-Jeans instability, 530
Rayleigh-Jeans Radiation Law, 321
realism, 55, 57, 59
recurrence theorem or paradox, 95–96, 352, 628–36, 644–46
REDTENBACHER, F., 205, 332
REE, F. H., 416
REEDER, C., 103
referees' reports, 210–11, 214–15, 219–25
refractive index, 78, 86, 243
REGNAULD, J., 664, 685, 687
REGNAULT, H. V., 78, 127, 163, 168, 328, 399–401, 506, 561, 576
REICHENBACH, H., 290, 549, 639
REILLY, C., 17
REILLY, J. G., 18
REINGANUM, M., 416, 432, 757
REINGOLD, N., 132
relativity, 90, 542
reversibility, 40, 571; electromagnetism, 644–45; paradox, 83, 95, 238–39, 241, 445, 602–7, 616–18, 622, 634
revolutions, scientific, 35–48, 51, 100, 310
REY, A., 549, 637
REYNOLDS, O., 7, 677, 708, 757–58; Maxwell, 224–27, 230; radiometer, 215–16, 229, 498; thermal transpiration, 85, 218, 221
RICHARZ, F., 758
RICHMANN, G. W., 476
RIEMANN, B., 289, 378, 379
RINALDINI, C., 473
RITTER, A., 582
RITTER, E., 30, 264, 397–400, 405
RIVIERE, C., 517
RO, S. T., 506
ROBERTS, W. O., 456
ROBIDA, K., 758
ROBIN, C. P., 663
ROBINSON, C., 109
ROBINSON, N. H., viii, 144
ROBISON, J., 477
RØMER, O., 13, 18
ROGER, J., 563

Book 1, pp. 1–300; Book 2, pp. 301–770.

INDEX

ROGERS, W. B., 66
ROGET, P. M., 33, 117
ROGOVSKY, E. A., 758
ROITI, A., 758–59
ROLLER, D., 33
romanticism, 51–56, 672
ROMÉ DE L'ISLE, J. B. L., 563
RONGE, G., 33, 167, 176, 596
RONKAR, J. E. J., 759
ROOSEVELT, T., 57
ROSENBERGER, F., 321, 324
ROSENBLUETH, A. W., 416
ROSENBLUETH, M. N., 416
ROSENFELD, L., 19, 65, 274–75, 421, 549–50, 653
ROSENTHAL, A., 80, 242
ROSSETTI, F., 510, 521
rotation of molecules, 140, 161, 172
ROUSE, H., 22, 425, 685
ROWLINSON, J. S., viii, 263, 269, 273, 411, 416, 532
Royal Society of London, 41, 115–19, 121, 140, 142, 157, 210, 229, 709, 711
RUBENS, H., 511, 512, 533
RUDWICK, M. J. S., 558
RÜCKER, A. W., 78, 759
RÜHLMANN, R., 89
RUFFNER, J. A., 470–71
RUMFORD, COUNT (B. Thompson), 25, 26, 28, 29, 31, 32, 33, 305, 306, 307, 308–9, 330, 334; heat transfer, 469, 473–75, 507
RUSSELL, A., 532
RUSSELL, B., 292
RUSSELL, H. N., 582
RUTHERFORD, E., 701
Rutherford–Bohr model, 90–91

SABINE, E., 190, 675
SAINT-ROBERT, P. DE, 50, 582
SACHS, R. G., 549
SAIGEY, E., 333
SAINT-VENANT, A. J. C. BARRÉ DE, 684
SAMBURSKY, S., 15
SANDARS, 226
SANDEMAN, R., 134
SANDRUCCI, A. A. L. G., 759
SANKEY, W. S., 34

SARTON, G., 164
SARTORI, L., 22
SATTERLY, J., 145
Saturn's rings, 189, 368
SAUNDERS, F. A., 521
SAUSSURE, H. B. DE, 473
scaling properties, 221 (*see also* dimensional analysis)
SCHAEFER, C., 511
SCHÄFER, K., 506
SCHAFFNER, K. F., 684
SCHAGRIN, M. L., 229, 522
SCHAMP, H. W., 542
SCHANCK, R. L., 49
SCHEELE, C. W., 308, 477
SCHELLING, F., 51–53, 397
SCHERER, A. N., 31
SCHLEGEL, F., 52
SCHLEGEL, R., 550
SCHLEIDEN, J. M., 55
SCHLEIERMACHER, A., 503, 504
SCHLÜNDER, E. U., 532
SCHMIDT, A., 198, 759
SCHMIDT, E., 467
SCHMIDT, G. J. L., 760
SCHMIDT, H., 637, 639
SCHNEEBELI, H., 517
SCHNEIDER, I., 176
SCHNEIDER, S. H., 511
SCHOENFLIES, A., 380
SCHOFIELD, R. E., 49, 66, 388
SCHRÖDINGER, E., 91, 294, 299, 639
SCHUETZ, 623
SCHULTZ, C. A. S., 662
SCHUSTER, A., 760; radiometer, 215–217, 228, 229; Waterston, 158
SCHWANN, T., 55
SCHWARZE, W., 503, 504
SCHWENDENER, S., 671
scientism, 60
SCOTT, G. D., 89
SCOTT, R., 582
SCOTT, W. L., 69, 113, 314, 333, 546, 579
SECCHI, A., 509
SEDDIG, M., 696
SEGUIN, M., 325, 580
SELIGMANN-LUI, 103
SELLMEIER, W., 760
SENGERS, J. V., viii, 256, 532, 534

Book 1, pp. 1–300; Book 2, pp. 301–770.

set theory, 79, 378–80, 631
SHAKESPEARE, W., 52
SHANKLAND, R. S., 685
SHAPIRO, A. E., 18
SHAPLEY, H., 18
SHARLIN, H. I., viii, 49, 50, 159, 566
SHERLOCK, T. T., 68
SHEYNIN, O. B., 550, 594, 595, 596
SHILLING, W. G., 356
SHIPLEY, A., 158
SHORTLEY, G., 18
SIEDENTOPF, H., 695
SIEMENS, W., 517
SILLIMAN, B., 332
SILLIMAN, R. H., 49, 210, 313
SIMON, C. M. E. T., 760
SIMPSON, T. K., 103, 597, 613
SINGER, C., 159
SLATER, J. C., 418
slip (of fluid along surface), 217–18, 229–30, 438, 676
SMITH, E. B., 563
SMITH, F. J., 442
SMITH, W. R., 15
SMOLUCHOWSKI, M. VON, 7, 431, 455, 527–29, 654, 708, 760; Brownian movement, 91, 101, 298, 649, 656, 686–92; Einstein, 686
SMYTH, A. L., 26
SNELDERS, H. A. M., 65, 180, 281
SODDY, F., 528
SOGIN, H. H., 533
solar system, 545–48, 554, 628
solids, 173; elasticity, 34, 64, 168
SOMERVILLE, M., 318, 324
SOMMERFELD, A., 96, 294
SONIN, N. Y., 760–61
SORET, L., 330, 332, 512
SOROKIN, P., 637
SOUBEIRAN, E., 320
sound, propagation of, 11, 31, 85–88, 90, 110, 123, 138, 141, 149–55, 171, 203–4, 208, 308, 506
SOUTH, J., 676
SPALLANZANI, L., 657, 661
SPARROW, W. J., 33
specific heats, 25, 28, 33, 91, 112; gases, 4, 139, 162–63, 173, 245, 278, 335, 619, 707; ratio, 71–73, 79, 84, 86, 96, 148, 175, 194–95, 203, 208, 337, 345, 353–54; speed of sound, 71, 85–86, 151, 152
specific inductive capacity, 78
spectroscopy, 79, 91, 354–55
speed of molecule, 69, 123, 138, 150, 174, 177, 668
SPENCER, H., 562, 591–92
SPENCER, J. B., 66, 397
spheres, hard or elastic, 72–74, 242, 358–59, 401–14, 440–41, 454, 504, 530
SPRINGER, G. S., 533
SPRONSEN, J. W. VAN, 102
STAIGMULLER, H. C. D., 761
STALLO, J. B., 63–65, 69, 280, 292, 293, 298
STAMBAUGH, J., 638
STANKEVICH, B. V., 761
STANLEY, W. F., 102
states of matter, 249–50, 272, 392–94
statistical mechanics, 71, 79, 84, 98, 241, 376, 411, 419–20, 651, 673–74, 677–78, 688–89
statistics, 56, 183, 544–45, 584, 587
STAUFFER, R. C., 65, 280
steam engines, 256, 305, 551, 554, 566–72
STEBBING, S., 66
STEFAN, J., 7, 84, 192, 231, 428, 441, 450, 642, 708, 761; diffusion, 85, 192–93, 404; heat conduction, 497, 500–4, 505, 515–16; radiation, 481, 482, 514–16; sound, 87, 202
Stefan–Boltzmann Radiation Law, 84, 88, 192, 244, 476, 481, 514–26, 542
STENBERG, V. A., 307
STEPHENSON, 124
STERN, O., 22
STEVENS, F. W., 421
STEVENS, W. L., 523
STEWART, B., 201, 362
STICKER, B., 563
STIEGLER, K., 595
stochastic, *see* random
STOKES, G. G., 219, 224, 226, 484; formula for resistance to motion of sphere, 675–79, 682, 692, 696–97; sound, 86–87, 467; viscosity, 190–91, 193, 424, 432–33, 437, 675–76, 684

STONEY, G. J., 76, 198–99, 201, 372, 482, 616, 708, 762–63; radiometer, 216–17, 224, 228
Stosszahlansatz, 627, 640
stress in rarefied gas, 217–18, 221
SUDARSHAN, E. C. G., 549
sun's heat or temperature, 140, 141–42, 508–10, 517
superconductivity, 4
surface tension, 75, 264
SUSLOV, G. K., 763
Sussex, Duke of, 121
SUTHERLAND, W., 7, 277, 432, 454, 708, 763; radiometer, 228, 498; viscosity, 428–30
SVARTHOLM, N., 468
SVEDBERG, T., 666, 681, 696–97, 698
SWIFT, J., 550
SWINBURNE, J., 583
Swiss science, 171, 709, 711
SZILY [von Nagy-Szigeth], C. [or K.], 232, 279, 593, 617, 764

TAIT, P. G., 142, 280, 333, 583, 764–65; Boltzmann, 357–58, 431; equation of state, 413–14; equipartition theorem, 357–59, 616; Forbes, 313; Hooke, 19; kinetic theory, 356, 411, 432, 625, 708; Maxwell, 226, 589; Mayer, 66, 323–24; mean free path, 427; Mohr, 323–24, 329; radiation, 517, 520; radiometer, 217; reversal of motions, 602; viscosity, 427; van der Waals, 415
TALBOT, G. R., 20, 129
TANAKA, M., 281
TANUKADATE, T., 442
TAYLOR, F. S., 26
TAYLOR, L. W., 159
technology (*see also* steam engines), 510–11
TEDESCO, G., 313
TEISSERENC DE BORT, L., 589
temperature, 20, 470–71, 478; absolute, 23–26, 111, 118, 479, 576, 578; discontinuity at surface, 217, 501, 528–29, 688; effect of gravity, 110; mixtures, 111–12, 114; molecular kinetic energy, 70, 113, 137, 166–67, 172, 316; momentum, 110–12

Book 1, pp. 1–300; Book 2, pp. 301–770.

TERESCHIN, S., 523, 532
TERLETSKII, Y. P., 640
TESKE, A., 693
THENARD, L. J., 312
THEOBALD, D. W., 164
thermal conductivity, *see* heat conduction
thermal diffusion, *see* diffusion
thermal expansion, *see* expansion
thermal transpiration, 218, 221, 225–26, 228
thermo-diffusion, 84
thermodynamics, 32, 33, 42, 55, 166–68, 205, 279, 297, 328, 330, 419, 546; Second Law, 62, 83, 95–96, 101, 182, 232, 244, 291, 352, 367–68, 518, 539–41, 543, 569–79, 589, 630, 631, 666; statistical interpretation, 63, 94, 291, 297, 443, 589–90, 605; word, 322
THIELE, J., 282
THIRION, J., 666
THOMPSON, B., *see* RUMFORD
THOMPSON, H. A., 533
THOMPSON, S. P., 566
THOMSON, J., 260–63, 266
THOMSON, J. J., 50, 66, 91, 299, 677, 699, 765–66
THOMSON, T., 33, 109, 314, 477, 534
THOMSON, W., *see* KELVIN
TILLOCH, A., 34
time, direction, 96, 240, 289–90, 551, 584
TIMERDING, H., 164
TIMIRYAZEV, K. A., 582
TOBEY, R. C., 582
TODD, A. C., 109
TODHUNTER, I., 34, 322
TÖDHEIDE, K., 507
TÖPLER, A. J. I., 243
TÖPLER, E., 766
TOEPLITZ, O., 449
TOLMAN, R. C., 612, 613, 614, 615, 639, 640, 654
TOPPER, D. R., 50, 66
TORRICELLI, E., barometer and air pressure, 9–12, 16, 389
TOULMIN, S., 36, 563
TOWNLEY, C., 12
TOWNLEY, J., 12
TOWNLEY, R., 12, 17, 18, 389
TRAILL, T. S., 304, 312, 477, 488

TRAVERS, M. W., 356
transport processes (*see also* diffusion, etc.), 73, 80–81, 84, 180, 278, 493
TREDGOLD, T., 766
TREMBLY, J., 20
TRENEER, A., 120
TRIBUS, M., 581
TRICKER, R. A. R., 50
TRUESDELL, C., vii, 18, 20, 22, 34, 132, 157, 383, 387, 396, 397, 425, 477, 498, 499, 532, 550, 580, 581, 597, 613; recurrence theorem, 638; sound, 87–88, 150, 155; transport theory, 432, 442, 506
TURNER, F. M., 594
TURNER, J., 188
TUTTLE, E. R., 269
TYNDALL, J., 34, 58–59, 198–200, 332, 585, 596; Clausius, 168; radiation, 204, 310, 488, 508–10, 511, 513–15

UBBINK, J. B., 534
UHLENBECK, G. E., 269, 411, 418, 685
ultraviolet catastrophe, 4, 306, 641
UMOV, N. A., 582
UNGER, F., 662
URE, A., 112, 482

vacuum, 9–12, 472, 474–76, 479, 526–27
valence theory of chemical bonds, 55
VAN BYLEVELT, J. S., 532
VAN DER PAS, P. W., 662
VAN DER PYL, A. W., 18
VANDERSLICE, J. T., 542
VAN DER WAALS, *see* WAALS, VAN DER
VAN LAAR, *see* Laar, van
VAN MELSEN, A. G., 15
VAN ORSTRAND, C. E., 523
VAN SPRONSEN, J. W., 102
VAN'T HOFF, J. H., 242, 254, 677–78
VAN ZEE, E., 33
VARGAFTIK, N., 532
velocity distribution function, 80–82, 85, 180, 233–37, 443–44, 450–53, 458–59, 587–89, 599–600; finite system, 188
VENART, J. E. S., 534
VERDET, 174
vibration theory of gases, 9, 19, 20, 200, 473

vibrations of molecules, 172
VICAIRE, E., 512
VIOLI, A., 766
VIOLLE, J., 227, 512, 517, 520, 523
VIRGO, S. E., 90
virial coefficients, 242, 400, 414–17, 461
virial theorem, 75, 181, 404–6
viscosity, bulk, 433, 684; dense gases, 460–62; liquids, 424–25, 436, 545, 675; word, 423
viscosity of gases, 11, 58, 73, 75, 78, 84, 189–91, 225, 402–3, 423–31, 434–42, 454–55, 675–76; experiments, 11, 191, 435–41; interatomic forces, 425, 429–30, 440–42, 454–55; effect on speed of sound, 86–87
VISWANATHAN, B., 263
VOGT, J. G., 638
VOIGT, W., 766
VOLTAIRE, 552
VOLTERRA, V., 447
VON BAER, K. E., 613
vortex atom, 92, 161, 206–7, 388
VRIES, H. DE, 254

WAALS, J. D. VAN DER, 7, 242, 250–54, 277, 393, 395, 706, 708, 766–67; Boltzmann, 414–16, 418; critical point, 77, 394; equation of state, 75, 77, 78, 264, 406–11; force, 270, 417–18; virial coefficients, 414–16
WAALS, J. D. VAN DER, JR., viii, 251, 256, 414, 416, 654
WAARD, C. DE, 16
WAGNER, C. K., 524
WAGNER, R., 532
WAINWRIGHT, T., 415
WAKEHAM, W., 506
WALD, F., 68, 103, 281, 291, 292, 638
WALDMANN, L., 230
WALKER, H. M., 595
WALLACE, A. R., 58
WALLING, H. F., 582
WALLIS, J., 41
WANG, S. C., 418
WARBURG, E., 84, 441; rarefied gases, 217–18, 501, 517, 528; specific heats of mercury, 79, 354
WARD, J., 68
WARTMANN, E., 35

WATANABE, S., 281
water, 256; vapor, 509
WATERSTON, G. (father of J. J.), 134
WATERSTON, G. (nephew of J. J.), 142–43
WATERSTON, J. J., 6, 21, 134–36, 141–43, 195, 403–4, 489, 588, 767–68; difficulty in getting theory published, 35, 71, 115–16, 140; equipartition theorem, 336–38; gravity, 48, 135; Herapath, 130, 154; kinetic theory, 70–72, 113, 137–40, 146–48, 175, 330, 422, 585, 708; size of molecule, 75, 142; reception of theory, 141, 158, 167, 206; sound, 86, 149, 152–54, 156; sun, 508; wave theory of heat, 146, 329–30
WATSON, H. W., 195, 354, 359, 360, 615, 622, 627, 768
WATT, J., 33, 256, 259, 554
wave theory of heat, 32, 44, 146, 171, 203, 303–34, 481
WEBER, E. H., 55, 666
WEBER, H. F., 520–21
WEBER, S., 503, 504, 534
WEBSTER, C., 12, 16, 17, 18
WEBSTER, T., 193, 322, 482
WEDDERBURN-MAXWELL, 229
WEDGWOOD, J., 480
WEIERSTRASS, K. T. W., 378
WEIMBERG, C. B., 296
WEINER, C., 323, 333, 582
WEINSTEIN, M. B., 768
WEIZSÄCKER, C. F. VON, 639
WELTER, J. J., 149, 152, 153
WENCK, J., 332
WEST, G. D., 228
WEST, W., 318
WESTERGAARD, H. L., 595
WESTFALL, R. S., 155
WETHERALD, R. T., 511
WETTERHAM, D., 68
WETZELS, W. D., 65
WHEELER, L. P., 421
WHEWELL, W., 32, 120, 318, 321
WHITE, L., 637
WHITEHEAD, A. N., 49
WHITROW, G. J., 549, 637
WHITROW, M., 8

WHITTAKER, E. T., 49, 50, 522, 684
WHYTE, L. L., 66
WIEDEMANN, E. [E. G.], 519, 769
WIEN, W., 306, 523, 647
WIENER, C., 664, 668, 685
Wiener [N.] process, 671, 683, 685, 693
WIGNER, E., 42
WILHELMY, L., 481–83, 484, 511, 520, 595
WILKIE, J. S., 672
WILLIAMS, D., 18
WILLIAMS, H. S., 313, 333
WILLIAMS, L. P., 31, 49, 50, 53, 66, 120, 323, 397
WILLIAMSON, D. E., 511
WILSON, E. B., 421
WILSON, G., 16
WILSON, W., 159
WILSON, W. E., 521, 523
WIND, C. H., 769
WINKELMAN, A., 84, 501–4, 522
WITZ, A., 489, 510
WOLF, A., 26, 33, 114, 313, 477
WOLF, E., 49
WOODRUFF, A. E., 228, 229, 522
WORTHINGTON, A. M., 102
WREN, C., 41
WRIGHT, C., 68
WRIGHT, C. A., 532
WRIGHT, P. G., 89, 129, 615
WRISBERG, H. A., 659, 661
WÜLLNER, A., 482, 514

YOUNG, C. A., 517
YOUNG, T., 32, 33, 43, 75, 117, 135, 213, 305, 392; heat, 309, 475
YVON-VILLARCEAU, A. J. F., 406, 769

ZAITSEVA, L. S., 506
ZANSTRA, H., 639
ZANTEDESCHE, F., 35
ZEMPLEN, G., 769
ZERMELO, E., 7, 94, 95–97, 99, 103, 239, 352, 448, 630, 632–35, 638, 644, 654, 694, 708, 769
ZEUNER, G., 205
ZOELLNER, J. K. F., 64
ZOUBOV, V. P., 114
ZSIGMONDY, R., 685, 695

Book 1, pp. 1–300; Book 2, pp. 301–770.

STUDIES IN STATISTICAL MECHANICS

Editors: J. de Boer and G. E. Uhlenbeck

CONTENTS OF VOLUME I (1962)

A. Problems of a Dynamical Theory in Statistical Physics, *N.N. Bogoliubov* — 1–118
B. The Theory of Linear Graphs with Applications to the Theory of the Virial Development of the Properties of Gases, *G.E. Uhlenbeck and G.W. Ford* — 119–211
C. Some Topics in Quantum Statistics: the Wigner Function and Transport Theory, *H. Mori, I. Oppenheim and J. Ross* — 213–298
D. A Study of Models in Non-Equilibrium Statistical Mechanics, *M. Dresden* — 299–343

CONTENTS OF VOLUME II (1964)

A. Imperfect Bose Gas, *Kerson Huang* — 1–106
B. A Critical Study of some Theories of the Liquid State including a Comparison with Experiment, *J.M.H. Levelt and E.G.D. Cohen* — 107–239
C. The Heat Conductivity and Viscosity of Polyatomic Gases, *C.S. Wang Chang, G.E. Uhlenbeck and J. de Boer* — 241–268

CONTENTS OF VOLUME III (1965)

A. Diagram Expansions in Quantum Statistical Mechanics, *C. Bloch* — 3–211
B. Construction Operator Formalism in Many Particle Systems, *J. de Boer* — 215–275
C. On the Theory of the Equation of State, *B. Kahn* — 281–382

CONTENTS OF VOLUME IV (1969)

The Maxwell Equations. Non-Relativistic and Relativistic Derivations from Electron Theory, *S. R. de Groot* — 1–179

CONTENTS OF VOLUME V (1970)

A. The Kinetic Theory of Gases, *C.S. Wang Chang and G. E. Uhlenbeck* — 1–100
B. The Dispersion of Sound in Monoatomic Gases, *J.D. Foch Jr. and G. W. Ford* — 101–231

SEP 14 1978